Introduction to Graph Theory

Douglas B. West

University of Illinois - Urbana

PRENTICE HALL
Upper Saddle River, NJ 07458

Library of Congress Cataloging-in-Publication Data

West, Douglas Brent
   Introduction to Graph Theory / Douglas B. West
     p. cm.
   Includes bibliographical references and index.
   ISBN 0-13-227828-6
   1. Graph Theory. I. Title
  QA166.W43 1996
  511'.5--dc20   95-24773
                                    CIP

Acquisitions editor: George Lobell
Production editor: Tina Trautz
Cover designer: Douglas West / Ed Scheinerman
Production coordinator: Alan Fischer

 © 1996 by Prentice-Hall, Inc.
Simon & Schuster / A Viacom Company
Upper Saddle River, NJ 07458

Printed in the United States of America

10 9 8 7 6 5 4 3 2 1

ISBN 0-13-227828-6

PRENTICE-HALL INTERNATIONAL (UK) LIMITED, *London*
PRENTICE-HALL OF AUSTRALIA PTY. LIMITED, *Sydney*
PRENTICE-HALL CANADA INC., *Toronto*
PRENTICE-HALL HISPANOAMERICANA S.A., *Toronto*
PRENTICE-HALL OF INDIA PRIVATE LIMITED, *New Delhi*
PRENTICE-HALL OF JAPAN INC., *Tokyo*
SIMON & SCHUSTER ASIA PTE. LTD., *Singapore*
EDITORA PRENTICE-HALL DO BRASIL, LTDA., *Rio de Janeiro*

*I dedicate this book
to Ching and Blake and to my parents,
with thanks
for their support and encouragement
and especially for their patience.*

Contents

Preface   xi

Chapter 1   Fundamental Concepts   1

1.1   Definitions and Examples   1
    What is a Graph?, 1
    Graphs as Models, 2
    Matrices and Isomorphism, 5
    Exercises, 11

1.2   Paths and Proofs   14
    Induction and Walks, 15
    Equivalences and Connected Graphs, 17
    Contradiction and Bipartite Graphs, 20
    Extremality, 22
    Exercises, 23

1.3   Vertex Degrees and Counting   25
    Counting and Bijections, 26
    The Pigeonhole Principle, 29
    Turán's Theorem, 32
    Exercises, 36

1.4   Degrees and Algorithmic Proof   40
    Algorithmic or Constructive Proof, 40
    Graphic Sequences, 42
    Degrees and Digraphs, 46
    Exercises, 47

Chapter 2   Trees and Distance   51

2.1   Basic properties   51
Properties of Trees, 51
Distance in Graphs, 54
Proving a Stronger Result, 55
Disjoint Spanning Trees, 58
Exercises, 59

2.2   Spanning Trees and Enumeration   63
Enumeration of Trees, 63
Spanning Trees in Graphs, 65
Decomposition and Graceful Labelings, 69
Exercises, 70

2.3   Optimization and Trees   73
Minimum Spanning Tree, 74
Shortest Paths, 75
Trees in Computer Science, 79
Exercises, 82

2.4   Eulerian Graphs and Digraphs   85
Eulerian Circuits, 85
Directed Graphs, 88
Applications, 91
Exercises, 94

Chapter 3   Matchings and Factors   98

3.1   Matchings in Bipartite Graphs   98
Maximum Matchings, 99
Hall's Matching Condition, 100
Min-Max Theorems, 102
Independent Sets in Bipartite Graphs, 103
Exercises, 105

3.2   Applications and Algorithms   109
Maximum Bipartite Matching, 109
Weighted Bipartite Matching, 111
Stable Matchings (optional), 116
Faster Bipartite Matching (optional), 118
Exercises, 120

3.3   Matchings in General Graphs   121
Tutte's 1-factor Theorem, 121
f-factors of Graphs, 125
Edmonds' Blossom Algorithm (optional), 127
Exercises, 131

Chapter 4 Connectivity and Paths 133

4.1 Cuts and Connectivity 133
Connectivity, 133
Edge-connectivity, 136
Blocks, 139
Exercises, 141

4.2 k-connected Graphs 144
2-connected Graphs, 144
Connectivity of Digraphs, 147
k-connected and k-edge-connected Graphs, 148
Applications of Menger's Theorem, 152
Exercises, 155

4.3 Network Flow Problems 158
Maximum Network Flow, 158
Integral Flows, 163
Supplies and Demands (optional), 166
Exercises, 169

Chapter 5 Graph Coloring 173

5.1 Vertex Colorings and Upper Bounds 173
Definitions and Examples, 173
Upper Bounds, 175
Brooks' Theorem, 178
Exercises, 180

5.2 Structure of k-chromatic Graphs 184
Graphs with Large Chromatic Number, 184
Critical Graphs, 186
Forced Subdivisions (optional), 188
Exercises, 190

5.3 Enumerative Aspects 193
Counting Proper Colorings, 194
Chordal Graphs, 198
A Hint of Perfect Graphs, 200
Counting Acyclic Orientations (optional), 202
Exercises, 203

Chapter 6 Edges and Cycles 206

6.1 Line Graphs and Edge-coloring 206
Edge-colorings, 207
Characterization of Line Graphs (optional), 212
Exercises, 215

6.2 Hamiltonian Cycles   218
      Necessary Conditions, 219
      Sufficient Conditions, 221
      Cycles in Directed Graphs (optional), 226
      Exercises, 227

6.3 Complexity (optional)   232
      Intractability, 232
      Heuristics and Bounds, 235
      NP-Completeness Proofs, 238
      Exercises, 245

Chapter 7 Planar Graphs   **247**

7.1 Embeddings and Euler's Formula   247
      Drawings in the Plane, 247
      Dual Graphs, 250
      Euler's Formula, 255
      Exercises, 257

7.2 Characterization of Planar Graphs   259
      Preparation for Kuratowski's Theorem, 260
      Convex Embeddings, 261
      Bridges and Planarity Testing (optional), 264
      Exercises, 267

7.3 Parameters of Planarity   269
      Coloring of Planar Graphs, 269
      Edge-colorings and Hamiltonian Cycles, 274
      Crossing Number, 277
      Surfaces of Higher Genus (optional), 280
      Exercises, 283

Chapter 8 Additional Topics (optional)   **288**

8.1 Perfect Graphs   288
      The Perfect Graph Theorem, 289
      Chordal Graphs Revisited, 293
      Other Classes of Perfect Graphs, 297
      Imperfect Graphs, 305
      The Strong Perfect Graph Conjecture, 312
      Exercises, 315

8.2 Matroids   320
      Hereditary Systems and Examples, 321
      Properties of Matroids, 326
      The Span Function and Duality, 330
      Minors and Planar Graphs, 336

Matroid Intersection, 340
Matroid Union, 343
Exercises, 347

8.3   Ramsey Theory   353
The Pigeonhole Principle Revisited, 353
Ramsey's Theorem, 355
Ramsey Numbers, 359
Graph Ramsey Theory, 361
Sperner's Lemma and Bandwidth, 364
Exercises, 369

8.4   More Extremal Problems   373
Encodings of Graphs, 374
Branchings and Gossip, 381
List Colorings and Choosability, 386
Partitions Using Paths and Cycles, 390
Circumference, 394
Exercises, 400

8.5   Random Graphs   404
Existence and Expectation, 405
Properties of Almost All Graphs, 409
Threshold Functions, 412
Evolution and Properties of Random Graphs, 415
Connectivity, Cliques, and Coloring, 420
Martingales, 423
Exercises, 429

8.6   Eigenvalues of Graphs   432
The Characteristic Polynomial, 433
Linear Algebra of Real Symmetric Matrices, 436
Eigenvalues and Graph Parameters, 439
Eigenvalues of Regular Graphs, 441
Eigenvalues and Expanders, 444
Strongly Regular Graphs, 446
Exercises, 449

Glossary of terms   **453**

Glossary of notation   **472**

References   **474**

Author Index   **497**

Subject Index   **502**

Preface

Graph theory is a delightful playground for the exploration of proof techniques in discrete mathematics, and its results have applications in many areas of the computing, social, and natural sciences. The design of this book permits usage in a one-semester introduction at the undergraduate or beginning graduate level, or in a patient two-semester introduction. No previous knowledge of graph theory is assumed. Many algorithms and applications are included, but the focus is on understanding the structure of graphs and the techniques used to analyze problems in graph theory.

Many textbooks have been written about graph theory. Due to its emphasis on both proofs and applications, the initial model for this book was the elegant text by J.A. Bondy and U.S.R. Murty, *Graph Theory with Applications* (Macmillan/North-Holland [1976]). Graph theory is still young, and no consensus has emerged on how the introductory material should be presented. Selection and order of topics, choice of proofs, objectives, and underlying themes are matters of lively debate. Revising this book dozens of times has taught me the difficulty of these decisions. This book is my contribution to the debate.

Features

Various features of this book facilitate students' efforts to understand the material. I include an early discussion of proof techniques, more than 850 exercises of varying difficulty, more than 300 illustrations, and many examples. I have tried to include the statements and illustrations that are needed in class to complete the flow of argument.

   This book contains much more material than other introductions
to graph theory. Collecting the advanced material as a final optional
chapter of "additional topics" permits usage at different levels. The
undergraduate introduction consists of the first seven chapters, leaving
Chapter 8 as topical reading for interested students. The first five sections summarize and illustrate proof techniques while developing fundamental properties of graphs. Undergraduate students with minimal
previous exposure to proofs find this discussion helpful as they begin to
write proofs of their own. With advanced students, the reminder of elementary techniques can be ignored. Advanced students also may have
previous exposure to graphs from a general course in combinatorics, discrete structures, or algorithms. A graduate course can treat most of
Chapters 1 and 2 as recommended reading, moving rapidly to Chapter 3
in class and reaching some topics of Chapter 8. Chapter 8 can also be
used as the bulk of a second course in graph theory.

   Most of the exercises require written proofs. Many undergradu-
ates begin graph theory with little practice at presenting explanations,
and this hinders their appreciation of graph theory and other mathematics. The intellectual discipline of justifying an argument is valuable
independently of mathematics; I hope that students will become com-
fortable with this. In writing solutions to exercises, students should be
careful in their use of language ("say what you mean"), and they should
be intellectually honest ("mean what you say"), which includes acknowl-
edging when they have left gaps.

   Although we select terminology in graph theory that suggests the
intended meaning, the large quantity of definitions remains an obstacle
to fluency. Mathematicians like to start with a clean collection of definitions, but students usually want to master one concept before receiving
the next. By request of instructors, I have postponed many definitions
to their first important application. For example, the definition of
strongly connected digraph first appears in Section 2.4 on Eulerian circuits, the definition of Cartesian product appears in 5.1 with coloring
problems, and the definition of line graph appears in 6.1.

   Many results in graph theory have several proofs; illustrating this
can increase students' flexibility in trying multiple approaches to a
problem. I include some alternative proofs as remarks and others as
exercises. Many exercises have hints. Exercises marked "(–)" or "(+)"
are easier or more difficult respectively than unmarked problems.
Those marked "(–)" may be suitable as exam problems. Exercises
marked "(!)" are especially valuable, instructive, or entertaining. Exercises that relate several concepts usually appear when the last is introduced. Many exercises are referenced in the text while discussing a relevant concept or result. Exercises in the current section are cited by
giving only the index of that exercise among the exercises of that section. Other cross-references are by Chapter.Section.Item.

Organization

I have sought a development that is intellectually coherent and displays a gradual (not monotonic) increase in difficulty of proofs and in algorithmic complexity. Most graph theorists agree that the König-Egerváry Theorem deserves an independent proof without network flow. Also, students find connectivity a more abstract concept than matching. I therefore treat matching before connectivity.

The gradual rise in difficulty puts Eulerian graphs early and Hamiltonian and planar graphs later. When students encounter the lack of good algorithms for coloring and Hamiltonian cycle problems, they may be curious about NP-completeness. Section 6.3 can be read to satisfy this curiosity; it also can be discussed after Chapter 7. Presentation of NP-completeness via formal languages can be technically abstract, so many students appreciate a more "nuts and bolts" discussion in the context of graph problems. Also, NP-completeness proofs illustrate the variety and usefulness of "graph transformation" arguments.

Turán's Theorem uses only elementary ideas and hence appears in Chapter 1. The applications that motivate Sperner's Lemma involve advanced material, so this waits until Chapter 8. Trees and distance appear together (Chapter 2), because various results about distance reduce to trees and because trees are related to Dijkstra's algorithm and to Eulerian circuits. Petersen's Theorem on 2-factors (Chapter 3) uses Eulerian circuits and bipartite matching. Menger's Theorem appears before network flow (Chapter 4), and separate applications are provided for network flow. The $k-1$-connectedness of k-color-critical graphs (Chapter 5) uses bipartite matching. Section 5.3 offers a brief introduction to perfect graphs, emphasizing chordal graphs. The main discussion of perfect graphs (with the proof of the Perfect Graph Theorem) appears in Chapter 8. Graph orientations appear in many exercises and examples, including the Gallai-Roy Theorem and Stanley's connection between the chromatic polynomial and acyclic orientations (Chapter 5). The proof given of Vizing's Theorem for simple graphs (Chapter 6) is algorithmic and short. The proof of Kuratowski's Theorem (Chapter 7) uses Thomassen's approach and fits into one energetic lecture.

Chapter 8 contains highlights of advanced material and is not intended for a standard undergraduate course. It assumes more sophistication than earlier chapters and is written more tersely. The sections are independent. Each selects appealing results from a large topic that merits a chapter of its own. Some of the sections become more difficult near the end; an instructor may prefer to sample early material in several sections rather than present one completely. I will treat advanced graph theory more thoroughly in *The Art of Combinatorics*, of which volumes I-II are devoted to graph theory, with matroids in Volume III and random graphs in Volume IV.

Design of Courses

I intend the 23 sections in Chapters 1-7 for a pace of roughly two sections per week, skipping optional material as needed to present a balanced coverage of topics. With beginning students, instructors may want to spend more time on Chapters 1-2. Some items in the text are explicitly labeled "optional".

For a slower one-semester course, the following items can be omitted without damage to continuity. 1.3: counting of subgraphs and even graphs. 1.4: the 2-switch theorem. 2.1: distance sums. 2.2: proof of the Matrix Tree Theorem. 2.3: Huffman coding. 2.4: directed Eulerian circuits and street-sweeping. 3.2: Hopcroft-Karp algorithm. 3.3: everything after Petersen's Theorem. 4.1: algorithm for blocks. 4.2: applications of Menger's Theorem. 4.3: supplies and demands. 5.1: proof of Brooks' Theorem. 5.2: everything after edge-connectivity of k-critical graphs. 5.3: inclusion-exclusion computation and acyclic orientations. 6.1: characterization of line graphs. 6.2: cycles in digraphs. 6.3: everything. 7.1: Jordan curve proof and outerplanar graphs. 7.2: bridges and planarity testing. 7.3: 4-color discussion and genus.

Courses that start in Chapter 3 may wish to include from the first two chapters such topics as Turán's Theorem, graphic sequences, the matrix tree theorem, Kruskal's algorithm, and algorithms for Eulerian circuits. Courses that introduce graph theory over two quarters can cover the first seven chapters thoroughly.

A one quarter course must aim for the highlights. 1.1: adjacency matrix and isomorphism. 1.2: all. 1.3: degree-sum formula and Turán's Theorem. 1.4: large bipartite subgraphs and Havel-Hakimi test. 2.1: through definition of distance. 2.2: through statement of Matrix Tree Theorem. 2.3: Kruskal's algorithm and possibly Dijkstra's algorithm. 2.4: Eulerian graph characterization and Chinese Postman Problem. 3.1: all. 3.2: none. 3.3: statement of Tutte's Theorem and proof of Petersen's results. 4.1: through definition of blocks, skipping Harary graphs. 4.2: through open ear decomposition, plus Menger's Theorem(s) (proving only the edge version). 4.3: duality between flows and cuts, statement of Max-flow = Min-cut. 5.1: through Szekeres-Wilf theorem. 5.2: Mycielski's construction. 5.3: through chromatic recurrence, plus perfection of chordal graphs. 6.1: through Vizing's Theorem. 6.2: through Ore's condition, plus the Chvátal-Erdös condition. 6.3: none. 7.1: non-planarity of K_5 and $K_{3,3}$, examples of dual graphs, and Euler's formula. 7.2: statement and examples of Kuratowski's Theorem and Tutte's Theorem. 7.3: the 5-color Theorem, Tait's Theorem, and Grinberg's Theorem.

Acknowledgments

This text has benefited from classroom testing of successively-improving pre-publication versions at many universities. Instructors who used the text on this experimental basis were, roughly in chronological order, Ed Scheinerman (Johns Hopkins), Kathryn Fraughnaugh (Colorado-Denver), Paul Weichsel / Paul Schupp / Xiaoyun Lu (Illinois), Dean Hoffman / Pete Johnson / Chris Rodger (Auburn), Dan Ullman (George Washington), Zevi Miller / Dan Pritikin (Miami-Ohio), David Matula (Southern Methodist), Pavol Hell (Simon Fraser), Grzegorz Kubicki (Louisville), Jeff Smith (Purdue), Ann Trenk (Wellesley), Ken Bogart (Dartmouth), Kirk Tolman (Brigham Young), Roger Eggleton (Illinois State), Herb Kasube (Bradley) and Jeff Dinitz (Vermont). Many of these (or their students) provided helpful suggestions. Other thought-provoking comments came from reviewers, including Paul Edelman, Renu Laskar, Gary MacGillivray, Joseph Neggers, Joseph Malkevitch, James Oxley, Sam Stueckle, and Barry Tesman. Reviewers for sections of Chapter 8 included Mike Albertson, Sanjoy Barvah, James Oxley, Chris Rodger, and Alan Tucker. I thank George Lobell, my editor at Prentice Hall, for his commitment to the project as it evolved and for finding diligent reviewers who suffered with early versions.

The cover illustration was produced by Ed Scheinerman using BRL-CAD, a product of the U.S. Army Ballistic Research Laboratory. Chris Hartman contributed vital assistance in preparing the bibliography and index (the former printed in TEX), writing processing scripts and tracking down elusive references. Many students contributed to a marathon session to gather data for the index, including Maria Axenovich, Nicole Henley, Andre Kundgen, Peter Kwok, Kevin Leuthold, John Jozwiak, Radhika Ramamurty, and Karl Schmidt. In addition to giving yeoman service on the index, Peter Kwok proofread many chapters and Andre Kundgen proofread the bibliography. Maria Muyot helped with the index and the glossary. I prepared the text and illustrations using the "groff" typesetting system, a product of the Free Software Foundation. Indispensable in solving many groff difficulties was Ted Harding of the University of Manchester Institute of Science and Technology.

I have tried to eliminate errors, but surely one remains. I welcome corrections and suggestions, including comments on topics, attributions of results, suggestions for exercises, notice of typographical errors or omissions from the glossary, etc. With enough readers and printings, we can get it all right.

Douglas B. West
west@math.uiuc.edu
Urbana, 1995

Chapter 1

Fundamental Concepts

1.1 Definitions and Examples

How can we lay cable at minimum cost to make every telephone reachable from every other? What is the fastest route from the national capital to each state capital? How can n jobs be filled by n people with maximum total utility? What is the maximum flow per unit time from source to sink in a network of pipes? How many layers does a computer chip need so that wires in the same layer don't cross? How can the season of a sports league be scheduled into the minimum number of weeks? In what order should a traveling salesman visit cities to minimize travel time? Can we color the regions of every map using four colors so that neighboring regions receive different colors? These and many other practical problems involve graph theory. In this book, we develop the theory of graphs and apply it to such problems.

WHAT IS A GRAPH?

1.1.1. Definition. A *simple graph* G with n *vertices* and m *edges* consists of a *vertex set* $V(G) = \{v_1, \ldots, v_n\}$ and an *edge set* $E(G) = \{e_1, \ldots, e_m\}$, where each edge is an unordered pair of vertices. We write uv for the edge $\{u, v\}$. If $uv \in E(G)$, then u and v are *adjacent*. We write $u \leftrightarrow v$ to mean "u is adjacent to v". The vertices contained in an edge e are its *endpoints*.

We visualize a graph on paper by assigning a point to each vertex and drawing a curve for each edge between the points representing its

1

endpoints. The graph drawn on the left below has four vertices and three edges. The terms "vertex" and "edge" arise from the vertices and edges of polyhedra in space, such as cubes or tetrahedra. Although we may visualize an edge as a curve, in the definition of a graph an edge is only a pair of vertices. An alternative model, illustrated on the right below, treats edges as ordered pairs.

1.1.2. Definition. A *simple directed graph* or *simple digraph* G consists of a *vertex set* $V(G)$ and an *edge set* $E(G)$, where each edge is an ordered pair of vertices. We write uv for the edge (u, v), with u being the *tail* and v being the *head*. We write $u \to v$ when $uv \in E(G)$, meaning "there is an edge from u to v".

The choice of head and tail for an edge of a directed graph assigns a "direction" to the edge, which we illustrate by drawing edges as arrows. This book emphasizes graphs, which have undirected edges. Occasionally we discuss analogous concepts or applications involving directed graphs; these are particularly important in Section 4.3.

For some applications, we consider more general models that allow repeated edges or edges with both endpoints the same. These are *multiple edges* and *loops*, respectively. For example, the graph with vertex set $\{u, v\}$ and edge set $\{uv, uv, vv\}$ has both a repeated edge and a loop. The term *simple graph* explicitly forbids multiple edges and loops, and the term *multigraph* explicitly allows them. Technically, "graph" indicates the most general model (multigraph), but we often use "graph" alone instead of "simple graph" after establishing a context where multiple edges and loops are irrelevant.

GRAPHS AS MODELS

Graphs arise in many settings, and these settings may suggest useful terminology for graphs.

1.1.3. Example. *Acquaintance relations and subgraphs.* A famous brain-teaser asks this question: Does every set of six people have three mutual acquaintances or three mutual strangers (Exercise 6)?

Since "acquaintance" is a symmetric relation, we can model it using a simple graph with a vertex for each person and an edge for each acquainted pair. The "nonacquaintance" relation on the same set produces another graph. The *complement* of a simple graph G, written $\overline{G}$, is a graph having the same vertex set as G, such that u, v are adjacent in $\overline{G}$ if and only if u, v are not adjacent in G. Below we draw a graph and its complement.

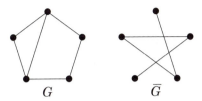

$$G \qquad\qquad \overline{G}$$

A *subgraph* of a graph G is a graph H such that $V(H) \subseteq V(G)$ and $E(H) \subseteq E(G)$; we write this as $H \subseteq G$ and say that "G contains H". An *induced subgraph* of G is a subgraph H such that every edge of G contained in $V(H)$ belongs to $E(H)$. The graph G drawn above has six subgraphs with five edges, but it has no induced subgraph with five edges. If H is an induced subgraph of G with vertex set S, then we write $H = G[S]$ and say that H is the subgraph of G "induced by S".

A *complete graph* or *clique* is a simple graph in which every pair of vertices forms an edge. A complete graph has many subgraphs that are not cliques, but every induced subgraph of a complete graph is a clique. The complement of a complete graph has no edges.

An *independent set* in a graph G is a vertex subset $S \subseteq V(G)$ such that the induced subgraph $G[S]$ has no edges. Our 6-person brainteaser asks whether every 6-vertex graph contains a clique or an independent set with three vertices. In the graph G drawn above, the largest clique and largest independent set have sizes 3 and 2, respectively. These values reverse in the complement $\overline{G}$, since cliques become independent sets (and vice versa) under complementation.

Terms defined: complement $\overline{G}$, subgraph $H \subseteq G$, induced subgraph $G[S]$, complete graph (clique), independent set. □

1.1.4. Example. *Job assignments and bipartite graphs.* Suppose we have m jobs and n people, and each person can do some of the jobs. Can we make assignments to fill the jobs? We model the available assignments by a graph having a vertex for each job and each person, putting job j adjacent to person p if p can do j.

A graph is *bipartite* if its vertex set can be partitioned into (at most) two independent sets. The graph of available person-job assignments is bipartite. Since a person can do only one job and we can assign a job to only one person, we seek m pairwise disjoint edges in the graph. Chapter 3 presents a test for whether this is possible.

A *complete bipartite graph*, illustrated below, is a bipartite graph

in which the edge set consists of all pairs having a vertex from each of
the two independent sets in the vertex partition. When the graph of
permitted job assignments is a complete bipartite graph, matching up
the vertices is easy, so we seek the "best" way to do it. Given numerical
weights on the edges that measure their desirability, perhaps the best
way to match up the vertices is the one where the selected edges have
maximum total weight. In Chapter 3, we develop an algorithm to find
the assignment with maximum weight.

   Terms defined: bipartite graph, complete bipartite graph. □

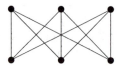

1.1.5. Example. *Scheduling and graph coloring.* Suppose we want to
schedule Senate committee meetings, with each committee assigned one
time period during the week. We cannot assign two committees to the
same time period if they have a common member. How many time periods do we need?

   We model this by creating a vertex for each committee, making
vertices adjacent if the corresponding committees have a common member. We must assign labels (time periods) to the vertices so the endpoints of each edge receive different labels; we want to use the fewest
labels. The labels have no numerical value, so we call them *colors*, and
the vertices receiving a particular label form a *color class*. The number
of colors needed is called the *chromatic number* of G, written $\chi(G)$; we
study this in Chapter 5. Since vertices of the same color must form an
independent set, $\chi(G)$ equals the minimum number of independent sets
that partition $V(G)$. This generalizes the notion of bipartite graphs. A
graph G is *k-partite* if $V(G)$ can be partitioned into k or fewer independent sets. The independent sets in a specified partition are *partite sets*
(or color classes). The graph drawn below has chromatic number 3 and
is 3-partite (also 4-partite, 5-partite, etc.).

   The most (in)famous problem in graph theory involves coloring. A
map is a partition of the plane into connected regions. Can we color the
regions of every map using at most four colors so that neighboring

regions have different colors? In each map, we introduce a vertex for each region and an edge for regions sharing a boundary. We are then asking whether every planar graph has chromatic number at most 4, where a graph is *planar* if it can be drawn in the plane without edge crossings. The graph above can be redrawn without crossings. We study planar graphs in Chapter 7.

Terms defined: color class, chromatic number $\chi(G)$, k-partite, partite set, planar. □

1.1.6. Example. *Road networks and connection.* We can model a road network by a graph having edges that correspond to road segments between intersections. We can assign edge weights to measure distance or travel time. In this context edges do represent physical links. We may want to know the shortest route from x to y.

Informally, we think of a *path* in a graph as an ordered list of distinct vertices $v_1, \ldots, v_n$ such that $v_{i-1}v_i$ is an edge for all $2 \leq i \leq n$. Similarly, we think of a *cycle* as an ordered list $v_1, \ldots, v_n$ such that all $v_{i-1}v_i$ and also $v_n v_1$ are edges. The first and last vertices of a path are its *endpoints*; a u,v-path is a path with endpoints u and v. We will define these concepts more precisely in Section 1.2.

To find the shortest route from x to y, we want to find the x,y-path having the least total weight among all x,y-paths in G. We solve this problem in Chapter 2. Similarly, in a network of n cities, we may want to visit all the cities and return home at least total cost. Using the costs as weights on the edges of the complete graph, we seek the n-vertex cycle with minimum total cost. This is the "Traveling Salesman Problem", discussed in Chapter 6.

In a road network or communication network, every site should be reachable from every other. A graph G is *connected* if it has a u,v-path for each pair $u,v \in V(G)$. The graph drawn below is not connected.

Terms defined: path, cycle, connected graph. □

MATRICES AND ISOMORPHISM

How can we specify a graph? We can list the vertices and edges, but there are other useful ways to encode this information.

1.1.7. Definition. Given a graph or digraph G with vertices indexed as $V(G) = \{v_1, \ldots, v_n\}$, the *adjacency matrix* of G, written $A(G)$, is the matrix in which entry a_{ij} is the number of copies of the edge $v_i v_j$ in G. The *degree* of a vertex is the number of non-loop edges containing it plus twice the number of loops containing it.

If vertex v belongs to edge e, then v and e are *incident*. The *incidence matrix* of a loopless graph G, written $M(G)$, has rows indexed by $V(G)$ and columns indexed by $E(G)$, with $m_{ij} = 1$ if vertex v_i belongs to edge e_j, otherwise $m_{ij} = 0$. If G is a digraph, then $m_{ij} = +1$ if v_i is the tail of e_j and $m_{ij} = -1$ if v_i is the head of e_j.

1.1.8. Remark. A graph may have many adjacency matrices; each ordering of the vertices determines an adjacency matrix. If G is a graph (not a digraph), then every adjacency matrix is symmetric ($a_{ij} = a_{ji}$ for all i, j). If G is a simple graph, then every entry in $A(G)$ is 0 or 1, with 0's on the diagonal. □

1.1.9. Example. Below we draw a simple graph G and a digraph H, together with the adjacency matrix and incidence matrix that result from the vertex ordering w, x, y, z and the edge ordering a, b, c. The adjacency matrix for the multigraph having two copies of each of these edges would be obtained by changing each 1 to a 2. □

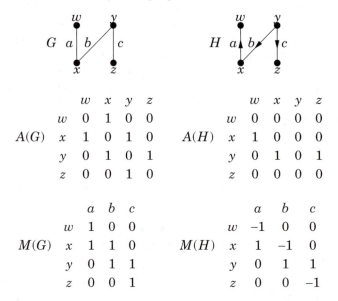

When we present an adjacency matrix for a graph, we are implicitly naming the vertices by the order of the rows; the ith vertex corresponds to the ith row and column. This provides names for the vertices. We cannot store a graph in a computer without naming the vertices. Nevertheless, we want to study properties (like "connected") that do not depend on the names of the vertices. If we can find a one-to-one correspondence (*bijection*) between $V(G)$ and $V(H)$ that preserves the adjacency relation, then G and H have the same structural properties.

1.1.10. Definition. An *isomorphism* from G to H is a bijection $f: V(G) \to V(H)$ such that $uv \in E(G)$ if and only if $f(u)f(v) \in E(H)$. We say "G *is isomorphic to* H", written $G \cong H$, if there is an isomorphism from G to H.

1.1.11. Remark. *Isomorphism and adjacency matrices.* When G is isomorphic to H, also H is isomorphic to G, so we may say "G and H are isomorphic" (to each other).[†] Since an adjacency matrix encodes the adjacency relation, we can also describe isomorphism using adjacency matrices. The graphs G and H are isomorphic if and only if we can apply a permutation to the rows of $A(G)$ and the same permutation to the columns of $A(G)$ to obtain $A(H)$. Permuting $A(G)$ in this way renumbers the vertices of G; instead of $v_1, \ldots, v_n$, the rows now correspond to $v_{\pi(1)}, \ldots, v_{\pi(n)}$ in order. If the new matrix is the matrix $A(H)$ that corresponds to the ordering $u_1, \ldots, u_n$ of $V(H)$, then the bijection that sends $v_{\pi(i)}$ to u_i for each i is an isomorphism from G to H. □

1.1.12. Example. The graphs G and H drawn below are 4-vertex paths. They are isomorphic by an isomorphism that maps w, y, z, x to a, b, c, d, respectively. Rewriting $A(G)$ by placing the rows in the order w, y, z, x and the columns also in that order yields $A(H)$. Another isomorphism maps x, z, y, w to a, b, c, d, respectively. □

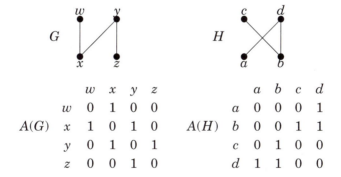

1.1.13. Definition. A *relation* R on a set S is a collection of ordered pairs from S. When R is a relation, we often write xRy to mean $(x, y) \in R$ and say that (x, y) *satisfies* R. An *equivalence relation* is a relation R that is reflexive (xRx for all $x \in S$), symmetric (xRy implies yRx), and transitive (xRy and yRz imply xRz).

[†]The adjective "isomorphic" applies only to pairs of graphs; "G is isomorphic" by itself has no meaning.

Adjacency is a relation on the vertices of a graph G; if G has P_3 as an induced subgraph, then the adjacency relation on $V(G)$ is *not* an equivalence relation. The set of pairs (G, H) such that G is isomorphic to H is the *isomorphism relation* on the set of graphs.

1.1.14. Proposition. Isomorphism is an equivalence relation.

Proof. The identity map on $V(G)$ is an isomorphism from G to itself, so $G \cong G$. If $f: V(G) \to V(H)$ is a isomorphism from G to H, then f^{-1} is an isomorphism from H to G, so $G \cong H$ implies $H \cong G$. If $f: V(F) \to V(G)$ and $g: V(G) \to V(H)$ are isomorphisms, then the composition $g \circ f$ is a bijection from $V(F)$ to $V(H)$ that preserves the adjacency relation and hence is an isomorphism from F to H (this means $F \cong G$ and $G \cong H$ imply $F \cong H$. Hence the isomorphism relation is reflexive, symmetric, and transitive. $\square$

An equivalence relation partitions objects into *equivalence classes*, where two elements satisfy the relation if and only if they lie in the same equivalence class.

1.1.15. Definition. An *isomorphism class* of graphs is an equivalence class of graphs under the isomorphism relation.

1.1.16. Remark. *"Unlabeled" graphs.* When discussing the structure of a graph G, we consider a fixed vertex set for G, but our comments apply to every graph isomorphic to G. For this reason, we sometimes use the informal expression "unlabeled graph" to mean an isomorphism class of graphs. When we draw a graph, its vertices are named by the physical points where we have located them, even if we gave them no other names. Hence a drawing of a graph is a representative of its isomorphism class. We often use the name "graph" for a drawing of a graph. When we redraw a graph to illustrate something about its structure, we have chosen a more convenient member of the isomorphism class.

When we know that two graphs are isomorphic, we often discuss them using the same name; this recognizes that we are making statements about the isomorphism class containing them. For this reason, we usually write $G = H$ instead of $G \cong H$. Similarly, when we say "G is a subgraph of H", we mean technically that G is isomorphic to a subgraph of H, or that H contains a "copy" of G.

Based on this treatment of isomorphism classes, we use K_n, P_n, C_n, respectively, to denote any graph that is a clique, path, or cycle with n vertices, without naming the vertices. Asking whether G "is" C_n is asking whether G is isomorphic to a cycle with n vertices. Similarly, we use $K_{r,s}$ to denote "the" complete bipartite graph with partite sets of sizes r and s. The picture in Example 1.1.4 shows $K_{3,3}$. $\square$

1.1.17. Example. *The number of n-vertex graphs.* Suppose X is a set of size n; X contains $\binom{n}{2} = n(n-1)/2$ unordered pairs.[†] We may include or omit each pair as an edge, so there are $2^{\binom{n}{2}}$ simple graphs with vertex set X. There are 64 simple graphs having a fixed set of four vertices, but these fall into only 11 isomorphism classes. These appear below in complementary pairs; only P_4 is isomorphic to its complement. The isomorphism classes have different sizes, so we cannot count the isomorphism classes of n-vertex simple graphs by dividing $2^{\binom{n}{2}}$ by the size of a class. □

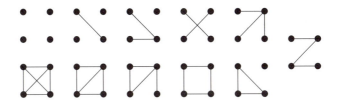

An isomorphism from G to H preserves the adjacency relation. Since structural properties of graphs are determined by their adjacency relations, we can prove that G and H are *not* isomorphic by finding some structural property of one that is not true of the other. If they have different vertex degrees, different sizes of the largest clique or smallest cycle, etc., then they cannot be isomorphic, because these properties are preserved by isomorphism. On the other hand, no known list of common structural properties implies that $G \cong H$; we must present a bijection $f: V(G) \to V(H)$ that preserves the adjacency relation.

1.1.18. Example. *Isomorphic or not?* In the graphs below, each vertex has degree 3, but the graphs are not pairwise isomorphic. Several are drawings of $K_{3,3}$, as shown by exhibiting isomorphisms. One contains C_3 and hence cannot be a drawing of $K_{3,3}$.

To prove $G \cong H$, we give names to the vertices, provide a bijection between the vertex sets, and verify that the adjacency relation is preserved. In the first drawing below, we could name the top row u, v, w and the bottom row x, y, z. In the second drawing, we could name the vertices 1,2,3,4,5,6 in order around the outside cycle. The bijection that sends u, v, w, x, y, z to $1, 3, 5, 2, 4, 6$, respectively, is an isomorphism from the first graph to the second. The map sending u, v, w, x, y, z to $6, 4, 2, 1, 3, 5$, respectively, is another isomorphism.

[†]The *binomial coefficient* $\binom{n}{k}$ is the number of k-element subsets of an n-element set.

Consider also the graphs below. Since they have many edges, we prefer to compare their complements; graphs G and H are isomorphic if and only if $\bar{G}$ and $\bar{H}$ are isomorphic (Exercise 10). The complement of one of these graphs is connected, but the complement of the other is disconnected, so they cannot be isomorphic. □

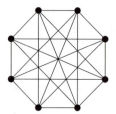 

1.1.19. Example. *The Petersen graph.* The *Petersen graph* is most commonly drawn as the graph on the left below. The graph is useful enough to have an entire book devoted to it (Holton-Sheehan [1993]). The other graphs below are also drawings of the Petersen graph. Exercise 19 requests proof that these are pairwise isomorphic. The Petersen graph has a simple description using the set S of 2-element subsets of a 5-element set. Let G be the graph with vertex set S in which two pairs form an edge if and only if they are disjoint as sets. The graph G is isomorphic to each graph below. This follows from labeling the vertices of each graph with the members of S such that the adjacency relation is disjointness. □

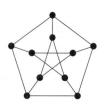  

1.1.20. Definition. An *automorphism* of G is a permutation of $V(G)$ that is an isomorphism from G to G. A graph G is *vertex-transitive* if for every pair $u, v \in V(G)$ there is an automorphism that maps u to v.

1.1.21. Example. *Automorphisms.* Let G be the path with vertex set $\{1, 2, 3, 4\}$ and edge set $\{12, 23, 34\}$. This graph G has two automorphisms: the identity permutation and the permutation that switches 1 with 4 and switches 2 with 3. Interchanging vertex 1 and vertex 2 is not an automorphism of G, although G is isomorphic to the graph H

with vertex set $\{1, 2, 3, 4\}$ and edge set $\{12, 13, 34\}$. The automorphisms of G are the permutations that can be applied simultaneously to the rows and columns of $A(G)$ without changing $A(G)$. In $K_{r,s}$, permuting the vertices of one independent set does not change the adjacency matrix, so $K_{r,s}$ has $r!s!$ automorphisms if $r \neq s$. On the other hand, $K_{t,t}$ has $2(t!)^2$ automorphisms.

If $n > 2$, then P_n is not vertex-transitive, since no automorphism can map a vertex of degree 1 into a vertex of degree 2. The complete bipartite graph $K_{r,s}$ is vertex-transitive if and only if $r = s$. Every cycle is vertex-transitive, and all graphs in Examples 1.1.18 and 1.1.19 are vertex-transitive. $\square$

We can prove a statement for every vertex in a vertex-transitive graph by proving it for one vertex. Vertex-transitivity guarantees that the graph "looks the same" from every vertex; every vertex plays the same role.

EXERCISES

Most of these exercises can be solved using ad hoc reasoning or case analysis. Other techniques of proof appear explicitly in the remainder of this chapter. A mark of "(–)" indicates that a problem is easier or shorter than most, while a "(+)" indicates that it is harder or longer than most, and a "(!)" indicates that it is particularly useful or instructive.

1.1.1. (–) Write down all possible adjacency matrices and incidence matrices for a 3-vertex path. Also write down an adjacency matrix for a path with six vertices and for a cycle with six vertices.

1.1.2. (–) Using rectangular blocks whose entries are all equal, write down an adjacency matrix for $K_{m,n}$.

1.1.3. (–) Prove that cutting opposite corner squares from an 8 by 8 checkerboard leaves a subboard that cannot be partitioned into rectangles consisting of two adjacent unit squares. Generalize this using bipartite graphs.

1.1.4. (–) Consider the following four families of graphs: $A = \{$paths$\}$, $B = \{$cycles$\}$, $C = \{$cliques$\}$, $D = \{$bipartite graphs$\}$. For each pair of these families, determine all isomorphism classes of graphs that belong to both families.

1.1.5. (–) Prove or disprove: If every vertex of a finite simple graph G has degree 2, then G is a cycle.

1.1.6. (!) Prove that at any party with six people, there are three mutual acquaintances or three mutual strangers.

1.1.7. (!) Suppose that G is a connected simple graph that is not a complete graph. Prove that every vertex of G belongs to some 3-vertex induced subgraph isomorphic to P_3.

1.1.8. Suppose that G is a simple graph having no vertex of degree 0 and no induced subgraph with two edges. Prove that G is a complete graph.

1.1.9. (+) Suppose that G is a simple graph having no vertex of degree 0 and no induced subgraph with three edges. Prove that G has at most four vertices.

1.1.10. (–) Prove that $G \cong H$ if and only if $\overline{G} \cong \overline{H}$.

1.1.11. (–) Determine the number of isomorphism classes of simple 7-vertex graphs in which every vertex has degree 4. (Hint: consider the complements.)

1.1.12. Among the graphs below, which pairs are isomorphic?

1.1.13. (!) In each class below, determine the smallest n such that there exist nonisomorphic n-vertex graphs having the same list of vertex degrees.
　　　(a) all (multi)graphs,　(b) loopless multigraphs,　(c) simple graphs.
(Hint: since each class contains the next, the answers form a nondecreasing triple. For (c), use the list of isomorphism classes in Example 1.1.17.)

1.1.14. Suppose G is a simple graph with adjacency matrix A and incidence matrix M. Prove that the degree of v_i is the ith diagonal entry in A^2 and in MM^T. What do the entries in position (i, j) of A^2 and MM^T say about G?

1.1.15. A simple graph isomorphic to its complement is *self-complementary*.
　　　a) Prove that the number of vertices in a self-complementary graph is congruent to 0 or 1 modulo 4.
　　　b) Construct a self-complementary n-vertex graph for every n congruent to 0 or 1 modulo 4. (Hint: for $n \equiv 0 \bmod 4$, generalize the structure of the self-complementary graph P_4 by placing the vertices into four groups. For $n \equiv 1 \bmod 4$, add one vertex to the graph constructed for $n - 1$.)

1.1.16. Characterize the pairs of positive integers m, n such that $K_{m,n}$ has two isomorphic subgraphs with each edge of $K_{m,n}$ in exactly one of them.

1.1.17. Prove that K_n has three pairwise-isomorphic subgraphs such that each edge of K_n appears in exactly one of them if and only if n is congruent to 0 or 1 modulo 3.

1.1.18. Find three pairwise-isomorphic connected subgraphs of the Petersen graph such that each edge of the Petersen graph appears in exactly one of them.

1.1.19. (–) *The Petersen graph.* Prove that the graphs drawn in Example 1.1.19 are pairwise isomorphic. (Comment: one can use disjoint vertex sets and establish the required bijections or give all the graphs the same vertex set, assigned so that the graphs have the same adjacency matrix).

1.1.20. (−) Is the graph drawn below isomorphic to the Petersen graph?

1.1.21. (!) *Automorphisms of the Petersen graph.* Let S be the collection of 2-element subsets of $\{1, \ldots, 5\}$. Use ab (or ba) to denote the element $\{a, b\}$ of S. Let G be the graph with vertex set S defined by $ab \leftrightarrow cd$ if and only if $\{a, b\} \cap \{c, d\} = \varnothing$.

a) Prove that the Petersen graph as drawn in Example 1.1.19 is isomorphic to G. Use this to conclude that the Petersen graph is vertex-transitive.

b) Prove that every automorphism of G maps the 5-cycle with vertices $12, 34, 51, 23, 45$ to a 5-cycle with vertices ab, cd, ea, bc, de determined by a permutation of $\{1, 2, 3, 4, 5\}$ taking elements 1,2,3,4,5 to a, b, c, d, e, respectively.

c) Prove that every permutation of $\{1, 2, 3, 4, 5\}$ specifies an automorphism of G in the way described above. Conclude that the Petersen graph has exactly 120 automorphisms.

1.1.22. Count the automorphisms of P_n, C_n, and K_n.

1.1.23. Construct a simple graph with six vertices that has only one automorphism. Construct a simple graph that has exactly three automorphisms.

1.1.24. *"Extra" symmetry in the Petersen graph.* Suppose that $P = (u_0, u_1, u_2, u_3)$ and $Q = (v_0, v_1, v_2, v_3)$ are 3-edge paths in the Petersen graph. Prove that there is exactly one automorphism of the Petersen graph that maps u_i into v_i for $i = 0, 1, 2, 3$. (Hint: use the disjointness description from Exercise 21.)

1.1.25. (!) Determine whether the graphs drawn below are isomorphic. (The graph on the left appears on the cover of Wilson-Watkins [1990].)

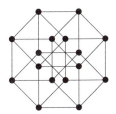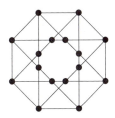

1.1.26. (+) *Edge-transitivity.* A graph G is *edge-transitive* if for all $e, f \in E(G)$ there is an automorphism of G that maps the endpoints of e to the endpoints of f (in either order). Prove that the graphs of Exercise 24 are vertex-transitive and edge-transitive. (Comment: the complete graphs, complete bipartite graphs, and Petersen graph (by Exercise 23) are edge-transitive.)

1.1.27. *Edge-transitive versus vertex-transitive.*

a) Suppose that G is obtained from K_n with $n \geq 3$ by replacing each edge of K_n with a path of length two (the new vertices have degree 2). Prove that G is edge-transitive but not vertex-transitive.

b) (+) Suppose that G is edge-transitive but not vertex-transitive and has no vertices of degree 0. Prove that G is bipartite.

c) Prove that the graph below is vertex-transitive but not edge-transitive.

1.2 Paths and Proofs

In the remainder of this chapter we illustrate basic proof techniques of discrete mathematics by developing elementary results about graphs. We begin with definitions of path-like objects. The informal definitions of path and cycle in Section 1.1 are adequate for most purposes, but it helps to view these as special cases of a more general concept. A tourist wandering in a city may want to allow vertex repetitions but avoid edge repetitions. A mail carrier delivers mail to houses on both sides of the street and hence traverses each edge twice.

1.2.1. Definition. A *walk* of *length* k is a sequence $v_0, e_1, v_1, e_1, \ldots,$ e_k, v_k of vertices and edges such that $e_i = v_{i-1}v_i$ for all i. A *trail* is a walk with no repeated edge. A *path* is a walk with no repeated vertex. A u,v-*walk* is a walk with first vertex u and last vertex v; these are its *endpoints*, and it is *closed* if $u = v$. A *cycle* is a closed trail of length at least one in which "first = last" is the only vertex repetition (a loop is a cycle of length 1).

Paths and trails are walks and hence have endpoints and lengths. A path or trail may have length 0, but a cycle cannot. We use the words "path" and "cycle" in three closely-related contexts: as a graph or subgraph, as a special case of a walk, and as a set of edges.

In a simple graph a walk is completely specified by its sequence of vertices; hence we usually describe a path or cycle (or walk) in a simple graph by its ordered list of vertices. Although we may list only vertices, the walk still consists of both vertices and edges.

We may start a cycle at any vertex. To emphasize the cyclic aspect of the ordering, we often list each vertex only once when naming a cycle.

1.2.2. Example. *Walks, paths, and cycles.* The graph below has a closed walk of length 12 that visits vertices in the order $(a, x, a, x, u, y, c, d, y, v, x, b, a)$. Omitting the first two steps yields a closed trail (no edge repetition). The edge set of this trail is the union of the edge sets of three pairwise edge-disjoint cycles. The u, v-trail u, y, c, d, y, v contains the edges of the u, v-path u, y, v. □

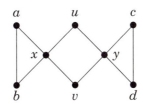

These definitions and remarks hold also for digraphs with each e_i being an *ordered* pair $v_{i-1}v_i$. In a path or cycle of a directed graph, successive edges "follow the arrows"; for example, the digraph consisting only of the edge xy has an x, y-path but no y, x-path.

1.2.3. Definition. A graph G is *connected* if it has a u, v-path for each pair $u, v \in V(G)$ (otherwise it is *disconnected*). If G has a u, v-path, then u is *connected to* v in G. The *connection relation* in a graph consists of the vertex pairs (u, v) such that u is connected to v.

Connectedness is a property of graphs, but the connection relation on vertices and the phrase "u is connected to v" are convenient when writing proofs. To specify the stronger condition $uv \in E(G)$, we say "u and v are adjacent" or "u and v are joined by an edge". We use "connected to" for the existence of a u, v-path and "joined to" for the existence of an edge uv.

INDUCTION AND WALKS

The technique of induction can be used to prove many statements that involve a positive integer parameter. The technique rests on the *Well Ordering Property*, which we ASSUME for the positive integers: every nonempty set of positive integers has a least element.

1.2.4. Theorem. (Induction Principle). If $P(n)$ is a statement with an integer parameter n and the following two conditions hold, then $P(n)$ is true for every positive n.
1) $P(1)$ is true.
2) For all $n \geq 1$, "$P(n)$ is true" implies "$P(n + 1)$ is true".

Proof. Suppose that conditions (1) and (2) hold. If $P(n)$ does not hold for every positive integer n, then $P(n)$ fails for some nonempty set of positive integers. By the Well Ordering Property, there is some least value for which $P(n)$ fails. By (1), this value cannot be 1. By (2), this value cannot be any integer larger than 1. □

For example, we can use induction on n to prove that $\Sigma_{i=1}^{n} i = \binom{n+1}{2}$ $= (n+1)n/2$. A common phrasing to prove $P(n)$ by induction is: 1) verify that $P(n)$ is true when $n = 1$, and 2) prove that if $P(n)$ is true when $n = k$, then $P(n)$ is also true when $n = k+1$. Verifying (1) is called the *basis step* of the proof; verifying (2) is called the *induction step*.

There are many other ways to phrase proofs by induction. We can start at 0 to prove a statement for nonnegative integers. We also have the equivalent principle stated next.

1.2.5. Theorem. (Strong Induction Principle). If $P(n)$ is a statement with an integer parameter n and the following two conditions hold, then $P(n)$ is true for every positive n.
1) $P(1)$ is true.
2) For all $n > 1$, "$P(k)$ is true for $1 \le k < n$" implies "$P(n)$ is true".

Proof. As in Theorem 1.2.4, there is no least value where $P(n)$ fails. □

We seldom distinguish between induction and strong induction. We call "$P(k)$ is true for all $k < n$" the *induction hypothesis*, because it is the hypothesis of the implication proved in the induction step.

Our first applications of induction are lemmas about walks and paths. Although the concatenation of a u,v-path and a v,w-path is a u,w-walk, it need not be a u,w-path. Nevertheless, it contains a u,w-path, and hence we can prove that G is connected by proving for all $u,v \in V(G)$ that G has a u,v-walk. Saying that a walk W *contains* a path P means that the vertices and edges of P appear as a subsequence of the vertices and edges of W; in order but not necessarily consecutive.

1.2.6. Lemma. If u and v are distinct vertices in G, then every u,v-walk in G contains a u,v-path.

Proof. We use (strong) induction on l to prove for all l the statement $P(l)$ that every u,v-walk of length l contains a u,v-path (see Exercise 3 for "ordinary" induction.) Suppose W is a u,v-walk of length l. If $l = 1$, then the only edge of W is uv, and W is itself a u,v-path.

For the induction step, suppose $l > 1$, and suppose the claim holds for walks of length less than l. If W has no repeated vertex, then W is itself a u,v-path. If W has a repeated vertex, then we can delete the edges and vertices in W between appearances of the repeated vertex to obtain a shorter u,v-walk contained in W. By the induction hypothesis, this walk contains a u, v-path, which appears in order in W. □

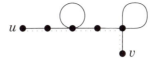

We drop the formal notation $P(n)$ for a statement being proved by induction on n. For the basis step in the next lemma, we explicitly consider multigraphs (Lemma 1.2.6 holds for multigraphs as well). A loop is a cycle of length one, and two distinct edges with the same endpoints form a cycle of length two. A walk is *odd* or *even* as its length is odd or even. We view a closed walk or a cycle as a cyclic arrangement that can start at any point; the next lemma requires this viewpoint.

1.2.7. Lemma. Every closed odd walk contains an odd cycle.

Proof. Suppose W is a closed odd walk; we use induction on the length l of W. If $l = 1$, then W is a loop, which is a cycle of length 1. For the induction step, suppose $l > 1$, and suppose the claim holds for walks of length less than l. If W has no repeated vertex (other than first=last), then W is itself a cycle of odd length. If vertex v is repeated in W, then we can partition W into two v, v-walks. Since the total length of W is odd, one of these is odd and the other is even. The odd one is shorter than W. By the induction hypothesis, it contains an odd cycle, which appears in order in W. □

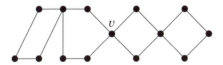

A closed walk of even length need not contain a cycle. Nevertheless, if edge uv appears *exactly once* in a closed walk, then Lemma 1.2.6 implies that the walk does contain a cycle, because deleting uv leaves a u, v-walk that contains a u, v-path but not uv itself.

EQUIVALENCES AND CONNECTED GRAPHS

If we prove that two conditions are equivalent, then in a situation where one of them holds we can also use the other. To prove "A if and only if B", we usually prove "A implies B" and its *converse* "B implies A". The statement "(not B) implies (not A)" is the *contrapositive* of "A implies B" and has the same meaning as "A implies B". Alternatively, using a chain of equivalences to prove "A if and only if B" proves both directions simultaneously.

1.2.8. Definition. The *components* of a graph G are its maximal connected subgraphs. A component (or graph) is *nontrivial* if it contains an edge. A *cut-edge* or *cut-vertex* of a graph is an edge or vertex whose deletion increases the number of components. The notation for the subgraph obtained by deleting an edge $e \in E(G)$ or $v \in V(G)$ (and the edges incident to v) is $G - e$ or $G - v$, respectively. Deleting a set $S \subseteq V(G)$ yields the subgraph induced by the remaining vertices; we write $G - S = G[\bar{S}]$, where $\bar{S} = V(G) - S$.

1.2.9. Remark. *Components and connection.* The connection relation is an equivalence relation on the vertices of a graph (Lemma 1.2.6 implies that the relation is transitive). The equivalence classes of this relation are the vertex sets of the components. Two components have no common vertex, and no edge has endpoints in different components. □

1.2.10. Lemma. A graph is connected if and only if for every partition of its vertices into two nonempty sets, there is an edge with endpoints in both sets.

Proof. Suppose G is connected. Given a partition of $V(G)$ into nonempty sets S, T, choose $u \in S$ and $v \in T$. Since G is connected, G has a u, v-path P. After its last vertex in S, P has an edge from S to T.

If G is disconnected and H is a component of G, then no edge has exactly one endpoint in H. This means that the partition S, T with $S = V(H)$ has no edge with endpoints in both sets. We have proved that if G is disconnected, then the partition condition fails. By the contrapositive, the partition condition implies that G is connected. □

1.2.11. Remark. *Adding or deleting cut-edges.* Adding an edge to G reduces the number of components by at most one, since the edge cannot have endpoints in more than two components of G. Similarly, deleting a cut-edge increases the number of components by exactly one. □

We next characterize cut-edges.

1.2.12. Lemma. An edge $e = xy$ is a cut-edge of a graph G if and only if $G - e$ has no x, y-path.

Proof. By using the contrapositive, the statement is equivalent to "e is not a cut-edge if and only if $G - e$ has an x, y-path." Deletion of e does not change any component not containing e, so it suffices to prove this statement for the component H of G that contains e. If e is not a cut-edge of H, then $H - e$ is connected and hence contains an x, y-path.

Conversely, suppose $H - e$ has an x, y-path Q; we prove that $H - e$ is connected. Choose $u, v \in V(H)$ arbitrarily. Since H is connected, H has a u, v-path P. If P does not contain e, then P also exists in $H - e$. If P contains e, then we form a u, v-walk in $H - e$ by following P until it reaches e, following Q instead of e to reach the other end of e, and then continuing on the rest of P to reach v. By Lemma 1.2.6, this u, v-walk in $H - e$ contains a u, v-path. Since u, v were chosen arbitrarily from $V(H)$, we have proved that $H - e$ is connected. □

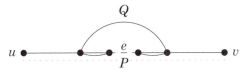

The picture shows the use of Lemma 1.2.6; Q uses edges of P (dotted) and the resulting u, v-walk (solid) is not a path. Lemma 1.2.12 has a useful corollary. We prove this characterization by a chain of equivalences; the two implications are proved simultaneously.

1.2.13. Corollary. An edge of an undirected graph is a cut-edge if and only if it belongs to no cycle.

Proof. By using the contrapositive, the statement is equivalent to "An edge belongs to a cycle if and only if it is not a cut-edge." We observe that $e = xy \in E(G)$ belongs to a cycle if and only if $G - e$ has an x, y-path, which by Lemma 1.2.12 is true if and only if e is not a cut-edge. □

We have introduced notation for deletion of vertices or edges. If G and H are connected graphs, we also have simple notation for the graph with components isomorphic to G and H. This is a special case of the operation of graph "union".

1.2.14. Definition. The *union* of graphs G and H, written $G \cup H$, has vertex set $V(G) \cup V(H)$ and edge set $E(G) \cup E(H)$. To specify the *disjoint union* with $V(G) \cap V(H) = \varnothing$, we write $G + H$. More generally, mG is the graph consisting of m pairwise disjoint copies of G. The *join* of G and H, written $G \vee H$, is obtained from $G + H$ by adding the edges $\{xy: x \in V(G), y \in V(H)\}$.

1.2.15. Example. *Sums and joins.* If G, H are connected graphs, then $G + H$ denotes the disconnected graph whose components are G and H. This notation is convenient when we have not named the vertices of G and H. In the illustration below, $G = P_4$ and $H = K_3$; on the left, we have chosen vertex sets that share two elements.

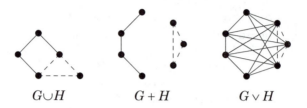

$$G \cup H \qquad G + H \qquad G \vee H$$

The graph mK_2 consists of m disjoint edges. We can express $K_{m,n}$ as $K_{m,n} = \overline{K_m + K_n}$ or as $K_{m,n} = (mK_1) \vee (nK_1)$. More generally, we always have $\overline{G + H} = \overline{G} \vee \overline{H}$. □

CONTRADICTION AND BIPARTITE GRAPHS

We already used proof by contradiction ("indirect proof") in justifying the Induction Principle and in our discussion of equivalences. The *conditional* statement "A implies B" is false only if A is true and B is false. The method of contradiction proves "A implies B" by showing that "A true and B false" is impossible.

1.2.16. Proposition. If v is a cut-vertex of a simple graph G, then v is not a cut-vertex of $\overline{G}$.

Proof. We assume that v is a cut-vertex of both G and $\overline{G}$ with vertex set V, and we derive a contradiction. Since v is a cut-vertex of G, there is a partition S, T of $V - \{v\}$ into nonempty sets such that G has no edge with endpoints in both S and T. Similarly, there is a partition S', T' of $V - \{v\}$ such that $\overline{G}$ has no edge with endpoints in both S and T.

This implies that $xy \notin G$ and $xy \notin \overline{G}$ if $x \in S \cap S'$ and $y \in T \cap T'$; hence these sets cannot both have vertices. Similarly, $S \cap T'$ and $T \cap S'$ cannot both have vertices. Making one of $\{S \cap S', T \cap T'\}$ and one of $\{S \cap T', T \cap S'\}$ empty makes one of S, T, S', T' empty, which contradicts our hypothesis about the partitions of $V - \{v\}$. Hence v cannot be a cut-vertex of both G and $\overline{G}$. □

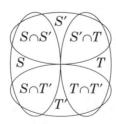

We could also prove this statement without contradiction. If v is a cut-vertex of G, then for each $x, y \in V - \{v\}$ we can build an explicit x, y-

path of length at most two (Exercise 14).

Characterizing a class **G** by a condition P requires proving that P is both *necessary* and *sufficient* for membership in **G**. Necessity means that membership in **G** requires property P, i.e. $G \in$ **G** *only if* G satisfies P. Sufficiency means that satisfying P guarantees membership in **G**, i.e. $G \in$ **G** *if* G satisfies P. We may speak of "necessity" and "sufficiency" of the condition instead of "only if" and "if" (when the class is named first and the condition is named second). Next we characterize the class of bipartite graphs by proving that absence of odd cycles is necessary and sufficient for membership in the class.

1.2.17. Theorem. A graph is bipartite if and only if it has no odd cycle.

Proof. *Necessity.* Suppose G is bipartite. Every walk in G alternates between the two color classes, so every return to the original class (including to the original vertex) happens after an even number of steps. Hence G has no odd cycle.

Sufficiency. Suppose G has no odd cycle. We prove that G is bipartite by partitioning the vertices of each component of G into two independent sets. Let u be a vertex in a component H of G. If G has u, v-walks of different parity for some $v \in V(H)$, then their concatenation is a closed walk of odd length. By Lemma 1.2.7, this contains an odd cycle, which contradicts the assumed absence of odd cycles. We can thus partition $V(H)$ into the set X of vertices reachable from u by even walks and the set Y of vertices reachable from u by odd walks. Each of X, Y is an independent set, since an edge v, v' within X or Y again creates a closed odd walk and contradicts the absence of odd cycles. □

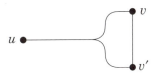

If G is a bipartite graph, a partition of $V(G)$ into two independent sets X, Y is a *bipartition* of G. If G is a disconnected bipartite graph, then there is more than one way to partition $V(G)$ into two independent sets. The statement "Let G be a bipartite graph with bipartition X, Y" specifies one such partition.

Technically, the sets of a *partition* are nonempty. For this reason the definition of k-partite graph allows vertex partitions using *at most k* independent sets. In particular, K_1 is a bipartite graph (and is k-partite for all k). We obtain the same results if we think of k-partite graphs as those whose vertex set can be covered by k independent sets that are allowed to be empty.

EXTREMALITY

The technique of extremality involves selecting an extremal example of a structure and using the lack of a "more extreme" example to gain extra leverage for the proof. The technique is well-suited to proving inequalities. For example, if we want to prove that a graph has a path with length at least l, then we may consider a longest path, because such a path satisfies the inequality if some path satisfies it.

1.2.18. Lemma. If G is a finite simple graph in which every vertex has degree at least k, then G contains a path of length at least k. If $k \geq 2$, then G also contains a cycle of length at least $k + 1$.

Proof. Since $V(G)$ is finite, we can select a longest path P in G. "Longest" implies that P cannot be extended, and hence every neighbor of an endpoint u of P also belongs to P. Since u has at least k neighbors, P must have at least k vertices other than u (because G is simple), so P has length at least k. If $k \geq 2$, then the edge from u to its farthest neighbor v along P completes a sufficiently long cycle with the u, v-portion of P. □

$$u \qquad\qquad\qquad v$$

1.2.19. Lemma. Suppose G is a finite graph with at least one edge. If G has no cycle, then G has a vertex of degree 1.

Proof. Because $V(G)$ is finite, every path in G is finite. Let e be an edge in G, and let P be a maximal path containing e ("maximal" means P is not contained in any longer path). Since P cannot be extended, every neighbor of an endpoint v of P belongs to P. To avoid creating a cycle, v must have no neighbor other than its neighbor along P. □

Lemma 1.2.19 is equivalent to "Every finite graph with minimum vertex degree at least 2 has a cycle." Finiteness is needed; if $V(G) = \mathbf{Z}$ and $E(G) = \{ij: |i - j| = 1\}$, then every vertex of G has degree 2, but G has no cycle. We will mention a few graphs with infinite vertex sets like $\mathbf{Z}$ or $\mathbb{R}^2$, but generally we discuss only finite graphs in this book.

Extremality is used in many ways. Given a connected graph G and disjoint sets $S, T \in V(G)$, we can obtain a path from S to T having only its endpoints in $S \cup T$ by choosing a *shortest* path from S to T (see Exercise 26). The components of a graph are the *maximal* connected subgraphs. We may choose maximal or maximum paths, vertices of minimum or maximum degree, maximal independent sets, first edge where two u, v-paths diverge, etc. Many such choices require the graph to be finite.

In Lemma 1.2.18, we chose a longest path because we were discussing length, but a maximal path would work as well. The adjective *maximum* means "maximum-sized," and *maximal* means "no larger one contains this one". For example, the graph $K_{1,m}$ has two maximal independent sets. The vertex of degree m by itself is a *maximal* independent set, but it is not a *maximum* independent set if $m > 1$. The vertices of degree 1 form the only maximum independent set. Every maximum(-sized) instance is a maximal instance, but the converse usually fails. The words have different meanings in any context involving containment and size, but not when describing numbers; "maximum vertex degree" and "maximal vertex degree" have the same meaning.

EXERCISES

Most problems in this book require proofs. Words like "construct", "show", "obtain", "determine", etc., explicitly state that proof is required. Disproof by providing a counterexample includes showing that it is a counterexample.

1.2.1. (–) Determine whether K_4 contains each of the following, by giving an example or a proof of non-existence.
   a) A walk that is not a trail.
   b) A trail that is not closed and is not a path.
   c) A closed trail that is not a cycle.

1.2.2. (–) Prove or disprove: If W is a closed walk in a simple graph, then W contains a cycle.

1.2.3. Use ordinary (not strong) induction to prove that every u, v-walk contains a u, v-path.

1.2.4. (!) Prove that the edge set of every closed trail can be partitioned into pairwise edge-disjoint cycles.

1.2.5. (!) Prove or disprove the following statements about simple graphs. (Comment: "distinct" does not mean "disjoint".)
   a) The union of the edge sets of distinct u, v-walks must contain a cycle.
   b) The union of the edge sets of distinct u, v-paths must contain a cycle.

1.2.6. Suppose edge e appears an odd number of times in a closed walk W. Prove that W contains a cycle using e.

1.2.7. (–) Suppose that T is a maximal trail in a graph G and that T is not a closed trail. Prove that the endpoints of T have odd degree.

1.2.8. (–) Suppose G has k components and H has l components. Determine the number of components of $G + H$ and the number of components of $G \vee H$.

1.2.9. Let G be the graph whose vertex set is the set of permutations of $\{1, \ldots, n\}$, with two permutations $a_1, \ldots, a_n$ and $b_1, \ldots, b_n$ adjacent if they differ by interchanging a pair of adjacent entries. Prove that G is connected.

1.2.10. Let G be the graph whose vertex set is the set of n-tuples with coordinates in $\{0, 1\}$, with x adjacent to y if x and y differ in two places. Determine the number of components of G.

1.2.11. (!) Prove that every graph with n vertices and k edges has at least $n - k$ components.

1.2.12. (–) Let G be the graph with vertex set $\{1, \ldots, 24\}$ in which i and j are adjacent if and only if their greatest common factor exceeds 1. Count the components, and find the maximum length of a path that is an induced subgraph.

1.2.13. (–) Suppose v is a vertex of a connected simple graph G. Prove that v has a neighbor in every component of $G - v$. Conclude that no graph has a cut-vertex of degree 1.

1.2.14. Suppose v is a cut-vertex of a graph G. Without using contradiction, prove directly that $\overline{G} - v$ is connected.

1.2.15. Suppose G is a graph with vertices $v_1, \ldots, v_n$, and let $H_i = G - v_i$ for $1 \le i \le n$. Prove that G is connected if and only if at least two of the graphs in $\{H_i\}$ are connected.

1.2.16. (–) In the graph below, find a bipartite subgraph with the maximum number of edges. Prove that no other bipartite subgraph has this many edges.

1.2.17. (–) Let G be the graph whose vertex set is the set of n-tuples with coordinates in $\{0, 1\}$, with x adjacent to y if x and y differ in one place. Determine whether G is bipartite.

1.2.18. (!) Let G be the graph whose vertices are the permutations of $\{1, \ldots, n\}$, with two permutations $a_1, \ldots, a_n$ and $b_1, \ldots, b_n$ adjacent if they differ by interchanging some pair of entries. Prove that G is bipartite. (Hint: for each permutation a, count the pairs i, j such that $i < j$ and $a_i > a_j$.)

1.2.19. Let G be a simple graph with vertices $v_1, \ldots, v_n$. Let A^k denote the kth power of the adjacency matrix of G under matrix multiplication. Prove that entry i, j of A^k is the number of v_i, v_j-walks of length k in G. Prove that G is bipartite if and only if, for some odd integer $r > n$, the diagonal entries of A^r are all 0. (Reminder: A walk is an *ordered* list of vertices and edges.)

1.2.20. (!) Prove that G is bipartite if and only if every subgraph H of G has an independent set consisting of at least half of $V(H)$.

1.2.21. (–) Prove that a finite directed graph contains a (directed) cycle if every vertex is the tail of at least one edge. (The same conclusion holds if every vertex is the head of at least one edge.)

1.2.22. (!) *The Odd Graph.* The graph O_k is a graph whose vertices are the k-element subsets of $\{1, 2, \ldots, 2k+1\}$, with two vertices being adjacent if they are disjoint sets. For example, O_2 is the Petersen graph. Determine whether O_k is bipartite. Prove that the shortest cycle in O_k has length 6 if $k \geq 3$.

1.2.23. (−) In the graph below, find all the maximal paths, maximal cliques, and maximal independent sets. Also find all the maximum paths, maximum cliques, and maximum independent sets.

1.2.24. (!) Prove that a finite graph having at least one edge contains at least two vertices that are not cut-vertices. (Hint: use extremality.)

1.2.25. (+) Suppose G does not have two vertices of degree 1 with a common neighbor. Prove that G has two adjacent vertices whose deletion does not separate G. (Hint: prove that the last two vertices of a longest path have this property.) (Lovász [1979, p269])

1.2.26. (!) Suppose that P and Q are two paths of maximum length in a connected graph G. Prove that P and Q have a common vertex.

1.2.27. Suppose G is a connected simple graph that does not have a 4-vertex induced subgraph that is a path or a cycle. Prove that G has a vertex adjacent to every other vertex. (Wolk) (Hint: consider a vertex of maximum degree.)

1.2.28. The *girth* of G is the length of a shortest cycle in G, with the girth of an acyclic graph being infinite. Prove that
   a) A k-regular graph of girth four has at least $2k$ vertices, and there exists exactly one unlabeled k-regular graph of girth four with $2k$ vertices.
   b) A k-regular graph of girth five has at least $k^2 + 1$ vertices, and there is such a graph with $k^2 + 1$ vertices if k is 2 or 3. (The graph is unique, but this need not be proved.)

1.3 Vertex Degrees and Counting

A graph *parameter* is a real-valued function on graphs. The vertex degrees are graph parameters as fundamental as the number of vertices and number of edges. We repeat the definition.

1.3.1. Definition. The *degree* of a vertex v in a graph G, written $d_G(v)$ or $d(v)$, is the number of non-loop edges containing v plus twice the number of loops containing v. The maximum degree is $\Delta(G)$; the minimum degree is $\delta(G)$. A graph G is *regular* if $\Delta(G) = \delta(G)$;

k-regular if $\Delta(G) = \delta(G) = k$. A vertex of degree k is *k-valent*. The *neighborhood* of v, written $N_G(v)$ or $N(v)$, is $\{x \in V(G): x \leftrightarrow v\}$; x is a *neighbor* of v if $x \in N(v)$. An *isolated vertex* has degree 0.

1.3.2. Definition. The *order* of a graph G, written $n(G)$, is the number of vertices in G. An *n-vertex graph* is a graph of order n. We use $e(G)$ to denote the number of edges in G, even though we also use e by itself to denote an edge.

The order of a graph makes sense only when the graph has finitely many vertices. This definition and many other statements in this book implicitly restrict discussion to finite graphs. We adopt this convention explicitly now: unless otherwise specified, all statements are made for finite graphs only.

COUNTING AND BIJECTIONS

One way to prove equality between two formulas is to show that they count the same set in two different ways. We use this first to prove a relationship between vertex degrees and edges.

1.3.3. Theorem. (Degree-Sum Formula). If G is a graph with vertex degrees $d_1, \ldots, d_n$, then $\Sigma d_i = 2e(G)$.

Proof. Summing the degrees counts each edge twice, since each edge has two ends and contributes to the degree of each endpoint. The set we have counted in two ways is the set of pairs (v, e) such that $v \in V(G)$, $e \in E(G)$, and $v \in e$; these are precisely the 1's in the incidence matrix $M(G)$ defined in Section 1.1. There are two 1's in $M(G)$ for each edge, so counting the 1's of $M(G)$ by columns gives $2e(G)$. There are $d(v)$ such 1's for vertex v, so counting the 1's by rows gives $\Sigma d(v)$. $\square$

By the degree-sum formula, the average vertex degree is $2e(G)/n(G)$, and hence $\delta(G) \leq 2e(G)/n(G) \leq \Delta(G)$. We list two other immediate corollaries.

1.3.4. Corollary. Every graph has an even number of vertices of odd degree. No graph of odd order is regular with odd degree. $\square$

1.3.5. Corollary. A k-regular graph with n vertices has $nk/2$ edges. $\square$

1.3.6. Example. *k-dimensional cubes.* There are 2^k k-tuples in which each position is 0 or 1; call this set S. The *k-dimensional cube* or *hypercube* is the graph Q_k with vertex set S in which two k-tuples are adjacent if and only if they differ in exactly one position. Below we illustrate Q_3. The hypercube is an architecture for parallel computers; processing units can communicate directly if they correspond to adjacent vertices in Q_k.

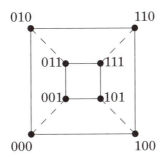

The *weight* of a 0,1-vector is the number of 1's. Every edge of Q_k consists of a vector of even weight and a vector of odd weight. Hence the vectors of even weight form an independent set, as do the vectors of odd weight, and Q_k is bipartite. Since each vector can be changed in k places, Q_k is k-regular; by Corollary 1.3.5, Q_k has $k2^{k-1}$ edges.

Deleting the dashed edges in the illustration leaves $2Q_2$. This suggests an inductive description of Q_k. The basis is the 1-vertex graph Q_0 (the unique binary vector of length 0). Given Q_{k-1}, we construct Q_k in two steps: 1) take two disjoint copies of Q_{k-1}; call them Q^0 and Q^1. 2) For each $v \in V(Q_{k-1})$, append a 0 to the end of the vector for v in Q^0 and a 1 to the end of the vector for v in Q^1, and add an edge consisting of these two vertices. □

We proved the degree-sum formula by a counting argument; after counting the edges containing each vertex, we divided by the number of times each edge was counted. This technique generalizes for larger subgraphs. Suppose we want to count the subgraphs of G isomorphic to H (the "copies" of H in G). Suppose J is a graph contained in H. If we count the copies of H containing each copy of J, then we will have counted the copies of H equally often, because each copy of H contains the same number of copies of J. The degree-sum formula does this for the case $H = K_2$, using $J = K_1$.

1.3.7. Proposition. Suppose J, H, G are graphs with $J \subseteq H \subseteq G$, and suppose H contains l copies of J. If G contains m copies of J, and the ith copy of J appears in k_i copies of H in G, then G contains $\Sigma_{i=1}^{m} k_i / l$ copies of H.

Proof. The sum $\Sigma_{i=1}^{m} k_i$ counts each copy of H l times. □

1.3.8. Example. *The Petersen graph has twelve 5-cycles.* Let G be the Petersen graph, $H = C_5$, and $J = P_3$. Each $v \in V(G)$ has degree 3, so there are $\binom{3}{2} = 3$ copies of J with v as the central vertex. Each copy of J has one central vertex, so there are $10 \cdot 3 = 30$ copies of J in G.

Each copy of J with vertices u, v, w can be extended at w in two ways to obtain P_4. Each P_4 appears in one 5-cycle, because nonadjacent vertices have exactly one common neighbor in G. Hence each copy of J appears in two copies of $H = C_5$. Since each 5-cycle contains five copies of J, Proposition 1.3.7 yields exactly $(30 \cdot 2)/5 = 12$ 5-cycles in G. $\square$

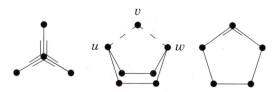

We can count a set by finding a one-to-one correspondence (a *bijection*) between it and a set of known size. For example, given vertices $v_1, \ldots, v_n$, each pair can be an edge or not. This establishes a bijection between the simple graphs with vertex set $v_1, \ldots, v_n$ and the subsets of an $\binom{n}{2}$-element set. Hence there are $2^{\binom{n}{2}}$ simple graphs with this vertex set. The next result uses this idea and the parity of vertex degrees.

1.3.9. Theorem. For $n \geq 1$, there are $2^{\binom{n-1}{2}}$ simple graphs with vertex set $\{v_1, \ldots, v_n\}$ such that every vertex degree is even.

Proof. Since $2^{\binom{n-1}{2}}$ is the number of simple graphs with vertex set $\{v_1, \ldots, v_{n-1}\}$, we establish a bijection to that set of graphs. Given a simple graph G with vertices $v_1, \ldots, v_{n-1}$, we form a new graph G' by adding a vertex v_n and making it adjacent to each vertex that has odd degree in G, as illustrated below. The vertices with odd degree in G have even degree in G'. Also, v_n itself has even degree because the number of vertices of odd degree in G is even. Conversely, deleting the vertex v_n from any graph on $\{v_1, \ldots, v_n\}$ with even degrees produces a graph on $\{v_1, \ldots, v_{n-1}\}$, and this is the inverse of the first procedure. We have established a one-to-one correspondence between the sets; hence they have the same size. $\square$

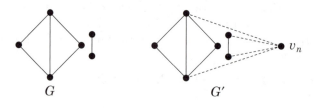

THE PIGEONHOLE PRINCIPLE

The pigeonhole principle is a simple notion that leads to elegant proofs and may reduce case analysis. It essentially says that any set of numbers has a number at least as large as the average.

1.3.10. Lemma. (Pigeonhole Principle). If a set consisting of more than kn objects is partitioned into n classes, then some class receives more than k objects.

Proof. The contrapositive states that if every class receives at most k objects, then a total of at most kn objects have been distributed. □

1.3.11. Proposition. Every simple graph with at least two vertices has two vertices of equal degree.

Proof. In a simple graph with n vertices, every vertex degree belongs to the set $\{0, \ldots, n-1\}$. If fewer than n values occur, then the pigeonhole principle yields the claim. Otherwise, both $n-1$ and 0 occur as vertex degrees. This is impossible; if one vertex is adjacent to all others, then there can be no isolated vertex. □

1.3.12. Proposition. If G is a simple n-vertex graph with $\delta(G) \geq (n-1)/2$, then G is connected.

Proof. Choose $u, v \in V(G)$. If $u \not\leftrightarrow v$, then at least $n-1$ edges join $\{u, v\}$ to the remaining vertices, since $\delta(G) \geq (n-1)/2$. There are $n-2$ other vertices, so the pigeonhole principle implies that one of them receives two of these edges. Since G is simple, this vertex is a common neighbor of u and v. We have proved that every pair of vertices is adjacent or has a common neighbor, so G is connected.

Alternatively, we could use set operations to prove that nonadjacent vertices have a common neighbor. If $u \not\leftrightarrow v$, then $|N(u) \cup N(v)| \leq n-2$. We compute $|N(u) \cap N(y)| = |N(u)| + |N(v)| - |N(u) \cup N(v)| \geq (n-1)/2 + (n-1)/2 - (n-2) = 1$. □

1.3.13. Example. *A best possible result.* The graph $K_{\lfloor n/2 \rfloor} + K_{\lceil n/2 \rceil}$ has two components.[†] Since it has minimum degree $\lfloor n/2 \rfloor - 1$ and yet is disconnected, the inequality in Proposition 1.3.12 cannot be improved. □

$K_{\lfloor n/2 \rfloor}$ $K_{\lceil n/2 \rceil}$

[†]We use $\lfloor x \rfloor$ and $\lceil x \rceil$ to denote the *floor* and *ceiling* of x; $\lfloor x \rfloor$ is the largest integer less than or equal to x, and $\lceil x \rceil$ is the least integer greater than or equal to x.

A result is *best possible* if the conclusion no longer holds when we weaken one of the conditions. We can also state the results above as "The minimum value of $\delta(G)$ that forces an n-vertex simple graph G to be connected is $\lfloor n/2 \rfloor$," or "The maximum value of $\delta(G)$ for a disconnected n-vertex simple graph is $\lfloor n/2 \rfloor - 1$." An *extremal problem* asks for an extreme value of some parameter over a class of objects. Proving that $\max_{x \in S} f(x) = \beta$ requires showing 1) that $f(x) \le \beta$ for all $x \in S$, and 2) that $f(x) = \beta$ for some $x \in S$. The proof of the upper bound must apply to every $x \in S$. We can prove that the bound is achievable by constructing an example and proving that it has the desired properties.

1.3.14. Example. *Covering cliques with bipartite subgraphs.* Consider an air traffic system with k airlines and n cities. Suppose that 1) direct service between two cities means round-trip direct service, and 2) each pair of cities has direct service from at least one airline. Suppose also that no airline can schedule a cycle through an odd number of cities. As a function of k, what is the maximum number of cities in the system?

The answer is 2^k. Phrased using graph theory, the question asks for the largest n such that K_n can be expressed as the union of k bipartite graphs, one for each airline. Suppose K_n is the union of bipartite graphs $G_1, \ldots, G_k$, with X_i, Y_i being the bipartition of G_i. We may assume that $X_i \cup Y_i = V(K_n)$, since adding an isolated vertex does not introduce an odd cycle. For each vertex v, we define a binary k-tuple a by setting $a_i = 0$ if $v \in X_i$ and $a_i = 1$ if $v \in Y_i$. If there are more than 2^k vertices, then by the pigeonhole principle some pair have the same k-tuple. Since these two k-tuples are equal in each coordinate, these vertices belong to the same partite set in each bipartite subgraph. Hence the edge between them belongs to none of the bipartite subgraphs, which contradicts the hypotheses that $K_n = \cup G_i$.

We have proved that $K_n = \cup G_i$ requires $n \le 2^k$. If $n = 2^k$, then we can express K_n as such a union. We assign distinct binary k-tuples to the n vertices and let $E(G_i)$ consist of all edges between vertices whose ith coordinate is 0 and vertices whose ith coordinate is 1. This constructs k bipartite subgraphs. Since distinct k-tuples differ in some coordinate, every edge belongs to some G_i, and we have constructed $G_1, \ldots, G_k$ such that $K_n = \cup G_i$ (shown below for $k = 2$). Hence the upper bound of 2^k is achievable. $\square$

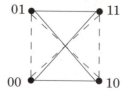

We cannot prove the upper bound in Example 1.3.14 by taking the successful construction with 2^k vertices (the "worst case") and showing that no vertex can be added. This does not consider all ways to build a clique with $2^k + 1$ vertices; it considers only those that contain the special construction on some 2^k vertices. To use this approach, we would have to prove that every expression of K_{2^k} as a union of k bipartite subgraphs has the form we presented.

A similar trap arises in inductive proofs. If, in the induction step, we grow an object with the new value of the parameter from a smaller object (where the induction hypothesis holds), then we must prove that all the objects of the new size have been considered.

1.3.15. Example. *The induction trap.* We consider an example where this error produces a **false** conclusion. By the degree-sum formula, every regular graph with odd degree has even order. The smallest 3-regular simple graph, K_4, is connected and has no cut-edge. Suppose we want to prove that every 3-regular simple connected graph has no cut-edge. Given a simple 3-regular graph G with $2k$ vertices, we can obtain a simple 3-regular graph G' with $2(k + 1)$ vertices (the next larger size) by "expansion": take two edges of G, replace them by paths of length 2 through new vertices, and add an edge joining the two new vertices. For example, $K_{3,3}$ arises from K_4 by expansion on two disjoint edges; expansion on two edges of K_4 with a common endpoint yields the other 3-regular 6-vertex simple graph.

If G is connected, then the graph G' obtained by an expansion is also connected: a path between old vertices that traversed an edge now subdivided has merely lengthened, and a path in G' with a new vertex as the desired endpoint can be obtained from a path in G to its neighbor. Also, if G is connected and has no cut-edge, then every edge of G lies on a cycle (Corollary 1.2.12). This remains true in G', even for the edges formed by subdividing. The edge involving the new vertices also lies on a cycle using a path in G between the edges that were subdivided. We have proved that if G is connected and has no cut-edge, then G' is connected and has no cut-edge.

We might think we have proved by induction on k that every 3-regular simple connected graph with $2k$ vertices has no cut-edge, but the graph below is a counterexample. The proof fails because we cannot build every 3-regular simple connected graph from K_4 by expansions. In Section 3.3 we will prove that every 3-regular simple graph with $2k$

vertices and no cut-edges has k pairwise disjoint edges, but even this we cannot prove using expansion, because expansion does not generate all 3-regular connected graphs that don't have cut-edges (Exercise 31). $\square$

Most inductive proofs in this book avoid the induction trap as follows. In the induction step to prove $P(n)$, we start with an arbitrary instance G having the new (larger) value n for the induction parameter. We obtain from this a smaller example G' that *satisfies the hypothesis of P*. The induction hypothesis then applies to G'; hence the conclusion of P holds for G'. We use this to obtain the conclusion of P for G. Verifying that the smaller example satisfies the hypothesis of P plays the same role as verifying that we have considered all the larger instances when we write the proof by starting with the smaller object.

TURÁN'S THEOREM

In politics and warfare, seldom do two enemies have a common enemy; usually two of the three will ally against the third. Given n factions, how many pairs of enemies can there be if no two enemies can have a common enemy? In the language of graphs, we are asking for the maximum number of edges in a simple n-vertex graph with no triangle (K_3). Bipartite graphs have no triangles, but non-bipartite graphs such as long odd cycles and the Petersen graph also have no triangles. Using extremality (choosing a vertex of maximum degree), we will prove that the solution is always a complete bipartite graph. For example, $K_{2,3}$ has more edges than C_5.

1.3.16. Definition. A graph G is *H-free* if H is not an induced subgraph of G.

1.3.17. Example. *The induction trap, again.* We might try to prove by induction that $K_{\lfloor n/2 \rfloor, \lceil n/2 \rceil}$ is the largest triangle-free simple graph with n vertices. In the induction step, we might say "suppose the claim is true when $n = k$, so that $K_{\lfloor k/2 \rfloor, \lceil k/2 \rceil}$ is the largest triangle-free graph with k vertices. When we add a vertex to $K_{\lfloor k/2 \rfloor, \lceil k/2 \rceil}$ to obtain a triangle-free graph with $k + 1$ vertices, we cannot make it adjacent to vertices from both partite sets. Hence we add the maximum number of edges by adding the new vertex to the partite set of size $\lfloor k/2 \rfloor$ and joining it to all vertices in the other class. This proves the claim when $n = k + 1$."

We fell into the induction trap! We did not consider all triangle-free graphs with $k + 1$ vertices, only those that contain the extremal graph $K_{\lfloor k/2 \rfloor, \lceil k/2 \rceil}$ on k vertices as an induced subgraph. Although it is true that every largest example with $k + 1$ vertices contains $K_{\lfloor k/2 \rfloor, \lceil k/2 \rceil}$ as an induced subgraph, we cannot use that before proving it. We have not eliminated the possibility that the largest example with $k + 1$ vertices arises by adding a new vertex of high degree to a non-maximal example with k vertices. (Exercise 36 suggests a correct proof by induction.) $\square$

1.3.18. Proposition. (Mantel [1907]) The maximum number of edges in an n-vertex triangle-free simple graph is $\lfloor n^2/4 \rfloor$.

Proof. Suppose G is an n-vertex triangle-free simple graph. Let $k = \Delta(G)$, and let x be a vertex of maximum degree in G. Since G has no triangles, there are no edges between neighbors of x. Hence summing the degrees of x and its non-neighbors counts at least one endpoint of every edge, and the total is at least $e(G)$. We sum over $n - k$ vertices, each having degree at most k; hence $e(G) \le (n - k)k$.

Observe that $(n - k)k$ counts the edges in $K_{n-k,k}$. If the partite sets differ in size by more than one, then moving a vertex from the larger set to the smaller set gains more edges than it loses. Hence the bound $(n - k)k$ is maximized for integer k when $k = \lfloor n/2 \rfloor$, and we conclude that $e(G) \le \lfloor n^2/4 \rfloor$. To prove that the bound is best possible, we exhibit a triangle-free graph with $\lfloor n^2/4 \rfloor$ edges: this is $K_{\lfloor n/2 \rfloor, \lceil n/2 \rceil}$. $\square$

This approach extends to solve the general problem of finding the largest n-vertex graph containing no complete subgraph with $r + 1$ vertices. The extremal graphs belong to a special family.

1.3.19. Definition. A *complete multipartite graph* is a graph whose vertices can be partitioned into sets so that uv is an edge if and only if u and v belong to different sets. Equivalently, G is a complete multipartite graph if and only if every component of $\overline{G}$ is a complete graph. When $k \ge 2$, we write $K_{n_1,\ldots,n_k}$ for the complete k-partite graph with partite sets of sizes $n_1, \ldots, n_k$ and complement $K_{n_1} + \cdots + K_{n_k}$.

Although K_n is a complete multipartite graph with one vertex in each color class, we use the notation K_n instead of $K_{1,\ldots,1}$. The only 1-partite graphs are independent sets; hence the notation for complete k-partite graphs applies only for $k \ge 2$. If G is a complete multipartite graph and the partite set containing v has size t, then $d_G(v) = n - t$. This makes it easy to count edges (Exercise 39).

1.3.20. Example. *The Turán graph.* The *Turán graph* $T_{n,r}$ is the complete r-partite graph with n vertices that has b parts of size $a + 1$ and $r - b$ parts of size a, where $a = \lfloor n/r \rfloor$ and $b = n - ra$. Turán proved that $T_{n,r}$ is the unique largest simple n-vertex graph with no $r + 1$-clique. No r-partite graph has an $r + 1$-clique, because each partite set contributes at most one vertex to a clique. □

1.3.21. Theorem. (Turán [1941]) Among the n-vertex simple graphs with no $r + 1$-clique, $T_{n,r}$ has the maximum number of edges.

Proof. We prove first that $T_{n,r}$ has the most edges among r-partite graphs. If H is a complete r-partite graph whose largest and smallest partite sets differ by more than one in size, then we can gain edges by moving a vertex v from the largest class to the smallest class. The edges not involving v are the same as before, but v now has more neighbors, gaining those in its old class and losing those in its new class. Hence we maximize the edge count by equalizing the sizes as in $T_{n,r}$.

It remains to prove that when G has no $r + 1$-clique, there is an r-partite graph H such that $V(H) = V(G)$ and $e(H) \geq e(G)$. We use induction on r. When $r = 1$, G and H have no edges. For the induction step, suppose $r > 1$. Let G be an n-vertex graph with no $r + 1$- clique, and let $x \in V(G)$ be a vertex of degree $k = \Delta(G)$. Let G' be the subgraph of G induced by the neighbors of x. Since x is adjacent to every vertex in G', the graph G' has no r-clique. The induction hypothesis yields an $r - 1$-partite graph H' with vertex set $N(x)$ such that $e(H') \geq e(G')$.

Let H be the graph formed from H' by joining all of $N(x)$ to all of $S = V(G) - N(x)$. Since S is an independent set, H is r-partite. We claim that $e(H) \geq e(G)$. By construction, $e(H) = e(H') + k(n - k)$. We also have $e(G) \leq e(G') + \Sigma_{v \in S} d_G(v)$, since the sum counts each edge of G once for each endpoint it has outside $V(G')$. Since $\Delta(G) = k$, we have $d_G(v) \leq k$ for each $v \in S$, and $|S| = n - k$. Hence $e(G) \leq e(G') + (n - k)k \leq e(H') + k(n - k) = e(H)$, as desired. □

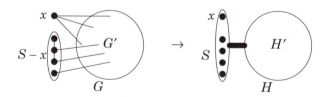

Exercises 38-44 pertain to Turán's Theorem, including the uniqueness of the extremal graph, alternative proofs, the value of $e(T_{n,r})$, and applications. Turán's theorem applies to extremal problems when some condition forbids cliques of a given order; we describe one geometric application, presented in Bondy-Murty [1976, p113-115].

1.3.22. Example. *Distant pairs of points.* Consider a circular city of diameter 1. We might want to place n police cars so as to maximize the number of pairs that are far apart, separated by distance at least $d = 1/\sqrt{2}$. If six cars occupy equally spaced points on a circle, then the only pairs that are not at least d apart are the consecutive pairs around the outside, so there are nine good pairs. If instead we place two cars each near the vertices of an equilateral triangle with sides of length $\sqrt{3}/2$, then only three pairs will be bad and twelve will be good. (This may not be the best criterion for locating police cars!) In general, if we place $\lceil n/3 \rceil$ or $\lfloor n/3 \rfloor$ cars near each vertex of this triangle, the good pairs correspond to edges of the tripartite Turán graph. We show next that this is the best construction. □

1.3.23. Theorem. If S is a set of n points in the plane with no pair more than distance 1 apart, then the maximum number of pairs of points more than distance $1/\sqrt{2}$ apart is $\lfloor n^2/3 \rfloor$.

Proof. Create a graph G with vertex set S in which vertices are adjacent when their distance exceeds $1/\sqrt{2}$. By Turán's Theorem and the construction above, it suffices to show that G has no K_4.

Among any four points, some triple must form an angle of at least $90°$: If the points form a convex quadrilateral, then the interior angles sum to $360°$, and one of them is at least $90°$. If one point is inside the triangle formed by the other three, then its rays to the others form three angles summing to $360°$, and one is at least $120°$.

Suppose G has a 4-clique corresponding to the points w, x, y, z, in which $\angle xyz \geq 90°$. Since the lengths of xy and yz exceed $1/\sqrt{2}$, xz is longer than the hypotenuse of a right triangle with legs of length $1/\sqrt{2}$. Hence the distance between x and z exceeds 1, which contradicts the hypothesis. □

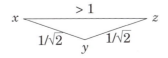

EXERCISES

A statement with a parameter must be proved for all values of the parameter; it cannot be proved by giving examples.

1.3.1. (–) Is it true that a graph having exactly two vertices of odd degree must contain a path from one to the other? Give a proof or a counterexample.

1.3.2. In a class with nine students, each student sends valentine cards to three others. Is it possible that each student receives cards from the same three students to whom he or she sent cards?

1.3.3. In a league with two divisions of 13 teams each, determine whether it is possible to schedule a season with each team playing nine games against teams within its division and four games against teams in the other division.

1.3.4. Suppose l, m, n are nonnegative integers with $l + m = n$. Find necessary and sufficient conditions on l, m, n such that there exists a connected simple n-vertex graph with l vertices of even degree and m vertices of odd degree.

1.3.5. Suppose C is a closed walk in a graph G, and let H be the subgraph of G consisting of the edges that appear an odd number of times in C. Prove that $d_H(v)$ is even for every $v \in V(G)$.

1.3.6. (–) Suppose H is a graph formed by deleting a vertex from a regular graph G. Describe (and justify) a method for obtaining G from H.

1.3.7. For each $k \geq 4$, determine the smallest n such that
   a) there is a simple k-regular graph with n vertices.
   b) there exist nonisomorphic simple k-regular graphs with n vertices.

1.3.8. (!) Suppose $k \geq 2$ and G is a k-regular bipartite graph with bipartition X, Y. Prove that $|X| = |Y|$. Prove also that G has no cut-edge.

1.3.9. (!) Prove that a finite graph with every vertex degree even has no cut-edge. For each $k \geq 1$, construct a $2k + 1$-regular simple graph having a cut-edge.

1.3.10. (+) A *mountain range* is a polygonal curve from $(a, 0)$ to $(b, 0)$ in the upper half-plane. Suppose A and B are at $(a, 0)$ and $(b, 0)$, respectively. Prove that A and B can meet by traveling on the mountain range in such a way that at all times their heights above the horizontal axis are the same. (Hint: define a graph to model the movements, and use the even parity of the number of vertices with odd degree.) (Communicated by D. Hoffman.)

1.3.11. (–) Prove that the inductive and direct definitions of the k-dimensional cube Q_k yield the same graph, and prove inductively that $e(Q_k) = k2^{k-1}$.

1.3.12. (–) Count the copies of P_3 in Q_k.

1.3.13. (–) Prove that the 3-dimensional cube Q_3 can be expressed as the union of edge-disjoint copies of $K_{1,3}$ and also as the union of edge-disjoint copies of P_4.

1.3.14. (–) Prove that the graph drawn below is isomorphic to $\overline{Q}_3$.

1.3.15. Determine the smallest simple bipartite graph that is not a subgraph of any cube Q_k.

1.3.16. (+) *Automorphisms of the k-dimensional cube Q_k.*
    a) Prove that the only subgraphs of Q_k isomorphic to Q_l are those induced by a set of vertices agreeing on some $k - l$ coordinates. (Hint: prove that vertices serving as antipodal vertices of Q_l differ in l coordinates.)
    b) Use part (a) to count the automorphisms of Q_k.

1.3.17. Prove that every cycle of length $2r$ in a hypercube is contained in a subcube of dimension at most r. Prove that this subcube is unique when $r = 2$ or $r = 3$, but that it need not be unique when $r = 4$.

1.3.18. *Cycles in the k-dimensional cube Q_k.*
    a) Prove that if (x, y, z) and (a, b, c) are two 3-vertex paths in Q_k, then there is an automorphism mapping x, y, z into a, b, c, respectively.
    b) Use part (a) and Prop. 1.3.7 to count the 4-cycles and the 6-cycles in Q_k.

1.3.19. Given $k \in \mathbb{N}$, let G be the subgraph of Q_{2k+1} induced by the vertices in which the number of ones and zeros differs by 1. Prove that G is regular, and compute $n(G)$, $e(G)$, and the length of the shortest cycle in G.

1.3.20. (!) Count the cycles of length n in K_n and the cycles of length $2n$ in $K_{n,n}$.

1.3.21. Count the 6-cycles in $K_{m,n}$.

1.3.22. Prove that the Petersen graph has exactly ten 6-cycles. (Hint: establish a bijection between the 6-cycles and the copies of $K_{1,3}$.)

1.3.23. (!) Use graphs and bijections (not algebra!) to prove that
    a) $\binom{n}{2} = \binom{k}{2} + k(n - k) + \binom{n-k}{2}$ for $0 \le k \le n$.
    b) If $\Sigma n_i = n$, then $\Sigma \binom{n_i}{2} \le \binom{n}{2}$.

1.3.24. (+) Suppose that G is a triangle-free simple n-vertex graph such that every pair of nonadjacent vertices has exactly two common neighbors.
    a) Prove that G is regular.
    b) Given that G is regular of degree k, prove that $n(G) = 1 + \binom{k+1}{2}$.

1.3.25. (!) Prove or disprove:
    a) Deleting a maximum degree vertex cannot increase the average degree.
    b) Deleting a minimum degree vertex cannot reduce the average degree.

1.3.26. (+) Suppose n, k are integers satisfying $1 < k < n - 1$ and $n \geq 4$. Suppose that G is a simple n-vertex graph and that every k-vertex induced subgraph of G has m edges.

a) Suppose G' is an induced subgraph of G with l vertices, where $l > k$. Prove that $e(G') = m\binom{l}{k}/\binom{l-2}{k-2}$.

b) Use (a) to prove that $G = K_n$ or $G = \overline{K}_n$. (Hint: use (a) to compute the entry in the adjacency matrix for the vertex pair uv; the formula is independent of the choice of u and v.)

1.3.27. Suppose G is a loopless graph, and $a = 2e(G)/n(G)$ is the average degree in G. Let $t(v)$ denote the average of the degrees of the neighbors of v. Prove that $t(v) \geq a$ for some $v \in V(G)$. Prove constructively that there are infinitely many graphs G such that $t(v) > a$ for all $v \in V(G)$.

1.3.28. Suppose S is a subset of the vertices in Q_k such that no pair of vertices in S is adjacent or has a common neighbor. Use the pigeonhole principle to prove that $|S| \leq \lfloor 2^k/(k+1) \rfloor$. Show that the bound is best possible when $k = 3$. (Comment: this bound is not best possible when $k = 4$.)

1.3.29. (!) *Covering cliques with bipartite subgraphs* (alternative proof for Example 1.3.14). Use induction on k (for both implications) to prove that K_n is a union of k bipartite graphs if and only if $n \leq 2^k$.

1.3.30. *Membership in cliques.*

a) Suppose a simple n-vertex graph G has an edge uv with $d(u) + d(v) = n + k$, where $k \geq 0$. Prove that uv belongs to at least k copies of K_3 in G.

b) Suppose H is a copy of K_r in a simple n-vertex graph G, and $\Sigma_{v \in H} d(v) > (r-1)n + k$. Prove that H appears in more than k copies of K_{r+1} in G.

1.3.31. *Expansion of 3-regular graphs* (see Example 1.3.15). For $n = 4k$, where $k \geq 2$, construct a connected 3-regular simple graph with n vertices that has no cut-edge but cannot be obtained from a smaller 3-regular simple graph by expansion. (Hint: G can be obtained from a smaller simple graph by expansion if and only if G has some edge to which the inverse "erasure" operation can be applied to obtain a simple graph smaller than G.)

1.3.32. (!) Consider a party attended by n married couples. Suppose no person shakes hands with his or her spouse, and the $2n - 1$ people other than the host shake hands with different numbers of people. With how many people does the hostess shake hands? (Hint: use induction. In the induction step, consider a general instance with n couples and extract an instance with $n - 1$ couples.)

1.3.33. (!) Suppose $n \geq 2$. Determine the disconnected n-vertex simple graphs that have the maximum number of edges.

1.3.34. (−) Determine the maximum number of edges in an n-vertex simple graph that has an independent set of size a.

1.3.35. Suppose G is a simple graph with $n > 3$ vertices.

a) Prove that if G has more than $n^2/4$ edges, then G has a vertex whose deletion leaves a graph with more than $(n-1)^2/4$ edges.

b) Use (a) to prove by induction that G contains a triangle if $e(G) > n^2/4$.

1.3.36. Prove that every n-vertex triangle-free simple graph with the maximum number of edges is isomorphic to $K_{\lfloor n/2 \rfloor, \lceil n/2 \rceil}$. (Hint: examine the proof of Proposition 1.3.18 more carefully.)

1.3.37. *Complete multipartite graphs.*
   a) Prove that a simple graph G is a complete multipartite graph if and only if G has no 3-vertex induced subgraph with one edge. (Hint: consider $\overline{G}$.)
   b) Use part (a) to prove that if a simple connected graph G has at least four vertices and is not a complete multipartite graph, then G has one of the graphs below as an induced subgraph.

1.3.38. (!) Use induction on r (without Turán's Theorem) to prove directly that every n-vertex simple graph with no $r+1$-clique has at most $(1-1/r)n^2/2$ edges.

1.3.39. (!) The Turán graph (Example 1.3.20) $T_{n,r}$ is the complete r-partite graph with b parts of size $a+1$ and $r-b$ parts of size a, where $a = \lfloor n/r \rfloor$ and $b = n - ra$.
   a) Prove that $e(T_{n,r}) = (1-1/r)n^2/2 - b(r-b)/(2r)$.
   b) Since $e(G)$ must be an integer, part (b) implies $e(T_{n,r}) \le \lfloor (1-1/r)n^2/2 \rfloor$. Determine the smallest value of r such that strict inequality occurs for some n. For this value of r, determine the values of n such that $e(T_{n,r}) < \lfloor (1-1/r)n^2/2 \rfloor$.

1.3.40. Compare the Turán graph $T_{n,r}$ with the graph $\overline{K}_a + K_{n-a}$ to prove directly that $e(T_{n,r}) = \binom{n-a}{2} + (r-1)\binom{a+1}{2}$.

1.3.41. (!) Suppose a graph G has n vertices and m edges. Use Turán's Theorem to give lower bounds on the number of vertices in the largest clique and the largest independent set in G.

1.3.42. Prove that among the n-vertex simple graphs with no $r+1$-clique, the Turán graph $T_{n,r}$ is the only one having the maximum number of edges. (Hint: examine the proof of Theorem 1.3.21 more carefully.)

1.3.43. (!) Each game of "bridge" involves two teams, each consisting of two "partners". Consider a club in which four players cannot play a game if any pair of them have previously been partners that night. Suppose 15 people come to play, but one decides to study graph theory instead. The other 14 people play until each person has been a partner with four of the others. Now the rule against having previous partners at the table makes it difficult to schedule games; the 14 players succeed in playing six more games before they can no longer find four players containing no pair of previous partners. Prove that if they can convince the graph theorist to play bridge, then at least one more game can be played (adapted from Bondy-Murty [1976, p111]).

1.3.44. Eighteen cellular-phone power stations are to be placed in a circular city of diameter four miles. Each station has a transmission range of six miles.

Prove that no matter where within the city the stations are placed, at least two of them will be able to transmit directly to at least five other stations. (adapted from Bondy-Murty [1976, p115])

1.3.45. (+) *Maximum size with no induced P_4.*
a) Suppose that G is a simple connected graph and $\overline{G}$ is disconnected. Prove that $e(G) \le \Delta(G)^2$, with equality only for $K_{\Delta(G),\Delta(G)}$.
b) Suppose that G is a simple connected graph with maximum degree D and no induced subgraph isomorphic to P_4. Prove that $e(G) \le D^2$. (Seinsche [1974], Chung-West [1993])

1.3.46. Partial analogue of Turán's Theorem for $K_{2,m}$.
a) Prove that if G is simple and $\Sigma_{v \in V(G)} \binom{d(v)}{2} > (m-1)\binom{n}{2}$, then G contains $K_{2,m}$. (Hint: view $K_{2,m}$ as two vertices with m common neighbors.)
b) Prove that $\Sigma_{v \in V(G)} \binom{d(v)}{2} \ge e(2e/n - 1)$, where G has e edges.
c) Use (a) and (b) to prove that a graph with more than $\frac{1}{2}(m-1)^{1/2}n^{3/2} + n/4$ edges contains $K_{2,m}$.
d) Application: Given n points in the plane, prove that the distance is exactly 1 for at most $\frac{1}{\sqrt{2}}n^{3/2} + n/4$ pairs. (Bondy-Murty [1976, p111-112])

1.4 Degrees and Algorithmic Proof

We continue our study of vertex degrees and proof techniques.

1.4.1. Definition. The *degree sequence* of a graph is the list of vertex degrees, usually written in nonincreasing order, as $d_1 \ge \cdots \ge d_n$. Some applications use nondecreasing order.

ALGORITHMIC OR CONSTRUCTIVE PROOF

We can prove that something exists by building it. Such proofs can be implemented as computer algorithms. A constructive proof requires more than stating an algorithm; we must also prove that the algorithm terminates and yields the desired result. This may involve induction, contradiction, finiteness, etc. Suppose we want to prove that every graph has a large bipartite subgraph.

1.4.2. Theorem. Every loopless graph G has a bipartite subgraph with at least $e(G)/2$ edges.

Proof. Suppose we start with an arbitrary partition of $V(G)$ into two sets X, Y. By including the edges having one endpoint in each set, we obtain a bipartite subgraph H with bipartition X, Y. If H contains fewer than half the edges of G incident to a vertex v, then v has more neighbors in its own class than in the other class, as illustrated below. By moving v to the other class, we gain more edges of G than we lose.

We make such a local switch in the bipartition as long as the current bipartite subgraph has a vertex that contributes fewer than half its edges. Each such switch increases the number of edges in the subgraph, so the process must terminate. When it terminates, we have $d_H(v) \geq d_G(v)/2$ for every $v \in V(G)$, and hence $e(H) \geq e(G)/2$ by the degree-sum formula. $\square$

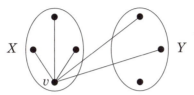

We could also use extremality: the bipartite subgraph H with the most edges has at least half the edges of G. Otherwise, we have $d_H(v) < d_G(v)/2$ for some $v \in V(G)$, and then switching v in the bipartition contradicts the choice of H. See Exercise 3 for a stronger result.

1.4.3. Example. *Local maximum.* The algorithm in Theorem 1.4.2 does not necessarily produce a bipartite subgraph with the most edges, merely one with at least half the edges. The graph below is 5-regular with 8 vertices and hence has 20 edges. The bipartition $X = \{a, b, c, d\}$ and $Y = \{e, f, g, h\}$ yields a bipartite subgraph with 12 edges, in which each vertex has degree 3. The algorithm terminates here; switching one vertex would pick up two edges but lose three. Nevertheless, the bipartition $X = \{a, b, g, h\}$ and $Y = \{c, d, e, f\}$ produces a 4-regular bipartite subgraph with 16 edges. An algorithm seeking a maximum example by local changes may get stuck in a local maximum. $\square$

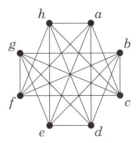

The proof of Theorem 1.4.2 illustrates one way to prove the existence of a desired configuration: define a sequence of changes to an arbitrary configuration that must terminate but can only terminate when the desired property occurs.

GRAPHIC SEQUENCES

Every graph has a degree sequence, but which sequences occur? Given nonnegative integers $d_1, \ldots, d_n$, can we determine whether there is a graph for which it is the degree sequence? The degree-sum formula implies that Σd_i must be even. This is a *necessary* condition, but it is not *sufficient* if we want a simple graph: (2,0,0) is not the degree sequence of any simple graph. We show first that the obvious necessary condition is sufficient when we allow loops and multiple edges.

1.4.4. Proposition. The nonnegative integers $d_1, \ldots, d_n$ are the vertex degrees of some multigraph if and only if Σd_i is even.

Proof. The degree-sum formula makes the condition necessary. Conversely, suppose Σd_i is even. We construct a multigraph with vertex set $v_1, \ldots, v_n$ and $d(v_i) = d_i$ for all i. Since Σd_i is even, the number of odd values is even. First form an arbitrary pairing of $\{v_i: d_i \text{ is odd}\}$, and establish an edge for each such pair. Now the remaining degree needed at each vertex is even and nonnegative; satisfy this demand for each i by placing $\lfloor d_i/2 \rfloor$ loops at v_i. $\square$

Here the sufficiency proof is an explicit construction. We could also prove that such a multigraph exists by induction on Σd_i or by induction on n (Exercise 10); there may be many workable choices for an induction parameter. The availability of loops makes the construction easy. If we forbid loops, then $(2, 0, 0)$ is no longer realizable. Exercise 16 develops a necessary and sufficient condition for realizability by a loopless multigraph.

Our discussion of large bipartite subgraphs suggests a connection between algorithmic proof and proof using induction or extremality. Proofs using induction or extremality often can be rephrased as recursive or iterative algorithms. Sometimes seek such a proof first and later convert it to an algorithm.

1.4.5. Definition. A *graphic sequence* is a list of nonnegative numbers that is the degree sequence of some simple graph. A simple graph with degree sequence d "realizes" d.

1.4.6. Example. *A recursive condition.* The list 2,0,0 is not graphic, but 2,2,1,1 is graphic, as is 1,0,1. The graph $K_2 + K_1$ realizes 1,0,1; if we add a new vertex adjacent to the isolated vertex and to one vertex of degree 1, then we obtain a graph with degree sequence 2,2,1,1 (illustrated below). Conversely, if we have a graph realizing 2,2,1,1 in which some vertex w of maximum degree is adjacent to vertices of degree 2 and 1, then we can delete w to obtain a graph with degree list 1,0,1.

These observations suggest a recursive test for graphic sequences. To test the sequence 33333221, we can seek a realization with a vertex y of degree 3 that has three neighbors of degree 3. Such a graph exists if and only if 2223221 is graphic (by deleting y). We reorder this as 3222221 and seek a realization having a vertex x of degree 3 with three neighbors of degree 2. Such a graph exists if and only if 111221 is graphic (by deleting x). We reorder this as 221111 and seek a realization having a vertex w of degree 2 with neighbors of degrees 2 and 1. Such a graph exists if and only if 10111 is graphic. Perhaps we recognize that this is indeed graphic. Beginning with a realization of 10111, we can insert w, x, y with the properties desired to obtain a realization of the original sequence 33333221. The realization is not unique. □

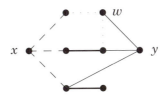

1.4.7. Theorem. (Havel [1955], Hakimi [1962]) For $n > 1$, the nonnegative integer list d of size n is graphic if and only if d' is graphic, where d' is the list of size $n - 1$ obtained from d by deleting its largest element Δ and subtracting 1 from its Δ next largest elements. The only 1-element graphic sequence is $d_1 = 0$.

Proof. For $n = 1$, the statement is trivial. For $n > 1$, we first prove that the condition is sufficient. Given d with $d_1 \geq \cdots \geq d_n$, and given a simple graph G' with degree sequence d', we add a new vertex adjacent to vertices in G' having degrees $d_2 - 1, \ldots, d_{\Delta+1} - 1$. These d_i are the Δ largest elements of d after (one copy of) Δ itself, but the numbers $d_2 - 1, \ldots, d_{\Delta+1} - 1$ need not be the Δ largest numbers in d'.

To prove necessity, we begin with a simple graph G realizing d and produce a simple graph G' realizing d'. Let w be a vertex of degree Δ in G. Let S be a set of Δ vertices in G having the "desired degrees" $d_2, \ldots, d_{\Delta+1}$. If $N(w) = S$, then we can delete w to obtain G'. Otherwise, some vertex of S is missing from $N(w)$. In this case, we will modify G to increase $|N(w) \cap S|$ without changing the degree of any vertex. Since $|N(w) \cap S|$ can increase at most Δ times, repeating this procedure converts an arbitrary G that realizes d into a graph G^* that realizes d and has $N(w) = S$. From G^* we then delete w to obtain the desired graph G' realizing d'.

If $N(w) \neq S$, then we can choose $x \in S$ and $z \notin S$ so that $w \leftrightarrow z$ and $w \not\leftrightarrow x$, since $d(w) = \Delta = |S|$. By the choice of S, $d(x) \geq d(z)$. We want to add wx and delete wz, but we must not change vertex degrees, so we must restore the degrees of x and z. It suffices to find a vertex y outside $T = \{x, z, w\}$ such that $y \leftrightarrow x$ and $y \not\leftrightarrow z$; if such a y exists, then we also delete xy and add zy (see illustration). Let ε be the number of copies of the edge xz (0 or 1). Now x has $d(x) - \varepsilon$ neighbors outside T, and z has $d(z) - 1 - \varepsilon$ neighbors outside T. Since $d(x) \geq d(z)$, the desired y outside T exists, and we can perform the desired switch. $\square$

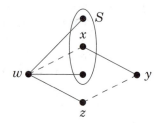

Theorem 1.4.7 tests a list of n numbers by testing a list of $n - 1$ numbers; hence it can be implemented as an iterative algorithm to test whether d is graphic. The necessary condition (Σd_i even) occurs implicitly in this characterization. Since $\Sigma d_i' = (\Sigma d_i) - 2\Delta$ and d' must have even sum to be realizable, the recursive condition implies that d must also have even sum.

In an algorithmic proof using "local change", we push an object toward a desired condition. This can also be phrased as proof by induction, where the induction parameter is the "distance" from the desired condition. In the proof above, this distance is the number of vertices in S that are missing from $N(w)$.

We used edge switches to transform an arbitrary graph with degree sequence d into a graph satisfying the desired condition. More generally, every simple graph with degree sequence d can be transformed by such switches into every other.

1.4.8. Definition. A *2-switch* is the replacement of a pair of edges xy and zw in a simple graph by the edges yz and wx, given that yz and wx did not appear in the graph originally.

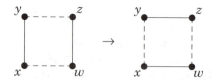

The dashed lines in this illustration indicate edges that do not belong to the graph. If $y \leftrightarrow z$ or $w \leftrightarrow x$, then the 2-switch above cannot be performed, because the resulting graph would not be simple. A 2-switch does not change the degree of any vertex. If some 2-switch turns H into H^*, then a 2-switch on the same four vertices turns H^* into H. Below we illustrate a sequence of two 2-switches that turns one graph into another with the same vertex degrees:

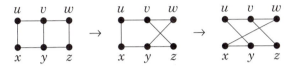

1.4.9. Theorem. If G and H are two simple graphs with vertex set V, then $d_G(v) = d_H(v)$ for every $v \in V$ if and only if there is a sequence of 2-switches that transforms G into H.

Proof. Every 2-switch leaves the degrees unchanged, so the condition is sufficient. Conversely, suppose $d_G(v) = d_H(v)$ for all $v \in V$; we prove the existence of the desired sequence by induction on the number of vertices, n. If $n \leq 3$, then for each vector $d_1, \ldots, d_n$ there is at most one simple graph with $d(v_i) = d_i$. Hence we can use $n = 3$ as a basis.

Suppose $n \geq 4$, and let w be a vertex of maximum degree, Δ. Let $S = \{v_1, \ldots, v_\Delta\}$ be a fixed set of vertices with the Δ highest degrees other than w. As we showed in proving the Havel-Hakimi Theorem, some sequence of 2-switches transforms G to a graph G^* such that $N_{G^*}(w) = S$, and some such sequence transforms H to a graph H^* such that $N_{H^*}(w) = S$.

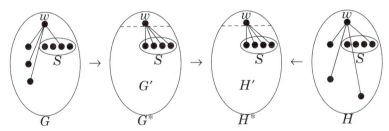

Since $N_{G^*}(w) = N_{H^*}(w)$, deleting w leaves simple graphs $G' = G^* - w$ and $H' = H^* - w$ with $d_{G'}(v) = d_{H'}(v)$ for every vertex v. By the induction hypothesis, some sequence of 2-switches transforms G' to H'. Since these don't involve w, and w has the same neighbors in G^* and H^*, applying this sequence to G^* transforms it to H^*. Hence we can transform G to H by performing the sequence that transforms G to G^*, then the sequence that transforms G^* to H^*, then in reverse order the (inverses of) 2-switches in the sequence that transforms H to H^*. $\square$

We could also prove this by induction on the number of edges appearing in exactly one of G and H, which is 0 if and only if they are already the same. In this approach, it suffices to find a 2-switch in G that makes it closer to H or a 2-switch in H that makes it closer to G.

DEGREES AND DIGRAPHS

The vertex degree notation for digraphs incorporates the distinction between heads and tails of edges.

1.4.10. Definition. Let v be a vertex in a directed graph. The *out-degree* $d^+(v)$ is the number of edges with tail v. The *in-degree* $d^-(v)$ is the number of edges with head v. The *out-neighborhood* or *successor set* $N^+(v)$ is $\{x \in V(G): v \to x\}$. The *in-neighborhood* or *predecessor set* $N^-(v)$ is $\{x \in V(G): x \to v\}$.

For digraphs we have a sequence of "degree pairs" $(d^+(v_i), d^-(v_i))$. Many graph results have analogues for digraphs. There are $2^{\binom{n}{2}}$ simple graphs with vertices $v_1, \ldots, v_n$. Similarly, there are 2^{n^2} digraphs with these vertices such that each *ordered* pair appears at most once as an edge. If we forbid loops and forbid having both $x \to y$ and $y \to x$, then only $3^{\binom{n}{2}}$ digraphs remain; these are closely related to (undirected) simple graphs.

1.4.11. Definition. An *orientation* of a simple graph G is a digraph D obtained from G by choosing an orientation ($x \to y$ or $y \to x$) for each edge $xy \in E(G)$. The result is an *oriented graph*, meaning a digraph with no loops and with no two edges having the same endpoints (i.e., at most one of $\{xy, yx\}$ is an edge).

1.4.12. Example. *Tournaments.* In an n-team league where each team plays every other exactly once, we can record the outcome of the season as a directed graph. For each pair u, v, we include the edge uv or vu depending on which team wins the match between them. At the end of the season we have an orientation of K_n, since we have oriented each vertex pair. The "score" of a team is its out-degree in this digraph; this equals the number of wins.

Since this is a model for "round-robin tournaments", an orientation of a complete graph is called a *tournament*. The number of tournaments with vertices $v_1, \ldots, v_n$, like the number of graphs, is $2^{\binom{n}{2}}$. The degree pairs of a tournament are determined by the out-degrees (the *score sequence*), since $d^+(v) + d^-(v) = n - 1$ for every vertex v. As shown below, a tournament may have more than one vertex with maximum out-degree, so there may be no clear "winner". □

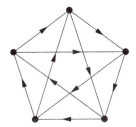

1.4.13. Proposition. (Landau [1953]) Every tournament has a *king*, defined to be a vertex from which every other vertex is reachable by a path of length at most 2.

Proof. Consider a vertex x in a tournament T. If x is not a king, then there exists a vertex y that is not reachable from x by a path of length at most 2. Since T is an orientation of K_n, every successor of x must be a successor of y. Also $y \to x$. Hence $d^+(y) > d^+(x)$. Now consider whether y is a king. Since T is finite, we cannot continue forever considering vertices of successively higher degree. This procedure must terminate, and it terminates only when we have found a king. □

We could also have phrased this proof using extremality, proving that every vertex of maximum out-degree in a tournament is a king. Exercises 29-31 ask further questions about kings in tournaments (see also Maurer [1980]).

EXERCISES

1.4.1. (–) Determine the maximum number of edges in a bipartite subgraph of P_n, of C_n, and of K_n.

1.4.2. (–) Prove or disprove: When the algorithm of Theorem 1.4.2 is applied to a bipartite graph, it finds the bipartite subgraph with the most edges.

1.4.3. Use induction on n to prove that every loopless graph G with at least one edge has a bipartite subgraph H such that H has *more* than half the edges of G.

1.4.4. Construct a sequence $\{G_n\}$ of graphs, with G_n having $2n$ vertices, such that $\lim_{n \to \infty} f_n = 1/2$, where f_n is the fraction of $E(G_n)$ belonging to the largest bipartite subgraph of G_n.

1.4.5. Prove that every loopless graph G has a k-partite subgraph H such that $e(H) \geq (1 - 1/k)e(G)$.

1.4.6. Determine the largest number of edges in a bipartite subgraph of the Petersen graph.

1.4.7. Determine the maximum number of edges in an n-vertex simple graph with k components.

1.4.8. (!) *Subgraphs with large minimum degree.* Suppose G is a finite graph with average vertex degree a (recall that $a = 2e(G)/n(G)$).
   a) Prove that $G - x$ has average degree at least a if and only if $d(x) \le a/2$.
   b) Use part (a) to give an algorithmic proof that if $a > 0$, then G has a subgraph with minimum degree greater than $a/2$.
   c) Using $K_{1,n-1}$, prove that the bound in part (b) is best possible.

1.4.9. Suppose G is an n-vertex simple graph with maximum degree $\lceil n/2 \rceil$ and minimum degree $\lfloor n/2 \rfloor - 1$. Does this imply that G is connected?

1.4.10. (−) Use induction (on n or on Σd_i) to prove that if $d_1, \dots, d_n$ are nonnegative integers and Σd_i is even, then there exists an n-vertex multigraph for which $d_1, \dots, d_n$ is the list of vertex degrees. (Comment: This requests an alternative proof of Proposition 1.4.4.)

1.4.11. (−) Which of the following sequences are graphic? Provide a construction or a proof of impossibility for each.

$$\text{a) } (5,5,4,3,2,2,2,1), \quad \text{c) } (5,5,5,3,2,2,1,1),$$
$$\text{b) } (5,5,4,4,2,2,1,1), \quad \text{d) } (5,5,5,4,2,1,1,1).$$

1.4.12. Suppose $d = (d_1, \dots, d_{2k})$ is defined by $d_{2i} = d_{2i-1} = i$ for $1 \le i \le k$. Prove that d is graphic. (Hint: do not use the Havel-Hakimi test.)

1.4.13. Suppose $G \cong \overline{G}$ and $n(G) \equiv 1 \bmod 4$. Prove that G has at least one vertex of degree $(n(G) - 1)/2$.

1.4.14. Suppose n is congruent to 0 or 1 modulo 4. Construct an n-vertex simple graph G with $\frac{1}{2} \binom{n}{2}$ edges such that $\Delta(G) - \delta(G) \le 1$.

1.4.15. *Construction of 3-regular simple graphs*
   a) Prove that a 2-switch can be performed by performing a sequence of expansions and erasures, where these are the operations defined in Example 1.3.15. (Caution: erasure is not allowed when it would produce multiple edges.)
   b) Use part (a) to prove that every 3-regular simple graph can be obtained from K_4 by a sequence of expansions and erasures. (Batagelj [1984])

1.4.16. (!) Let $d = (d_1, \dots, d_n)$ be an n-tuple of integers such that $d_1 \ge \cdots \ge d_n \ge 0$. Prove that there exists a loopless multigraph with degree sequence d if and only if Σd_i is even and $d_1 \le d_2 + \cdots + d_n$. (Hakimi [1962])

1.4.17. (!) Let $d_1 \le \cdots \le d_n$ be the vertex degrees of a simple graph G. Prove that G is connected if $d_k \ge k$ for every k with $k \le n - 1 - d_n$. (Hint: consider a component that omits some vertex of maximum degree.)

1.4.18. (!) Prove that if the nonnegative integers $d_1 \ge \cdots \ge d_n$ are the degree sequence of a simple graph, then Σd_i is even and $\Sigma_{i=1}^{k} d_i \le k(k-1) + \Sigma_{i=k+1}^{n} \min\{k, d_i\}$ for $1 \le k \le n$. (Hint: do not try to use induction. Comment: Erdős-Gallai [1960] proved that this condition is also sufficient.)

1.4.19. A simple graph G is a *split graph* if $V(G)$ can be partitioned into Q and S such that Q induces a clique and S is an independent set. Let $d_1 \geq \cdots \geq d_n$ be the degree sequence of a simple graph G, and let m be the largest value of k such that $d_k \geq k - 1$. Prove that G is a split graph if and only if $\Sigma_{i=1}^m d_i = m(m-1) + \Sigma_{i=m+1}^n d_i$. (Hammer-Simeone [1977])

1.4.20. (+) Suppose $a_1 < \cdots < a_k$ are distinct positive integers. Prove that there is a simple graph with $a_k + 1$ vertices whose *set* of distinct vertex degrees is $a_1, \ldots, a_k$. (Hint: use induction on k to construct such a graph.) (Kapoor-Polimeni-Wall [1977])

1.4.21. (–) Prove that the pairs $\{(d_i^+, d_i^-)\}_{i=1}^n$ of nonnegative integers are the in,out-degree pairs of some directed multigraph (loops and multiple edges allowed) if and only if $\Sigma d_i^+ = \Sigma d_i^-$.

1.4.22. (–) For each $n \geq 1$, prove or disprove: There is no n-vertex digraph without loops or multiple edges such that the out-degrees of the vertices are distinct and the in-degrees of the vertices are distinct.

1.4.23. (–) Prove that there is an n-vertex tournament with in-degree equal to out-degree at every vertex if and only if n is odd.

1.4.24. Determine the minimum n such that there is a pair of nonisomorphic n-vertex tournaments with the same list of out-degrees.

1.4.25. (!) Suppose $p_1 \geq \cdots \geq p_m$ and $q_1 \geq \cdots \geq q_n$ are sequences of nonnegative integers. The pair (p, q) is *bigraphic* if and only if there is a simple bipartite graph in which $p_1, \ldots, p_m$ are the vertex degrees for one partite set and $q_1, \ldots, q_n$ are the degrees for the other.

a) Prove that (p, q) is bigraphic if and only if (p', q') is bigraphic, where (p', q') is obtained from (p, q) by deleting the largest element p_1 from p and subtracting one from each of the p_1 largest elements of q.

b) Suppose G and H are two simple bipartite graphs each having vertex bipartition $V = X \cup Y$. Prove that $d_G(v) = d_H(v)$ for all $v \in V$ if and only if there is a sequence of 2-switches that transforms G into H without ever changing the bipartition (each 2-switch replaces two edges joining X and Y by two other edges joining X and Y).

1.4.26. Suppose A and B are two m by n matrices with entries in $\{0, 1\}$. Prove that if A and B have the same vector of row sums and have the same vector of column sums, then A can be transformed into B by a sequence of steps in which the 0's and 1's are interchanged in a 2 by 2 permutation submatrix (given two rows and two columns, the submatrices $\binom{0\,1}{1\,0}$ and $\binom{1\,0}{0\,1}$ can be substituted for each other). Interpret this conclusion in the context of digraphs. (Ryser [1957])

1.4.27. (!) Suppose G and H are two tournaments on a vertex set V. Prove that $d_G^+(v) = d_H^+(v)$ for all $v \in V$ if and only if G can be turned into H by a sequence of direction-reversals on cycles of length 3. (Hint: Use induction on the number of edges oriented differently in G and H, and consider a vertex incident to a maximum number of such edges.) Ryser [1964]

1.4.28. (+) Suppose that $0 \le p_1 \le \cdots \le p_n$ is a nondecreasing sequence of integers. Prove that there exists a tournament with out-degrees $p_1, \ldots, p_n$ if and only if $\Sigma_{i=1}^{k} p_i \ge \binom{k}{2}$ for $1 \le k < n$ and $\Sigma_{i=1}^{n} p_i = \binom{n}{2}$. (Landau [1953])

1.4.29. By Proposition 1.4.13, every tournament has a king. Use this to prove that every tournament having no vertex with in-degree 0 has at least two kings.

1.4.30. Consider a vertex x in a tournament T. If x has in-degree 0, call x a king and stop. Otherwise, delete x and its successors (out-neighbors) and repeat the iterative step on the remaining tournament. Prove that this algorithm produces a king.

1.4.31. Prove that if n is odd, then there is an n-vertex tournament in which every player is a king. Prove that there is no such tournament when $n = 4$.

1.4.32. Prove that for every digraph D there is a set S such that $D[S]$ has no edges but every vertex is reachable from S by a path of length at most 2 (this generalizes Theorem 1.4.13). (Hint: use strong induction on $n(D)$.) (Chvátal-Lovász [1974])

1.4.33. Suppose G is a tournament and L_0 lists its vertices in some order. If y immediately follows x in L_0 but $y \to x$ in G, then yx is a *reverse edge*. We are permitted to switch x and y in the order when yx is a reverse edge (this may increase the number of reverse edges). Suppose a sequence $L_0, L_1, \cdots$ is produced by successively switching one reverse edge in the current order. Prove that this process always reaches a list having no reverse edges. Determine the maximum number of steps to termination. (Comment: in the special case where the vertices correspond to numbers and each edge points to the higher number of the pair, the result says that successively switching two adjacent numbers that are out of order always eventually sorts the list.) (Locke [1995])

1.4.34. A directed graph is *unipathic* if for every pair of vertices x, y there is at most one (directed) x, y-path. Let T_n be the tournament on n vertices with the edge between v_i and v_j directed toward the vertex with larger index. What is the maximum number of edges in a unipathic subgraph of T_n? How many unipathic subgraphs are there with the maximum number of edges? (Hint: Use Turán's Theorem.) (Maurer-Rabinovitch-Trotter [1980])

Chapter 2

Trees and Distance

2.1 Basic Properties

The word "tree" suggests branching out from a root and never completing a cycle. Trees as graphs have many applications, especially in data storage and communication (including computation of distances).

2.1.1. Definition. A graph having no cycle is *acyclic*. A *forest* is an acyclic graph; a *tree* is a connected acyclic graph. A *leaf* (or *pendant vertex*) is a vertex of degree 1. A *spanning subgraph* of G is a subgraph with vertex set $V(G)$. A *spanning tree* is a spanning subgraph that is a tree.

If G has a u, v-path, then the *distance* from u to v, written $d_G(u, v)$ or simply $d(u, v)$, is the least length of a u, v-path. If G has no such path, then $d(u, v) = \infty$.

A spanning subgraph of G need not be connected, and a connected subgraph of G need not be a spanning subgraph. For example, the subgraph with vertex set $V(G)$ and edge set $\varnothing$ is spanning but not connected (if $n(G) > 1$), and the subgraph consisting of a single edge is connected but not spanning (if $n(G) > 2$).

PROPERTIES OF TREES

Trees have many equivalent characterizations, any of which could be taken as the definition. Such characterizations are useful because we need only verify that a graph satisfies any one of them to prove that it is a tree, after which we can use all the other properties.

We first prove that deleting a leaf from a tree yields a smaller tree. This implies that every tree with more than one vertex can be grown from a smaller tree by adding a vertex of degree 1. This facilitates inductive proofs for trees by allowing the induction step to grow an $n + 1$-vertex tree by adding a vertex to an arbitrary n-vertex tree without falling into the induction trap.

2.1.2. Lemma. Every finite tree with at least two vertices has at least two leaves. Deleting a leaf from an n-vertex tree produces a tree with $n - 1$ vertices.

Proof. Every connected graph with at least two vertices has an edge. In an acyclic graph, the endpoints of a maximum path have only one neighbor on the path and therefore have degree 1. Hence the endpoints of a maximum path provide the two desired leaves.

Suppose v is a leaf of a tree G, and let $G' = G - v$. If $u, w \in V(G')$, then no u, w-path P in G can pass through the vertex v of degree 1, so P is also present in G'. Hence G' is connected. Since a vertex deletion cannot create a cycle, G' is also acyclic. We conclude that G' is a tree with $n - 1$ vertices. $\square$

2.1.3. Theorem. For an n-vertex simple graph G (with $n \geq 1$), the following are equivalent (and characterize the trees with n vertices).
A) G is connected and has no cycles.
B) G is connected and has $n - 1$ edges.
C) G has $n - 1$ edges and no cycles.
D) For every pair $u, v \in V(G)$, G has exactly one u, v-path.

Proof. We first demonstrate the equivalence of A,B,C, by proving that any two of {connected, acyclic, $n - 1$ edges} implies the third.

A $\Rightarrow$ B,C. We use induction on n. For $n = 1$, an acyclic 1-vertex graph has no edge. For the induction step, suppose $n > 1$, and suppose the implication holds for graphs with fewer than n vertices. Given G, Lemma 2.1.2 provides a leaf v and states that $G' = G - v$ is acyclic and connected. Applying the induction hypothesis to G' yields $e(G') = n - 2$, and hence $e(G) = n - 1$.

B $\Rightarrow$ A,C. Delete edges from cycles of G one by one until the resulting graph G' is acyclic. Since no edge of a cycle is a cut-edge (Corollary 1.2.13), G' is connected. By the paragraph above, G' has $n - 1$ edges. Since this equals $e(G)$, no edges were deleted, and G itself is acyclic.

C $\Rightarrow$ A,B. Suppose G has k components with orders $n_1, \ldots, n_k$. Since G has no cycles, each component satisfies property A, and by the first paragraph the ith component has $n_i - 1$ edges. Summing this over all components yields $e(G) = \Sigma(n_i - 1) = n - k$. We are given $e(G) = n - 1$, so $k = 1$, and G is connected.

A $\Rightarrow$ D. Since G is connected, G has at least one u,v-path for each pair $u,v \in V(G)$. Suppose G has distinct u,v-paths P and Q. Let $e = xy$ be an edge in P but not in Q. The concatenation of P with the reverse of Q is a closed walk in which e appears exactly once. Hence $(P \cup Q) - e$ is an x,y-walk not containing e. By Lemma 1.2.6, this contains an x,y-path, which completes a cycle with e and contradicts the hypothesis that G is acyclic. Hence G has exactly one u,v-path.

D $\Rightarrow$ A. If there is a u,v-path for every $u,v \in V(G)$, then G is connected. If G has a cycle C, then G has two paths between any pair of vertices on C. $\square$

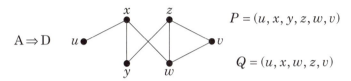

$$P = (u, x, y, z, w, v)$$

A $\Rightarrow$ D

$$Q = (u, x, w, z, v)$$

2.1.4. Remark. *Every connected graph contains a spanning tree.* As in the proof of B $\Rightarrow$ A,C above, we can begin with a connected graph G and iteratively delete an edge from a cycle until we obtain a spanning subgraph that is acyclic. Since an edge of a cycle is not a cut-edge, the graph remains connected. $\square$

The next result illustrates induction using deletion of a leaf.

2.1.5. Proposition. If T is a tree with k edges and G is a simple graph with $\delta(G) \geq k$, then T is a subgraph of G.

Proof. We use induction on k. For the basis step $k = 0$, note that every simple graph contains K_1. For the induction step, suppose $k > 0$, and suppose the claim holds for trees with fewer than k edges. Since $k > 0$, Lemma 2.1.2 allows us to choose a leaf v with neighbor u in T and consider the smaller tree $T' = T - v$. By the induction hypothesis, G contains T' as a subgraph, since $\delta(G) \geq k > k - 1$. Let x be the vertex in this copy of T' that corresponds to u (see illustration). Because T' has only $k - 1$ vertices other than u, x has a neighbor y in G that does not appear in this copy of T'. Adding the edge xy to correspond to uv expands this copy of T' in G into a copy of T. $\square$

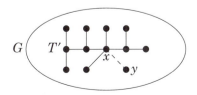

Since a tree has no cycles, Corollary 1.2.13 implies that every edge of a tree is a cut-edge. Since a tree has a unique path linking each pair of vertices, adding any edge creates exactly one cycle. We next apply these observations. We use subtraction and addition as operations involving sets to indicate deletion and inclusion of edges.

2.1.6. Proposition. If T, T' are two spanning trees of a connected graph G and $e \in E(T) - E(T')$, then there is an edge $e' \in E(T') - E(T)$ such that $T - e + e'$ is a spanning tree of G.

Proof. Since e is a cut-edge of T, we may let U, U' be the vertex sets of the components of $T - e$. Since T' is connected, it contains a path P between the endpoints of e. Since P connects vertices in U and U', it has at least one edge with endpoints in both sets. Any such edge can be used as e', since it reconnects $T - e$ without introducing a cycle. $\square$

2.1.7. Proposition. If T, T' are two spanning trees of a connected graph G and $e \in E(T) - E(T')$, then there is an edge $e' \in E(T') - E(T)$ such that $T' + e - e'$ is a spanning tree of G.

Proof. Adding e to T' creates a unique cycle C. The edges of this cycle other than e cannot all belong to T, since T has no cycle. If we delete any edge $e' \in E(T') - E(T)$ from C, we obtain a connected acyclic graph $T' + e - e'$. $\square$

With the proper choice of $e' \in E(T) - E(T')$, these two conclusions hold simultaneously (Exercise 20).

DISTANCE IN GRAPHS

In a connected graph, how far apart can two vertices be?

2.1.8. Definition. The *diameter* of a graph G is $\max_{u,v \in V(G)} d(u,v)$. The *eccentricity* of a vertex u, written $\varepsilon(u)$, is $\max_{v \in V(G)} d(u,v)$. The *radius* of G equals $\min_{u \in V(G)} \varepsilon(u)$. A *center* of G is a vertex of minimum eccentricity.

In a disconnected graph, the diameter and radius (and the eccentricity of every vertex) are infinite, because distance between vertices in different components is infinite. Every finite connected graph has at least one center, and diameter is related to eccentricity by $\text{diam} G = \max_{u \in V(G)} \varepsilon(u)$. The use of "diameter" in graph theory is motivated by the use of diameter in geometry, where it equals the greatest distance between two vertices in a set. In the graph below, we have labeled each vertex with its eccentricity; this graph has only one center.

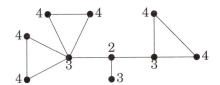

Every path in a tree is the shortest (the only!) path between its endpoints, so the diameter of a tree is the length of its longest path. We next describe the centers of trees. The proof uses deletion of *all* leaves to obtain a subtree; this sometimes provides a cleaner inductive proof than deleting one leaf.

2.1.9. Theorem. (Jordan [1869]) A tree has exactly one center or has two adjacent centers.

Proof. We use induction on the number of vertices; the statement is trivial for trees with one or two vertices. For $n > 2$, let T be an arbitrary n-vertex tree. Form T' by deleting every leaf of T; note that T' is also a tree. For any vertex u in a tree, every vertex at maximum distance from u is a leaf (take the path between v and the farthest vertex x; if x is not a leaf, then it has another neighbor, which is farther from v). Since all the leaves have been removed and no path between two other vertices uses a leaf, $\varepsilon_{T'}(u) = \varepsilon_T(u) - 1$ for every $u \in V(T')$. Also, the eccentricity of a leaf in T is greater than the eccentricity of its neighbor in T. Hence the vertices minimizing $\varepsilon_T(u)$ are the same as the vertices minimizing $\varepsilon_{T'}(u)$. By the induction hypothesis, for T' they consist of a single vertex or two adjacent vertices. $\square$

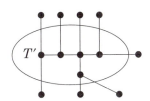

PROVING A STRONGER RESULT

It may be easiest to prove that an integer is nonzero by proving that it is odd. It may be easiest to prove a statement about trees by proving it more generally for forests or for all bipartite graphs. Forests are acyclic graphs that need not be connected; every component of a forest is a tree. When we prove a statement about forests by induction, we

can delete an arbitrary vertex or edge in the induction step. In a proof
by induction for trees, if we delete an edge or a path from a tree, we no
longer have a tree and cannot directly apply the induction hypothesis;
we can only apply it to each component of what remains.

2.1.10. Theorem. If G is a tree with $2k \geq 0$ vertices of odd degree, then
$E(G)$ is the union of k pairwise edge-disjoint paths.

Proof. We prove the claim for every forest G, using induction on k.
Basis step: if $k = 0$, then G has no leaf and hence no edge. For the induction step, suppose $k > 0$, and suppose each forest with $2k - 2$ vertices
of odd degree has a decomposition into $k - 1$ paths. Since $k > 0$, some
component of G is a tree with at least two vertices. This component has
at least two leaves; let P be a path connecting two leaves. Deleting $E(P)$
changes the parity of the vertex degree only for the endpoints of P; it
makes them even. Hence $G - E(P)$ is a forest with $2k - 2$ vertices of odd
degree. By the induction hypothesis, $G - E(P)$ is the union of $k - 1$ pairwise edge-disjoint paths; together with P, these paths partition $E(G)$. □

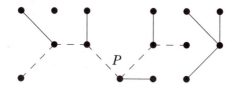

Sometimes proving a result *requires* proving a stronger result.

2.1.11. Example. *Sum of distances from a leaf of a tree.* The sum of the
distances from a leaf of the path P_n to all other vertices is $\sum_{i=0}^{n-1} i = \binom{n}{2}$,
and intuition suggests that no other n-vertex tree has a leaf with larger
distance sum. Suppose we want to prove that the sum of the distances
to other vertices is at most $\binom{n}{2}$ for any leaf v of an n-vertex tree. If we
want to use induction, the distances from v can be expressed in terms of
the distances from its neighbor w within the tree $T - v$, but w need not
be a leaf in $T - v$. Thus it is hard to prove this theorem without proving
the stronger result that the bound holds for *every* vertex of a tree. □

2.1.12. Theorem. If u is a vertex of an n-vertex tree G, then
$$\sum_{v \in V(G)} d(u, v) \leq \binom{n}{2}.$$

Proof. We use induction on n; the result holds trivially for $n = 2$. Suppose $n > 2$. The graph $T - u$ is a forest with components $T_1, \ldots, T_k$,
where $k \geq 1$. Because T is connected, u has a neighbor in each T_i;
because T has no cycles, u has exactly one neighbor v_i in each T_i. If
$v \in V(T_i)$, then the unique u, v-path in T passes through v_i, and we have

$d_T(u,v) = 1 + d_{T_i}(v_i, v)$. Letting $n_i = n(T_i)$, we obtain $\Sigma_{v \in V(T_i)} d_T(u,v) = n_i + \Sigma_{v \in V(T_i)} d_{T_i}(v_i, v)$.

By the induction hypothesis, $\Sigma_{v \in V(T_i)} d_{T_i}(v_i, v) \le \binom{n_i}{2}$. If we sum the formula for distances from u over all the components of $T - u$, we obtain $\Sigma_{v \in V(T)} d_T(u,v) \le (n-1) + \Sigma_i \binom{n_i}{2}$. Now observe that $\Sigma \binom{n_i}{2} \le \binom{m}{2}$ whenever $\Sigma n_i = m$, because the right side counts the edges in K_m and the left side counts the edges in a subgraph of K_m (a disjoint union of cliques). Hence we have $\Sigma_{v \in V(T)} d_T(u,v) \le (n-1) + \binom{n-1}{2} = \binom{n}{2}$. $\square$

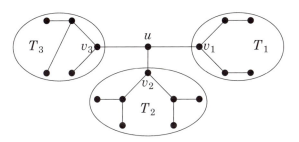

When we want to prove $P(n)$ by induction on n, we may find that more information than $P(n-1)$ alone is needed about the smaller object to prove $P(n)$ for the larger object. For example, in Example 2.1.11 we needed the statement to hold for all vertices of the smaller tree, not only leaves. If we include these extra conclusions in the statement to be proved, forming a stronger statement $Q(n-1)$, then the induction step will prove $P(n)$ from $Q(n-1)$, but it must also prove the stronger statement $Q(n)$ to complete the induction step for statement Q. This is one type of "proving a stronger result"; it has the special name of "loading the induction hypothesis".

"Reducing to a special case" is in some sense the reverse technique. We can reduce a theorem about arbitary graphs to the case of connected graphs by proving that the result can be obtained for arbitrary graphs from its validity for connected graphs. Similarly, some questions on general graphs can be reduced to the case of trees.

2.1.13. Lemma. If H is a subgraph of G, then $d_G(u,v) \le d_H(u,v)$.

Proof. Every u,v-path in H appears also in G (G may have additional u,v-paths that are shorter than any in H). $\square$

2.1.14. Corollary. If u is a vertex of a connected graph G, then
$$\Sigma_{v \in V(G)} d(u,v) \le \binom{n(G)}{2}.$$

Proof. Let T be a spanning tree of G. By Lemma 2.1.13, $d_T(u,v) \ge d_G(u,v)$. Since we already know the bound for the special case of trees, we have $\Sigma_{v \in V(G)} d_G(u,v) \le \Sigma_{v \in V(G)} d_T(u,v) \le \binom{n(G)}{2}$. $\square$

The sum of the distances over all pairs of distinct vertices in a graph G is the *Wiener index* $W(G) = \Sigma_{u,v \in V(G)} d(u,v)$. Assigning vertices for the atoms and edges for the atomic bonds, we can use graphs to study molecules. Used originally by Wiener to study the boiling point of paraffin, the Wiener index for graphs has been shown to correlate with many chemical properties of the corresponding molecules. Exercise 29 explores the extreme values of the Wiener index on trees.

DISJOINT SPANNING TREES

We have seen that every connected graph has a spanning tree. Edge-disjoint spanning trees provide alternate communication protocols in the event that an edge in the primary tree fails. Tutte [1961] and Nash-Williams [1961] independently characterized graphs having k pairwise edge-disjoint spanning trees (see Exercise 40).

We describe one application of edge-disjoint spanning trees. David Gale devised a game marketed under the name "Bridg-it" (copyright 1960 by Hassenfeld Bros., Inc. - "Hasbro Toys"). Each of two players owns a rectangular grid of posts. The players move alternately, at each move joining a pair of friendly posts by a unit-length bridge. In the picture below, Player 1's posts are solid; Player 2's are hollow. The object of Player 1 is to construct a bridge from the left to the right; Player 2 wants one from the top to the bottom.

Player 1 can start with an arbitrary move and then follow the strategy of Player 2, making an arbitrary move if the strategy for Player 2 ever calls for a bridge in the location of the first move. If the strategy for Player 2 leads to a win, Player 1 wins instead, because extra moves cannot hurt. Hence Player 2 cannot have a winning strategy. Using ideas of planarity, one can show that no ties are possible; therefore Player 1 must have a winning strategy. Here we give an explicit strategy that Player 1 can use to win. (The argument holds more generally in the context of "matroids"---see Theorem 8.2.44.)

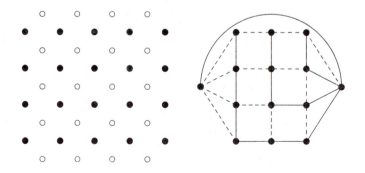

2.1.15. Theorem. Player 1 has a winning strategy in Bridg-it.

Proof. We form a graph of the potential connections for Player 1. Posts on the same end are equivalent, so we collect the (solid) posts from the end columns as single vertices. We add an auxiliary edge between the ends. The picture suggests that this graph is the union of two edge-disjoint spanning trees; we omit a technical description of the two trees.

Together, the two trees contain edge-disjoint paths between the goal vertices. Since the auxiliary edge doesn't really exist, we can pretend Player 2 moved first and took that edge. A move by Player 2 cuts one edge e in the graph and makes it no longer available. This cuts one of the trees into two components. By Lemma 2.1.6, some edge e' from the other tree reconnects the cut tree.

Player 1 chooses such an edge e'. This makes e' uncuttable, in effect putting e' in both spanning trees. By making e' a double edge with one copy in each tree, we retain a multigraph consisting of two edge-disjoint spanning trees. Since Player 2 cannot cut a double edge, Player 2 cannot cut both trees, and Player 1 can always defend. The process terminates only when only double edges remain. These double edges form a spanning tree held by Player 1 in actual bridges, so Player 1 has constructed a path between the desired ends. □

EXERCISES

2.1.1. (–) For each k, list the isomorphism classes of trees with at most six vertices that have maximum degree k. Do the same for diameter k. (Include justification that there are no others.)

2.1.2. (–) *Characterization of trees.*
   a) Prove that a graph G is a tree if and only if G is connected and for every $e \in E(G)$, $G - e$ is not connected.
   b) Prove that a simple graph G is a tree if and only if for every $e \in E(\overline{G})$, adding the edge e to G creates exactly one cycle.

2.1.3. (–) Prove that a graph G is a tree if and only if G is loopless and has exactly one spanning tree.

2.1.4. (–) Prove that every tree with maximum degree $\Delta > 1$ has at least Δ vertices of degree 1. Show that this is best possible by constructing an n-vertex tree with exactly Δ leaves, for each choice of n, Δ with $n > \Delta \geq 2$.

2.1.5. (–) Prove or disprove: If n_i denotes the number of vertices of degree i in a tree T, then $\Sigma i n_i$ depends only on the number of vertices in T.

2.1.6. (–) A saturated hydrocarbon is a molecule formed from k carbon atoms and l hydrogen atoms by adding bonds between atoms such that each carbon atom is in four bonds, each hydrogen atom is in one bond, and no sequence of bonds forms a cycle of atoms. Prove that $l = 2k + 2$. (Bondy-Murty [1976, p27])

2.1.7. (–) Suppose T is a tree with average degree a. Determine $n(T)$.

2.1.8. (–) Suppose T is a tree with exactly $k-1$ vertices that are not leaves, with one having degree i for each $2 \le i \le k$. Determine $n(T)$.

2.1.9. (–) Suppose T is a tree in which every vertex has degree 1 or degree k. Determine the values of $n(T)$ for which this is possible.

2.1.10. (–) Prove that every nontrivial tree has at least two maximal independent sets, with equality only for stars. (Note: maximal $\ne$ maximum.)

2.1.11. (!) Suppose G is an n-vertex graph such that every graph obtained by deleting one vertex of G is a tree. Determine the number of edges in G, and use this to determine G itself.

2.1.12. (!) Let $d_1, \ldots, d_n$ be positive integers. Prove that there exists a tree with vertex degrees $d_1, \ldots, d_n$ if and only if $\Sigma d_i = 2n - 2$.

2.1.13. Suppose $d_1 \ge \cdots \ge d_n$ are nonnegative integers. Prove that there exists a connected multigraph (loops and multiple edges allowed) with degree sequence $d_1, \ldots, d_n$ if and only if Σd_i is even, $d_n \ge 1$, and $\Sigma d_i \ge 2n - 2$. (Hint: consider a realization with the minimum number of components.) Is the statement true for simple graphs?

2.1.14. Suppose T is a tree in which every vertex adjacent to a leaf has degree at least 3. Prove that T has some pair of leaves with a common neighbor.

2.1.15. Prove that a simple connected graph having exactly two vertices that are not cut vertices is a path.

2.1.16. Let e be an edge in a connected graph G. Prove that e is a cut-edge of G if and only if e belongs to every spanning tree of G. Prove that e is a loop if and only if e belongs to no spanning tree of G.

2.1.17. (!) Suppose G is a connected n-vertex graph. Prove that G has exactly one cycle if and only if G has exactly n edges.

2.1.18. Suppose $k \ge 1$, T is a tree with $k + 1$ edges, and G is a simple graph with average degree at least $2k$. Use Exercise 1.4.8b to prove that $T \subseteq G$.

2.1.19. Suppose T is a tree of even order. Prove that T has exactly one subgraph in which every vertex has odd degree.

2.1.20. (!) Suppose that T, T' are two spanning trees of a connected graph G and that $e \in E(T) - E(T')$. Prove that there is an edge $e' \in E(T') - E(T)$ such that $T' + e - e'$ and $T - e + e'$ are both spanning trees of G.

2.1.21. (!) A subgraph H of an undirected graph G is a *parity subgraph* if $d_H(v) \equiv d_G(v) \pmod 2$ for all $v \in V(G)$. Prove that every spanning tree of G contains a parity subgraph of G. (Hint: Use a constructive or an inductive proof. If the induction parameter is chosen wisely, the proof is quite short.)

2.1.22. (!) Suppose G is a tree with k leaves. Prove that G is the union of paths $P_1, \ldots, P_{\lceil k/2 \rceil}$ such that $P_i \cap P_j \ne \varnothing$ for all $i \ne j$. (Ando-Kaneko-Gervacio [1996])

2.1.23. (–) Suppose G is a graph. Prove that a maximal subgraph of G that is a forest consists of a spanning tree from each component of G.

2.1.24. Prove that each property below characterizes forests.
a) Every induced subgraph has a vertex of degree at most 1.
b) Every connected subgraph is an induced subgraph.

2.1.25. (–) Suppose the processors in a parallel computer are labeled by binary k-tuples, with pairs able to communicate directly if and only their k-tuples are adjacent in the k-dimensional cube Q_k. Suppose a processor with address u wants to send a message to the processor with address v. How can it determine where to send the message as the first step on a shortest path to v?

2.1.26. (!) Suppose G is a connected graph. Let G' be a new graph having one vertex for each spanning tree of G, with two vertices t, t' in G' forming an edge if and only if the corresponding trees T, T' have exactly $n(G) - 2$ common edges. Prove that G' is connected. Give a formula for $d_{G'}(t, t')$ in terms of the corresponding trees T, T' in G.

2.1.27. *Centers of trees.* Suppose T is a tree.
a) Give a noninductive proof that T has one center or two adjacent centers.
b) Prove that T has one center if and only if diameter$(T) = 2$ radius(T).
c) Use (a) to prove that if $n(T)$ is odd, then every automorphism of T maps some vertex to itself.

2.1.28. Given $x \in V(G)$, let $s(x) = \Sigma_{v \in V(G)} d(x, v)$. Prove that if G is a tree and $y, z \in N(x)$, then $2s(x) < s(y) + s(z)$. Conclude that $s(x)$ attains its minimum at one vertex or at two adjacent vertices. (Comment: vertices minimizing $s(x)$ are called *barycenters* of G.)

2.1.29. Given a tree T, let $W(T)$ denote the sum of $d(x, y)$ over all $\binom{n}{2}$ pairs of vertices x, y (the Wiener index). Determine the maximum and minimum values of $W(T)$ for trees on n vertices. Prove that only one n-vertex tree achieves the maximum and only one achieves the minimum (only one isomorphism class).

2.1.30. Suppose S is a tree with leaves $\{x_1, \ldots, x_k\}$, and T is a tree with leaves $\{y_1, \ldots, y_k\}$. Suppose also that $d_S(x_i, x_j) = d_T(y_i, y_j)$ for each pair i, j. Prove that S and T are isomorphic.

2.1.31. (–) Count the isomorphism classes of n-vertex trees with diameter 3.

2.1.32. Suppose G is a tree with n vertices, k leaves, and maximum degree k. Determine the maximum and minimum possible values of the diameter of G.

2.1.33. (–) Given a simple graph G, define G' to be the simple graph on the same vertex set such that $xy \in E(G')$ if and only if x and y are adjacent in G or have a common neighbor in G. Prove that diam$(G') = \lceil \text{diam}(G)/2 \rceil$.

2.1.34. (!) Prove that diam$G \geq 3$ implies diam$\overline{G} \leq 3$. Apply this to conclude that diam$G \geq 4$ implies diam$\overline{G} \leq 2$. (Hint: vertices are connected by a path of length 2 if and only if they have a common neighbor.)

2.1.35. Suppose G has diameter d and maximum degree k. Prove that $n(G) \le 1 + [(k-1)^d - 1]k/(k-2)$. (Comment: equality holds for the Petersen graph.)

2.1.36. *Diameter and radius.*

a) Prove that the distance function $d(u,v)$ on pairs of vertices of a graph satisfies the triangle inequality: $d(u,v) + d(v,w) \ge d(u,w)$.

b) Use part (a) to prove that $\mathrm{diam}G$ is at most twice the radius of G.

c) Given positive integers r, d with $r \le d \le 2r$, construct a simple graph with radius r and diameter d. (Hint: build a suitable graph with one cycle.)

2.1.37. Suppose G is a connected graph that is not a tree. Prove that G has a cycle of length at most $2\mathrm{diam}(G) + 1$. For each $k \in \mathbb{N}$, show that this is best possible by exhibiting a graph with diameter k and minimum cycle length $2k + 1$.

2.1.38. (+) Suppose G is a connected graph of order n and minimum degree k, with $k \ge 2$ and $n - 2 \ge 2(k+1)$. Prove that $\mathrm{diam}G \le 3(n-2)/(k+1) - 1$, and provide examples achieving this bound when $n - 2$ is a multiple of $k + 1$. (Moon [1965]) (Comment: for extra credit, it is possible to achieve equality using a k-regular graph when $k + 1$ divides $n - 2$, though separate constructions are needed for even and odd k.)

2.1.39. Suppose $F_1, \ldots, F_m$ are forests whose union is G. Prove that $m \ge \max_{H \subseteq G} \lceil \frac{e(H)}{n(H)-1} \rceil$. (Comment: Nash-Williams [1964] and Edmonds [1965b] proved that this bound is always achieveable - Corollary 8.2.56.)

2.1.40. (!) Prove that the following is a necessary condition for the existence of k pairwise edge-disjoint spanning trees in G: for any partition of the vertices of G into r parts, there are at least $k(r-1)$ edges of G whose endpoints are in different parts of the partition. (Comment: Corollary 8.2.58 shows that this condition is also sufficient - Tutte [1961], Nash-Williams [1961], Edmonds [1965c].)

2.1.41. Can the graph drawn below be expressed as the union of edge-disjoint spanning trees? As the union of isomorphic edge-disjoint spanning trees?

2.1.42. (+) Prove the Helly property for subtrees of a tree. In other words, prove that if $G_1, \ldots, G_k$ are pairwise-intersecting subtrees of a tree G, then G has a vertex that belongs to all of $G_1, \ldots, G_k$. (Hint: use induction on k.)

2.1.43. (+) Prove that a simple graph G is a forest if and only if for every pairwise intersecting family of paths in G, the paths have a common vertex. (Hint: for sufficiency, use induction on the size of the family of paths.)

2.1.44. (+) Prove that every n-vertex tree other than $K_{1,n-1}$ is (isomorphic to) a subgraph of its complement. (Hint: prove a stronger result about the placement of two edge-disjoint copies of T in a clique of the same order.)

2.1.45. (+) Let S be an n-element set, and let $\mathbf{A} = \{A_1, \ldots, A_n\}$ be a collection of n distinct subsets of S. Prove that S has an element x such that the sets $A_1 \cup \{x\}, \ldots, A_n \cup \{x\}$ are distinct. (Hint: define a graph with vertices $a_1, \ldots, a_n$ such that $a_i \leftrightarrow a_j$ if and only if A_i and A_j differ by a single element. Use that element as a label on the edge; if $a_i a_j$ has label y, then y cannot be the desired element x. Prove that there is a forest containing all the labels that occur on edges, and use this to obtain the desired x.) (Bondy [1972])

2.2 Spanning Trees and Enumeration

We have seen that there are $2^{\binom{n}{2}}$ simple graphs with vertex set $\{1, \ldots, n\}$. Counting the trees with this vertex set is harder but still yields to bijective arguments. We can also count spanning trees in arbitrary graphs and consider structural aspects of trees.

ENUMERATION OF TREES

There are n^{n-2} trees having a fixed set of n vertices; this is *Cayley's formula*. Prüfer, Kirchhoff, Polya, Renyi, and others found proofs. J.W. Moon [1970] wrote a book about enumeration of classes of trees.

2.2.1. Example. *Listing of small trees.* With a vertex set of size one or two, only one tree can be formed. With three vertices there is still only one isomorphism class, but the adjacency matrix is determined by which vertex is the center, and there are three trees. Given a set of four vertices, there are four stars and twelve paths; 16 trees in total. □

2.2.2. Definition. When n is a natural number, $[n]$ is the set of natural numbers $\{1, \ldots, n\}$.

2.2.3. Theorem. (Cayley's Formula [1889]). There are n^{n-2} trees with vertex set $[n]$.

Proof. (Prüfer [1918]). There are n^{n-2} sequences of length $n-2$ with entries from $[n]$; we establish a bijection between the set of trees and this set of sequences. To compute the Prüfer sequence $f(T)$ for a labeled tree T, iteratively delete the leaf with smallest label and append the label of its *neighbor* to the sequence. After $n-2$ iterations a single edge remains and we have produced a sequence $f(T)$ of length $n-2$. The sequence corresponding to the tree below is 744171, and the edge that remains is $\{1, 8\}$.

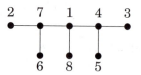

To prove that f is a bijection, we prove that a sequence can only come from one tree and that every sequence arises in this way. Suppose at each step in computing f we mark the deleted leaf as "finished". Let S denote the set of leaves of the remaining tree; these are the unfinished vertices whose labels do not appear in the remainder of the sequence. The next leaf deleted is the least number in S. Hence we can retrieve T from the sequence $a = f(T)$ as follows. Begin with the vertex set $[n]$ and no edges. At the ith step, let x be the label in position i of a. Let y be the smallest label that does not appear in positions after i and has not been marked "finished". Add the edge yx and mark y "finished". After $n-2$ steps, join the two unfinished vertices by an edge.

We have proved that if these $n-1$ edges form a tree T, then $a = f(T)$, because we determined the edge that must have been deleted from T at each stage. To see that the edges do form a tree, note that we begin with a graph (the trivial graph) in which every component has one unfinished vertex. At each step we add an edge joining unfinished vertices in distinct components and mark one finished; this reduces the number of components by one and leaves one unfinished vertex in each component. The last edge joins the two remaining components. Hence we produce a graph with $n-1$ edges and one component: a tree. We have proved that our reverse procedure is f^{-1}. $\square$

Cayley approached the problem algebraically, using a generating function to enumerate the labeled trees by their vertex degrees. Prüfer's bijective proof also provides this information.

2.2.4. Corollary. The number of trees with vertex set $[n]$ in which vertices $1, \ldots, n$ have degrees $d_1, \ldots, d_n$, respectively, is $\dfrac{(n-2)!}{\Pi(d_i - 1)!}$.

Proof. When we delete vertex x from T while constructing the Prüfer sequence, all neighbors of x except one have already been deleted. We recorded x once for each such neighbor, and after its deletion x never appears again. If x remains at the end, then one edge incident to x also remains. In each case, x appears $d(x) - 1$ times in the sequence.

Therefore, we count trees with each i having degree d_i by counting sequences of length $n-2$ having $d_i - 1$ copies of i for each i. If we assign subscripts to the copies of each i to distinguish them, then there are $(n-2)!$ sequences. Since the copies of each i are in fact indistinguishable, we have counted each desired arrangement $\Pi(d_i - 1)!$ times, once for each way to order the subscripts on each type of label. $\square$

2.2.5. Example. *Trees with fixed degrees.* Consider trees with vertices $\{1, 2, 3, 4, 5, 6, 7\}$ that have degrees $(3,1,2,1,3,1,1)$, respectively. We compute $\frac{(n-2)!}{\Pi(d_i-1)!} = 30$; the trees are suggested below. There are six ways to complete the first tree (pick from the remaining four vertices the two adjacent to vertex 1) and twelve ways to complete each of the others (pick the neighbor of vertex 3 from the remaining four, and then pick the neighbor of the central vertex from the remaining three). □

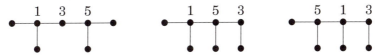

When $\Sigma n_i = n$, the quantity $\frac{n!}{\Pi n_i!}$ is called the *multinomial coefficient* $\binom{n}{n_1, \ldots, n_k}$, because it is the coefficient of $\Pi(x_i^{n_i})$ in the expansion of $(\Sigma_{i=1}^k x_i)^n$. The contributions to the coefficient of this term correspond to n-tuples that are arrangements of the n letters consisting of n_i letters of type i. When we set $x_i = 1$ for all i, this tells us that the total number of n-tuples formed from k types of letters, over all possible multiplicities, is k^n, which agrees with Cayley's formula.

SPANNING TREES IN GRAPHS

Cayley's formula also follows from the more general Matrix Tree Theorem implicit in earlier work of Kirchhoff [1847]. This provides a formula that counts the spanning subtrees of any multigraph G; Cayley's formula results when $G = K_n$ (Exercise 15). We first describe a recursive way to count spanning trees in a graph, by separately counting those that contain a particular edge e and those that omit e.

2.2.6. Definition. If e is an edge of G, then *contracting* e means replacing both endpoints of e by a single vertex whose incident edges are all edges that were incident to the endpoints of e, except e itself. The notation for the graph obtained by contracting e is $G \cdot e$.

Visually, we think of contracting e as shrinking e to a single point. Contracting an edge can produce multiple edges. To count spanning trees correctly, we must keep the multiple edges (the example below shows why), but in other applications of contraction the multiple edges may be irrelevant. When counting spanning trees, we may discard loops that arise in contraction, because no spanning tree can contain a loop. The recurrence applies to all multigraphs.

2.2.7. Proposition. If $\tau(G)$ denotes the number of spanning trees of a graph G and $e \in E(G)$, then $\tau(G) = \tau(G - e) + \tau(G \cdot e)$.

Proof. The spanning trees of G that omit e are precisely the spanning trees of $G - e$. The number of spanning trees that contain e is $\tau(G \cdot e)$, because there is a natural bijection between spanning trees of $G \cdot e$ and spanning trees of G that contain e. Contracting e in a spanning tree of G that contains e yields a spanning tree of $G \cdot e$. The other edges maintain their identity under contraction, so no two trees collapse onto the same spanning tree of $G \cdot e$ via this operation. Furthermore, each spanning tree of $G \cdot e$ arises in this way. Hence the map is a bijection. $\square$

2.2.8. Example. *A step in the recurrence.* The graphs on the right each have four spanning trees, so the recurrence for spanning trees implies that the graph on the left has eight spanning trees. $\square$

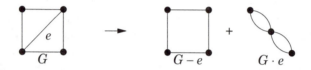

Computation using the recurrence requires initial conditions for graphs with no edges. If one vertex remains, there is one spanning tree. If more than one vertex remains, there is no spanning tree. If the computer follows the recurrence by deleting or contracting every edge, it computes $2^{e(G)}$ terms. This can be reduced by deleting loops and by recognizing special multigraphs G where we know $\tau(G)$.

2.2.9. Remark. If G is a connected multigraph with no cycle other than repeated edges, then $\tau(G)$ is the product of the edge multiplicities. A disconnected multigraph has no spanning trees. $\square$

Despite such reductions, the recursive computation is impractical for large graphs. The Matrix Tree Theorem computes $\tau(G)$ by a determinant. Determinants of n by n matrices can be computed using fewer than n^3 operations (for large n), which is much faster than $2^{e(G)}$. We can delete loops before the computation since they don't affect spanning trees. The proof of the Matrix Tree Theorem requires matrix multiplication (and determinants).

2.2.10. Example. *A Matrix Tree computation.* The Matrix Tree Theorem below instructs us to form a matrix with the vertex degrees on the diagonal, subtract the adjacency matrix, delete a row and a column, and take the determinant. For the graph $K_4 - e$ in the example above, the vertex degrees are 3,3,2,2, so we form the matrix on the left below and take the determinant of the matrix in the middle. The result is the number of spanning trees! □

$$\begin{pmatrix} 3 & -1 & -1 & -1 \\ -1 & 3 & -1 & -1 \\ -1 & -1 & 2 & 0 \\ -1 & -1 & 0 & 2 \end{pmatrix} \quad \rightarrow \quad \begin{pmatrix} 3 & -1 & -1 \\ -1 & 2 & 0 \\ -1 & 0 & 2 \end{pmatrix} \quad \rightarrow \quad 8$$

2.2.11. Theorem. (Matrix Tree Theorem). Let A be the adjacency matrix of a loopless (multi)graph G; entry (i, j) is the number of edges of the form $v_i v_j$. Let D be a diagonal matrix with $d_{ii} = d_G(v_i)$, and let $Q = D - A$. For any s, t, $\tau(G)$ equals $(-1)^{s+t}$ times the determinant of the matrix obtained by deleting row s and column t of Q.

Proof. We prove this only when $s = t$. Let Q^* be the matrix obtained by deleting row t and column t of Q. (The general statement follows from this by Exercise 18.)

Step 1. If G' is an orientation of G, and M is the incidence matrix of G', then $Q = MM^T$. If the directed edges are $e_1, \ldots, e_m$, then the entries of M are $m_{ij} = 1$ if v_i is the tail of e_j, $m_{ij} = -1$ if v_i is the head of e_j, and $m_{ij} = 0$ if v_i does not belong to e_j. Since every entry in the n by n matrix MM^T is the dot product of rows of M, off-diagonal entries in the product count -1 for every edge of G between the two vertices, and diagonal entries count vertex degrees.

$$M = \begin{array}{c} 1 \\ 2 \\ 3 \\ 4 \end{array} \begin{pmatrix} \overset{a}{-1} & \overset{b}{1} & \overset{c}{1} & \overset{d}{0} & \overset{e}{0} \\ 0 & 0 & -1 & -1 & 0 \\ 0 & 0 & 0 & 1 & -1 \\ 1 & -1 & 0 & 0 & 1 \end{pmatrix} \qquad Q = \begin{pmatrix} 3 & -1 & 0 & -2 \\ -1 & 2 & -1 & 0 \\ 0 & -1 & 2 & -1 \\ -2 & 0 & -1 & 3 \end{pmatrix}$$

Step 2. If B is an $(n-1) \times (n-1)$ submatrix of M, then $\det B = 0$ if the corresponding $n-1$ edges contain a cycle, and $\det B = \pm 1$ if they form a spanning tree of G. If the edges corresponding to the columns contain a cycle C, then the columns sum to the zero vector when weighted with $+1$ or -1 according as the directed edge is followed forward or backward when following the cycle. This equation of dependence implies $\det B = 0$.

For the other case, we use induction on n. For $n = 1$, by convention a 0×0 matrix has determinant 1. Suppose $n > 1$, and let T be the spanning tree whose edges are the columns of B. Since T has at least two leaves, B contains a row corresponding to a leaf x of T. This row has only one nonzero entry in B. When computing the determinant by expanding along that row, the only submatrix B' given nonzero weight in the expansion corresponds to the spanning subtree of $G - x$ obtained by deleting x and its incident edge from T. Since B' is an $(n-2) \times (n-2)$ submatrix of the incidence matrix for an orientation of $G - x$, the induction hypothesis implies that the determinant of B' is ± 1, and multiplying it by ± 1 gives the same result for B.

Step 3. Computation of $\det Q^*$. Let M^* be the matrix obtaining by deleting row t of M, so $Q^* = M^*(M^*)^T$. We may assume $m \geq n - 1$, else both sides have determinant 0 and there are no spanning subtrees. The Binet-Cauchy formula expresses the determinant of a product of matrices, not necessarily square, in terms of the determinants of submatrices of the factors. In particular, if $m \geq p$, A is a $p \times m$ matrix, and B is an $m \times p$ matrix, then $\det AB = \Sigma_S \det A_S \det B_S$, where the summation runs over all $S \subseteq [m]$ consisting of p indices, A_S is the submatrix of A having the columns indexed by S, and B_S is the submatrix of B having the rows indexed by S (Exercise 19). When we apply the Binet-Cauchy formula to $Q^* = M^*(M^*)^T$, the submatrix A_S is an $(n-1) \times (n-1)$ submatrix of M as discussed in Step 2, and $B_S = A_S^T$. Hence the summation counts $1 = (\pm 1)^2$ for each set of $n - 1$ edges corresponding to a spanning tree and 0 for each other set of $n - 1$ edges. $\square$

Tutte extended this theorem to directed graphs. His theorem reduces to the Matrix Tree Theorem when the digraph is symmetric; a digraph is *symmetric* if its adjacency matrix is symmetric.

2.2.12. Definition. A *branching* or *out-tree* is an orientation of a tree having a root of in-degree 0 and all other vertices of indegree 1. An *in-tree* is an out-tree with its edges reversed. Given a digraph G, let $Q^- = D^- - A'$ and $Q^+ = D^+ - A'$, where D^-, D^+ are the diagonal matrices of in-degrees and out-degrees in G, and the i, j-entry of A' is the number of edges from v_j to v_i.

2.2.13. Theorem. (Directed Matrix Tree Theorem - Tutte [1948]) In a digraph, with Q^- and Q^+ defined as above, the number of out-trees (in-trees) rooted at v_i is the value of any cofactor in the ith row of Q^- (ith column of Q^+). $\square$

2.2.14. Example. The digraph below has two out-trees rooted at 1 and two in-trees rooted at 3. The determinants behave as claimed. $\square$

$$Q^+ = \begin{pmatrix} 2 & 0 & 0 \\ -1 & 1 & 0 \\ -1 & -1 & 0 \end{pmatrix} \qquad Q^- = \begin{pmatrix} 0 & 0 & 0 \\ -1 & 1 & 0 \\ -1 & -1 & 2 \end{pmatrix}$$

DECOMPOSITION AND GRACEFUL LABELINGS

A *decomposition* of a graph G is a partition of $E(G)$ into pairwise edge-disjoint subgraphs. We can always decompose G into single edges, so we might ask whether we can decompose G into isomorphic copies of a larger tree T. This requires that $e(G)$ be a multiple of $e(T)$; is that also sufficient? If G is regular, the answer is "maybe". Häggkvist conjectured that if G is a $2m$-regular graph and T is a tree with m edges, then $E(G)$ can be partitioned into $n(G)$ copies of T. Even the "simplest" case when G is a clique is still open and notorious.

2.2.15. Conjecture. (Ringel [1964]) If T is a fixed tree with m edges, then K_{2m+1} can be decomposed into $2m+1$ copies of T. □

Attempts to prove Ringel's conjecture have focused on a stronger conjecture about trees, called the *Graceful Tree Conjecture*. This conjecture implies Ringel's conjecture and a similar statement about decomposing cliques of even order (Exercise 25).

2.2.16. Conjecture. (Graceful Tree Conjecture - Kotzig, Ringel [1964]) If T is a tree with m edges, then the vertices of T can be given the distinct numbers $0, \ldots, m$ in such a way that the edge-differences are $\{1, \ldots, m\}$. Such a numbering is called a *graceful labeling*. □

2.2.17. Theorem. If T is a tree with m edges that has a graceful labeling, then K_{2m+1} can be decomposed into $2m+1$ copies of T.

Proof. View the vertices of K_{2m+1} as the congruence classes modulo $2m+1$. The *displacement* between two congruence classes is the number of unit moves needed to get from one to the other; the maximum displacement between two congruence classes modulo $2m+1$ is m. The edges of K_{2m+1} consist of m "displacement classes", each of size $2m+1$.

From a graceful labeling of T, we define copies of T in K_{2m+1} for $0 \le k \le 2m$. In the kth copy, the vertices are $k, \ldots, k+m \bmod 2m+1$, with $k+i$ adjacent to $k+j$ if and only if i is adjacent to j in the graceful labeling. The 0th copy of T looks just like the graceful labeling and has

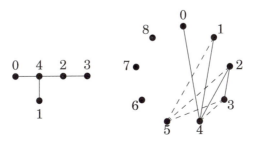

one edge with each displacement. Moving to the next copy shifts each edge to the next edge in its displacement class. Hence the $2m + 1$ copies of T cycle through the $2m + 1$ edges from each displacement class, without repetitions, and these $2m + 1$ copies of T decompose K_{2m+1}. $\square$

Graceful labelings are known to exist for some types of trees. In some ways, the *stars* $(K_{1,n-1})$ and the paths (P_n) are the simplest trees; stars minimize the diameter and paths minimize the maximum degree. We can obtain more general trees than stars by considering diameter at most k, for some fixed k. To generalize paths, we permit the addition of edges incident to a path, obtaining a class that includes the stars and the paths and has graceful labelings.

2.2.18. Example. *Graceful labeling of caterpillars.* A *caterpillar* is a tree having a path that contains at least one vertex of every edge (it can be taken to be a path of maximum length). The illustration shows a caterpillar with a graceful labeling and a tree that is not a caterpillar. Every caterpillar has a graceful labeling (Exercise 29). A *lobster* is a tree having a path from which every vertex has distance at most 2 (caterpillars with longer legs); it is not yet known whether all lobsters are graceful. $\square$

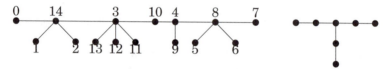

2.2.19. Theorem. The following conditions on a tree G are equivalent and characterize the class of *caterpillars*.
   A) G has a path incident to every edge.
   B) Every vertex of G has at most two non-leaf neighbors.
   C) G does not contain the tree on the right above.

Proof. Let G' denote the tree obtained from G by deleting each leaf of G. Condition A states that G' is a path, which is equivalent to $\Delta(G') \leq 2$. Since the non-leaf neighbors of each non-leaf vertex remain in G', $\Delta G' \leq 2$ is also equivalent to condition B, and we have proved A $\Leftrightarrow$ B. For B $\Leftrightarrow$ C, G has a vertex with three non-leaf neighbors if and only if G has the forbidden subtree. $\square$

EXERCISES

2.2.1. (–) Prove that the n-vertex graph $K_1 \vee C_{n-1}$ has a spanning tree with diameter k for each $k \in \{2, \ldots, n-1\}$.

2.2.2. Use the Prüfer correspondence to count the trees with vertex set $[n]$ that have $n-2$ leaves and the trees that have 2 leaves.

2.2.3. Use Cayley's Formula to prove that the graph obtained from K_n by deleting an edge has $(n-2)n^{n-3}$ spanning trees.

2.2.4. Let $S(m,r)$ denote the number of partitions of an m-element set into r nonempty subsets. In terms of these numbers, count the trees with vertex set $\{v_1,\ldots,v_n\}$ that have exactly k leaves. (Rényi [1959])

2.2.5. Suppose G has m spanning trees. Let G' be the multigraph obtained by replacing each edge of G with k copies of that edge. Let G'' be the graph obtained by replacing each edge $uv \in E(G)$ with a u,v-path of length k through $k-1$ new vertices. Determine $\tau(G')$ and $\tau(G'')$.

2.2.6. Compute $\tau(K_{2,m})$. Also compute the number of isomorphism classes of spanning trees of $K_{2,m}$.

2.2.7. (+) Determine $\tau(K_{3,m})$.

2.2.8. *Spanning trees in $K_{n,n}$.* Consider a copy of $K_{n,n}$ with bipartite sets $x_1,\ldots,x_n$ and $y_1,\ldots,y_n$. For each spanning tree T of $K_{n,n}$, form a sequence $f(T)$ of pairs of integers (written vertically) as follows: Let u,v be the least-indexed leaves of the remaining subtree that occur in X and Y. Add the pair $\binom{a}{b}$ to the sequence, where a is the index of the neighbor of u and b is the index of the neighbor of v. Delete $\{u,v\}$ and iterate until $n-2$ pairs are generated and one edge remains. Part (a) shows that f is well-defined.

　　a) Prove that every spanning tree of $K_{n,n}$ has a leaf in each partite set.

　　b) Prove that f is a bijection from the set of spanning trees of $K_{n,n}$ to the set of $n-1$-sequences of pairs of elements from $[n]$. Conclude that $K_{n,n}$ has n^{2n-2} spanning trees. (Pritikin [1994])

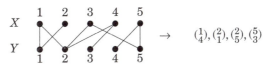

2.2.9. (+) Given K_n with vertex set $[n]$, let $f(r,s)$ be the number of spanning trees of the clique that have partite sets of sizes r and s (with $r+s=n$). Prove that $f(r,s) = \binom{r+s}{s}s^{r-1}r^{s-1}$ if $r \neq s$. What is the formula when $r=s$? (Hint: first show that the Prüfer sequence for such a tree will have $r-1$ of its terms from the set of s integers and $s-1$ of its terms from the set of r integers.) (Scoins [1962], Glicksman [1963])

2.2.10. Let G_n be the graph on $2n$ vertices and $3n-2$ edges pictured below, for $n \geq 1$. Determine $\tau(G_n)$.

2.2.11. (–) Use Proposition 2.2.7 and Remark 2.2.9 to count the spanning trees in $K_1 \vee C_4$.

2.2.12. Let a_n be the number of spanning trees in $K_1 \vee P_n$, for $n \geq 1$. For examples $a_1 = 1$, $a_2 = 3$, $a_3 = 8$. Prove for $n > 1$ that $a_n = a_{n-1} + 1 + \sum_{i=1}^{n-1} a_i$.

2.2.13. Give a combinatorial proof that the number t_n of trees with vertex set $[n]$ vertices satisfies the recurrence $t_n = \sum_{k=1}^{n-1} k\binom{n-2}{k-1} t_k t_{n-k}$. (Hint: In a tree with vertex set $[n]$, cut the edge incident to vertex n on the path from n to 1. Comment: since $t_n = n^{n-2}$, this provides a combinatorial proof of the identity $n^{n-2} = \sum_{k=1}^{n-1} \binom{n-2}{k-1} k^{k-1} (n-k)^{n-k-2}$.) (see Dziobek [1917], Lovász [1979, p219])

2.2.14. (–) Let G be the multigraph below. Use the matrix tree theorem to find a matrix whose determinant is $\tau(G)$. Compute $\tau(G)$.

2.2.15. Use the Matrix Tree Theorem to prove Cayley's formula.

2.2.16. Use the Matrix Tree Theorem to determine the number of spanning trees in $K_{r,s}$. (Lovász [1979, p223])

2.2.17. A matrix is *totally unimodular* if every square submatrix has determinant in $\{0, 1, -1\}$. Prove that the incidence matrix of a simple graph is totally unimodular if and only if the graph is bipartite. (Reminder: the incidence matrix of a simple graph has two +1's in each column).

2.2.18. (+) Given a matrix A, let b_{ij} equal $(-1)^{i+j}$ times the matrix obtained by deleting row i and column j of A. Let $\mathrm{Adj}A$ be the matrix whose entry in position i, j is b_{ji}. The definition of the determinant by expansion along rows of A yields $A(\mathrm{Adj}A) = (\det A)I$. Use this formula to prove that if the columns of A sum to the 0 vector, then b_{ij} is independent of j. (Comment: With the next exercise, this completes the proof of the Matrix Tree Theorem.)

2.2.19. (+) Let $C = AB$, where A and B are $n \times m$ and $m \times n$ matrices. Given $S \subseteq [m]$, let A_S be the $n \times n$ matrix whose columns are the columns of A indexed by S, and let B_S be the $n \times n$ matrix whose rows are the rows of B indexed by S. Prove the Cauchy-Binet Formula: $\det C = \sum_S \det A_S \det B_S$, where the summation extends over all n-element subsets of $[m]$. (Hint: consider the matrix equation $\begin{pmatrix} I_m & 0 \\ A & I_n \end{pmatrix}\begin{pmatrix} -I_m & B \\ A & 0 \end{pmatrix} = \begin{pmatrix} -I_m & B \\ 0 & AB \end{pmatrix}$.)

2.2.20. (–) Prove that a 3-regular graph with more than six vertices cannot be decomposed into three paths.

2.2.21. Suppose G is a 3-regular graph. Prove G has a decomposition into copies of $K_{1,3}$ if and only if G is bipartite.

2.2.22. Prove that no 3-regular graph has a decomposition into copies of P_5.

2.2.23. Prove that $K_{2m-1,2m}$ has a decomposition into m spanning paths.

2.2.24. Suppose G is an n-vertex simple graph having a decomposition into k spanning trees. Suppose also that $\Delta(G) = \delta(G) + 1$. Determine the degree sequence of G.

2.2.25. Prove that if the Graceful Tree Conjecture is true and T is a tree with m edges, then K_{2m} can be decomposed into $2m - 1$ copies of T. (Hint: use the proof of Theorem 2.2.17 for a tree with $m - 1$ edges.)

2.2.26. Let $d_1, \ldots, d_n$ be positive integers. Prove directly that there exists a caterpillar with vertex degrees $d_1, \ldots, d_n$ if and only if $\Sigma d_i = 2n - 2$. (Comment: in light of Exercise 2.1.12, the next exercise provides a different proof of this statement.)

2.2.27. Use the forbidden subtree characterization of caterpillars (Theorem 2.2.19) to prove that every tree can be transformed into a caterpillar with the same degree sequence by a succession of "cut and paste" operations. Such an operation consists of deleting an edge of the tree and adding another edge to reconnect the two components.

2.2.28. A bipartite graph is *drawn on a channel* if the vertices of one partite set are placed on one line in the plane (in some order) and the vertices of the other partite set are placed on a line parallel to it and the edges are drawn as straight-line segments between them. Prove that a connected graph G can be drawn on a channel without edge crossings if and only if G is a caterpillar.

2.2.29. An *up/down labeling* is a graceful labeling for which there exists a *critical value* α such that every edge joins vertices with labels above and below α. Prove that every caterpillar has an up/down labeling. Prove that the 7-vertex tree that is not a caterpillar has no up/down-labeling.

2.2.30. (+) Prove that the number of isomorphism classes of n-vertex caterpillars is $2^{n-4} + 2^{\lfloor n/2 \rfloor - 2}$ if $n \geq 3$. (Harary-Schwenk [1973], Kimble-Schwenk [1981])

2.3 Optimization and Trees

Now we seek the "best" spanning tree. This may be very difficult, especially with weights assigned to the edges, but the problem is surprisingly easy when "best" means "minimum total weight". We use *weighted graph* to mean a graph with weights assigned to the edges.

2.3.1. Definition. A *good* algorithm is one whose number of computational steps is always bounded by a polynomial function of the size of the input. An algorithm *runs in time* $O(f(n))$ ("order" $f(n)$) if there exist constants a, c such that the number of computational steps used is bounded by $c|f(n)|$ for all inputs of size at least a.

For graphs, we take the order $n(G)$ and size $e(G)$ as measuring the size of the input. Most problems we study in the first half of this book have good algorithms, so technical notions of complexity (such as "NP-completeness" in Section 6.3) need not trouble us yet.

MINIMUM SPANNING TREE

Each edge in a connected graph of possible communication links has a weight recording its length or cost. All spanning trees have $n - 1$ edges, but we seek a spanning tree to minimize the sum of the edge weights. This problem was considered as early as Borůvka [1926]. A *greedy* approach always chooses the cheapest remaining edge that does not complete a cycle with the edges already chosen (ties are broken arbitrarily). Since every acyclic subgraph with $n - 1$ edges is a tree, this produces a spanning tree when we have chosen $n - 1$ edges.

This is *Kruskal's Algorithm*. The term "greedy" refers generally to a locally optimal heuristic; we may consider greedy algorithms in various contexts. Few greedy algorithms work, but Kruskal's algorithm always produces the cheapest spanning tree.

2.3.2. Example. *Application of Kruskal's algorithm.* The choices in Kruskal's algorithm use only the order of the weights, not their magnitude. In the graph below, we have labeled the edges in increasing order of weight to emphasize the order of examination of edges.

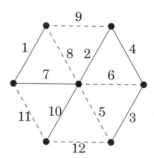

In a computer, the weights appear in a matrix, with huge weight on "unavailable" edges. Edges of equal weight may be examined in any order; the resulting trees have the same cost. The algorithm begins with a forest of n isolated vertices. Each successive selected edge combines two components. In this example, the four cheapest edges are selected, but then we cannot take the fifth or sixth. □

2.3.3. Theorem. (Kruskal [1956]). In a connected weighted graph G, Kruskal's Algorithm constructs a minimum-weight spanning tree.

Proof. Let T be the tree built by Kruskal's Algorithm, and let T^* be a minimum spanning tree. If $T \neq T^*$, let e be the first edge chosen for T that is not in T^*. Adding e to T^* creates one cycle, which contains an edge $e' \notin E(T)$ since T has no cycle. Now $T^* + e - e'$ is a spanning tree. Since T^* contains e' and all the edges of T chosen before e, both e' and e are available when the algorithm chooses e, and hence $w(e) \leq w(e')$. Thus $T^* + e - e'$ is a spanning tree with weight at most T^* that contains a longer initial segment of T. Since T is finite, iterating this switch leads to a minimum-weight spanning tree that contains all of T. (Phrased extremally, we have proved that the minimum spanning tree that agrees with T for the longest initial segment is T itself.) □

2.3.4. Example. *Implementation and analysis of Kruskal's Algorithm.* First we sort the m edges by weight, which can be done using $O(m \log m)$ pairwise comparisons among m numbers. While building the tree, we label each vertex by the component containing it in the current tree. We accept the next cheapest edge if its endpoints have different labels. In that case, we merge the two components by assigning the lower of the two labels to every vertex having the higher label. If we initially assign label i to the component consisting of vertex v_i, then the label on v_i changes at most $i - 1$ times, and altogether there are at most $\binom{n}{2}$ changes. If we keep track of the sizes of the components and always merge the smaller component into the larger, then the number of changes is $O(n \log n)$. In this case the time for processing large graphs is determined by the time to sort m numbers. □

Both Borůvka [1926] and Jarník [1930] posed and solved the minimum spanning tree problem. Borůvka's algorithm picks the next edge by considering the cheapest edge leaving each component of the current forest. Later improvements use clever data structures to implement the algorithms more efficiently. Fast versions appear in Tarjan [1984] for when the edges are previously sorted and in Gabow-Galil-Spencer-Tarjan [1986] for when they are not. Thorough discussion and further references appear in Ahuja-Magnanti-Orlin [1993, Chapter 13]. For more recent developments, see Karger-Klein-Tarjan [1995].

SHORTEST PATHS

Given a road map with distances specified between junctions, we may ask "what is the quickest way from here to there?" We may also want to know the shortest route to every other location from a particular location, such as our home town. This requires finding shortest paths from one specified vertex to all other vertices in a weighted graph, where the edge weights correspond to nonnegative distances between junctions. Together, these paths will form a spanning tree.

Dijkstra's Algorithm (discovered by Dijkstra [1959] and by Whiting and Hillier [1960]) solves this problem quickly. The approach rests on the following: if P is a shortest u, z-path and P contains v, then the u, v-portion of P is a shortest u, v-path. This suggests that we should determine optimal routes from u to every other vertex z in increasing order of the distance $d(u, z)$. We maintain a current tentative distance from u to each vertex z. We confirm the smallest tentative distance as a correct distance and use this to update the remaining tentative distances. The details are more complicated than for Kruskal's Algorithm, so we give a more formal presentation.

2.3.5. Algorithm. *Dijkstra's Algorithm* (to compute distances from u).
Input: A weighted graph (or digraph) and starting vertex u. The weight of edge xy is $w(xy)$; let $w(xy) = \infty$ if xy is not an edge.
Idea: Maintain the set S of vertices to which the shortest route from u is known, enlarging S to include all vertices. To do this, maintain also a tentative distance $t(z)$ from u to each $z \notin S$; this is the length of the shortest u, z-path yet found.
Initialization: Set $S = \{u\}$; $d(u, u) = 0$; $t(z) = w(uz)$ for $z \neq u$.
Iteration: Select the vertex v outside S such that $t(v) = \min_{z \notin S} t(z)$. Add v to S. Explore edges from v to update tentative distances: for each edge vz with $z \notin S$, update $t(z)$ to $\min\{t(z), d(u, v) + w(vz)\}$.

The iteration continues until $S = V(G)$ or until $t(z) = \infty$ for every $z \notin S$. In the latter case, no vertex is selectable; the remaining vertices are unreachable from u and have infinite distance from u. □

2.3.6. Example. *Application of Dijkstra's algorithm.* In the weighted graph below, shortest paths from u are found to the other vertices in the order a, b, c, d, e, with distances 1,3,5,6,8, respectively. To reconstruct

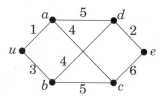

the paths, we need only know the final edge on which each shortest path arrives at its destination, because the earlier portion of a shortest u, z-path that reaches z on the edge vz is a shortest u, v-path. The algorithm can maintain this information by recording the identity of the "selected vertex" whenever the tentative distance to z is updated. When

z is selected, the vertex that was recorded when $t(z)$ was last updated is the predecessor of z on the u, z-path of length $d(u, z)$. In this example, the final edges on the paths to a, b, c, d, e generated by the algorithm are ua, ub, ac, ad, de, respectively, and these are the edges of the spanning tree generated from u. $\square$

As described, Dijkstra's algorithm works equally well for directed graphs, generating a branching from u (an out-tree rooted at u) if every vertex is reachable from u. The proof works for graphs and for digraphs without change.

2.3.7. Theorem. Dijkstra's algorithm computes $d(u, z)$ for every $z \in V(G)$.

Proof. We prove the stronger statement that at each step of the algorithm, 1) the distance from u that has been confirmed for each $v \in S$ is $d(u, v)$, and 2) each finite $t(z)$ for $z \notin S$ is the least length of a u, z-path reaching z directly from S. We prove this by induction on $|S|$; this is an example of "loading the induction hypothesis". The basis step follows from the initialization: $k = 1$, $S = \{u\}$, $d(u, u) = 0$, and there is a path of finite length reaching z from S if and only if uz is an edge, in which case $t(z) = w(uz)$.

For the induction step, suppose that when $|S| = k$, the various claims about S are true. Let v be a vertex among $z \notin S$ such that the tentative distance $t(z)$ is smallest. We first argue that $d(u, v) = t(v)$. A shortest u, v-path must exit S before reaching v. The induction hypothesis states that the length of the shortest path going directly to v from S is $t(v)$. The induction hypothesis and choice of v also guarantee that a path visiting any vertex outside S and later reaching v has length at least $t(v)$. Hence $d(u, v) = t(v)$. Before we update, the shortest u, z-path reaching z directly from S has length $t(z)$ (∞ if no such path has been found). When we add v to S, we must also consider paths reaching z from v. Since we have now computed $d(u, v)$, the shortest such path has length $d(u, v) + w(vz)$, and we compare this with the previous value of $t(z)$ to update $t(z)$. We have verified that each claim being proved in the induction step holds for the new set $S \cup \{v\}$ of size $k + 1$. $\square$

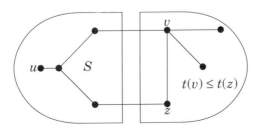

The algorithm maintains the condition that $d(u,x) \le t(z)$ for all $x \in S$ and $z \notin S$; hence it selects vertices in nondecreasing order of distance from u. The special case when G is unweighted is *Breadth-First Search* from u. In this case, both the algorithm and the proof (Exercise 11) have simpler descriptions.

2.3.8. Algorithm. *Breadth-First Search (BFS).*
Input: An unweighted graph (or digraph) and a start vertex u.
Idea: Maintain a set R of vertices that have been reached but not searched and a set S of vertices that have been searched. The set R is maintained as a First-In First-Out list (queue) so that the first vertices found are the first vertices explored.
Initialization: $R = \{u\}$, $S = \varnothing$, $d(u,u) = 0$.
Iteration: As long as $R \ne \varnothing$, we search from the first vertex v of R. The neighbors of v not in R or S are added to the back of R and assigned distance $d(u,v) + 1$, and then v is removed from the front of R and placed in S. □

The largest distance from a vertex u to another vertex is the eccentricity $\varepsilon(u)$. Hence we can compute diam$G$ by running Breadth-First Search from each vertex. Other search strategies have other applications. In *Depth-First Search* (DFS), we explore always from the most recently discovered vertex that has unexplored edges (this is also called *backtracking*). In contrast, BFS explores from the oldest vertex, so the difference between DFS and BFS is that in DFS the set R is maintained as a Last-In First-Out "stack" rather than as a queue.

2.3.9. Example. *Depth-First Search.* In the graph below, one depth-first search from u finds the vertices in the order u, a, b, c, d, e, f, g. For both BFS and DFS, the vertex order depends on the order of exploring edges from a searched vertex. Section 4.1 has an application of depth-first search. □

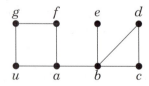

A breath-first or depth-first search from u generates a tree rooted at u; each time we discover a new vertex v, we include the edge between v and the vertex from which we reached v. This grows a tree that becomes a spanning tree of the component containing u. The usefulness of depth-first search stems from a fundamental property of the resulting spanning tree.

2.3.10. Lemma. If T is a spanning tree of a connected graph G grown by a depth-first search from u, then every edge of G not in T consists of two vertices v, w such that v lies on the u, w-path in T.

Proof. Suppose vw is an edge of G, with v encountered before w in the depth-first search. Because vw is an edge, we cannot finish v before w is added to T. Hence w appears somewhere in the subtree formed before finishing v, and the path from w to u contains v. $\square$

TREES IN COMPUTER SCIENCE

Most applications of trees in computer science use rooted trees.

2.3.11. Definition. A *rooted tree* distinguishes one vertex r as *root*. For each vertex v, let $P(v)$ be the unique v, r-path. The *parent* of v is the neighbor of v on $P(v)$. The *children* of v are its other neighbors. The *ancestors* of v are the vertices of $P(v) - v$. The *descendants* of v are the vertices u such that $P(u)$ contains v (including the children of v). The *leaves* are the vertices with no children (non-root vertices of degree 1). A *rooted plane tree* or *planted tree* is a rooted tree with a left-to-right ordering specified for the children of each vertex.

When running BFS or DFS from u, we usually view the resulting tree T as rooted at u. In this language, Lemma 2.3.9 states that every edge of G outside a spanning tree T formed by a depth-first search joins two vertices such that one is an ancestor of the other.

2.3.12. Definition. A *binary tree* is a rooted plane tree in which each vertex has at most two children, and each child of a vertex is designated as its *left child* or *right child*. The subtrees rooted at the children of the root are the *left subtree* and the *right subtree* of the tree. A *k-ary tree* allows each vertex up to k children.

In some applications of binary trees, non-leaves must have exactly two children (Exercise 2.3.18). Binary trees permit storage of data for efficient access. When we associating each data item with a leaf of a rooted binary tree, we can access items by searching from the root if we can tell at each non-leaf which subtree contains the desired leaf. We do this by associating with each leaf the 0,1-sequence that encodes the sequence of left steps or right steps on the path to it from the root. The length of the search is the length of this path. Given probabilities for the n items to be accessed, we want to associate them with the n leaves of a binary tree to minimize the expected search length.

Similarly, given large computer files and limited disk storage, we want to encode the characters as bit-sequences to minimize the total length. Given the frequencies of the characters (or messages), we can divide the frequencies by the total number of characters to obtain probabilities $\{p_i\}$, and then this reduces to the first problem: we want to assign binary code words to minimize the average message length. The length of code words may vary, so we need a way to recognize the end of the current code word. If no code word is a prefix of another code word, then the current word ends at the first (and only) bit such that the sequence since the end of the previous word is a code word.

The prefix-free condition ensures that the code words correspond to the leaves of a binary tree, where the code for a leaf is the 0,1-sequence generated by the path from the root by letting 0 represent traveling to a left child and letting 1 represent traveling to a right child. The expected length of a message is $\Sigma p_i l_i$, where l_i is the length of the code word (path from root) assigned to the ith message word. Constructing the optimal code is surprisingly easy.

2.3.13. Algorithm. *Huffman [1952].* The input is a discrete probability distribution $\{p_i\}$ on n message words; the output is a prefix-free code. If $n = 2$, the algorithm assigns code word 1 to one message word and 0 to the other. If $n > 2$, the algorithm combines the two least likely items into a single item whose probability is the sum of their probabilities, calls itself recursively to find the code for the resulting set of $n - 1$ items, and replaces the resulting code word for the combined item with its two extensions by 1 and 0, assigned to the two least likely items. □

2.3.14. Example. *Huffman coding.* Suppose the frequencies of eight messages are 5,1,1,7,8,2,3,6. The algorithm combines items according to the tree on the left below, combining items by working from the bottom up. First the two items of weight 1 combine to form one of weight 2. Now this and the original item of weight 2 are the cheapest, combining to form one of weight 4. The 3 and 4 now combine, after which the cheapest elements are the original items of weights 5 and 6. The remaining combinations in order are $5 + 6 = 11$, $7 + 7 = 14$, $8 + 11 = 19$, $14 + 19 = 33$. From the drawing of this tree on the right, we can choose code words. Corresponding to the original order of the items, the code words are 100, 00000, 00001, 01, 11, 0001, 001, and 101. The expected length of this code is $\Sigma p_i l_i = 90/33 < 3$, while the expected length of a code using only the eight words of length 3 would be 3. □

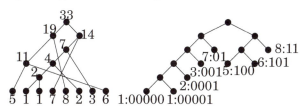

2.3.15. Theorem. Given a probability distribution $\{p_i\}$ on n words, the Huffman algorithm produces the prefix-free code with minimum expected length.

Proof. We use induction on n. For $n = 2$, we must send a bit to send a message, and the algorithm produces code length 1; this completes the basis step. For the induction step, suppose $n > 2$, and suppose the algorithm computes the optimal code when given a distribution for $n - 1$ items. Every binary tree with n leaves corresponds to a code for n messages. For any fixed tree with n leaves, we can minimize the expected length by greedily assigning the messages with probabilities $p_1 \geq \cdots \geq p_n$ to leaves in increasing order of depth. Thus in an optimal code we may assume that the two least likely messages are assigned to leaves of greatest depth (every leaf at maximum depth has another leaf as its sibling). We may further assume that the least likely messages appear as siblings at greatest depth, since permuting the items associated with leaves at a given depth does not change the expected length.

Suppose T is an optimal tree, with the least likely p_n and p_{n-1} located as sibling leaves at greatest depth. Let T' be the tree obtained from T by deleting these leaves. Let $\{q_i\}$ be the probability distribution obtained by replacing $\{p_{n-1}, p_n\}$ by $q_{n-1} = p_{n-1} + p_n$. The tree T' yields a code for $\{q_i\}$. The expected length for T is the expected length for T' plus q_{n-1}, since if k is the depth of the leaf assigned q_{n-1}, we lose kq_{n-1} and gain $(k + 1)(p_{n-1} + p_n)$ in moving from T' to T. This is true for each choice of T', so it is best to use the tree T' that is optimal for $\{q_i\}$. By the induction hypothesis, the optimal choice for T' is obtained by applying Huffman's algorithm to $\{q_i\}$. Since the replacement of $\{p_{n-1}, p_n\}$ by q_{n-1} is the first step of Huffman's algorithm for $\{p_i\}$, we conclude that Huffman's algorithm generates the optimal tree T for $\{p_i\}$. □

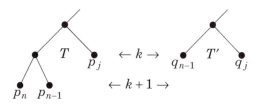

Huffman's algorithm computes an optimal prefix-free code, and its expected length is close to the optimum over all types of binary codes. Shannon [1948] proved that for every code with binary digits, the expected length is at least the *entropy* of the discrete probability distribution $\{p_i\}$, defined to be $-\Sigma p_i \lg p_i$ (Exercise 19). When each p_i is a power of 1/2, the Huffman code meets this bound exactly (Exercise 18).

EXERCISES

2.3.1. (–) Prove or disprove: If T is a minimum-weight spanning tree of a weighted graph G, then the u, v-path in T is a minimum-weight u, v-path in G.

2.3.2. (–) There are five cities in a network. The cost of building a road directly between i and j is the entry a_{ij} in the matrix below. An infinite entry indicates that there is a mountain in the way and the road cannot be built. Determine the least cost of making all the cities reachable from each other.

$$\begin{pmatrix} 0 & 3 & 5 & 11 & 9 \\ 3 & 0 & 3 & 9 & 8 \\ 5 & 3 & 0 & \infty & 10 \\ 11 & 9 & \infty & 0 & 7 \\ 9 & 8 & 10 & 7 & 0 \end{pmatrix}$$

2.3.3. In the graph $K_1 \vee C_4$, assign the weights $(1, 1, 2, 2, 3, 3, 4, 4)$ to the edges in two ways: one way so that the mimimum-weight spanning tree is unique, and another way so that the minimum-weight spanning tree is not unique.

2.3.4. Suppose G is a weighted graph in which the edge weights are distinct. Without using Kruskal's algorithm, prove that G has only one minimum-weight spanning tree.

2.3.5. Suppose F is a spanning forest of a connected weighted graph G. Among all the edges of G having endpoints in different components of F, let e be one of minimum weight. Prove that among all the spanning trees of G that contain F, there is one of minimum weight that contains e. Use this to give another proof that Kruskal's algorithm works.

2.3.6. (!) Prim's Algorithm grows a spanning tree from an arbitrary vertex of a weighted G, iteratively adding the cheapest edge between a vertex already absorbed and a vertex not yet absorbed, finishing when the other $n - 1$ vertices of G have been absorbed. (Ties are broken arbitrarily.) Prove that Prim's Algorithm produces a minimum-weight spanning tree of G. (discovered independently in Jarník [1930], Prim [1957], Dijkstra [1959]).

2.3.7. A *minimax* or *bottleneck* spanning tree is a spanning tree in which the maximum weight of the edges is as small as possible. Prove that every minimum-weight spanning tree is a bottleneck spanning tree.

2.3.8. Suppose T is a minimum-weight spanning tree in a weighted connected graph G. Prove that T omits some heaviest edge from every cycle in G.

2.3.9. (!) Given a connected weighted graph, iteratively delete a heaviest non-cut-edge until the resulting graph is acyclic. Prove that the subgraph remaining is a minimum-weight spanning tree.

2.3.10. (!) Suppose T is a minimum-weight spanning tree in G, and T' is another spanning tree in G. Prove that T' can be transformed into T by a sequence of steps that exchange one edge of T' for one edge of T, such that the edge set is always a spanning tree and the total weight never increases.

2.3.11. Suppose we seek a minimum-weight spanning path greedily: Iteratively select the edge of least weight so that the edges selected so far form a disjoint union of paths. When $n - 1$ edges have been chosen, the result is a spanning path. Prove that this algorithm always gives a minimum-weight spanning path, or give an infinite family of counterexamples where it fails.

2.3.12. (–) There are five cities in a network. The travel time for traveling directly from i to j is the entry a_{ij} in the matrix below. The matrix is not symmetric (use directed graphs), and $a_{ij} = \infty$ indicates that there is no direct route. Determine the least travel time and quickest route from i to j for each pair i, j.

$$\begin{pmatrix} 0 & 10 & 20 & \infty & 17 \\ 7 & 0 & 5 & 22 & 33 \\ 14 & 13 & 0 & 15 & 27 \\ 30 & \infty & 17 & 0 & 10 \\ \infty & 15 & 12 & 8 & 0 \end{pmatrix}$$

2.3.13. Given a starting vertex u in an unweighted graph or digraph G, prove directly (without Dijkstra's algorithm) that the BFS algorithm computes $d(u, z)$ for every $z \in V(G)$.

2.3.14. *Minimum diameter spanning tree.* An MDST is a spanning tree in which the maximum length of a path is as small as possible. Intuition suggests that running Dijkstra's algorithm from a vertex of minimum eccentricity (a center) will produce an MDST, but this may fail.

a) Construct a 6-vertex example of an unweighted graph (edge weights all equal 1) in which Dijkstra's algorithm can be run from some vertex of minimum eccentricity and produce a spanning tree that does not have minimum diameter. (Note: when there are multiple candidates with the same distance from the root, or multiple ways to reach the new vertex with minimum distance, the choice in Dijkstra's algorithm can be made arbitrarily.)

b) Construct a 4-vertex example of a weighted graph such that Dijkstra's algorithm cannot produce an MDST when run from any vertex.

2.3.15. Develop an efficient algorithm that, given a graph as input, decides whether the graph is bipartite. The graph is given by adjacency matrix or lists of vertices and their neighbors. The algorithm should not need to look at any edge more than twice.

2.3.16. Let $f(G)$ denote the maximum number of leaves in a spanning tree of G. Form H from a cyclic sequence of $3m$ cliques by making every vertex adjacent to each vertex in the clique before it and the clique after it. Let the clique sizes be

$k/2, k/2, 1, k/2, k/2, 1, \cdots$, so H is k-regular. Determine $f(H)$. (Comment: It is conjectured that H minimizes f over all n-vertex graphs with minimum degree at least k. The conjecture is known to be true for $k \leq 5$.)

2.3.17. Suppose G is an n-vertex rooted plane tree in which every vertex has 0 or k children. Given k, for what values of n is this possible?

2.3.18. Find a recurrence relation to count the binary trees with $n + 1$ leaves (here each non-leaf vertex has exactly two children, and the left-to-right order of children matters). When $n = 2$, the possibilities are the two trees below.

2.3.19. Find a recurrence relation for the number of rooted plane trees with n vertices. (As in a rooted binary tree, the subtrees obtained by deleting the root of a rooted plane tree are distinguished by their order from left to right.)

2.3.20. (–) Compute a code with minimum expected length for a set of ten messages whose relative frequencies are 1,2,3,4,5,5,6,7,8,9. What is the expected length of a message in this optimal code?

2.3.21. (–) In the game of *Scrabble*, the relative frequencies of letters are as in the table below. This is not the true distribution in English; the number of S's is reduced, for example, to improve the game. There are also two blanks, making the total 100 letters; treat blanks as an additional character. Pretend that these are the actual relative frequencies in English text, and compute the prefix-free code of minimum expected length for transmitting messages. Give your answer by listing the relative frequency for each length of code word. Compute the expected length of the code (per character). (Comment: Straight ASCII coding uses five bits per letter, so this code will beat that. Of course, it is unfair to ASCII that we have not included codes for punctuation.)

A	B	C	D	E	F	G	H	I	J	K	L	M	N	O	P	Q	R	S	T	U	V	W	X	Y	Z	∅
9	2	2	4	12	2	3	2	9	1	1	4	2	6	8	2	1	6	4	6	4	2	2	1	2	1	2

2.3.22. Suppose that n messages occur with probabilities $p_1, \ldots, p_n$ and that each p_i is a power of 1/2 (each $p_i \geq 0$ and $\Sigma p_i = 1$).

a) Prove that the two least likely messages have equal probability.

b) Prove that the expected message length of the Huffman code for this distribution is $-\Sigma p_i \lg p_i$.

2.3.23. (+) Suppose that n messages occur with probabilities $p_1, \ldots, p_n$ and that the words are assigned distinct binary code words. Prove that for every code, the expected length of a code word with respect to this distribution is at least $-\Sigma p_i \lg p_i$. (Shannon [1948]) (Hint: use induction on n.)

2.4 Eulerian Graphs and Digraphs

Some say that graph theory was born in the city of Königsberg, located on the banks of the Pregel. The river surrounded the island of Kneiphof, and there were seven bridges linking the four land masses of the city, as illustrated on the left below. It seems that the residents wanted to know whether it was possible to take a stroll from home, cross every bridge exactly once, and return home. The problem reduces to traversing the multigraph drawn on the right, where the vertices represent land masses and the edges represent bridges. We want to know when a multigraph contains a single closed trail traversing all its edges; we use the term *circuit* to have the same meaning as "closed trail".

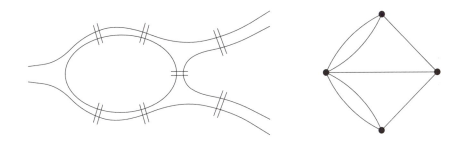

EULERIAN CIRCUITS

The Swiss mathematician Leonhard Euler (pronounced "oiler") observed in 1736 that the desired stroll did not exist. A circuit contributes twice to the degree of a vertex for each visit, so a multigraph traversable in this way must have each vertex degree even (and must have all edges in the same component). All vertex degrees in the graph above are odd, so it fails this necessary condition. Euler stated that the condition is also sufficient, and in his honor we say that a graph whose edges belong to a single circuit is *Eulerian*. Euler's paper, which appeared in 1741, did not give a proof that the obvious necessary condition is sufficient. Hierholzer [1873] gave the first published proof; the diagram on the right above for the multigraph model did not appear until 1894 (see Wilson [1986] for a discussion of the historical record).

In this section, we allow loops and multiple edges because our focus is on traversing edges. We also use *odd vertex* or *even vertex* to indicate the parity of a vertex degree, and we say that a (multi)graph is *even* if its vertices are all even. Recall that a *nontrivial* graph is a graph having at least one edge.

2.4.1. Lemma. Every maximal trail in an even graph is a closed trail.

Proof. Let T be a maximal trail. If T is not closed, then T has an odd number of edges incident to the final vertex v, but then there is another edge incident to v that is not in T and can be used to extend T. □

2.4.2. Theorem. A finite graph G is Eulerian (its edges are traversable by a single circuit) if and only if all its vertex degrees are even and all its edges belong to a single component.

Proof. We have seen that the conditions are necessary. Conversely, suppose that G satisfies them and has at least one edge. Let T be a maximal trail in G. By Lemma 2.4.1, T is closed. If T does not contain $E(G)$, let $G' = G - E(T)$. Since T has even degree at every vertex, G' is an even graph. Because G has only one nontrivial component, there is a path in G from every edge to every other edge. Hence some edge e of G' is incident to some vertex v of T.

Let T' be a maximal trail in G' beginning from v along e. By Lemma 2.4.1, T' is closed. Hence we may detour from T along T' when we reach v (and then complete T) to obtain a longer closed trail than T. Replace T by this trail and repeat the argument. Since $E(G)$ is finite, this process must end, and it can only end by constructing an Eulerian circuit. □

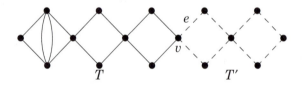

We obtained a short proof by employing extremality (the use of maximal trails). Exercises 3 and 7 request alternative proofs.

Given a figure G to be drawn on paper, how many times must we pick up and move the pen (or plotter) to do it? This is the minimum number of pairwise edge-disjoint trails whose union is $E(G)$. We may reduce the problem to connected graphs, since the number of trails needed to draw G is the sum of the number needed to draw each component. Recall that every circuit is a closed trail. The graph on the left below has four odd vertices and can be decomposed into two trails. Adding the dashed edges on the right makes it Eulerian.

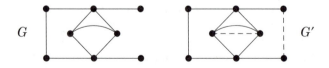

2.4.3. Theorem. For a connected nontrivial graph with $2k$ odd vertices, the minimum number of pairwise edge-disjoint trails covering the edges is max $\{k, 1\}$.

Proof. A trail contributes even degree to every vertex, except that a non-closed trail contributes odd degree to its endpoints. Therefore, a partition of the edges into trails must have some trail ending at each odd vertex. Since each trail has (at most) two ends, this requires at least k trails. It also requires at least one trail since $e(G) > 0$, and we have proved that one trail suffices if $k = 0$.

To prove that k trails suffice when $k > 0$, pair up the odd vertices in G arbitrarily and form G' by adding a copy of each pair as an edge, as illustrated above. The resulting G' is connected, even, and has an Eulerian circuit, because Theorem 2.4.2 allows multiple edges. As we traverse the circuit, we start a new trail in G each time we traverse an edge of $G' - E(G)$. This yields k edge-disjoint trails partitioning $E(G)$. $\square$

We could also prove this by induction (Exercise 6); transforming the graph and using Theorem 2.4.2 avoids induction. When $k = 2$, we obtain a single trail using all of $E(G)$; this is an *Eulerian trail*.

The proof of Theorem 2.4.2 provides an algorithm that constructs an Eulerian circuit. It iteratively merges circuits into the current circuit until all edges are absorbed. The next algorithm approaches the "draw this figure" problem more directly, building a circuit one edge at a time without backtracking. An Eulerian circuit can start anywhere; an Eulerian trail must start at an odd vertex. Neither traverses an edge whose deletion cuts the remaining graph into two nontrivial components, because it could not return to pick up the stranded edges. This necessary condition is sufficient; if we avoid such edges, we can complete the traversal.

2.4.4. Algorithm. (Fleury's Algorithm - constructing Eulerian trails).
Input: A graph G with one nontrivial component and at most two odd vertices.
Initialization: Start at a vertex that has odd degree unless G is even, in which case start at any vertex.
Iteration: From the current vertex, traverse any remaining edge whose deletion from the remaining graph does not leave a graph with two non-trivial components. Stop when all edges have been traversed. $\square$

2.4.5. Theorem. If G has one nontrivial component and at most two odd vertices, then Fleury's algorithm constructs an Eulerian trail.

Proof. We use induction on $e(G)$. The claim is immediate for $e(G) = 1$. Suppose $e(G) > 1$, and suppose the claim holds for graphs with $e(G) - 1$ edges. If G is even, then G has no cut-edge, since deleting it would leave subgraphs with one odd vertex (Exercise 1.3.9). Hence starting from some vertex v along an edge vu leaves a graph with two odd vertices, and by the induction hypothesis the algorithm completes the circuit of G by completing a u,v-trail in $G-vu$.

Suppose G has two odd vertices u,v. If $d(u)=1$, then traversing ux leaves one nontrivial component. If $d(u)>1$, let P be a u,v-path, and let ux be an edge incident to u but not on P. Since u and v are connected in $G-ux$, ux cannot be a cut-edge, because x would be the only odd vertex in its component of $G-ux$ (see illustration). Hence again u has an incident edge ux whose traversal leaves a graph $G-ux$ having at most one nontrivial component and at most two odd vertices.

If $x=v$, then $G-ux$ is even. If $x \neq v$, then the vertices of $G-ux$ are even except for x and v (this includes the possibility $x=u$). Hence the induction hypothesis applies, and the algorithm completes an Eulerian u,v-trail of G by traversing an Eulerian x,v-trail of $G-ux$. $\square$

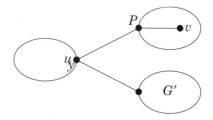

Fleury's Algorithm (see Lucas [1921]) seems to be the oldest algorithm for constructing Eulerian trails directly (Exercise 7 describes another). It provides another proof of the characterization of Eulerian graphs. To make the proof algorithmic, we must *find* a noncut edge incident to u. We can run BFS or DFS search to test whether u and x are in the same component of $G-ux$. Because we need many such tests, our proof of Theorem 2.4.2 yields a faster algorithm.

DIRECTED GRAPHS

A circuit in a digraph traverses edges from tail to head; an *Eulerian* circuit traverses every edge. This requires a condition analogous to connectedness for graphs.

2.4.6. Definition. A digraph G is *strongly connected* or *strong* if for every ordered pair $u,v \in V(G)$ there is a u,v-path in G.

Each entrance to a vertex is followed by a departure, so an Eulerian digraph must also have $d^+(u)=d^-(u)$ for all $u \in V(G)$. This is also sufficient, if the edges of G belong to one strong component. The proofs are similar to those for undirected graphs (Exercise 10). We present a constructive proof, analogous to Fleury's algorithm. It needs only one search computation to create a tree at the beginning. After that, the information needed to construct the circuit is present in the tree.

2.4.7. Algorithm. (Directed Eulerian circuit).
Input: A digraph G that is an orientation of a connected graph and has $d^+(u) = d^-(u)$ for all $u \in V(G)$.
Step 1: Choose a vertex $v \in V(G)$. Let G' be the digraph obtained from G by reversing direction on each edge. Search G' to construct a tree T' consisting of paths from v to all other vertices (e.g., use BFS or DFS).
Step 2: Let T be the reversal of T'; T contains a u, v-path in G for each $u \in V(G)$. Specify an arbitrary ordering of the edges that leave each vertex u, except that for $u \neq v$ the edge leaving u in T must come last.
Step 3: Construct an Eulerian circuit from v as follows: whenever u is the current vertex, exit along the next unused edge in the ordering specified for edges leaving u. □

2.4.8. Example. *Directed Eulerian circuit.* In the digraph below, the solid edges indicate an "in-tree" T of paths into v. When we follow the Eulerian circuit starting with edge 1, we avoid using edges of T until we have no choice. Assuming that the ordering at v places 1 before 10 before 15, the algorithm traverses the edges in the order indicated. □

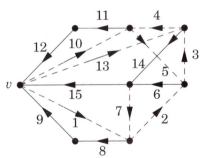

2.4.9. Theorem. If G is an orientation of a multigraph with one non-trivial component and $d^+(u) = d^-(u)$ for all $u \in V(G)$, then the algorithm above constructs an Eulerian circuit of G.

Proof. First we construct T'. The search (BFS) from v reaches a new vertex at each step. Otherwise, all edges between the current reached set R and the remaining vertices enter R. Each edge within R contributes once to the in-degree and once to the out-degree of vertices in R; each edge entering R contributes only to the in-degree. Hence the sum of in-degrees in R exceeds the sum of out-degrees, which contradicts our hypothesis that $d^+(u) = d^-(u)$ for each vertex u.

Now we use T to build a trail as directed. The trail can only end at v, because when we enter a vertex $u \neq v$, the edge leaving v in T remains, since $d^+(u) = d^-(u)$. We can end at v only if we have used all edges leaving v, and hence also all edges entering v. Since we cannot use an edge of T until it is the only remaining edge leaving its tail, we cannot use all edges entering v until we have finished all the other vertices, since T contains a path from each vertex to v. $\square$

2.4.10. Example. *Directed Eulerian circuit.* In the digraph below, every in-tree to v contains the edges uv, yz, wx, exactly one of $\{zu, zv\}$, and exactly one of $\{xy, xz\}$. There are four in-trees to v. For each in-tree, the number of legal sets of orderings of the edges leaving the vertices is $\Pi(d_i - 1)! = (0!)^3(1!)^3 = 1$. Hence we can generate one Eulerian circuit for each in-tree, starting along the edge $e = vw$ from v. The four in-trees and the corresponding circuits appear below. $\square$

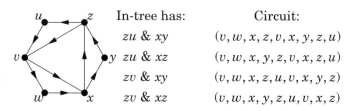

In-tree has:	Circuit:
zu & xy	$(v, w, x, z, v, x, y, z, u)$
zu & xz	$(v, w, x, y, z, v, x, z, u)$
zv & xy	$(v, w, x, z, u, v, x, y, z)$
zv & xz	$(v, w, x, y, z, u, v, x, z)$

Two Eulerian circuits are the same if the successive pairs of edges are the same. From each in-tree to v, Algorithm 2.4.7 can generate $\Pi_{u \in V(G)}(d^+(u) - 1)!$ different Eulerian circuits. The last out-edge is fixed by the tree for vertices other than v, and since we consider only the cyclic order of the edges we may also choose a particular edge e to start the ordering of edges leaving v. Any change in the exit orderings at vertices specifies at some point different choices for the next edge, so the circuits are distinct. Similarly, the circuits obtained from distinct T's are distinct. Hence we have generated $c\Pi_{u \in V(G)}(d^+(u) - 1)!$ distinct Eulerian circuits, where c is the number of in-trees to v.

In fact, these are all the Eulerian circuits. This provides a combinatorial proof that the number of in-trees to each vertex of an Eulerian digraph is the same. Since the graph obtained by reversing all the edges has the same number of Eulerian circuits, the number of out-trees from any vertex also has this same value, c. The value c can be computed using the Directed Matrix Tree Theorem (Theorem 2.2.13).

2.4.11. Theorem. (van Aardenne-Ehrenfest and de Bruijn [1951]). In an Eulerian digraph with $d_i = d^+(v_i) = d^-(v_i)$ the number of Eulerian circuits is $c\Pi_{i=1}^{n}(d_i - 1)!$, where c equals the number of in-trees to or out-trees from any node.

Proof. We have observed that Algorithm 2.4.7 generates this many distinct Eulerian circuits using in-trees to vertex v, with e being the first edge in the exit ordering at v. We need only show that each Eulerian circuit arises in this way. To generate an in-tree from an Eulerian circuit, follow the circuit from e, and record the edges leaving each particular vertex in the order used. The collection of edges that are last in these orderings forms an in-tree T to v, because it departs each edge other than v exactly once. Furthermore, this circuit is the circuit obtained from T and these exit orderings by Algorithm 2.4.7. □

APPLICATIONS

2.4.12. Application. *The Chinese Postman Problem.* Suppose a mail carrier traverses all edges in a road network, starting and ending at the same vertex. The edges have nonnegative weights representing distance or time. We seek a closed walk of minimum total length that uses all the edges. This is called the *Chinese Postman Problem* in honor of the Chinese mathematician Guan Meigu [1962], who proposed it.

If every vertex is even, the graph G is Eulerian and the answer is the sum of the weights. Otherwise, we must repeat edges. Every traversal is an Eulerian circuit of a multigraph obtained by duplicating edges of G. Finding the shortest traversal is equivalent to finding the minimum total weight of edges whose duplication will make all vertices even. We say "duplication" because we need not use an edge more than twice. If we use an edge three or more times in making all vertices even, then deleting two of those copies will leave all vertices even. There may be many ways to choose the duplicated edges. □

2.4.13. Example. In the example below, the eight outer vertices are odd. If we match them around the outside to make the degrees even, the extra cost is 4+4+4+4 = 16 or 1+7+7+1 = 16. We can do better by using all the vertical edges, which total only 10. □

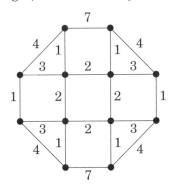

The example illustrates that adding an edge from an odd vertex to an even vertex makes the even vertex odd. We must continue adding edges until we complete a trail to an odd vertex. The duplicated edges must consist of a collection of trails that pair up the odd vertices. We may restrict our attention to paths pairing up the odd vertices (Exercise 17), though the paths may need to intersect.

Edmonds and Johnson [1973] described a way to solve the Chinese Postman problem. If there are only two odd vertices, then we can use Dijkstra's algorithm (Section 2.3) to find the shortest path between them and solve the problem. If there are $2k$ odd vertices, then we can use the shortest path algorithm to find the shortest paths connecting each pair of odd vertices. We can use these lengths as weights on the edges of K_{2k}, and then our problem is to find the minimum total weight of k edges that pair up these $2k$ vertices. This is a weighted version of the maximum matching problem discussed in Section 3.3. An exposition appears in Gibbons [1985, p163-165].

2.4.14. Application. *deBruijn Cycles*. There are 2^n binary strings of length n. Is there a cyclic arrangement of 2^n binary digits such that the 2^n strings of n consecutive digits are all distinct? This can be used to test the position of a rotating drum, as observed by Good [1946]. Suppose the drum has 2^n rotational positions, and a strip around the surface is partitioned into 2^n portions that can be coded 0 or 1. Taps reading n consecutive portions can determine the position of the drum if the specified cyclic arrangement exists. For $n = 4$ we can use (0000111101100101), shown below.

We can construct such an arrangement using Eulerian digraphs. Define a digraph D_n whose vertices are the $n - 1$-digit binary sequences. Place an edge from sequence a to sequence b if the last $n - 2$ digits of a agree with the first $n - 2$ digits of b. Label the edge with the last digit of b. For each sequence a, there are two digits we can append to obtain a new sequence, and hence there are two edges leaving each vertex. Similarly, because there are two digits we could drop from a sequence to obtain a, there are two edges entering each vertex. Hence D_n is Eulerian and has 2^n edges; D_4 appears below. □

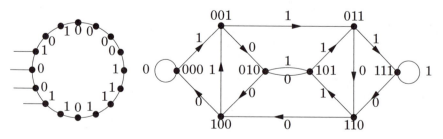

2.4.15. Theorem. The edge labels in any Eulerian circuit of D_n form a cyclic arrangement in which the consecutive segments of length n are the 2^n distinct binary vectors.

Proof. Suppose we are traversing an Eulerian circuit C in D_n and sit at a vertex whose sequence is $a = a_1, \ldots, a_{n-1}$. The previous $n-1$ edge labels, looking backward, must have been $a_{n-1}, \ldots, a_1$ in order, because the label on an edge entering a vertex agrees with the last digit of the sequence at the vertex. If C next traverses an edge with label a_n, then the subsequence ending there is $a_1, \ldots, a_n$. Since the 2^{n-1} vertex labels are distinct, and the two edges leaving each vertex have distinct labels, we have 2^n distinct subsequences determined from C in this way. $\square$

The graph D_n is the *deBruijn graph* of order n on an alphabet of size 2. It is useful for other purposes, because it has few edges (only twice the number of vertices) and yet its diameter is very small compared to its number of vertices. We can reach any desired vertex in $n-1$ steps by introducing the desired sequence in order from wherever we currently sit. Hence the diameter is $n-1$, even though there are 2^{n-1} vertices.

2.4.16. Application. *The Street-Sweeping Problem* (optional). In traversing a road network, it may be important to consider the direction of travel along segments of road, such as when sweeping the curb. The curb must be traversed in the direction that traffic flows, so a two-way street contributes a pair of oppositely directed edges, and a one-way street contributes two edges in the same direction. We consider a simple version of the street-sweeping problem, discussed in more detail in Roberts [1978], following the results of Tucker and Bodin [1976].

In New York City, parking is prohibited from some streetcurbs each weekday to allow for street sweeping. This defines a *sweep subgraph* G of the full digraph of curbs; G consists of the edges available for sweeping, directed to follow traffic. The question is how to sweep G while minimizing the *deadheading* time - the time when no sweeping is being done. If in-degree equals out-degree at each vertex of the subgraph, then no deadheading is needed. Otherwise, we must duplicate edges or add edges from the full digraph to obtain an Eulerian superdigraph of the sweep digraph. Each edge e in the full graph H has a deadheading time $t(e)$.

Let X be the set of vertices with in-degree exceeding out-degree, and set $\sigma(x) = d_G^-(x) - d_G^+(x)$ for $x \in X$. Let Y be the set of vertices with out-degree exceeding in-degree, and set $\partial(y) = d_G^-(y) - d_G^+(y)$ for $y \in Y$. Note that $\Sigma_{x \in X} \sigma(x) = \Sigma_{y \in Y} \partial(y)$. The Eulerian superdigraph must add $\sigma(x)$ edges with tails at $x \in X$ and $\partial(y)$ edges with heads at $y \in Y$. Since we must finish with a supergraph having degrees in balance, we can think of the additions as paths from X to Y. The cost of adding a path from x to y is the distance from x to y in H, which can be found by Dijkstra's Algorithm.

We now have the *Transportation Problem*. Given supplies $\sigma(x)$ for $x \in X$ and demands $\partial(y)$ for $y \in Y$ and costs $c(xy)$ for sending a unit from x to y, with $\Sigma\sigma(x) = \Sigma\partial(y)$, the problem is to satisfy the demands at least total cost. A version of the Transportation Problem was introduced by Kantorovich in 1939; the form above was introduced (with a constructive solution) by Hitchcock [1941] (see also Koopmans [1947]). The problem is discussed at length in Ford-Fulkerson [1962, p93-130]. In Section 3.2 we will solve the Transportation Problem for nonnegative integer supplies and demands. A more general solution appears in Section 4.3. □

EXERCISES

2.4.1. (–) Prove or disprove: There is no connected Eulerian simple graph that has an even number of vertices and an odd number of edges.

2.4.2. (–) Prove or disprove: If G is an Eulerian graph with edges e, f that share a vertex, then G has an Eulerian circuit in which e, f appear consecutively.

2.4.3. Use induction on the number of edges to prove the characterization of Eulerian graphs.

2.4.4. Use ordinary induction on k or on $e(G)$ (one by one) to prove that the edges of a connected graph having $2k$ odd vertices can be partitioned into k trails if $k > 0$. Does this remain true without the connectedness hypothesis?

2.4.5. (–) Two Eulerian circuits are equivalent if they have the same pairs of consecutive edges, viewed cyclically. For example, a cycle has only one equivalence class of Eulerian circuits. How many equivalence classes of Eulerian circuits are there in the graph drawn below?

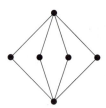

2.4.6. (+) Use induction on k to prove that the edges of a connected simple graph with $2k$ edges can be partitioned into paths of length 2. Does the conclusion remain true if the hypothesis of connectedness is omitted?

2.4.7. *Tucker's Algorithm.* Suppose G is a connected even graph. At each vertex, partition the incident edges arbitrarily into pairs (each edge appears in a pair for each of its endpoints). Starting along a given edge, form a trail by leaving each subsequent vertex along the edge paired with the edge most recently used to enter it, ending with the first return to the original vertex. This decomposes G into closed trails. As long as there is more than one trail in the decomposition, find two trails with a common vertex and splice them together into a longer trail by changing the pairing at the common vertex. End with an Eulerian circuit. Prove that the statements made about this procedure are correct and that it produces an Eulerian circuit. (Tucker [1976])

2.4.8. Say that a vertex v in a graph G is *randomly Eulerian* if every trail beginning at v can be extended to form an Eulerian circuit of G. (The graphs below have exactly one and exactly two randomly Eulerian vertices, respectively.) Prove the following statements about an Eulerian graph G.

   a) $v \in V(G)$ is randomly Eulerian if and only if $G - v$ is a forest. (Ore [1951])

   b) If v is randomly Eulerian, then $d(v) = \Delta(G)$. (Bäbler [1953])

   c) Every vertex of G is randomly Eulerian if and only if G is a cycle.

   d) If G is not a cycle, then G has at most two randomly Eulerian vertices. (adapted from Chartrand-Lesniak [1986, p61])

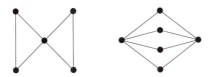

2.4.9. (+) *Additional characterization of Eulerian graphs.*

   a) Prove that if G is Eulerian and $G' = G - uv$, then G' has an odd number of u, v-trails that visit v only at the end. Prove also that the number of the trails in this list that are not paths is even. (Toida [1973])

   b) Suppose v has odd degree and for each edge e incident to v, $c(e)$ is the number of cycles containing e. Use $\Sigma_e c(e)$ to prove that $c(e)$ is even for some e incident to v. (McKee [1984])

   c) Use part (a) and part (b) to conclude that a nontrivial connected graph is Eulerian if and only if every edge belongs to an odd number of cycles.

2.4.10. (−) Prove that a digraph has a single circuit containing each edge if and only if $d^-(v) = d^+(v)$ for every vertex v and the underlying undirected graph has only one nontrivial component. Determine (with proof), the necessary and sufficient conditions for a digraph to have an Eulerian trail. (Good [1946])

2.4.11. (−) Let D be a digraph with $d^-(v) = d^+(v)$ for every vertex v, except that $d^+(x) - d^-(x) = k = d^-(y) - d^+(y)$. Use the characterization of Eulerian digraphs to prove that D contains k pairwise edge-disjoint x, y-paths.

2.4.12. Prove that every undirected graph G has an orientation D such that $|d^+(v) - d^-(v)| \le 1$ for every vertex $v \in V(G)$.

2.4.13. Prove or disprove: Every graph G has an orientation such that for every $S \subseteq V(G)$, the number of edges entering S and leaving S differ by at most one.

2.4.14. *Orientations and P_3-decomposition.*
a) Prove that every connected graph has an orientation in which the number of vertices with odd out-degree is at most 1. (Rotman [1991])
b) Use part (a) to conclude that a connected graph with an even number of edges can be decomposed into paths with two edges.

2.4.15. (–) Solve the Chinese Postman Problem in the k-dimensional cube Q_k under the condition that every edge has weight 1.

2.4.16. Every morning the Lazy Postman takes the bus to the Post Office. Starting there, he arranges his route so that he ends at home to go back to sleep as quickly as possible (**NOT** back at the post office). Below is a map of the streets along which he must deliver mail, giving the number of minutes required to walk each block, whether delivering or not. P denotes the post office and H denotes home. What must the edges traveled more than once satisfy? How many times will each edge be traversed in the optimal route?

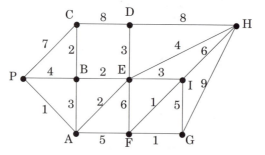

2.4.17. (–) Explain why the optimal trails pairing up odd vertices in an optimal solution to the Chinese Postman Problem may be assumed to be paths. Construct a weighted graph with four odd vertices where the optimal solution to the Chinese Postman Problem requires duplicating the edges on two paths that have a common vertex.

2.4.18. (–) Arrange seven 0's and seven 1's cyclically so that the 14 strings of four consecutive bits are all the 4-digit binary strings other than 0101 and 1010.

2.4.19. (–) *DeBruijn sequence for any alphabet and length.* Let A be an alphabet of size k. Prove that there exists a cyclic arrangement of k^l characters chosen from A such that the k^l strings of length l in the sequence are all distinct. (Good [1946], Rees [1946])

2.4.20. (–) Suppose S is an alphabet of m letters. Must there be a cyclic arrangement C of $m^4 - m$ letters from S such that all four-letter strings of consecutive letters in C are different and contain at least two distinct letters? Give a construction procedure or a counterexample.

2.4.21. (!) *Explicit deBruijn cycle.* Use the results of this section to prove that the following explicit algorithm generates a binary deBruijn cycle of length 2^n:

start with n 0's. Subsequently, append a 1 if doing so does not repeat a previous string of length n, otherwise append a 0. (Comment: for $n = 4$, the resulting cycle appears in the illustration accompanying Application 2.4.14.)

2.4.22. (!) *Tarry's Algorithm* (as presented by D.G. Hoffman). Consider a castle with finitely many rooms and finitely many corridors. Each corridor has two ends, with a door at each end into a room. Each room has doors, each of which leads to a corridor. Any room can be reached from any other by traversing corridors and rooms. Initially, no doors have marks. A robot powered up in some room must explore the castle using the following rules.

    1) After entering a corridor, traverse it and enter the room at the other end.

    2) Upon entering a room with all doors unmarked, mark I on the door of entry.

    3) In a room with an unmarked door, mark O on an unmarked door and use it.

    4) In a room with all doors marked, exit via a door not marked O if one exists.

    5) In a room with all doors marked O, stop.

Prove that the robot eventually stops after traversing every corridor exactly twice, once in each direction. (Hint: use induction on the number of corridors, proving also that the process ends in the starting room. Comment: all decisions are completely local; the robot sees nothing other than the current room or corridor. Tarry's Algorithm [1895] and others are described by König [1936, p35-56] and by Fleischner [1983, 1991].)

2.4.23. (!) Suppose that G is a graph and D is an orientation of G that is strongly connected. Prove that if G has an odd cycle, then D has an odd cycle. (Hint: consider each pair $\{v_i, v_{i+1}\}$ in an odd cycle $(v_1, \ldots, v_k)$ of G.)

2.4.24. (+) Given a strong digraph D, let $f(D)$ be the length of the shortest closed walk visiting every vertex. Prove that the maximum value of $f(D)$ over all strong digraphs with n vertices is $\lfloor (n+1)^2/4 \rfloor$ if $n \geq 2$. (Cull [1980])

Chapter 3

Matchings and Factors

3.1 Matchings in Bipartite Graphs

Within a set of people, each person is compatible with some of the others; under what conditions can we pair the people as compatible roommates? Many applications of graphs involve such pairings. In Section 1.1 we considered the case of jobs and applicants, asking whether all the jobs can be competently filled. Bipartite graphs have a natural vertex partition into two sets, and we want to know whether the edges can pair up the two sets. In the roommate question, the graph need not be bipartite.

3.1.1. Definition. A *matching* in an undirected graph G is a set of pairwise disjoint edges. The vertices belonging to the edges of a matching are *saturated* by the matching; the others are *unsaturated*. If a matching saturates every vertex of G, then it is a *perfect matching* or *complete matching*.

3.1.2. Example. *Matchings in familiar graphs.* $K_{n,n}$ has $n!$ complete matchings. Let $X = \{x_1, \ldots, x_n\}$ and $Y = \{y_1, \ldots, y_n\}$ be the partite sets. We can use X and Y to index the rows and columns of a matrix. Recording a 1 in position i, j for each edge $x_i y_j$ in a matching M establishes a bijection between complete matchings in $K_{n,n}$ and permutation matrices.

Since it has odd order, K_{2n+1} has no perfect matching. The number f_n of complete matchings in K_{2n} is the number of ways to pair up $2n$ distinct people. There are $2n - 1$ choices for the partner of v_{2n}, and for each of these there are f_{n-1} ways to complete the matching. Hence

$f_n = (2n - 1)f_{n-1}$ for $n \geq 1$, and with $f_0 = 1$ we can verify $f_n = \Pi_{i=0}^{n-1}(2n - 1 - 2i)$ by induction. There is also a bijective proof. Given any ordering of $2n$ people, we can pair up the first two, the next two, etc. From the $(2n)!$ orderings, we obtain each matching $2^n n!$ times, so the total number of matchings is $f_n = (2n)!/(2^n n!)$.

The usual drawing of the Petersen graph exhibits a complete matching between the vertices of the inner and outer 5-cycles. Counting the complete matchings requires some effort (Exercise 7). The inductive construction of the hypercube Q_k readily yields complete matchings, but again counting them is difficult (Exercise 8). □

MAXIMUM MATCHINGS

To seek a large matching, we could iteratively select an edge disjoint from those previously selected. This yields a maximal matching, but it need not yield a maximum matching. A maximal matching cannot be enlarged, because its edges are incident to all others. A maximum matching is a matching of maximum size.

3.1.3. Example. *Maximal $\neq$ maximum.* The smallest graph having a maximal matching that is not a maximum matching is P_4. If we take the middle edge, them we can add no other, but the two pendant edges form a larger matching. Below we show this for P_6. □

3.1.4. Definition. Given a matching M, an *M-alternating path* is a path that alternates between edges in M and edges not in M. An M-alternating path P that begins and ends at M-unsaturated vertices is an *M-augmenting* path; replacing $M \cap E(P)$ by $E(P) - M$ produces a new matching M' with one more edge than M.

Maximum matchings are those that have no augmenting paths. We prove this by examining the subgraph formed from the union of two matchings by deleting the common edges. We can define this operation for any two graphs with the same vertex set:

3.1.5. Definition. If G and H are graphs with vertex set V, then the *symmetric difference* $G \triangle H$ is the graph with vertex set V whose edges are all those edges appearing in exactly one of G and H. We also use this notation for sets of edges; in particular, if M and M' are matchings, then $M \triangle M' = (M \cup M') - (M \cap M')$.

3.1.6. Theorem. (Berge [1957]) A matching M in a graph G is a maximum matching in G if and only if G has no M-augmenting path.

Proof. We have noted that an M-augmenting path produces a larger matching. For the converse, suppose that G has a matching M' larger than M; we want to construct an M-augmenting path. Let F be the spanning subgraph of G with $E(F) = M \triangle M'$. Since M and M' are matchings, every vertex has at most one incident edge in each of them, and F has maximum degree at most 2.

Since $\Delta(F) \le 2$, F consists of disjoint paths and cycles. Furthermore, every path or cycle in F alternates between edges of M and edges of M'. This implies that each cycle in F has even length. Since $|M'| > |M|$, F has a component with more edges of M' than of M. Such a component can only be a path that starts and ends with an edge of M'; every such path in F is an M-augmenting path in G. □

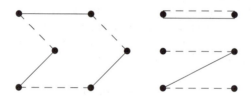

HALL'S MATCHING CONDITION

In the example of jobs and applicants, there may be many more applicants than jobs, so we may be able to fill the jobs without using all the applicants. Hence when G is a bipartite graph with bipartition X, Y we may ask whether G has a matching that saturates X; we call this a matching of X *into* Y.

If M saturates X, then for every $S \subseteq X$ there must be at least $|S|$ vertices that have neighbors in S, because the vertices matched to S must be chosen from that set. We use $N_G(S)$ or simply $N(S)$ to denote the set of vertices having a neighbor in S. Our observation is that $|N(S)| \ge |S|$ is a necessary condition. Hall proved that this obvious necessary condition is also sufficient.

3.1.7. Theorem. (P. Hall [1935]) If G is a bipartite graph with bipartition X, Y, then G has a matching of X into Y if and only if $|N(S)| \ge |S|$ for all $S \subseteq X$.

Proof. Necessity was observed above. For sufficiency, suppose $|N(S)| \ge |S|$ for all $S \subseteq X$, and consider a maximum matching M. If M does not saturate X, we will find a set S that violates the hypothesis.

Suppose u is an M-unsaturated vertex of X. Let S and T be the sets of vertices in X and Y, respectively, that are reachable from u by M-alternating paths.

We claim that M matches T with $S - u$. The M-alternating paths from u reach Y by edges not in M and reach X by edges in M. Since there is no M-augmenting path, every vertex of T is M-saturated, meaning that an alternating path reaching $y \in T$ extends via M to a vertex of S. Furthermore, every vertex of S other than u is reached via an edge in M from a vertex in T. Hence these edges of M establish a bijection between T and $S - u$, and we have $|T| = |S - u|$.

The matching between T and $S - u$ implies $T \subseteq N(S)$. In fact, $T = N(S)$. An edge between S and a vertex $y \in Y - T$ would be an edge not in M; this would create an M-alternating path to y, which contradicts $y \notin T$. With $T = N(S)$, we have $|N(S)| = |T| = |S| - 1 < |S|$, contradicting the hypothesis of the theorem. $\square$

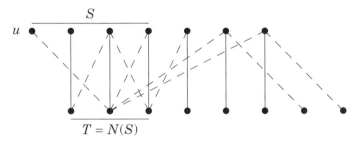

There are other proofs that Hall's condition is sufficient. A later proof by M. Hall [1948] gives a lower bound on the number of matchings when the condition is satisfied, as a function of the degrees of the vertices in X. We treat the problem algorithmically in Section 3.2.

When the sets of the bipartition have the same size, Hall's Theorem is the *Marriage Theorem*, proved originally by Frobenius [1917]. The name arises from the scenario of a symmetric compatibility relation between a set of n men and a set of n women. If also every man is compatible with k women and every woman is compatible with k men, then there must be a complete matching using compatible pairs.

3.1.8. Corollary. Every k-regular bipartite multigraph (with $k > 0$) has a complete matching.

Proof. Suppose the bipartition is X, Y. Counting the edges by endpoints in X and endpoints in Y shows that $k|X| = k|Y|$, so regularity implies that $|X| = |Y|$. Hence it suffices to show that Hall's condition holds; a matching saturating X will be a complete matching. Consider $S \subseteq X$, and suppose there are m edges between S and $N(S)$. Since G is k-regular, $m = k|S|$. Since these m edges are incident to $N(S)$, we have

$m \le k|N(S)|$.   Hence $k|S| \le k|N(S)|$, so $|N(S)| \ge |S|$.   Having   chosen
$S \subseteq X$ arbitrarily, we have established Hall's condition. $\square$

One can also prove this by contradiction, starting with the
assumption that G has no perfect matching, which is requires a set
$S \subseteq X$ such that $|N(S)| < |S|$. The argument that leads to a contradiction is essentially a rewording of the direct proof given above.

MIN-MAX THEOREMS

If G does not have a complete matching, we can prove that M is a
maximum matching by proving that G has no M-augmenting path.
Exploring all M-alternating paths to seek an augmentation would take
a long time. We would rather find an explicit structure in G that forbids
a matching larger than M. A "dual" optimization problem may provide
a short proof that the answer is optimal.

3.1.9. Definition. A *vertex cover* of G is a set S of vertices such that S
contains at least one endpoint of every edge of G. The vertices in
S "cover" the edges of G.

If our graph represents a road network (with straight roads and no
isolated vertices), then we can interpret the problem of finding a minimum vertex cover as the problem of installing the minimum number of
policeman in order to observe the entire road network.
Since no two edges of a matching can be covered by a single vertex,
the size of every vertex cover is at least the size of every matching.
Therefore, exhibiting a matching and a vertex cover of the same size
PROVES that each is optimal. We can find such an equality for every
bipartite graph, but not for every graph.

3.1.10. Example. *Matchings and vertex covers.* On the left below we
indicate a matching and a vertex cover of size 2. Presenting a matching
and a vertex cover of the same size PROVES that both are optimal,
since the smallest cover is at least as large as the largest matching. As
illustrated on the right, these sizes differ by one for an odd cycle. The
difference can be arbitrarily large (Exercise 3.3.5). $\square$

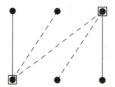

A *min-max relation* is a theorem stating equality between the answers to a minimization problem and a maximization problem over a class of instances. The König-Egerváry Theorem is such a relation for matching and vertex covering in bipartite graphs:

3.1.11. Theorem. (König [1931], Egerváry [1931]) If G is a bipartite graph, then the maximum size of a matching in G equals the minimum size of a vertex cover of G.

Proof. Suppose G has bipartition X, Y. Since distinct vertices must be used to cover the edges of a matching, we have $|U| \geq |M|$ whenever U is a vertex cover and M is a matching in G. Given a minimum vertex cover U of G, we construct a matching of size $|U|$ to prove that equality can always be achieved. Let $R = U \cap X$ and $T = U \cap Y$. Let H, H' be the subgraphs of G induced by $R \cup (Y - T)$ and $T \cup X - R$. We use Hall's Theorem to show that H has a complete matching of R into $Y - T$ and H' has a complete matching of T into $X - R$. Since these subgraphs are disjoint, the two matchings together form a matching of size $|U|$ in G.

Since $R \cup T$ is a vertex cover, G has no edge from $Y - T$ to $X - R$. Suppose $S \subseteq R$, and consider $N_H(S) \subseteq Y - T$. If $|N_H(S)| < |S|$, then we can substitute $N_H(S)$ for S in U and obtain a smaller vertex cover, since $N_H(S)$ covers all edges incident to S that are not covered by T. The minimality of U thus implies that Hall's condition holds in H, and hence H has a complete matching of R into $Y - T$. Applying the same argument to H' yields the rest of the matching. $\square$

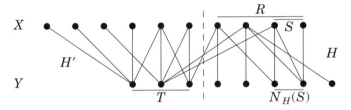

INDEPENDENT SETS IN BIPARTITE GRAPHS

We now turn from independent sets of edges to independent sets of vertices. The *independence number* of a graph is the maximum size of an independent set of vertices. The independence number of a bipartite graph does *not* always equal the size of a partite set:

Just as no vertex can cover two edges of a matching, so no edge can contain two vertices of an independent set. For independence number we again have a dual covering problem:

3.1.12. Definition. An *edge cover* of G is a set of edges that cover the vertices of G (only graphs without isolated vertices have edge covers). For the independence and covering problems we have defined, we use the following notation:

$$\begin{array}{ll} \text{maximum size of independent set} & \alpha(G) \\ \text{maximum size of matching} & \alpha'(G) \\ \text{minimum size of vertex cover} & \beta(G) \\ \text{minimum size of edge cover} & \beta'(G) \end{array}$$

This notation treats the answers to these optimization problems on graphs as graph parameters, like the order, size, maximum degree, diameter, etc. Our use of $\alpha'(G)$ to count a set of edges suggests a relationship with the parameter $\alpha(G)$ that counts a set of vertices. We discuss this relationship in Section 6.1.

We use $\beta(G)$ for minimum vertex cover due to its interaction with maximum matching. The "prime" goes on $\beta'(G)$ rather than on $\beta(G)$ because $\beta(G)$ counts a set of vertices and $\beta'(G)$ counts a set of edges.

In this notation, the König-Egerváry Theorem states that $\alpha'(G) = \beta(G)$ for every bipartite graph G. We will prove also that $\alpha(G) = \beta'(G)$ for bipartite graphs without isolated vertices; we have already observed that $\beta'(G) \geq \alpha(G)$.

3.1.13. Lemma. In a graph G, $S \subseteq V(G)$ is an independent set if and only if $\bar{S}$ is a vertex cover, and hence $\alpha(G) + \beta(G) = n(G)$.

Proof. If S is an independent set, then there are no edges within S, so every edge is incident to at least one vertex of $\bar{S}$. Conversely, if $\bar{S}$ covers all the edges, then there are no edges between vertices of S. Hence any maximum independent set is the complement of a minimum vertex cover, and $\alpha(G) + \beta(G) = n(G)$. □

For matchings and edge coverings, the relationship is more complicated, since the edges of G omitted by a matching need not form an edge cover of G. Nevertheless, a similar formula holds.

3.1.14. Theorem. (Gallai [1959]) If G has no isolated vertices, then $\alpha'(G) + \beta'(G) = n(G)$.

Proof. If we use a maximum matching M to construct an edge cover of size $n(G) - |M|$, then the smallest edge cover is no bigger than this, and we conclude $\beta'(G) \leq n(G) - \alpha'(G)$. Conversely, if we use a minimum edge cover L to construct a matching of size $n(G) - |L|$, then the largest matching is no smaller than this, and we conclude $\alpha'(G) \geq n(G) - \beta'(G)$. These two inequalities complete the proof.

Let M be a maximum matching in G. We can construct an edge cover of G by using M and adding one edge incident to each unsaturated vertex. The total number of edges used is $n(G) - |M|$, as desired; we save $|M|$ edges because each edge of the matching covers two vertices instead of one.

For the other inequality, let L be a minimum edge cover. If the endpoints of some edge e in L belongs to other edges in L, then e is not needed in the cover. Hence L consists of k disjoint stars, for some k. Since L has one edge for each vertex that is not a center of its stars, we have $|L| = n(G) - k$. We can form a matching M of size $k = n(G) - |L|$ by choosing one edge arbitrarily from each star in L. $\square$

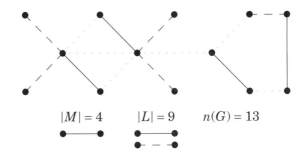

$$|M| = 4 \qquad |L| = 9 \qquad n(G) = 13$$

These two results yield what we call "König's Other Theorem".

3.1.15. Corollary. (König [1916]) If G is a bipartite graph with no isolated vertices, then $\alpha(G) = \beta'(G)$ (max |independent set| = min |edge cover|).

Proof. The two preceding results imply $\alpha(G) + \beta(G) = \alpha'(G) + \beta'(G)$, and then we subtract the König-Egerváry min-max relation $\alpha'(G) = \beta(G)$. $\square$

EXERCISES

3.1.1. (–) Prove that every tree has at most one perfect matching.

3.1.2. Given a maximal matching M in a graph G, prove that $|M| \geq \alpha'(G)/2$.

3.1.3. (–) Suppose S is a set of vertices in a simple graph G that is saturated by some matching. Prove that S is saturated by some maximum matching.

3.1.4. Suppose that M, M' are matchings in a bipartite graph G with bipartition X, Y. Suppose that $S \subseteq X$ is saturated by M and that $T \subseteq Y$ is saturated by M'. Prove that G contains a matching that saturates $S \cup T$.

3.1.5. Suppose M and N are matchings in a graph G, and $|M| > |N|$. Prove that there exist matchings M' and N' in G such that $|M'| = |M| - 1$, $|N'| = |N| + 1$, and M', N' have the same union and intersection (as edge sets) as M, N.

3.1.6. Derive a formula for the number of matchings in $K_{n,n}$ that do not match x_i to y_i for any i, and for the number of matchings in K_{2n} that do not match x_{2i-1} to x_{2i} for any i. Use the inclusion-exclusion principle to obtain a summation or derive a recurrence; no simple closed formulas are available.

3.1.7. *Matchings in the Petersen graph.*
   a) Prove that deleting any perfect matching from the Petersen graph leaves the subgraph $C_5 + C_5$.
   b) Use symmetry to count the 5-cycles in the Petersen graph.
   c) Use (a) and (b) to count the perfect matchings in the Petersen graph.

3.1.8. *Matchings in k-dimensional cubes.*
   a) $(-)$ Prove that Q_k has a perfect matching.
   b) Count the perfect matchings in Q_3. (Hint: prove first that every perfect matching in Q_k has an even number of edges in each direction, and that therefore no perfect matching in Q_3 has an edge in each direction.)
   c) Prove that Q_k has at least $2^{(2^{k-2})}$ perfect matchings if $k \geq 2$.

3.1.9. (!) Two people play a game on a graph G by alternately selecting distinct vertices $v_1, v_2, \cdots$ forming a path. The last player able to select a vertex wins. Prove that the second player has a winning strategy if G has a perfect matching, and the first player has a winning strategy if G has no perfect matching. (Hint: For the second part, the first player should start by picking a vertex omitted by some maximum matching.)

3.1.10. Determine the minimum size of a maximal matching in the cycle C_n.

3.1.11. (!) Let $\mathbf{A} = (A_1, \ldots, A_m)$ be a collection of subsets of a set Y. A *system of distinct representatives* (SDR) for $\mathbf{A}$ is a set of distinct elements $a_1, \ldots, a_m$ in Y such that $a_i \in A_i$. Prove that $\mathbf{A}$ has an SDR if and only if $|\cup_{i \in S} A_i| \geq |S|$ for every $S \subseteq \{1, \ldots, m\}$. (Hint: transform this to a graph problem).

3.1.12. Prove that a bipartite graph G has a complete matching (1-factor) if and only if $|N(S)| \geq |S|$ for all $S \subseteq V(G)$, and present an infinite class of examples to prove that this characterization does not hold for all graphs.

3.1.13. (+) *Alternative proof of Hall's Theorem.* Consider a bipartite graph G with bipartition X, Y, and suppose $|N(S)| \geq |S|$ for every $S \subseteq X$. Use induction on $|X|$ to prove that G has a matching that saturates X. (Hint: First consider the case where $|N(S)| > |S|$ for every proper subset S of X. When this does not hold, consider a minimal nonempty $T \subseteq X$ such that $|N(T)| = |T|$.)

3.1.14. A *permutation matrix* P is a 0,1-matrix having exactly one 1 in each row and column. Suppose A is a 0,1-matrix having exactly k 1's in each row and column. Prove that A can be expressed as the sum of k permutation matrices.

3.1.15. (!) A *doubly stochastic matrix* Q is a nonnegative real matrix in which every row and every column sums to 1. Prove that a doubly stochastic matrix Q

can be expressed as a convex combination of permutation matrices, meaning that $Q = c_1 P_1 + \cdots + c_m P_m$, where $c_1, \ldots, c_m$ are nonnegative real numbers summing to 1 and $P_1, \ldots, P_m$ are permutation matrices. For example,

$$\begin{pmatrix} 1/2 & 1/3 & 1/6 \\ 0 & 1/6 & 5/6 \\ 1/2 & 1/2 & 0 \end{pmatrix} = \frac{1}{2}\begin{pmatrix} 1 & 0 & 0 \\ 0 & 0 & 1 \\ 0 & 1 & 0 \end{pmatrix} + \frac{1}{3}\begin{pmatrix} 0 & 1 & 0 \\ 0 & 0 & 1 \\ 1 & 0 & 0 \end{pmatrix} + \frac{1}{6}\begin{pmatrix} 0 & 0 & 1 \\ 0 & 1 & 0 \\ 1 & 0 & 0 \end{pmatrix}$$

(Hint: Suppose Q is a nonnegative real matrix in which every row and every column sums to t. Use induction on the number of nonzero entries in Q to prove that Q is a linear combination of permutation matrices with nonnegative coefficients summing to t.)

3.1.16. (!) *A generalization of Tic-Tac-Toe.* A *positional game* consists of a set $X = x_1, \ldots, x_n$ of positions and a collection $W_1, \ldots, W_m$ of winning sets of positions (Tic-Tac-Toe has 9 positions and 8 winning sets of positions). Two players alternately choose positions; a player wins by collecting a winning set. Suppose each winning set has at least a positions and each position appears in at most b winning sets. Prove that Player 2 can force a draw if $a \ge 2b$. (Hint: form a bipartite graph with bipartition X, Y, where $Y = \{w_1, \ldots, w_m\} \cup \{w_1', \ldots, w_m'\}$, and put $x_i \leftrightarrow w_j$ and $x_i \leftrightarrow w_j'$ whenever $x_i \in W_j$. Is there a matching of size $2m$? If so, how can Player 2 use it? Note: as a consequence, Player 2 can force a draw in d-dimensional Tic-Tac-Toe if the sides are long enough.)

3.1.17. (!) Exhibit a perfect matching in the graph drawn below or give a short proof that it has none. (Lovász-Plummer [1986, p7])

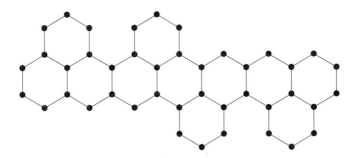

3.1.18. (–) Prove that every bipartite graph G has a matching of size at least $e(G)/\Delta(G)$.

3.1.19. (–) Suppose T is a tree with n vertices, and suppose k is the maximum size of an independent set in T. Determine $\alpha'(T)$.

3.1.20. (–) Determine the maximum number of edges in a simple bipartite graph that contains no matching with k edges and no star with l edges. (Isaak)

3.1.21. (!) Use the König-Egerváry Theorem to prove that any subgraph of $K_{n,n}$ with more than $(k-1)n$ edges has a matching of size at least k.

3.1.22. Use the König-Egerváry Theorem to prove Hall's Theorem.

3.1.23. Suppose G is a bipartite graph with bipartition X, Y. Use a graph transformation to prove that the maximum size of a matching in G is $|X| - \max_{S \subseteq X}(|S| - |N(S)|)$. (Hint: form a bipartite supergraph G' of G such that G' has a complete matching if and only if G has a matching of the desired size, and prove that G' satisfies Hall's condition. (Ore [1955])

3.1.24. (+) On a certain island with n married couples, every couple consists of a hunter and a farmer. The Ministry of Hunting divides the island into n hunting ranges of equal size. The Ministry of Agriculture independently divides it into n farming ranges of equal size. The Ministry of Marriage insists that the hunting range and farming range assigned to each couple must overlap. To everyone's surprise, this turns out to be possible. The Ministry of Religion declares it a miracle. Prove that this is not a miracle by showing that the ranges can always be matched up so that each couple's ranges share area at least $4/(n+1)^2$, where each range has area 1. Prove also that no larger common area can be guaranteed. (Marcus-Ree [1959], Floyd [1990])

3.1.25. Let G be a bipartite graph with bipartition X, Y. Prove that G is $(k+1)K_2$-free if and only if each $S \subseteq X$ has a subset of size at most k with neighborhood $N(S)$. (Liu-Zhou [1996])

3.1.26. Let G be a subgraph of $K_{m,m}$ that has a complete matching. Prove that G has at most $\binom{m}{2}$ edges that belong to no complete matching. Construct examples to show that this is best possible for every m.

3.1.27. (!) Suppose G is a bipartite graph of order $2m$. Prove that $\alpha(G) = m$ if and only if G has a perfect matching.

3.1.28. (!) Suppose G is regular. Prove that $\alpha(G) \leq n(G)/2$.

3.1.29. (!) Suppose G is a tree. Prove that $\alpha(G) \geq n(G)/2$, with equality if and only if G has a perfect matching.

3.1.30. A connected n-vertex graph has exactly one cycle if and only if it has exactly n edges (Exercise 2.1.17). Suppose G is such a graph, with cycle C. Prove that $\alpha(G) \geq \lfloor n(G)/2 \rfloor$, with equality if and only if $G - V(C)$ has a perfect matching. (Hint: Exercise 29 may be assumed.)

3.1.31. Prove that G is bipartite if and only if $\alpha(H) = \beta'(H)$ for every subgraph H of G with no isolated vertices.

3.1.32. (+) *The middle-levels graph.* Let G_k be the bipartite graph whose vertices are the k-element and $k+1$-element subsets of $[2k+1]$, with two vertices adjacent if one differs from the other by the addition or removal of a single element (G_k is an induced subgraph of the hypercube Q_{2k+1}). Given a vertex $X = \{x_1, \ldots, x_k\}$ with $0 = x_0 < x_1 < \cdots < x_k$, let t be the largest i for which $2i - x_i$ is maximized. Prove that the set of edges formed by joining X to $X \cup \{x_t + 1\}$ is a complete matching of G_k. For example, when $X = \{1, 3, 4, 7\}$, we have $t = 3$ and match X to $\{1, 3, 4, 5, 7\}$. When $X = \{3, 5, 7, 9\}$, we have $t = 0$ and match X to $\{1, 3, 5, 7, 9\}$. (White-Williamson [1977])

3.1.33. *Consequences of Gallai's Theorem.* Suppose M is a maximal matching and L is a minimal edge cover in a graph having no isolated vertices.

a) Prove that M is a maximum matching if and only if M is contained in a minimum edge cover.

b) Prove that L is a minimum edge cover if and only if L contains a maximum matching. (Norman-Rabin [1959], Gallai [1959])

3.1.34. (+) An edge e of a graph G is *critical* if $\alpha(G - e) > \alpha(G)$. Suppose that xy and xz are critical edges in G and that $y \not\leftrightarrow z$. Prove that G contains an odd cycle as an induced subgraph. (Hartman [1995a]) (Hint: Let Y, Z be maximum stable sets in $G - xy$ and $G - xz$, respectively. Let $H = G[Y \triangle Z]$. Use König's Theorem (Corollary 3.1.15) to prove that every component of H has the same number of vertices from Y and Z. Conclude that y and z belong to the same component of H. Comment: Markossian and Karapetian [1984] proved a more general result by a more difficult proof.)

3.2 Applications and Algorithms

MAXIMUM BIPARTITE MATCHING

The augmenting-path characterization of maximum matchings leads to an algorithm to find maximum matchings. We iteratively seek augmenting paths to enlarge the current matching by one edge at a time. In a bipartite graph, if we don't find an augmenting path, we will find a vertex cover with the same size as the current matching, thereby proving that the current matching has maximum size. This yields both an algorithm to solve the maximum matching problem and an algorithmic proof of the König-Egerváry Theorem.

Given a matching M in a bipartite graph G, we can search for an M-augmenting path from each M-unsaturated vertex. We need only consider the unsaturated vertices in one partite set, because every augmenting path has an end in each partite set. We can search from all the unsaturated vertices in one partite set simultaneously by carefully maintaining the information we have discovered during the search. Starting with a matching of size 0, $\alpha'(G)$ applications of the Augmenting Path Algorithm produce a maximum matching.

3.2.1. Algorithm. (Augmenting Path Algorithm).
Input: A bipartite graph G with bipartition X, Y, a matching M in G, and the set U of all M-unsaturated vertices in X.
Idea: Explore M-alternating paths from U, letting $S \subseteq X$ and $T \subseteq Y$ be the sets of vertices reached. *Mark* vertices of S that have been explored for extending paths. For each $x \in (S \cup T) - U$, record the vertex before

x on some M-alternating path from U.

Initialization: $S = U$ and $T = \varnothing$.

Iteration: If S has no unmarked vertex, stop and report $T \cup (X - S)$ as a minimum cover and M as a maximum matching. Otherwise, select an unmarked $x \in S$. To explore x, consider each $y \in N(x)$ such that $xy \notin M$. If y is unsaturated, terminate and trace back from y to report an M-augmenting path from U to y. Otherwise, y is matched to some $w \in X$ by M. In this case, include y in T (reached from x) and include w in S (reached from y). After exploring all such edges incident to x, mark x and iterate. $\square$

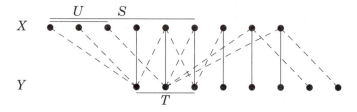

When exploring x in the iterative step, we may include y in T and w in S even though we have done so previously. This saves us the trouble of testing whether $y \in T$ each time we reach y. We still retrieve an M-augmenting path when we reach an unsaturated vertex of Y; the change in the vertex recorded for y affects which path we report but not the existence of such a path.

3.2.2. Theorem. Repeated application of the Augmenting Path Algorithm to a bipartite graph produces a matching and a vertex cover of the same size.

Proof. We need only verify that the Augmenting Path Algorithm produces an M-augmenting path or a vertex cover of size $|M|$. Let X, Y, U, S, T be the sets used in the algorithm, viewed when the algorithm terminates. If the algorithm produces an M-augmenting path, we are finished. Otherwise, it terminates by marking all vertices of S and claiming that $R = T \cup (X - S)$ is a vertex cover of size $|M|$. We must prove that R is a vertex cover and has size $|M|$.

An M-alternating path from U can enter X only on an edge of M; hence every vertex of $S - U$ is matched via M to a vertex of T, and there is no edge of M from S to $Y - T$. Once an M-alternating path reaches $x \in S$, it can continue along any unsaturated edge, and exploring x places all neighbors of x along unsaturated edges into T. Since the algorithm marks all of x before terminating, there is no unsaturated edge from S to $Y - T$. Hence there is no edge at all from S to $Y - T$, and R is a vertex cover.

Since the algorithm terminates without finding an M-augmenting path, every vertex of T is saturated, and this means each $y \in T$ is matched via M to a vertex of S. Since $U \subseteq S$, also each vertex of $X - S$ is saturated, and the edges of M incident to $X - S$ cannot involve T. Hence they are different from the edges saturating T, and we find that M has at least $|T| + |X - S|$ edges. Since there cannot be a matching larger than a vertex cover, we have $|M| = |T| + |X - S| = |R|$. □

We can evaluate the performance of this algorithm for maximum matching by counting the operations it may perform on an n-vertex bipartite graph G. Since matchings have at most $n/2$ edges, we apply the Augmenting Path Algorithm at most $n/2$ times. In each iteration, we search from a vertex of X at most once, before we mark it. Hence the number of operations in each iteration is bounded by a multiple of $e(G)$. This is at most quadratic in n, so the total number of operations is bounded by a multiple of n^3. At the end of this section, we describe a faster algorithm using a multiple of $n^{2.5}$ operations. The augmenting path characterization of maximum matchings also leads to a good algorithm for finding maximum matchings in general graphs; we postpone discussion of this to Section 3.3.

WEIGHTED BIPARTITE MATCHING

Our results on maximum matching generalize to weighted bipartite graphs. Given nonnegative edge weights, we seek the matching of maximum total weight. By assigning weight 0 to unavailable edges, we may assume that $G = K_{n,n}$. We solve both the maximization problem and a dual problem.

3.2.3. Example. *Weighted bipartite matching and its dual.* Suppose a farming company owns n farms and n processing plants, with each farm capable of producing corn to the capacity of one processing plant. The profit that results from sending the output of farm i to plant j is w_{ij}. This gives us a weighted bipartite graph with partite sets $X = \{x_1, \ldots, x_n\}$ and $Y = \{y_1, \ldots, y_n\}$; the weight on the edge $x_i y_j$ is w_{ij}. The company wants to find the matching with maximum total weight.

The government claims that too much corn is being produced, so it will pay the company not to process any corn. The government will pay u_i if the company doesn't use farm i and v_j if it doesn't use plant j. If $u_i + v_j < w_{ij}$, then the company makes more by using the edge $x_i y_j$ than by taking the government payments for those vertices. In order to offer enough to stop all production, the government must offer amounts such that $u_i + v_j \geq w_{ij}$ for all i, j. The government wants to find such values for $\{u_i\}$ and $\{v_j\}$ to minimize $\Sigma u_i + \Sigma v_j$. □

3.2.4. Definition. A *transversal* of an n by n matrix A consists of n positions: one in each row and each column. Finding a transversal of A with maximum sum is the *Assignment Problem*. This is the matrix formulation of the *maximum weighted matching* problem, where A is the matrix A of weights w_{ij} assigned to the edges $x_i y_j$ of $K_{n,n}$ and we seek a complete matching M with maximum total weight $w(M)$. Given the weights $\{w_{ij}\}$, a *(weighted) cover* is a choice of labels $\{u_i\}$ and $\{v_j\}$ such that $u_i + v_j \geq w_{ij}$ for all i, j. The *cost* $c(u,v)$ of a cover u, v is $\Sigma u_i + \Sigma v_j$. The *minimum weighted cover* problem is the problem of finding a cover of minimum cost.

This minimization problem generalizes the vertex cover problem in bipartite graphs: we must place large enough labels on the vertices so that the weight on each edge is "covered". Consider the special case where the weight on each edge is 0 or 1 and we use only integer labels. We use label 0 or 1 on each vertex, and the vertices receiving 1 form a vertex cover of the graph formed by the edges with weight 1. The next lemma expresses the "duality" of the weighted problems.

3.2.5. Lemma. If M is a complete matching in a weighted bipartite graph G and u, v is a cover, then $c(u,v) \geq w(M)$. Furthermore, $c(u,v) = w(M)$ if and only if M consists of edges $x_i y_j$ such that $u_i + v_j = w_{ij}$. In this case, M is a maximum weight matching and u, v is a minimum weight cover.

Proof. Since the edges in a matching M are disjoint, summing the constraints $u_i + v_j \geq w_{ij}$ that arise from its edges yields $c(u,v) \geq w(M)$ for every cover u, v. Furthermore, if $c(u,v) = w(M)$, then equality must hold in each of the n summands $u_i + v_j \geq w_{ij}$. Finally, since weight is bounded by cost for every matching and every cover, $c(u,v) = w(M)$ implies that there is no matching with weight greater than $c(u,v)$ and no cover with cost less than $w(M)$. $\square$

The observation that equality between weight and cost occurs only using edges covered with equality leads us to an algorithm. Let $G_{u,v}$ denote the *equality subgraph* for the cover u, v; this is the spanning subgraph of $K_{n,n}$ containing each edge $x_i y_j$ such that $u_i + v_j = w_{ij}$. If $G_{u,v}$ has a perfect matching, then the matching has weight $\Sigma u_i + \Sigma v_j$, and by Lemma 3.2.5 we have the optimal matching and cover. Otherwise, we will change the cover u, v.

Consider the structure of $G_{u,v}$ when the Augmenting Path Algorithm terminates; S and T are the subsets of X and Y reachable by M-alternating paths from the set U of unsaturated vertices in X. To seek a larger matching, we change u, v so as to keep M and the edges of M-alternating paths from U in the equality subgraph, plus introduce an edge from S to $Y - T$ in the hope of creating an M-augmenting path. We reduce u_i by a constant ε for $x_i \in S$, and we increase v_j by ε for $y_j \in T$. This preserves $u_i + v_j = w_{ij}$ on the desired edges.

A cover must have $u_i + v_j - w_{ij} \geq 0$ for all i, j. As in Theorem 3.2.2, the edges $x_i y_j$ with $x_i \in S$ and $y_j \in Y - T$ are not in $G_{u,v}$. Hence the "excess" for these edges is positive, and our proposed change reduces the excess by ε. We choose $\varepsilon = \min \{u_i + v_j - w_{ij} \colon x_i \in S, \ y_j \in Y - T\}$ to maintain feasibility in all constraints and introduce an edge between S and $Y - T$. Then we find a complete matching in the new equality subgraph. The resulting algorithm was named the *Hungarian Algorithm* by Kuhn, honoring the contributions of König and Egerváry.

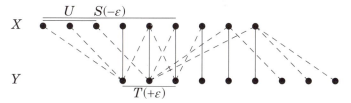

3.2.6. Algorithm. (Hungarian Algorithm - Kuhn [1955], Munkres [1957]).

Input: A matrix of weights on the edges of $K_{n,n}$ with bipartition X, Y.

Idea: Maintain a cover u, v, iteratively reducing the cost of the cover until the equality subgraph $G_{u,v}$ has a perfect matching.

Initialization: Let u, v be a feasible labeling, such as $u_i = \max_j w_{ij}$ and $v_j = 0$, and find a maximum matching M in $G_{u,v}$.

Iteration. If M is a complete matching, stop and report M as maximum weight matching. Otherwise, let U be the set of M-unsaturated vertices in X. Let S be the set of vertices in X and Y the set of vertices in Y that are reachable by M-alternating paths from U. Let

$$\varepsilon = \min \{u_i + v_j - w_{ij} \colon x_i \in S, \ y_j \in Y - T\}.$$

Decrease u_i by ε for all $x_i \in S$, and increase v_j by ε for all $y_j \in T$. If the new equality subgraph G' contains an M-augmenting path, replace M by a maximum matching in G' and iterate. Otherwise, iterate without changing M. $\square$

3.2.7. Theorem. The Hungarian Algorithm finds a maximum weight matching and a minimum cost cover.

Proof. The algorithm begins with a cover. Each iteration of the algorithm produces a cover, terminating only when the equality subgraph has a complete matching, which guarantees that the current matching and cover have equal value. If u, v is the current cover, let u', v' denote the new list of numbers assigned to the vertices. Because ε is the minimum of a set of positive numbers, $\varepsilon > 0$.

We verify first that u', v' is also a cover. Because we have changed the labels only on vertices of S and T, we have $u_i' + v_j' = u_i + v_j$ if $x_i \in S$ and $y_j \in T$, or if $x_i \in X - S$ and $y_j \in Y - T$. If $x_i \in X - S$ and $y_j \in T$, then $u_i' + v_j' = u_i + v_j + \varepsilon$, and the weight remains covered on such edges. If $x_i \in S$ and $y_j \in Y - T$, then $u_i' + v_j' = u_i + v_j - \varepsilon$, but $u_i + v_j - w_{ij} \geq \varepsilon$ for such edges, and again the weight remains covered. By the choice of ε, some edge from S to $Y - T$ enters the equality subgraph.

The algorithm can terminate only when the equality subgraph has a complete matching, so it suffices to show that the algorithm must terminate within $n^2/2$ iterations. Because the edges of M remain in G', the size of the current matching never decreases. When it remains unchanged, we claim that $|T|$ increases. Hence the size of the matching must increase within n iterations, and it increases at most $n/2$ times.

To prove that $|T|$ increases, observe that all M-alternating paths from U in $G_{u,v}$ lie in $G_{u,v}[S \cup T]$. Since all edges between S and T remain in G', every vertex reachable from U by an M-alternating path in $G_{u,v}$ is also reachable in G'. In G' we keep the same matching and hence the same set of unsaturated vertices. Now the introduction of an edge from S to $Y - T$ increases the number of vertices of Y reachable by M-alternating paths from U. $\square$

3.2.8. Remark. Although we did not use it in the proof, the cost of the cover decreases with each iteration of the Hungarian Algorithm. Since we increase labels on T and decrease labels on S, the net change in the cost of the cover is $\varepsilon(|T| - |S|)$. Because $|T| + |X - S| = |M| < n = |X|$ when we iterate, we have $|T| < |S|$, and the new cover costs less.

When the weights w_{ij} are rational, we have more flexibility in choosing S and T at each iteration. In this case there is an integer d such that each weight is a multiple of $1/d$. If the labels in u, v are multiples of $1/d$, then ε and the new labels are multiples of $1/d$, and $c(u, v)$ changes by a multiple of $1/d$. This remains true whenever $T \cup (X - S)$ is a vertex cover of $G_{u,v}$. Since we begin with a matching of finite weight, and every iteration reduces $c(u, v)$ by a multiple of $1/d$, there are finitely many iterations before we obtain a minimum cover. $\square$

Visualization in terms of bipartite graphs explains why the algorithm works, but visual computation with a changing $G_{u,v}$ is awkward. Hence we compute with matrices. The initial weights form a matrix A with w_{ij} in position i, j. We associate the vertices (and the variables u, v) with the rows and columns, which serve as X and Y, respectively. We subtract w_{ij} from $u_i + v_j$ to obtain the "excess" matrix $c_{ij} = u_i + v_j - w_{ij}$. The edges of the equality subgraph correspond to 0's in the excess matrix. A matching in $G_{u,v}$ corresponds to a set of 0's in the excess matrix having no two in any row or column. We know we have found a maximum matching in $G_{u,v}$ when we have a vertex cover of the same size, which corresponds to a collection of rows and columns that together cover all the 0's in the excess matrix.

3.2.9. Example. *Solving the Assignment Problem.* The first matrix below is a matrix of weights. The others display a cover and the corresponding excess matrix. We underscore entries in the excess matrix to mark a maximum matching M of $G_{u,v}$, which appears as solid edges in the equality subgraph drawn for the first two excess matrices.

We find S and T by exploring M-alternating paths from the unsaturated rows. From a reached row x_i, we can reach a column in which row x_i has a 0 not in M. From a reached column y_j, we can reach a row in which the column y_j has a 0 in M. Alternatively, since the weights are rational, Remark 3.2.8 allows us to use any $T \subseteq Y$ and $S \subseteq X$ such that $T \cup (X - S)$ covers all the 0's of the excess matrix. It may be easier to find rows and columns covering the 0's than to determine precisely which are rows and columns are reachable from the unsaturated rows.

In this example, the choice of $T \cup S$ as indicated in the first iteration enlarges S and T but not the matching. The second iteration produces a perfect matching. If in the first iteration, we use the last three columns as a vertex cover of $G_{u,v}$ (i.e. $S = X$), we obtain a larger matching immediately. The value of the optimal solution is unique, but the solution itself is not; this example has many maximum weight matchings and many minimum labelings, but all have total weight 31. $\square$

$$
\begin{pmatrix}
4 & 1 & 6 & 2 & 3 \\
5 & 0 & 3 & 7 & 6 \\
2 & 3 & 4 & 5 & 8 \\
3 & 4 & 6 & 3 & 4 \\
4 & 6 & 5 & 8 & 6
\end{pmatrix}
\rightarrow
\begin{array}{r}
 \\ 6 \\ 7 \\ 8 \\ 6 \\ 8
\end{array}
\begin{array}{c}
0\;\;0\;\;0\;\;0\;\;0 \\
\begin{pmatrix}
2 & 5 & 0 & 4 & 3 \\
2 & 7 & 4 & 0 & 1 \\
6 & 5 & 4 & 3 & 0 \\
3 & 2 & 0 & 3 & 2 \\
4 & 2 & 3 & 0 & 2
\end{pmatrix} \\
\;\;\;\;\;T\;\;\;\;T
\end{array}
\begin{array}{l}
S \\ S \\ \\ S \\ S
\end{array}
\qquad \text{Step 0}
$$

Step 0 (bipartite graph): X vertices labeled $S\;S\;\;\;S\;S$; Y vertices below with $T\;T$ beneath the third and fourth.

$\rightarrow$

Step 1 (bipartite graph): X vertices labeled $S\;S\;S\;S\;S$; Y vertices below with $T\;T\;T$ beneath the third, fourth, and fifth.

$$
\rightarrow \text{Step 1}
\begin{array}{r}
 \\ 5 \\ 6 \\ 8 \\ 5 \\ 7
\end{array}
\begin{array}{c}
0\;\;0\;\;1\;\;1\;\;0 \\
\begin{pmatrix}
1 & 4 & 0 & 4 & 2 \\
1 & 6 & 4 & 0 & 0 \\
6 & 5 & 5 & 4 & 0 \\
2 & 1 & 0 & 3 & 1 \\
3 & 1 & 3 & 0 & 1
\end{pmatrix} \\
\;\;\;\;\;T\;\;\;T\;\;\;T
\end{array}
\begin{array}{l}
S \\ S \\ S \\ S \\ S
\end{array}
\rightarrow
\begin{array}{r}
 \\ 4 \\ 5 \\ 7 \\ 4 \\ 6
\end{array}
\begin{array}{c}
0\;\;0\;\;2\;\;2\;\;1 \\
\begin{pmatrix}
0 & 3 & 0 & 4 & 2 \\
0 & 5 & 4 & 0 & 0 \\
5 & 4 & 5 & 4 & 0 \\
1 & 0 & 0 & 3 & 1 \\
2 & 0 & 3 & 0 & 1
\end{pmatrix}
\end{array}
$$

3.2.10. Example. *Solving the Transportation Problem (optional).* At the end of Chapter 2, we defined the Transportation Problem for application to the Street-Sweeping Problem. Given supplies $\sigma(x)$ for $x \in X$, demands $\delta(y)$ for $y \in Y$, and costs $c(xy)$ for sending a unit from x to y, with $\Sigma\sigma(x) = \Sigma\delta(y)$, the Transportation Problem is the problem of satisfying the demands with the least total cost. This is easy to solve using the Assignment Problem when the supplies and demands are integers.

Define a matrix with $\Sigma\sigma(x)$ rows and columns. For each $x \in X$, we have $\sigma(x)$ rows. For each $y \in Y$, we have $\delta(y)$ columns. If row i belongs to x and column j belongs to y, then the weight for position (i, j) is $M - c(xy)$, where $M = \max_{x,y} c(xy)$. Now we solve the Assignment Problem to find a maximum weight matching, which corresponds to a minimum cost solution to the Transportation Problem. Some details remain to prove that this works, such as showing that the shipments in the Transportation Problem break into unit transfers in some optimal solution when the supplies and demands are integers. □

STABLE MATCHINGS (optional)

Instead of optimizing total weight for a matching, we may try to optimize using preferences. For example, suppose we have a collection of n men and n women and want to establish a collection of "stable" marriages. Given societal habits, a collection of marriages is *stable* if and only if there is no man x and woman a such that x prefers a to his current partner and a prefers x to her current partner. Otherwise the matching is "unstable"; x and a will leave their current partners and switch to each other.

3.2.11. Example. Given men x, y, z, w, women a, b, c, d, and preferences listed below, the matching $\{xa, yb, zd, wc\}$ is a stable matching. □

$$
\begin{array}{ll}
\text{Men } \{x, y, z, w\} & \text{Women } \{a, b, c, d\} \\
x \colon a > b > c > d & a \colon z > x > y > w \\
y \colon a > c > b > d & b \colon y > w > x > z \\
z \colon c > d > a > b & c \colon w > x > y > z \\
w \colon c > b > a > d & d \colon x > y > z > w
\end{array}
$$

In their paper "College admissions and the stability of marriage", Gale and Shapley proved that a stable matching always exists and can be found using a relatively simple algorithm. There is an asymmetry in the algorithm; the women instead of the men could do the proposing. We will say more about this difference later. The algorithm below generates the matching of Example 3.2.11.

3.2.12. Algorithm. (Gale-Shapley Proposal Algorithm)
Input: Preference rankings by each of n men and n women.
Idea: Produce a stable matching using proposals by maintaining information about who has proposed to whom and who has rejected whom.
Iteration: Each unmatched man proposes to the highest woman on his list who has not previously rejected him and is not yet matched. If each woman receives exactly one proposal, stop with these being the remaining confirmed matches. Otherwise, at least one woman receives at least two proposals. Every woman receiving more than one proposal rejects all but the highest on her list. Every woman receiving a proposal says "maybe" to the most attractive proposal received. □

3.2.13. Theorem. (Gale-Shapley [1962]) The Proposal Algorithm produces a stable matching.

Proof. The algorithm terminates (with some matching), because on each nonterminal iteration, the total length of the lists of potential mates for the men decreases. This can happen only n^2 times.

The key observation is that the sequence of proposals made by each man is nonincreasing in his preference list, and the sequence of men to whom a woman says "maybe" is nondecreasing in her preference list, culminating in the man assigned. (Men propose repeatedly to the same woman until rejected or assigned.)

If the result is not stable, then there exist x matched to b and y matched to a such that a prefers x to y and x prefers a to b. By the key observation, x never proposed to a during the algorithm, since a received a mate less desirable than x. The key observation also implies that x would never have proposed to b without earlier proposing to a. This contradiction confirms the stability of the result. □

The asymmetry of the proposal algorithm suggests a question: Which sex is happier using this algorithm? When the first choices of the men are distinct, they all get their first choice, and the women are stuck with whomever proposed. The precise statement of "the men are happier" is this: If we instead run the algorithm by having the women propose according to their lists, then every woman winds up at least as happy as in the original algorithm, and every man winds up at least as unhappy. In Example 3.2.11, running the algorithm with women proposing immediately yields the matching $\{xd, yb, ca, wc\}$, in which all women are matched to their first choices. In fact, of all possible stable matchings, every man is happiest under the male-proposal algorithm, and every woman is happiest under the female-proposal algorithm (Exercise 8). Societal habits thus favor men.

The algorithm is used in another setting. Each year, the new graduates of medical schools submit a preference list of hospitals where they would like to serve as residents. The hospitals have their own list of preferences, where we can think of a hospital with multiple openings as several hospitals with the same preference list. Who is happier with the outcome? Since the medical organizations run the algorithm, they do the proposing, and therefore they are happier. The distinction is even clearer in another setting. When graduating students apply for jobs, they have their own lists of preferences, but it is the employers who are making the proposals, called "job offers". Amazingly, chaos in the market for residents (then called interns) forced hospitals to devise and implement the algorithm ten years before the Gale-Shapley paper introduced and solved the problem.

There may be stable matchings other than those found by the two versions of the proposal algorithm. If the cost of assigning each person to that person's ith choice is i, we may seek a "fair" stable matching by finding the stable matching that minimizes the total cost of the assignments. Such an assignment can be found as an application of network flows (Chapter 4). Knuth [1976] and Gusfield and Irving [1989] published books on the subject of stable marriages, the latter containing all we have mentioned and many other aspects of the problem (including the stable roommates problem - Exercise 9).

FASTER BIPARTITE MATCHING (optional)

We began this section with a maximum matching algorithm for bipartite graphs. The running time of the algorithm can be improved by seeking augmenting paths in a clever order; when short augmenting paths are available, we needn't explore as many edges to find one. Using a breadth-first search simultaneously from all the unsaturated vertices of X, we can find many paths of the same length with one examination of the edge set. Hopcroft and Karp [1975] proved that subsequent augmentations must use longer paths, so the searches can be grouped in phases finding paths of the same lengths. They combined these ideas to show that few phases are needed, enabling maximum matchings in bipartite graphs to be found in $O(n^{2.5})$ time.

First observe that if M is a matching of size r and M^* is a matching of size $s > r$, then there exist at least $s - r$ vertex-disjoint M-augmenting paths, because at least this many such paths can be found in $M \triangle M^*$. We use this to prove the next lemma; it implies that the sequence of path lengths in successive shortest augmentations is nondecreasing. Here we treat paths as sets of edges, and cardinality indicates number of edges.

3.2.14. Lemma. If P is a shortest M-augmenting path and P' is $M \triangle P$-augmenting, then $|P'| \geq |P| + |P \cap P'|$ (treating P as an edge set).

Proof. Note that $M \triangle P$ is the matching obtained by using P to augment M. Let N be the next matching, $N = (M \triangle P) \triangle P'$. Since $|N| = |M| + 2$, the remark above guarantees that $M \triangle N$ contains disjoint M-augmenting paths P_1, P_2, and each is as long as P since P is a shortest augmentation.

Since N is obtained from M by switching the edges in P and then switching the edges in P', an edge is in exactly one of M, N if and only if it is in exactly one of P, P'. Hence $M \triangle N = P \triangle P'$, which yields $|P \triangle P'| \geq |P_1| + |P_2| \geq 2|P|$. Combining this with $|P \triangle P'| = |P| + |P'| - |P \cap P'|$, we obtain $|P'| \geq |P| + |P \cap P'|$. $\square$

3.2.15. Lemma. If $P_1, P_2, \cdots$ is a sequence of successive shortest augmentations, then the augmentations of the same length are vertex-disjoint paths.

Proof. Proof by contradiction. Let P_k, P_l with $l > k$ be a closest pair in the sequence that have the same size but are not vertex-disjoint. Let M' be the matching arising after the augmentations $P_1, \ldots, P_k$. By the choice of l, the paths $P_{k+1}, \ldots, P_l$ are vertex-disjoint. Hence P_l is an M'-augmenting path, and the preceding lemma implies $|P_l| \geq |P_k| + |P_l \cap P_k|$. Since $|P_l| = |P_k|$, there is no common edge. If there is no common edge, there is no common vertex, because each vertex of P_k is saturated in M' using an edge of P_k, and any vertex of an M'-augmenting path P_l that is saturated in M' must contribute its saturated edge to P_l. $\square$

3.2.16. Theorem. (Hopcroft-Karp [1975]) The breadth-first phased maximum matching algorithm runs in $O(n^{2.5})$ time.

Proof. By the lemmas, when we search simultaneously from all the unsaturated vertices of X for the shortest augmenting paths, we find a set of vertex-disjoint paths such that after these augmentations all other augmenting paths are longer. Hence all augmentations of the same length can be found by a single examination of the edges, so each such phase runs in time $O(n^2)$. It suffices to show there are at most $2\lfloor \sqrt{n} \rfloor + 2$ phases.

List the augmenting paths as $P_1, \ldots, P_s$ in order by length. Because the paths of the same length are vertex-disjoint, each P_{i+1} is an augmenting path for the matching M_i formed by using the first i paths in the sequence. It suffices to prove the more general statement that whenever $P_1, \ldots, P_s$ are successive shortest augmenting paths that build a maximum matching, the number of distinct lengths among these paths is at most $2\lfloor \sqrt{s} \rfloor + 2$.

Let $r = \lfloor s - \sqrt{s} \rfloor$. Because $|M_r| = r$ and the maximum matching has size s, we have observed that there are at least $s - r$ vertex-disjoint M_r-augmenting paths. The shortest of these paths uses at most

$\lfloor r/(s-r) \rfloor$ edges from M_r. Hence $|P_{r+1}| \leq 2\lfloor r/(s-r) \rfloor + 1$. Since $\lfloor r/(s-r) \rfloor = \lfloor \lfloor s - \sqrt{s} \rfloor / \lceil \sqrt{s} \rceil \rfloor \leq \lfloor \sqrt{s} \rfloor$, the paths up to P_r provide all but the last $\lceil \sqrt{s} \rceil$ augmentations using length at most $2\lfloor \sqrt{s} \rfloor + 1$. There are at most $\lfloor \sqrt{s} \rfloor + 1$ distinct odd integers up to this value, and even if the last $\lceil \sqrt{s} \rceil$ paths have distinct lengths, they provide at most $\lfloor \sqrt{s} \rfloor + 1$ additional lengths, so altogether we use at most $2\lfloor \sqrt{s} \rfloor + 2$ distinct lengths. □

This approach was further improved Even and Tarjan [1975] to run in time $O(\sqrt{n}m)$, where m is the number of edges in the graph. Their algorithm also solves a more general problem.

EXERCISES

3.2.1. (–) Using non-negative edge weights, construct a weighted graph with four vertices in which the matching of maximum weight is not a matching of maximum size.

3.2.2. Find a transversal of maximum total sum (weight) in the matrices below. Prove that there is no larger weight transversal by exhibiting a solution to the dual problem. Explain why this proves that there is no larger transversal.

$$
\begin{array}{ccc}
\begin{matrix}
4\,4\,4\,3\,6 \\
1\,1\,4\,3\,4 \\
1\,4\,5\,3\,5 \\
5\,6\,4\,7\,9 \\
5\,3\,6\,8\,3
\end{matrix}
&
\begin{matrix}
7\,8\,9\,8\,7 \\
8\,7\,6\,7\,6 \\
9\,6\,5\,4\,6 \\
8\,5\,7\,6\,4 \\
7\,6\,5\,5\,5
\end{matrix}
&
\begin{matrix}
1\,2\,3\,4\,5 \\
6\,7\,8\,7\,2 \\
1\,3\,4\,4\,5 \\
3\,6\,2\,8\,7 \\
4\,1\,3\,5\,4
\end{matrix}
\end{array}
$$

3.2.3. Find a minimum-weight transversal in the matrix below. (Hint: use a transformation of the problem.)

$$
\begin{pmatrix}
4 & 5 & 8 & 10 & 11 \\
7 & 6 & 5 & 7 & 4 \\
8 & 5 & 12 & 9 & 6 \\
6 & 6 & 13 & 10 & 7 \\
4 & 5 & 7 & 9 & 8
\end{pmatrix}
$$

3.2.4. (!) *The Bus Driver Problem.* Suppose bus drivers are paid overtime for the time by which their routes in a day exceed t. Suppose there are n bus drivers, n morning routes with durations $x_1, \ldots, x_n$, and n afternoon routes with durations $y_1, \ldots, y_n$, and the objective is to assign one morning run and one afternoon run to each driver to minimize the total amount of overtime. Express this as a weighted matching problem, and prove that the best solution is to give the ith longest morning route and ith shortest afternoon route to the same driver, for each i. (R.B. Potts)

3.2.5. Suppose the weights in the matrix A have the form $w_{ij} = a_i b_j$, where $a_1, \ldots, a_n$ are numbers associated with the rows and $b_1, \ldots, b_n$ are weights associated with the columns. Determine the maximum weight of a transversal of A. What happens when $w_{ij} = a_i + b_j$? (Hint: in each case, guess the general pattern by examining the solution when $n = 2$.)

3.2.6. (–) Give an example of the stable matching problem with two men and two women in which there is more than one stable matching.

3.2.7. (–) Determine the stable matchings that result when the Proposal Algorithm is run with the men proposing and with the women proposing, given the preference orders listed below.

Men $\{u, v, w, x, y, z\}$	Women $\{a, b, c, d, e, f\}$
$u: a > b > d > c > f > e$	$a: z > x > y > u > v > w$
$v: a > b > c > f > e > d$	$b: y > z > w > x > v > u$
$w: c > b > d > a > f > e$	$c: v > x > w > y > u > z$
$x: c > a > d > b > e > f$	$d: w > y > u > x > z > v$
$y: c > d > a > b > f > e$	$e: u > v > x > w > y > z$
$z: d > e > f > c > b > a$	$f: u > w > x > v > z > y$

3.2.8. Prove that among all stable matchings, *every* man is happiest in the matching produced by the Gale-Shapley Proposal Algorithm with men proposing. (Hint: use induction on the number of rounds to prove that no man is ever rejected by a woman who is assigned to him in a stable matching.)

3.2.9. In the Stable Roommates Problem, each of $2n$ people has a preference ordering on the other $2n - 1$ people. A stable matching is a complete matching in which no pair prefers each other to their current roommates. Prove that the Stable Roommates Problem defined by the preference orderings below has no stable matching. (Gale-Shapley [1962])

$$a: b > c > d$$
$$b: c > a > d$$
$$c: a > b > d$$
$$d: a > b > c$$

3.3 Matchings in General Graphs

We will expand our discussion of matchings in arbitrary graphs to consider somewhat more general subgraphs.

3.3.1. Definition. A *factor* of G is a spanning subgraph of G. A *k-factor* is a spanning k-regular subgraph (a perfect matching is a 1-*factor*). An *odd component* of a graph is a component of odd order; the number of odd components of H is $o(H)$.

TUTTE'S 1-FACTOR THEOREM

Tutte found a necessary and sufficient condition for an arbitrary graph to have a 1-factor. If G has a 1-factor and we consider a set $S \subseteq V(G)$, then every odd component of $G - S$ has a vertex matched to something outside it, which can only belong to S.

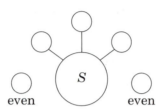

Since these must be distinct vertices of S, we conclude that $o(G - S) \leq |S|$. Tutte proved that this obvious necessary condition is also sufficient. Many proofs of this have appeared; we present the proof by Lovász using the ideas of symmetric difference and extremality.

3.3.2. Theorem. (Tutte [1947]) A graph G has a 1-factor if and only if $o(G - S) \leq |S|$ for every $S \subseteq V(G)$.

Proof. (Lovász [1975]). As noted above, Tutte's condition is necessary; we prove sufficiency. Tutte's condition is preserved by addition of edges: if $G' = G + e$ and $S \subseteq V(G)$, then $o(G' - S) \leq o(G - S)$, because when the addition of e combines two components of $G - S$ into one, the number of components that have odd order does not increase. Therefore, it suffices to consider a simple graph G such that G satisfies Tutte's condition, G has no 1-factor, and adding any edge to G creates a 1-factor. We will obtain a contradiction in every case by constructing a 1-factor in G. This implies that every graph satisfying Tutte's condition has a 1-factor.

By considering $S = \emptyset$, we know that $n(G)$ is even, since a graph of odd order must have a component of odd order. Let U be the set of vertices in G that have no non-neighbors. Suppose $G - U$ consists of disjoint complete graphs; we build a 1-factor for such a G. The vertices in each component of $G - U$ can be paired up arbitrarily, with one left over in the odd components. Since $o(G - U) \leq |U|$ and each vertex of U is adjacent to all of $G - U$, we can match these leftover vertices arbitrarily to vertices of U to complete a 1-factor.

This leaves the case where $G - U$ is not a disjoint union of cliques. Here $G - U$ contains two nonadjacent vertices x, z with a common neighbor y. Furthermore, $G - U$ has another vertex w not adjacent to y, since $y \notin U$. By the maximality of G, adding any edge to G produces a 1-factor; let M_1 and M_2 be 1-factors in $G + xz$ and $G + yw$. It suffices to show that in $M_1 \cup M_2$ we can find a 1-factor avoiding xz and yw, because this will be a 1-factor in G.

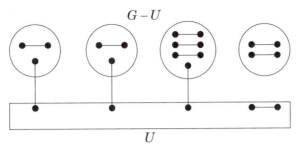

Let F be the set of edges that belong to exactly one of M_1 and M_2; note that F contains xz and yw. Since every vertex of G has degree 1 in each of M_1 and M_2, every vertex of G has degree 0 or 2 in F. Hence F is a collection of disjoint even cycles and isolated vertices. Let C be the cycle of F containing xz. If C does not also contain yw, then the desired 1-factor consists of the edges of M_2 from C and all of M_1 not in C. If C contains both yw and xz, as illustrated below, then we use the edge yx or the edge yz to obtain a matching of $V(C)$ using only edges of G (avoiding both xz and yw). We use yx if $d_C(y,x)$ is odd; we use yz if $d_C(y,z)$ is odd (as illustrated). The remaining vertices of C form two paths of even order; we use the edges of M_1 in one of these paths and the edges of M_2 in the other to produce a matching in C that does not use xz or yw. Combined with M_1 or M_2 outside C, we have a perfect matching of G. $\square$

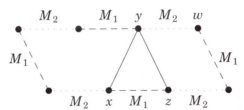

There may be a gap in size between the largest matching and the smallest vertex cover (Exercise 5). Nevertheless, another minimization problem yields a min-max relation for maximum matching in general graphs. The proof uses a graph transformation argument.

3.3.3. Remark. *Parity.* For any graph G and $S \subseteq V(G)$, the difference $o(G-S)-|S|$ has the same parity as $n(G)$. Hence $o(G-S)$ exceeds $|S|$ by at least two for some S if $n(G)$ is even and G has no 1-factor. $\square$

3.3.4. Corollary. (Berge [1958]). The largest number of vertices in a matching in G is $\min_{S \subseteq V(G)} \{n - d(S)\}$, where $d(S) = o(G-S)-|S|$.

Proof. Given $S \subseteq V(G)$, at most $|S|$ edges can match vertices of S to vertices in odd components of $G-S$, so every matching has at least

$o(G - S) - |S|$ unsaturated vertices. We want to achieve this bound.

Let $d = \max\{o(G - S) - |S|: S \subseteq V(G)\}$. The case $S = \varnothing$ implies $d \geq 0$. Let $G' = G \vee K_d$. Since $d(S)$ has the same parity as $n(G)$ for each S, we know that $n(G')$ is even. If G' satisfies Tutte's condition, then we can obtain a matching of the desired size in G from a complete matching in G', because deleting the d added vertices eliminates edges that saturate at most d vertices of G.

The condition $o(G' - S') \leq |S'|$ holds for $S' = \varnothing$ because $n(G')$ is even. If S' is nonempty but does not contain all of K_d, then $G' - S'$ has only one component, and $1 \leq |S'|$. Finally, if $K_d \subseteq S'$, let $S = S' - V(K_d)$. We have $G' - S' = G - S$, so $o(G' - S') = o(G - S) \leq |S| + d = |S'|$, and G' indeed satisfies Tutte's condition. $\square$

This corollary guarantees that there is a PROOF that a maximum matching is maximum by exhibiting a vertex set S whose deletion leaves the desired number of odd components. This is easier than proving there is no M-augmenting path, but finding S may be hard.

Most applications of Tutte's Theorem involve showing that some other condition implies Tutte's condition and hence guarantees a 1-factor. Some applications were proved by other means long before Tutte's Theorem was available.

3.3.5. Corollary. (Petersen [1891]) Every 3-regular graph with no cut-edge has a 1-factor.

Proof. Suppose G is 3-regular and has no cut-edge. We prove that each set $S \subseteq V(G)$ satisfies Tutte's condition. We count the edges between S and the odd components of $G - S$. Since G is 3-regular, each vertex of S is incident to at most three such edges. If each odd component H of $G - S$ is incident to at least three such edges, then $3o(G - S) \leq 3|S|$ and hence $o(G - S) \leq |S|$, as desired. The number of edges between S and H cannot be 1, since G has no cut-edge. It also cannot be even, because then the sum of the vertex degrees in H would be odd. Hence there are at least three edges from H to S, as desired. $\square$

The Petersen graph was used originally to show that this theorem is best possible; the Petersen graph satisfies the hypothesis but does not contain two edge-disjoint 1-factors. Petersen also proved a sufficient condition for 2-factors in regular multigraphs. The proof uses only Eulerian circuits and a corollary of Hall's Theorem. A connected $2k$-

regular graph is Eulerian and therefore is a closed trail (Theorem 2.4.2) and it easy to show that every closed trail can be partitioned into edge-disjoint cycles (Exercise 1.2.4). Petersen proved that when the graph is regular, these cycles can be organized into 2-factors.

3.3.6. Theorem. (Petersen [1891]) Every regular multigraph of even degree has a 2-factor.

Proof. Suppose G is $2k$-regular with vertices $v_1, \ldots, v_n$. Every component of G is Eulerian, with some Eulerian circuit C. For each component, define a bipartite graph H with vertices $u_1, \ldots, u_n$ and $w_1, \ldots, w_n$ by putting $u_i \leftrightarrow w_j$ if v_j immediately follows v_i somewhere on C. Because C leaves and enters each vertex k times, H is k-regular.

Every regular bipartite graph has a 1-factor (Corollary 3.1.8). A 1-factor in H designates one edge "leaving" v_i (incident to u_i in H) and one edge "entering" v_i (incident to w_i in H). Together, these edges form a 2-regular spanning subgraph of G (a 2-factor). $\square$

3.3.7. Example. *Construction of a 2-factor.* Consider the Eulerian circuit in $G = K_5$ that successively visits 1231425435. The corresponding bipartite graph H is on the right. For the 1-factor whose u, w-pairs are 12, 43, 25, 31, 54, the resulting 2-factor is the cycle (1,2,5,4,3). The remaining edges form another 1-factor, which converts back to the 2-factor (1,4,2,3,5) that remains in G. $\square$

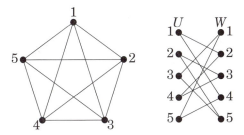

f-FACTORS OF GRAPHS

A factor is an arbitrary spanning subgraph of G; we ask existence questions about factors of special types. A k-factor is a k-regular factor; we have been studying 1-factors and 2-factors. More generally, we can try to specify the degree at each vertex. That is, given a function $f \colon V(G) \to \mathbb{N}$, we ask whether G has a subgraph H such that $d_H(v) = f(v)$ for all $v \in V(G)$. Such a subgraph H is an *f-factor* of G.

Multiple edges don't affect the existence of 1-factors, but they can affect the existence of f-factors. Tutte [1952] proved a necessary and

sufficient condition for a multigraph to have an f-factor. The original
proof was quite difficult; later Tutte reduced it to checking for a 1-factor
in a related graph. We describe the construction of this related graph.
This is a beautiful example of transforming a graph problem into a simpler graph problem.

We may assume that $f(w) \le d(w)$ for every vertex w; otherwise we
can say immediately that G has no f-factor. Given this, we construct a
graph H that has a 1-factor if and only if G has an f-factor. Let
$e(w) = d(w) - f(w)$ be the "excess" degree at w. To construct H, replace
each vertex v by a bipartite graph $K_{d(v),e(v)}$, where $A(v)$ is the partite set
of size $d(v)$ and $B(v)$ is the part of size $e(v)$. For each edge vw in G, join
one vertex of $A(v)$ to one vertex of $A(w)$, in such a way that every vertex
in every $A(v)$ gets exactly one of these edges. The figure below (from
notes by Chvátal) illustrates a typical multigraph G and the corresponding simple graph H we construct when f is the function indicated
by the labels on $V(G)$.

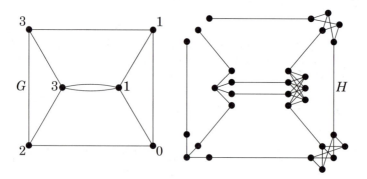

3.3.8. Theorem. A graph G has an f-factor if and only if the graph H
  constructed above from G and f has a 1-factor.

Proof. If G has an f-factor, the corresponding edges in H leave $e(v)$
vertices of $A(v)$ unmatched; match them arbitrarily to the vertices of
$B(v)$ to obtain a 1-factor of H. Similarly, deletion of the edges involving
B-vertices in a 1-factor of H leaves edges that collapse to an f-factor of
G when the remaining $f(v)$ vertices of each $A(v)$ are merged. Hence H
has a 1-factor if and only if G has an f-factor. $\square$

The necessary and sufficient condition $(o(H - S) \le |S|$ for all $S)$ for
a 1-factor in the derived graph H of Theorem 3.3.8 can be translated
into a necessary and sufficient condition for an f-factor in G. This condition has several applications. The question of whether $d_1, \dots, d_n$ is a
graphic sequence is the question of whether K_n has an f-factor with
$f(v_i) = d_i$ for all i. See Exercise 16 for further discussion.

Given an algorithm to check for 1-factors, The correspondence in Theorem 3.3.8 provides an algorithmic test for an f-factor. More generally, we may seek an algorithm for finding maximum matchings in graphs; we discuss this next.

EDMONDS' BLOSSOM ALGORITHM (optional)

Berge's Theorem (Theorem 3.1.6) states that a matching M is a maximum matching if and only if there is no M-augmenting path. Hence we can build a maximum matching by iteratively finding augmenting paths. Since we can only augment $n/2$ times, we obtain a good algorithm if the search for an augmenting path does not take too long. Edmonds [1965a] presented the first such algorithm in his famous paper "Paths, trees, and flowers".

In bipartite graphs, we can search quickly for augmenting paths (Algorithm 3.2.1) because we can limit the exploration of paths. When we search for M-augmenting paths from a particular vertex u, vertices in the same partite set as u are reached only via edges in M (saturated edges), and vertices in the opposite partite set are reached only via edges not in M (unsaturated edges). For this reason, we extend our search from a given vertex at most once. This property fails in graphs with odd cycles, in which alternating paths from an unsaturated vertex may reach a vertex x along saturated or along unsaturated edges.

3.3.9. Example. In the graph below with a matching M indicated by solid edges, a search for shortest M-augmenting paths from u will reach x via the unsaturated edge ax. If we do not also consider a longer path reaching x via a saturated edge, we will miss the augmenting path u, v, a, b, c, d, x, y. $\square$

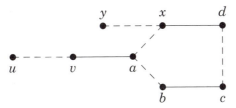

We describe Edmonds' solution to this difficulty. If an exploration of M-alternating paths from u reaches a vertex x by an unsaturated edge in one path and by a saturated edge in another path, then x belongs to an odd cycle. Alternating paths from u can split apart only when the next edge is unsaturated (leaving vertex a in Example 3.3.9); when the next edge is saturated there is only one choice for it. From the vertex where the paths diverge, the path reaching x on an unsaturated edge has odd length, and the path reaching it on a saturated edge has even length. Together, they form an odd cycle.

3.3.10. Definition. Let M be a matching in a graph G, and let u be an M-unsaturated vertex. A *flower* is the union of two M-alternating paths from u that reach a vertex x on steps of opposite parity (having not done so earlier). The *stem* of the flower is the maximal common initial path (of nonnegative even length). The *blossom* of the flower is the odd cycle obtained by deleting the stem.

In Example 3.3.9, the flower is the entire graph except y, the stem is the path u, v, a), and the blossom is the 5-cycle. This horticultural terminology evokes the use of *tree* for the structures generated by most search procedures.

Blossoms do not impede our search. For each vertex z in a blossom, there is an M-alternating u, z-path that reaches z on a saturated edge, found by traveling the proper direction around the blossom to reach z from the stem. We therefore can continue our search along any unsaturated edge from the blossom to a vertex not yet reached. Example 3.3.9 shows such an extension that immediately reaches an unsaturated vertex and completes an M-augmenting path.

Conversely, we cannot leave the blossom along a saturated edge. The effect of these two observations is that we can view the entire blossom as a single "supervertex" reached along the saturated edge at the end of the stem. We can search from all vertices of the supervertex blossom simultaneously along unsaturated edges.

We implement this consolidation by contracting the edges of a blossom B when we find it. The result is a new saturated vertex b incident to the last (saturated) edge of the stem. Its other incident edges are the unsaturated edges joining vertices of B to vertices outside B. We can explore from b in the usual way to extend our search. We may later find a blossom containing b; blossoms may contain earlier blossoms. If we find an M-alternating path in the contracted graph from u to an unsaturated vertex x, then we can undo the contractions to obtain an M-augmenting path to x in the original graph.

Except for the treatment of blossoms, the approach is that of Algorithm 3.2.1 for exploring M-alternating paths. In the corresponding phrasing, T is the set of vertices of the current graph reached along unsaturated edges, and S is the set of vertices reached along saturated edges. The vertices that arise by contracting blossoms belong to S.

3.3.11. Example. The graph on the left below exhibits a matching M in solid edges. We search from the unsaturated vertex u for an M-augmenting path. We first explore the unsaturated edges incident to u, reaching a and b. Since a and b are saturated, we immediately extend the paths along the edges ac and bd. Now $S = \{u, c, d\}$. If we next choose to explore from c, we find its neighbors e, f along unsaturated edges. Since these are saturated, we extend the paths along the edge ef, discovering the blossom with vertex set $\{c, e, f\}$. We contract the blossom to obtain the new vertex C, changing S to $\{u, C, d\}$. We have arrived at the second graph below.

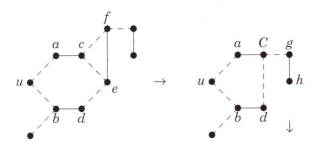

Suppose we now explore from the vertex $C \in S$. Unsaturated edges take us to g and to d. Since g is saturated by the edge gh, we place h in S. Since d is already in S, we have found another blossom. The paths reaching d are u, b, d and u, a, C, d. We contract the blossom, obtaining the new vertex B and the graph on the right below, with $S = \{U, h\}$. We next explore from h, finding nothing new (if we exhaust S without reaching an unsaturated vertex, then there is no M-augmenting path from u). Finally, we explore from U, reaching the unsaturated vertex x.

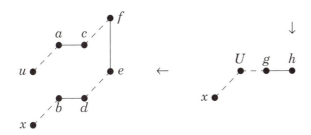

If we record the edge reaching each vertex, then we can extract from the search an M-augmenting u, x-path. We reached x from U, so we expand the blossom back into $\{u, a, C, d, b\}$ and recall that x is reached from U along bx. The path in the blossom U that reaches b on a saturated edge ends with C, d, b. Since C is a blossom in the original graph, we expand C back into $\{c, f, e\}$ and recall that d is reached from C along the unsaturated edge ed. Now the path (from the "base" of C) that reaches e along a saturated edge is c, f, e. Finally, c was reached from a and a from u, so we extract the full M-augmenting path u, a, c, f, e, d, b, x. □

We summarize the steps of the algorithm, glossing over the details of implementation.

3.3.12. Algorithm. (Edmonds' Blossom Algorithm - sketch).
Input: A graph G, a matching M in G, an M-unsaturated vertex u.
Idea: Explore M-alternating paths from U, letting S be the set of vertices reached along edges of M and T be the set of vertices reached along edges not in M. Mark vertices of S that have been explored for extensions, and record for each vertex of $S \cup T$ the vertex from which it was reached. When blossoms are found, contract them.
Initialization: $S = \{u\}$ and $T = \varnothing$.
Iteration: If S has no unmarked vertex, stop and report that there is no M-augmenting path from u. Otherwise, select an unmarked $v \in S$. To explore v, consider each $y \in N(x)$ such that $xy \notin M$. If y is unsaturated, stop and trace back from y (expanding blossoms as needed) to report an M-augmenting u, y-path. Otherwise, y is matched to some w by M. If $w \notin S$, include y in T (reached from v) and include w in S (reached from y). If $w \in S$ (or if w is a neighbor of v), then a blossom has been found. Contract the blossom, replacing its vertices in S and T by a single new vertex in S to continue the search in the smaller graph. After exploring all such edges incident to v, mark v and iterate. □

We cannot search from all unsaturated vertices simultaneously as in Algorithm 3.2.1, because the behavior of a blossom depends on where the stem reaches the blossom. Nevertheless, if we find no M-augmenting path from u, then we can delete u from the graph and ignore it when searching for M-augmenting paths from other vertices. This relies on Exercise 14.

Edmonds' original algorithm runs in time $O(n^4)$. The implementation given in Ahuja-Magnanti-Orlin [1993, p483-494] runs in time $O(n^3)$. This requires 1) appropriate data structures to represent the blossoms and to process contractions, and 2) careful analysis of the number of contractions that can be performed, the time spent exploring edges, and the time spent contracting and expanding blossoms.

The first algorithm solving the maximum matching problem in less than cubic time was the $O(n^{5/2})$ algorithm in Even-Kariv [1975]. The best algorithm now known runs in time $O(n^{1/2}m)$ for a graph with n vertices and m edges (this is faster than $O(n^{5/2})$ for sparse graphs). The algorithm is rather complicated and appears in Micali-Vazirani [1980], with a complete proof in Vazirani [1994].

We have not discussed the weighted matching problem for general graphs. Edmonds [1965d] found an algorithm for this, which was implemented in time $O(n^3)$ by Gabow [1975] and by Lawler [1976]. Faster algorithms appear in Gabow [1990] and in Gabow-Tarjan [1989].

EXERCISES

3.3.1. (–) In the graph drawn below, exhibit a k-factor for each k in $\{0, 1, 2, 3, 4\}$.

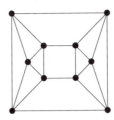

3.3.2. (–) Prove that the edges of a k-regular bipartite graph can be partitioned into r-factors if and only if r divides k.

3.3.3. (!) Prove that a tree T has a perfect matching if and only if $o(T - v) = 1$ for every $v \in V(T)$. (Chungphaisan)

3.3.4. For each $k > 1$, present a k-regular simple graph that has no 1-factor, and prove that it has no 1-factor.

3.3.5. (!) Suppose the maximum size of a matching in a simple graph G is k (i.e., $\alpha'(G) = k$). Determine, with proof, the maximum possible value of the vertex cover number $\beta(G)$ as a function of k. (This requires giving an upper bound on $\beta(G)$ and an example to achieve it, for each k.)

3.3.6. Draw a connected 3-regular graph that has a 1-factor and has a cut-vertex. Prove that if the edges of G can be partitioned into 1-factors, then G has no cut-vertex.

3.3.7. Prove that the edges of a 3-regular graph having no cut-edge can be partitioned into copies of P_4.

3.3.8. (!) *Generalization of Petersen's Theorem.* Suppose G is a connected k-regular graph of even order, and suppose that G remains connected whenever any set of fewer than $k - 1$ edges is deleted. Prove that G has a 1-factor.

3.3.9. (!) Let G be the graph below. Find a maximum matching in G and in $G + xy$, and for each use a dual problem to give a short proof that it has no larger matching. (Hint: the minimum vertex cover in $G + xy$ has size eight.)

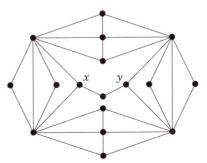

3.3.10. (!) Suppose G is a bipartite graph with bipartition X, Y. Let H be the graph obtained from G by adding one vertex to Y if $n(G)$ is odd and then adding the edges of a clique on the vertices of Y.

a) Prove that G has a matching of size $|X|$ if and only if H has a 1-factor.

b) Prove that if G satisfies Hall's Condition ($|N(S)| \geq |S|$ for all $S \subseteq X$), then H satisfies Tutte's Condition ($o(H - T) \leq |T|$ for all $T \subseteq V(H)$).

c) Use (a) and (b) to obtain Hall's Theorem from Tutte's Theorem.

3.3.11. Suppose G is a $K_{1,3}$-free connected graph of even order. Prove that G has a 1-factor. (Hint: prove the stronger result that the last edge in a longest path belongs to a 1-factor in G.) (Sumner [1974a], Las Vergnas [1975])

3.3.12. (+) Suppose n is even. Determine the maximum number of edges in an n-vertex graph with no 1-factor. (Erdős-Gallai [1961]) (Hint: Describe the structure of maximal graphs having no 1-factor. Count the edges by a quadratic polynomial whose argument is a parameter of the structure.)

3.3.13. Prove that G is $(k + 1)K_2$-free if and only if for each bipartite induced subgraph of G with partite sets X, Y, the set X has a subset of size at most k whose neighborhood contains $N(X) \cap Y$. (Liu-Zhou [1996])

3.3.14. Suppose M is a matching in a graph G, and u is an M-unsaturated vertex. Prove that if G has no M-augmenting path that starts at u, then u is unsaturated in some maximum matching in G.

3.3.15. f-*solubility.* Let $S = \mathbb{N} \cup \{0\}$. Given $f: V(G) \to S$, a graph G is f-*soluble* if it has an edge-weighting $w: E(G) \to S$ such that $\Sigma_{u \in N(v)} w(uv) = f(v)$ for every $v \in V(G)$.

a) Prove that G has an f-factor if and only if the graph H obtained from G by subdividing each edge twice and defining f to be 1 on the new vertices is f-soluble. (This reduces testing for an f-factor to testing f-solubility.)

Given G and an $f: E(G) \to S$, construct a graph H (with proof) such that G is f-soluble if and only if H has a 1-factor.

3.3.16. (+) *Tutte's f-factor condition and graphic sequences.* Suppose f assigns nonnegative integers to vertices of G. If S, T are disjoint subsets of $V(G)$, let $q(S, T)$ denote the number of components Q of $G - S - T$ such that $e(Q, T) + \Sigma_{v \in V(Q)} f(v)$ is odd, where $e(Q, T)$ is the number of edges from Q to T. Tutte [1952, 1954] proved that G has an f-factor if and only if

$$q(S, T) + \Sigma_{v \in T}(f(v) - d_{G-S}(v)) \leq \Sigma_{v \in S} f(v)$$

for all choices of disjoint subsets $S, T \subset V$.

a) *The Parity Lemma.* Let $\delta(S, T) = f(S) - f(T) + \Sigma_{v \in T} d_{G-S}(v) - q(S, T)$ be the *deficiency* of S, T with respect to f. Prove that $\delta(S, T)$ has the same parity as $f(V)$ for any disjoint sets $S, T \subseteq V(G)$.

b) Use the f-factor theorem and the Parity Lemma to prove that the nonnegative integers $d_1 \geq \cdots \geq d_n$ are the degree sequence of a simple graph if and only if Σd_i is even and $\Sigma_{i=1}^{k} d_i \leq k(k - 1) + \Sigma_{i=k+1}^{n} \min\{k, d_i\}$ for $1 \leq k \leq n$. (Erdős-Gallai [1960])

Chapter 4

Connectivity and Paths

4.1 Cuts and Connectivity

A good communication network is hard to disrupt. We want to preserve network service by ensuring that the graph (or digraph) of possible transmissions remains connected even when some vertices or edges fail. When communication links are expensive, we want to achieve these goals with few edges. Loops are irrelevant for connection, so we may assume that the graphs and digraphs of this chapter have no loops, especially when considering degree conditions.

CONNECTIVITY

Consider the deletion of vertices from graphs.

4.1.1. Definition. A *separating set* or *vertex cut* of a graph G is a set $S \subseteq V(G)$ such that $G - S$ has more than one component. A graph G is *k-connected* if every vertex cut has at least k vertices. The *connectivity* of G, written $\kappa(G)$, is the minimum size of a vertex cut (equivalently, $\kappa(G)$ is the maximum k such that G is k-connected).

4.1.2. Example. *Connectivity of K_n and $K_{m,n}$.* A clique has no separating set. We adopt the convention that $\kappa(K_n) = n - 1$ so that our results about connectivity will extend to cliques.

Consider $K_{m,n}$ with bipartition X, Y. Every induced subgraph of $K_{m,n}$ that has at least one vertex from X and from Y is connected. Hence every separating set of $K_{m,n}$ contains X or Y. Since X and Y

themselves are separating sets (unless their deletion leaves K_1), we have $\kappa(K_{m,n}) = \min\{m,n\}$. The connectivity of $K_{3,3}$ is 3, and $K_{3,3}$ is 1-connected, 2-connected, and 3-connected, but not 4-connected. $\square$

A graph has connectivity 0 if and only if it is disconnected. In general, the condition of connectivity k has no nice characterization, but in Section 4.2 we will see nice characterizations of the condition of k-connectedness.

4.1.3. Example. *Connectivity of the k-dimensional cube Q_k.* The cube Q_k is k-regular. We can cut Q_k by deleting the neighbors of a vertex, so $\kappa(Q_k) \le k$. To prove that $\kappa(Q_k) = k$, we show that every vertex cut of Q_k has at least k vertices.

We use induction on k. For $k = 0, 1$, Q_k is a clique with connectivity k; this completes the basis. For the induction step, suppose $k \ge 2$, and suppose $\kappa(Q_{k-1}) = k - 1$. We know that Q_k is obtained from two copies Q, Q' of Q_{k-1} by adding a matching joining corresponding vertices in Q and Q'. Let S be an arbitrary vertex cut in Q_k.

If $Q - S$ is connected and $Q' - S$ is connected, then $Q_k - S$ is also connected unless S deletes at least one endpoint of every matched pair. This requires $|S| \ge 2^{k-1}$, but $2^{k-1} \ge k$ for $k \ge 2$. Hence we may assume that $Q - S$ is disconnected, which means that S has at least $k - 1$ vertices in Q, by the induction hypothesis. If S contains no vertices of Q', then $Q' - S$ is connected and all vertices of $Q - S$ have neighbors in $Q' - S$, so $Q_k - S$ is connected. Hence S must also contain a vertex of Q', yielding $|S| \ge k$, as desired. $\square$

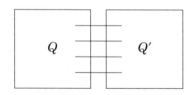

When G is not a clique, deleting the neighbors of a vertex separates G, so $\kappa(G) \le \delta(G)$. Equality need not hold; $2K_m$ has minimum degree $m - 1$ but connectivity 0. Since connectivity k requires $\delta(G) \ge k$, it also requires at least $\lceil kn/2 \rceil$ edges. This is achievable.

4.1.4. Example. *The k-connected Harary graph $H_{k,n}$.* Place n vertices in circular order, and suppose $k < n$. If $k = 2r$, form $H_{k,n}$ by making each vertex adjacent to the nearest r vertices in each direction around the circle. If $k = 2r + 1$ and n is even, form $H_{k,n}$ by making each vertex adjacent to the nearest r vertices in each direction and to the vertex opposite it on the circle. In each case, $H_{k,n}$ is k-regular.

If $k = 2r + 1$ and n is odd, index the vertices by the integers modulo n. Construct $H_{k,n}$ from $H_{2r,n}$ by adding the edges $i \leftrightarrow i + (n+1)/2$ for $0 \le i \le (n-1)/2$. The graphs $H_{4,8}$, $H_{5,8}$, and $H_{5,9}$ appear below. □

4.1.5. Theorem. (Harary [1962a]) $\kappa(H_{k,n}) = k$, and hence the minimum number of edges in a k-connected graph on n vertices is $\lceil kn/2 \rceil$.

Proof. We prove only the even case $k = 2r$, leaving the odd case as Exercise 6. Let $G = H_{2r,n}$. Since $\delta(G) = 2r$, it suffices to prove $\kappa(G) \ge 2r$. Choose any $S \subseteq V(G)$ with $|S| < 2r$; we prove that $G - S$ is connected. Choose any pair $u, v \in V(G) - S$. Deleting u, v from the original circular arrangement of vertices leaves two maximal segments A, B of consecutive vertices. In $G - S$, we have the potential for traveling from u to v in a clockwise or a counterclockwise direction, through A or B. Since $|S| < 2r$, the pigeonhole principle implies that S has fewer than r vertices in A or in B. Since each vertex has edges to the next r vertices in a particular direction, we can find a u, v-path in $G - S$ via the subset A or B in which S has fewer than r vertices. □

Harary's construction determines the degree conditions that *allow* a graph to be k-connected. Exercise 15 determines the degree conditions that *force* a simple graph to be k-connected. Because it considers vertex deletions, the connectivity of a multigraph is the same as the connectivity of the simple graph obtained by deleting extra copies of multiple edges (and deleting loops). Hence we discuss degree conditions that imply k-connectedness only for the class of simple graphs.

Direct proofs of $\kappa(G) = k$ consider a vertex cut S and prove that $|S| \ge k$ or consider a set S with fewer than k vertices and prove that $G - S$ is connected. The indirect approach assumes a cut of size less than k and obtains a contradiction. The indirect proof may be easier to find, but the direct proof may be clearer to state. It helps to observe that if $k < n(G)$ and G has a vertex cut of size less than k, then G has a vertex cut of size $k - 1$ (first delete the cut, then continue to delete vertices until $k - 1$ are gone, always leaving a vertex in each of two components). Finally, proving $\kappa(G) = k$ also requires presenting a vertex cut of size k; that is usually the easy part.

EDGE-CONNECTIVITY

Perhaps our transmitters are secure and never fail, but our communication links are subject to noise or other disruptions. In this situation, we want to make it hard to disconnect our graph by deleting edges.

4.1.6. Definition. A *disconnecting set* of edges is a set $F \subseteq E(G)$ such that $G - F$ has more than one component. Given $S, T \subseteq V(G)$, the notation $[S, T]$ specifies the set of edges having one endpoint in S and the other in T. An *edge cut* is an edge set of the form $[S, \bar{S}]$, where S is a nonempty proper subset of $V(G)$. A graph is *k-edge-connected* if every disconnecting set has at least k edges. The *edge-connectivity* of G, written $\kappa'(G)$, is the minimum size of a disconnecting set (equivalently, $\kappa'(G)$ is the maximum k such that G is k-edge-connected).

The notation for edge-connectivity continues our convention of using a "prime" for an edge parameter analogous to a vertex parameter. Using the same base letter emphasizes the analogy and avoids the confusion of using many different letters -- and running out of them.

4.1.7. Remark. *Disconnecting set vs. edge cut.* Every edge cut is a disconnecting set, since $G - [S, \bar{S}]$ has no path from S to $\bar{S}$. The converse is false. In K_3, the set of three edges is a disconnecting set but not an edge cut. In $K_{3,3}$, every set of seven edges is a disconnecting set, but none is an edge cut (Exercise 7).

If $n(G) > 1$, then every minimal disconnecting set of edges is an edge cut. If $G - F$ has more than one component for some $F \subseteq E(G)$, then we have deleted all edges having one endpoint in some component H of $G - F$. Hence F contains the edge cut $[V(H), \overline{V(H)}]$, and F is not a minimal disconnecting set unless $F = [V(H), \overline{V(H)}]$. □

Deleting one endpoint of each edge in an edge cut F deletes every edge of F. Hence we expect that $\kappa(G) \le \kappa'(G)$ always holds. To prove this, we must be careful not to delete the only vertex of a component of $G - F$ and thereby produce a connected subgraph. The inequality $\kappa(K_n) \le \kappa'(K_n)$ holds by our convention that $\kappa(K_n) = n - 1$.

4.1.8. Theorem. $\kappa(G) \le \kappa'(G) \le \delta(G)$.

Proof. The edges incident to a vertex v of minimum degree form a disconnecting set; hence $\kappa'(G) \le \delta(G)$. It remains only to show $\kappa(G) \le \kappa'(G)$. Because we define $\kappa(K_n) = n - 1$, we have $\kappa(G) \le n(G) - 1$ for all G. Suppose $n(G) > 1$ and $[S, \bar{S}]$ is a minimum edge cut, having size $k = \kappa'(G)$. If every vertex of S is adjacent to every vertex of $\bar{S}$, then $k = |S||\bar{S}| \ge$

$n(G) - 1$, and the inequality holds.

Hence we may assume there exist $x \in S$, $y \in \bar{S}$ with $x \not\leftrightarrow y$. Now, let T be the vertex set consisting of all neighbors of x in $\bar{S}$ and all vertices of $S - \{x\}$ that have neighbors in $\bar{S}$ (illustrated below). Because x and y belong to different components of $G - T$, T is a separating set. Since T contains the endpoints in $\bar{S}$ of distinct edges involving x and the endpoints in S of distinct edges not involving x, we have $|T| \leq |[S, \bar{S}]| = k$. Hence $\kappa(G) \leq \kappa'(G)$. $\square$

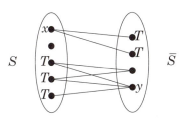

Because connectivity equals minimum degree for cliques, complete bipartite graphs, hypercubes, and Harary graphs, the inequality of Theorem 4.2.8 implies also that edge-connectivity equals minimum degree for these graphs. Although the set of edges incident to a vertex of minimum degree is always an edge cut, it need not be a minimum edge cut. The situation $\kappa'(G) < \delta(G)$ is precisely the situation where no minimum edge cut disconnects a single vertex from the rest of the graph.

4.1.9. Example. *Possibility of $\kappa < \kappa' < \delta$*. For graph G below, $\kappa(G) = 1$, $\kappa'(G) = 2$, and $\delta(G) = 3$. No minimum edge cut isolates a vertex. $\square$

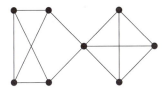

When $\kappa' < \delta$, a minimum edge cut $[S, \bar{S}]$ does not isolate a vertex, and in fact the set S must be much larger than a single element. This follows from a simple relationship between the size of an edge cut and the size of the graph induced by one side of the cut.

4.1.10. Proposition. If S is a nonempty proper subset of the vertices of a graph G, then $|[S, \bar{S}]| = [\sum_{v \in S} d(v)] - 2e(G[S])$.

Proof. Each edge in $G[S]$ contributes twice to $\sum_{v \in S} d(v)$, and each edge in $[S, \bar{S}]$ contributes once. This counts all the contributions, so $|[S, \bar{S}]| + 2e(G[S]) = \sum_{v \in S} d(v)$. $\square$

4.1.11. Corollary. If S is a nonempty proper subset of the vertices of a graph G such that $\|[S, \bar{S}]\| < \delta(G)$, then $|S| > \delta(G)$.

Proof. If $|S| = 1$, then $\|[S, \bar{S}]\| \geq \delta(G)$, so we may assume $|S| > 1$. By Proposition 4.1.10, we have $\delta(G) > \Sigma_{v \in S} d(v) - 2e(G[S])$. Since $d(v) \geq \delta(G)$ and $2e(G[S]) \leq |S|(|S| - 1)$, this becomes $\delta(G) > |S|\delta(G) - |S|(|S| - 1)$. Since $|S| > 1$, we can combine the terms involving $\delta(G)$ and cancel $|S| - 1$ to obtain $|S| > \delta(G)$. $\square$

An edge cut is a set of edges. An edge cut may contain another edge cut as a subset. For example, $K_{1,2}$ has three edge cuts, but one of them contains the other two. The minimal non-empty edge cuts of a graph have useful structural properties and a special name:

4.1.12. Definition. A *bond* is a minimal non-empty edge cut; a minimal set of edges whose deletion increases the number of components.

Here "minimal" means that no proper nonempty subset of these edges is also an edge cut. Bonds in connected graphs are characterized by a condition on the sets in the vertex partition defining the edge cut.

4.1.13. Proposition. If G is a connected graph and S is a nonempty proper subset of $V(G)$, then the edge cut $F = [S, \bar{S}]$ is a bond if and only if the induced subgraphs $G[S]$ and $G[\bar{S}]$ are connected.

Proof. If $G[S]$ and $G[\bar{S}]$ are connected, then F is a bond, because deleting $F' \subset F$ leaves the two components of $G - F$ plus at least one edge between them, making $G - F'$ connected. For the converse, suppose (by symmetry) that $G[S]$ has more than one component, so S has a partition into A, B with no edges between A and B. Now the edge cuts $[A, \bar{A}]$ and $[B, \bar{B}]$ are contained in F, so F is not a bond. $\square$

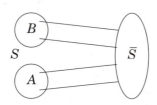

In Chapter 1, we introduced the terms "cut-vertex" and "cut-edge" for vertex cuts and edge cuts of size 1 whose deletion increases the number of components. Some authors have used *articulation point* to mean cut-vertex, and some authors have used *isthmus* or *bridge* to mean cut-edge. A disconnected graph may have a cut-vertex or a cut-edge. Every graph with connectivity 1 other than K_2 has a cut-vertex; these are sometimes called *separable graphs*.

BLOCKS

A connected graph with no cut-vertex need not be 2-connected, since it can be K_1 or K_2. The notion of a connected graph with no cut-vertex provides a useful decomposition of graphs.

4.1.14. Definition. A *block* of a graph G is a maximal connected subgraph of G that has no cut-vertex. If G itself is connected and has no cut-vertex, then G is a block.

4.1.15. Example. *Blocks.* If H is a block of G, then H as a graph has no cut-vertex, but H may contain vertices that are cut-vertices of G. For example, the graph drawn below has five blocks; three copies of K_2, one of K_3, and one subgraph that is neither a cycle nor a clique.

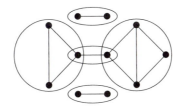

4.1.16. Remark. *Properties of blocks.* An edge of a cycle cannot itself be a block, because it belongs to a larger subgraph having no cut-vertex. Hence an edge is a block of G if and only if it is a cut-edge of G; the blocks of a tree are its edges. If a block has more than two vertices, then it is 2-connected. The blocks of a graph are its isolated vertices, its cut-edges, and its maximal 2-connected subgraphs. □

4.1.17. Proposition. Two blocks in a graph share at most one vertex.

Proof. A single vertex deletion cannot disconnect either block. If blocks B_1, B_2 share two vertices, then after deleting any single vertex x there remains a path within B_i from every vertex remaining in B_i to each vertex of $(B_1 \cap B_2) - x$. Hence $B_1 \cup B_2$ is a subgraph with no cut-vertex, which contradicts the maximality of the original blocks. □

Hence the blocks of a graph partition its edge set. When two blocks of G share a vertex, it must be a cut-vertex of G. The interaction between blocks and cut-vertices is described by a special graph.

4.1.18. Definition. The *block-cutpoint graph* of a graph G is a bipartite graph H in which one partite set consists of the cut-vertices of G, and the other has a vertex b_i for each block B_i of G. We include vb_i as an edge of H if and only if $v \in B_i$.

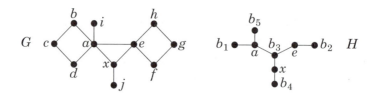

If G is connected, then H is a tree (Exercise 28) whose leaves are blocks of G. Hence a graph G with connectivity 1 has at least two blocks (called *leaf blocks*) that contain exactly one cut-vertex of G. Blocks can be found using depth-first search. By Lemma 2.3.10, every edge outside a tree found by DFS joins two vertices such that one is an ancestor of the other.

4.1.19. Algorithm. *Computing the blocks of G.*
Input: A connected graph G. (The blocks of a graph are the blocks of its components, which can be found by depth-first search, so we may assume G is connected.)
Idea: Build a depth-first search tree T of G, discarding portions of T as blocks are identified. Maintain one vertex called ACTIVE.
Initialization: Pick a root $x \in V(H)$; make x ACTIVE; set $T = \{x\}$.
Iteration: Let v denote the current active vertex.
1) If v has an unexplored incident edge vw, then
      1A) If $w \notin V(T)$; add vw to T, mark vw explored, make w ACTIVE.
      1B) If $w \in V(T)$, then w is an ancestor of v; mark vw explored.
2) If v has no more unexplored incident edges, then
      2A) If $v \neq x$, and w is the parent of v, make w ACTIVE. If no vertex in the current subtree T' rooted at v has an explored edge to an ancestor above w, then $V(T') \cup \{w\}$ is the vertex set of a block; record this information and delete $V(T')$ from T.
      2B) If $v = x$, terminate. □

4.1.20. Example. *Finding blocks.* For the graph below, one depth-first traversal from x visits the other vertices in the order $a, b, c, d, e, f, g,$ h, i, j. We find blocks in the order $\{a, b, c, d\}$, $\{e, f, g, h\}$, $\{a, i\}$, $\{x, a, e\}$, $\{x, j\}$. After finding each block, we delete the vertices other than the highest. Exercise 30 requests a proof of correctness. □

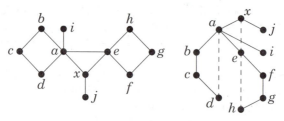

Recall that a digraph is *strongly connected* (or *strong*) if it has a path from each vertex to every other vertex (following the arrows).

4.1.21. Definition. The *strong components* of a digraph D are the maximal strongly connected subgraphs of D.

A digraph D is strongly connected if and only if it has only one strong component. The strong components of a digraph behave very much like the blocks of a graph, as discussed in Exercises 31-33.

EXERCISES

4.1.1. (–) Determine $\kappa(G)$, $\kappa'(G)$, and $\delta(G)$ for each graph G drawn below. For each k, which graphs are k-connected? Which are k-edge-connected?

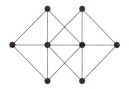 

4.1.2. (–) Prove that a graph G is k-connected if and only if $G \vee K_r$ is $k+r$-connected.

4.1.3. (–) Given a simple connected graph G, obtain G' by adding an edge joining every pair of vertices whose distance in G is 2. Prove that G' is 2-connected.

4.1.4. (–) Give a counterexample to the following statement, add a hypothesis to correct it, and prove the corrected statement: If e is a cut-edge of G, then at least one vertex of e is a cut-vertex of G.

4.1.5. (–) Let k, l, m be integers with $0 \le k \le l \le m$. Construct a graph $G_{k,l,m}$ with $\kappa(G_{k,l,m}) = k$, $\kappa'(G_{k,l,m}) = l$, and $\delta(G_{k,l,m}) = m$. (Chartrand-Harary [1968])

4.1.6. Suppose n is even and k is odd. Let G be the k-regular simple graph formed by placing n vertices on a circle and making each vertex adjacent to the opposite vertex and to the $(k-1)/2$ nearest vertices in each direction. Prove that $\kappa(G) = k$. (Harary [1962a])

4.1.7. Consider $K_{m,n}$ with bipartition X, Y. Suppose S consists of a members of X and b members of Y.
    a) Compute $\|[S, \bar{S}]\|$ in terms of a, b, m, n.
    b) Use part (a) to prove numerically that $\kappa'(K_{m,n}) = \min\{m, n\}$.
    c) Prove that every set of seven edges in $K_{3,3}$ is a disconnecting set, but no set of seven edges is an edge cut.

4.1.8. (!) Prove that $\kappa'(G) = \kappa(G)$ if G is a 3-regular simple graph. Find (with proof) the smallest 3-regular simple graph having connectivity 1.

4.1.9. (!) Use Proposition 4.1.10 and Exercise 8 to prove that the Petersen graph is 3-connected.

4.1.10. Use Proposition 4.1.10 to prove that the Petersen graph has an edge cut of size m if and only if $3 \le m \le 12$. (Hint: consider $|[S, \bar{S}]|$ for $1 \le |S| \le 5$.

4.1.11. Suppose G is a 3-regular graph with at most 10 vertices. Use Corollary 4.1.11 to prove that if G is not 3-edge-connected, then G has a triangle. Show that this result is best possible by exhibiting a 3-regular bipartite graph with 12 vertices that is not 3-edge-connected.

4.1.12. Prove that $\kappa(G) = \delta(G)$ if G is simple and $\delta(G) \ge n(G) - 2$. Prove that this is best possible for each $n \ge 4$ by constructing an n-vertex graph with minimum degree $n - 3$ and connectivity less than $n - 3$.

4.1.13. (!) Suppose G is a simple n-vertex graph with $\delta(G) \ge n/2 - 1$. Prove that G is k-connected for all k with $k \le 2\delta(G) + 2 - n$. Prove that this is best possible for all $\delta \ge n/2 - 1$ by constructing an n-vertex graph with minimum degree δ that is not k-connected for $k = 2\delta + 3 - n$. (Comment: Proposition 1.3.5 is the special case of this when $\delta(G) = (n - 1)/2$.)

4.1.14. (+) Suppose G is a simple n-vertex graph with $n \ge k + l$ and $\delta(G) \ge \frac{n + l(k - 2)}{l + 1}$. Prove that if $G - S$ has more than l components, then $|S| \ge k$. Prove that the hypothesis on $\delta(G)$ is best possible whenever $n \ge k + l$, by constructing an appropriate n-vertex graph with minimum degree $\lfloor \frac{n + l(k - 2) - 1}{l + 1} \rfloor$. (Comment: this generalizes the preceding problem.)

4.1.15. (!) *Sufficient condition for $k + 1$-connected graphs.* (Bondy [1969])
a) Suppose G is a simple n-vertex graph with vertex degrees $d_1 \le \cdots \le d_n$. Prove that if $d_j \ge j + k$ whenever $j \le n - 1 - d_{n-k}$, then G is $k + 1$-connected. (Comment: Exercise 1.4.13 is the special case of this when $k = 0$.)
b) Suppose $0 \le j, k \le n$. Construct an n-vertex graph G such that $\kappa(G) \le k$ and G has j vertices of degree $j + k - 1$, has $n - j - k$ vertices of degree $n - j - 1$, and has k vertices of degree $n - 1$. Explain in what sense this shows that part (a) is best possible.

4.1.16. (!) Suppose that G is an r-connected graph of even order having no $K_{1,r+1}$ as an induced subgraph. Prove that G has a 1-factor. (Sumner [1974b])

4.1.17. (!) *Degree conditions for $\kappa' = \delta$.* Suppose that G is a simple n-vertex graph. Use Corollary 4.1.11 to prove the following statements.
a) If $\delta(G) \ge \lfloor n/2 \rfloor$, then $\kappa'(G) = \delta(G)$. Prove this best possible by constructing for each $n \ge 3$ a simple n-vertex graph with $\delta(G) = \lfloor n/2 \rfloor - 1$ and $\kappa'(G) < \delta(G)$.
b) If $d(x) + d(y) \ge n - 1$ whenever $x \not\leftrightarrow y$, then $\kappa'(G) = \delta(G)$. Prove this best possible by constructing for each $n \ge 4$ and $\delta(G) = m \le n/2 - 1$ an n-vertex graph G with $\kappa'(G) < \delta(G) = m$ in which $d(x) + d(y) \ge n - 2$ whenever $x \not\leftrightarrow y$.

4.1.18. (!) $\kappa'(G) = \delta(G)$ *for diameter 2.* Suppose that G is a simple graph with diameter 2 and that $[S, \bar{S}]$ is a minimum edge cut with $|S| \le |\bar{S}|$.
a) Prove that every vertex of S has a neighbor in $\bar{S}$.
b) Use (a) and Corollary 4.1.11 to prove that $\kappa'(G) = \delta(G)$. (Plesník [1975])

4.1.19. (!) Suppose $F \subseteq E(G)$. Prove that F is an edge cut if and only if F contains an even number of edges from every cycle in G. (Hint: for sufficiency, consider the components of $G - F$.)

4.1.20. (!) Suppose that $[S, \bar{S}]$ is an edge cut in an undirected graph G. Prove that there is a set of pairwise edge-disjoint bonds whose union (as edge sets) is $[S, \bar{S}]$. (This holds trivially if $[S, \bar{S}]$ is itself a bond.)

4.1.21. (!) Prove that the symmetric difference of two edge cuts is an edge cut. (Comment: By this and Exercise 20, the symmetric difference of two bonds is an edge-disjoint union of bonds. This property holds also for cycles. See Section 8.2 for a more general context.)

4.1.22. (!) Suppose H is a spanning subgraph of a connected graph G. Prove that H is a spanning tree of G if and only if the subgraph $H^* = G - E(H)$ is a maximal subgraph that contains no bond of G. (Comment: we already know that H is a spanning tree of G if and only if H is a maximal subgraph containing no cycle. See Section 8.2 for a more general context.)

4.1.23. (−) Let G be the graph with vertex set $\{1, \ldots, 11\}$ in which $i \leftrightarrow j$ if and only if i and j have a common factor bigger than 1. Determine the blocks of G.

4.1.24. (−) Give a formula for the number of spanning trees of a graph in terms of the number of spanning trees of each of its blocks.

4.1.25. A *cactus* is a connected graph in which every block is an edge or a cycle. Prove that the maximum number of edges in a simple n-vertex cactus is $\lfloor 3(n - 1)/2 \rfloor$. (Hint: $\lfloor x \rfloor + \lfloor y \rfloor \leq \lfloor x + y \rfloor$.)

4.1.26. Prove that every vertex of G has even degree if and only if every block of G is Eulerian.

4.1.27. Prove that a connected graph is k-edge-connected if and only if each of its blocks is k-edge-connected.

4.1.28. (!) *The block-cutpoint graph* (see Definition 4.1.18). Let H be the block-cutpoint graph of a graph G that has a cut-vertex. (Harary-Prins [1966])
    a) Prove that H is a forest.
    b) Prove that G has at least two blocks that contain one cut-vertex of G.
    c) Prove that G has exactly $k + \Sigma_{v \in V(G)}(b(v) - 1)$ blocks, where k is the number of components of G and $b(v)$ is the number of blocks containing v.
    d) Prove that every graph has fewer cut-vertices than blocks.

4.1.29. Suppose that H and H' are distinct maximal k-connected subgraphs of a graph G. Prove that H and H' have at most $k - 1$ vertices in common. (Harary-Kodama [1964])

4.1.30. Prove that Algorithm 4.1.19 correctly computes blocks of graphs.

4.1.31. (−) Let R be the relation on the vertex set of a digraph D defined by $(u, v) \in R$ if u is connected to v and v is connected to u in D. Prove that R is an equivalence relation. Prove that the equivalence classes of R are the vertex sets of the strong components of G.

4.1.32. *Strong components.*

a) Prove that two maximal strongly connected subgraphs of a directed graph cannot share any vertices.

b) Let $D_1, \ldots, D_k$ be the strong components of a digraph D. The *condensation* of D is the digraph D* with vertices $v_1, \ldots, v_k$ such that $v_i \rightarrow v_j$ if and only if D has an edge from D_i to D_j. Prove that the condensation of D has no cycle.

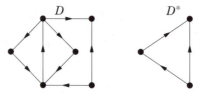

4.1.33. Develop an algorithm to compute the strong components of a digraph. Prove that it works. (Hint: Model the algorithm on Algorithm 4.1.19).

4.2 k-connected Graphs

A communication network is fault-tolerant if it has alternative paths between vertices; the more disjoint paths, the better. In this section, we prove that this alternative measure of connection is essentially the same as k-connectedness. In Section 1.2, we proved that a graph G is connected if and only if every partition of $V(G)$ into two non-empty sets yields an edge with an endpoint in each set. In other words, we proved that each pair of vertices is connected by a path if and only if G is 1-edge-connected. Here we generalize this characterization to k-edge-connected graphs and to k-connected graphs.

2-CONNECTED GRAPHS

We begin by characterizing 2-connected graphs. Two paths are *internally-disjoint* if neither contains a non-endpoint vertex of the other.

4.2.1. Theorem. (Whitney [1932]) An undirected graph G having at least three vertices is 2-connected if and only if each pair $u, v \in V(G)$ is connected by a pair of internally-disjoint u, v-paths in G.

Proof. When G has internally-disjoint u, v-paths, deletion of one vertex cannot separate u from v. Since this is given for every pair u, v, the condition is sufficient. For the converse, suppose that G is 2-connected. We prove by induction on $d(u, v)$ that G has two internally-disjoint u, v-paths. When $d(u, v) = 1$, the graph $G - uv$ is connected, since $\kappa'(G) \geq$

$\kappa(G) = 2$. A u, v-path in $G - uv$ is internally-disjoint in G from the u, v-path consisting of the edge uv itself.

For the induction step, we consider $d(u, v) = k > 1$ and assume that G has internally-disjoint x, y-paths whenever $1 \le d(x, y) < k$. Let w be the vertex before v on a shortest u, v-path. We have $d(u, w) = k - 1$, and hence by the induction hypothesis G has internally-disjoint u, w-paths P and Q. Since $G - w$ is connected, $G - w$ contains a u, v-path R. If this path avoids P or Q, we are finished, but R may share internal vertices with both P and Q. Let x be the last vertex of R belonging to $P \cup Q$. By symmetry, we may assume $x \in P$. We combine the u, x-subpath of P with the x, v-subpath of R to obtain a u, v-path internally-disjoint from $Q \cup wv$. $\square$

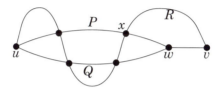

4.2.2. Lemma. (Expansion Lemma). If G is a k-connected graph, and G' is obtained from G by adding a new vertex y adjacent at least k vertices of G, then G' is k-connected.

Proof. Suppose S is a separating set of G'. If $y \in S$, then $S - \{y\}$ separates G, so $|S| \ge k + 1$. If $y \notin S$ and $N(y) \subseteq S$, then $|S| \ge k$. Otherwise, S must separate G, and again $|S| \ge k$. $\square$

4.2.3. Theorem. If $n(G) \ge 3$, then the following conditions are equivalent (and characterize 2-connected graphs).
   A) G is connected and has no cut-vertex.
   B) For all $x, y \in V(G)$, there are internally-disjoint x, y-paths.
   C) For all $x, y \in V(G)$, there is a cycle through x and y.
   D) $\delta(G) \ge 1$, and every pair of edges in G lie on a common cycle.

Proof. Theorem 4.2.1 is the equivalence of A and B. Cycles containing x and y correspond to pairs of internally-disjoint x, y-paths, so B $\Leftrightarrow$ C. For D $\Rightarrow$ C, apply D to edges incident to the desired x and y.

For A,C $\Rightarrow$ D, suppose G is 2-connected and $uv, xy \in E(G)$. Add to G the vertices w with neighborhood $\{u, v\}$ and z with neighborhood $\{x, y\}$. By the Expansion Lemma, the resulting graph G' is 2-connected, and hence w, z lie on a common cycle C in G'. Since w, z each have degree 2, this cycle must contain the paths u, w, v and x, z, y but not uv or xy. Replace the paths u, w, v and x, z, y in C by the edges uv and xy to obtain the desired cycle in G. $\square$

4.2.4. Definition. *Subdividing an edge uv* of an undirected graph G is the operation of deleting uv and adding a path u, w, v through a new vertex w.

4.2.5. Corollary. If G is 2-connected, then the graph G' obtained by subdividing an edge of G is 2-connected.

Proof. Suppose G' arises by adding w to subdivide uv. We prove that any two edges of G' lie on a cycle. For any pair not including uw or vw, we use the cycle guaranteed in G, unless it uses uv, in which case we modify it to pass through w between u and v. For a pair consisting of xy and one of $\{uw, wv\}$, we modify the cycle resulting from uv and xy in G; this also takes care of $\{uw, wv\}$. $\square$

Constructive characterizations or decomposition procedures can lead to algorithms for a class of graphs. The class of 2-connected graphs has a characterization that expresses the construction of such graphs from a cycle.

4.2.6. Definition. A *path addition* to G is the addition to G of a path of length $l \geq 1$ between two vertices of G, introducing $l - 1$ new vertices; the added path is an *ear*. An *ear decomposition* is a partition of $E(G)$ into sets $P_0, P_1, \ldots, P_k$ such that $C = P_0$ is a cycle, and P_i for $i \geq 1$ is a path addition to the graph formed by $P_0, \ldots, P_{i-1}$.

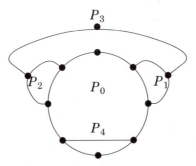

4.2.7. Theorem. (Whitney [1932]). A graph is 2-connected if and only if it has an ear decomposition. Furthermore, every cycle in a 2-connected graph is the initial cycle in some ear decomposition.

Proof. *Sufficiency.* Since cycles are 2-connected, it suffices to show that path addition preserves 2-connectedness. Let u, v be the endpoints of an ear P to be added to a 2-connected graph G. Adding an edge cannot reduce connectivity, so $G + uv$ is 2-connected. A succession of edge subdivisions converts $G + uv$ into $G \cup P$; by Corollary 4.2.5, each subdivision preserves 2-connectedness.

Necessity. Given a 2-connected graph G, we build an ear decomposition of G from a cycle C in G. Let $G_0 = C$. Suppose we have built up a subgraph G_i by adding ears. If $G_i \neq G$, then we can choose an edge uv of $G - E(G_i)$ and an edge $xy \in E(G_i)$. Because G is 2-connected, uv and xy lie on a common cycle C'. Let P be the path in C' that contains uv and exactly two vertices of G_i, one at each end of P. Now P is an ear that can be added to G_i to obtain a larger subgraph G_{i+1}. The process ends when all of G has been absorbed. $\square$

Every 2-connected graph is also 2-edge-connected, but the converse does not hold, so decomposition of 2-edge-connected graphs needs a more general operation. The proof is simpler.

4.2.8. Definition. A *closed-ear decomposition* of G is a partition of $E(G)$ into sets $P_0, P_1, \dots, P_k$ such that P_0 is a cycle and P_i for $i \geq 1$ is either an open ear (a path addition to $P_0 \cup \cdots \cup P_{i-1}$) or a closed ear (a cycle with exactly one vertex in $P_0 \cup \cdots \cup P_{i-1}$).

4.2.9. Theorem. A graph is 2-edge-connected if and only if it has a closed-ear decomposition, and every cycle in a 2-edge-connected graph is the initial cycle in some ear decomposition.

Proof. *Sufficiency.* Cut-edges are the edges not on cycles, so a connected graph is 2-edge-connected if and only if every edge lies on a cycle. Starting from a cycle, it suffices to show that adding an ear P to a 2-edge-connected G preserves 2-edge-connectedness. The edges in G already lie in cycles. Since G is connected, it contains a path between the endpoints of P (which may be the same point). The union of this path with P is a cycle in $G \cup P$ containing all the edges of P.

Necessity. Suppose G is 2-edge-connected and P_0 is a cycle in G. Suppose we have constructed a closed-ear decomposition $G_i = P_0 \cup \cdots \cup P_i$ of a subgraph of G. If $G_i \neq G$, then G has an edge $uv \notin G_i$ with $u \in V(G_i)$, since G is connected. Since G is 2-edge-connected, uv lies on a cycle C. Follow C until it returns to $V(G_i)$; use this as an open or closed ear P_{i+1} to enlarge the subgraph. The process ends only by absorbing all of G. $\square$

CONNECTIVITY OF DIGRAPHS

Our results about k-connected and k-edge-connected graphs will apply as well for digraphs, so we now introduce the corresponding concepts for digraphs.

4.2.10. Definition. A *separating set* or *vertex cut* of a digraph D is a set $S \subseteq V(D)$ such that $D - S$ is not strongly connected. A digraph is *k-connected* if every vertex cut has at least k vertices. The minimum size of a vertex cut is the *connectivity* $\kappa(D)$.

For $S, T \subseteq V(D)$, let $[S, T]$ be the set of edges from S to T. An *edge cut* is the set $[S, \bar{S}]$ for some $\varnothing \neq S \subset V(D)$. A digraph is *k-edge-connected* if every edge cut has at least k edges. The minimum size of an edge cut is the *edge-connectivity* $\kappa'(D)$.

4.2.11. Remark. Because $\|[S, \bar{S}]\|$ is the number of edges leaving S, we can restate the definition of edge-connectivity as follows: A graph or digraph G is k-edge-connected if and only if for every nonempty proper vertex subset S, there are at least k edges in G leaving S. □

Strong digraphs are similar to 2-edge-connected graphs.

4.2.12. Lemma. Adding a (directed) ear to a strong digraph produces a larger strong digraph.

Proof. As remarked above, a digraph is strong if and only if it has an edge leaving every non-empty vertex subset. If we add an open ear or closed ear P to a strong digraph D, then for every set S such that $\varnothing \subset S \subset V(D)$ we already have an edge leaving S and going to $V(D) - S$. We need only consider sets that don't intersect $V(D)$ and sets that contain all of $V(D)$ but not all of $V(P)$. For every such set, there is an edge leaving it along P. □

4.2.13. Example. *The One-Way Street Problem.* When can the streets in a road network all be made one-way without making any location unreachable from some other location? This is the question of when a graph has a strong orientation. Robbins [1939] proved that G has a strong orientation if and only if it is 2-edge-connected.

If G is disconnected, some vertices cannot reach others in any orientation. If G has a cut-edge, oriented from x to y in an orientation of G, then y cannot reach x in this orientation. Hence the condition is necessary. For sufficiency, we use a closed-ear decomposition of G. We orient the initial cycle consistently to obtain a strong digraph. As we add each new ear and direct it consistently, Lemma 4.2.12 guarantees that we still have a strong digraph. □

k-CONNECTED AND k-EDGE-CONNECTED GRAPHS

We have introduced two measures of well connected graphs: invulnerability to deletions and multiplicity of alternative communication paths. Extending Whitney's Theorem, we show that the two notions of *k*-connected graphs are the same (similarly for *k*-edge-connected). We first discuss the "local" situation considering x, y-paths for a fixed pair $x, y \in V(G)$. These definitions hold both for graphs and for digraphs.

4.2.14. Definition. Given $x, y \in V(G)$, a set $S \subseteq V(G) - \{x, y\}$ is an x, y-*separating set* if $G - S$ has no x, y-path. If $xy \notin E(G)$, the *local connectivity* $\kappa(x, y)$ is the minimum size of an x, y-separating set. The *local path-multiplicity* $\lambda(x, y)$ is the maximum number of pairwise internally-disjoint x, y-paths. The *path-multiplicity* $\lambda(G)$ is the maximum k such that G has k pairwise internally-disjoint x, y-paths for all $x, y \in V(G)$.

Separating G makes some vertex unreachable from another, so $\kappa(G) = \min\{\kappa(x, y): x, y \notin E(G)\}$. Similarly, $\lambda(G) = \min_{x,y \in V(G)} \lambda(x, y)$. Menger's original theorem stated that $\kappa(x, y) = \lambda(x, y)$; the global equality $\kappa(G) = \lambda(G)$ and analogous results for edge-connectivity were observed by others, but all are considered forms of Menger's Theorem. At least 15 proofs of Menger's original theorem have been published, many of which prove stronger results.[†]

We start with the local edge version. The argument is a distillation of that used for Ford-Fulkerson Theorem in Section 4.3. Again the definitions hold both for graphs and for digraphs.

4.2.15. Definition. Given $x, y \in V(G)$, a set $F \subseteq E(G)$ is an x, y-*cut* if $G - F$ has no x, y-path. The *local edge-connectivity* $\kappa'(x, y)$ is the minimum size of an x, y-cut. The *local edge-path-multiplicity* $\lambda'(x, y)$ is the maximum number of pairwise edge-disjoint x, y-paths. The *edge-path-multiplicity* $\lambda'(G)$ is the maximum k such that G has k pairwise edge-disjoint x, y-paths for all $x, y \in V(G)$.

4.2.16. Theorem. (Edge-Local Menger Theorem - Elias-Feinstein-Shannon [1956], Ford-Fulkerson [1956]) If x and y are distinct vertices in a graph or digraph G, then $\lambda'(x, y) = \kappa'(x, y)$.

Proof. Suppose $\lambda'(x, y) = k$, and let $\{P_1, \ldots, P_k\}$ be a set of k pairwise edge-disjoint x, y-paths. Every x, y-cut has an edge from each P_i, so $\kappa'(x, y) \geq \lambda'(x, y)$. To prove that equality holds, we define a set S containing x but not y, and we prove that only k edges leave S. Deletion of these edges makes y unreachable from S and proves $\kappa'(x, y) = k$.

Let $\mathbf{P} = \cup_{i=1}^{k} E(P_i)$, and let $G' = G - \mathbf{P}$; the dashed edges in each illustration are $E(G')$. Let S be the set of vertices that can be reached from x by a *legal sequence* of steps, where each legal step moves along an edge of G' or backwards along an edge of some P_i (to the preceding vertex of P_i). No edge uv of G' leaves S, because a legal sequence to u

---

[†]Menger's original argument had a gap that was later repaired by König.

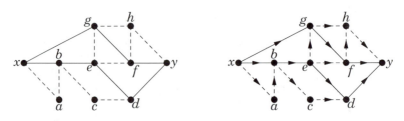

would extend to v and put v into S. No path P_i leaves S and later returns to S along an edge uv, because a legal sequence to v would extend to u by stepping back along P_i from v to u, putting u into S. Hence at most k edges leave S in G; up to one on each P_i.

To show that $y \notin S$, suppose otherwise. Let Q be the set of edges in a legal sequence reaching y (in the illustration, $k = 2$ and $x, a, b, c, d,$ e, f, g, h, y is a legal sequence reaching y). We claim that $\mathbf{P}' = \mathbf{P} \triangle Q$ contains $k + 1$ pairwise edge-disjoint paths in G. To prove this, we process the legal sequence one step at a time, maintaining $k + 1$ pairwise edge-disjoint paths in G, k of them from x to y and one an unfinished path originating at x. Suppose the claim is true through the current step of Q, and let the next step be from u to v.

Case 1: $vu \notin \mathbf{P}$. Extend the unfinished path.

Case 2: $vu \in \mathbf{P}$ and vu belongs to a current x, y-path, P. Replace P with the path P' consisting of the current unfinished path from x to v followed by the v, y-portion of P. The x, u-portion of P becomes the current unfinished path.

Case 3: $vu \in \mathbf{P}$ but vu is not on a current x, y-path. In this case, the step moves back along the current unfinished path. Discard the intervening v, v-portion of the unfinished path, retaining the initial x, v-portion as the current unfinished path.

When the last step of Q reaches y, it completes the unfinished path into an additional x, y-path, which contradicts $\lambda'(x, y) = k$. $\square$

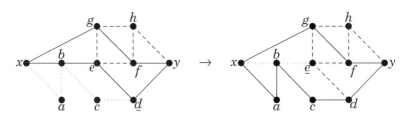

This argument works equally well for graphs and for digraphs, as long as direction along P_i is reversed when following Q. The behavior of vertex b in the illustration shows that this argument does not prove $\kappa(x, y) = \lambda(x, y)$. Nevertheless, we can obtain the remaining local equalities from the edge-local Menger Theorem for digraphs.

4.2.17. Theorem. (Local Menger Theorem for digraphs) In a digraph D with $x \not\to y$, the maximum number $\lambda(x, y)$ of pairwise internally-disjoint x, y-paths equals the minimum number $\kappa(x, y)$ of vertices whose deletion breaks all x, y-paths.

Proof. Every x, y-separating set has a vertex from each path in a minimum set of pairwise internally-disjoint paths, so $\kappa(x, y) \geq \lambda(x, y)$. To prove that equality holds, we transform the digraph D in such a way that internally disjoint x, y-paths in D become edge-disjoint x, y-paths in a new digraph D', and then we apply Theorem 4.2.16 to D'. Form D' from D by splitting each $v \in V(D) - \{x, y\}$ into two vertices v^+ and v^- with a single internal edge $v^- v^+$. Let R be the set of internal edges. For each edge $uv \in E(D)$, include $n(D)$ copies of the edge $u^+ v^-$ in D'.

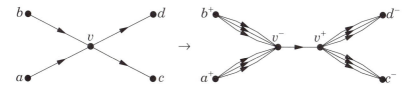

Since x, y-paths alternate between internal edges and copies of original edges, edge-disjoint x, y-paths in D' cannot use two copies of the same original edge. Furthermore, since the internal edge from each vertex of D can be used only once, edge-disjoint x, y-paths in D' become internally disjoint x, y-paths in D when each internal edge $v^- v^+$ is shrunk back into the original vertex v. Applying this to a maximum set of pairwise edge-disjoint x, y-paths in D' yields $\lambda_D(x, y) \geq \lambda'_{D'}(x, y)$.

The extra copies of original edges of D ensure that minimum x, y-cuts in D' use only internal edges. A minimum x, y-cut F that uses one copy of an edge uv uses all copies of uv; otherwise replacing that copy yields an x, y-path that can be rerouted using a copy of uv not in F to yield an x, y-path in $D' - F$. This implies that every minimal x, y-cut containing an edge outside R has at least $n(D)$ edges. The entire set R, on the other hand, has only $n(D) - 2$ edges and cuts all x, y-paths, so R contains every minimum x, y-cut. Let F be a minimum x, y-cut in D'. Since every x, y-path in D extends into an x, y-path in D', the set $S = \{v : v^- v^+ \in F\}$ is an x, y-separating set of size $\kappa'_{D'}(x, y)$ in D. Hence $\kappa_D(x, y) \leq \kappa'_{D'}(x, y)$.

We have proved that $\lambda'_{D'}(x, y) \leq \lambda_D(x, y) \leq \kappa_D(x, y) \leq \kappa'_{D'}(x, y)$. By Theorem 4.2.16, the two outer quantities are equal, and hence also the two inner quantities are equal. □

We have observed that the proof of Theorem 4.2.16 holds both for graphs and for digraphs. Only one local version of Menger's Theorem remains, for local connectivity in graphs, and it follows from Theorem 4.2.17 by a simple additional transformation.

4.2.18. Corollary. (Local Menger Theorem - Menger [1927]) If $x, y \in V(G)$ and $x \leftrightarrow y$, then $\lambda(x, y) = \kappa(x, y)$.

Proof. Let D be the digraph obtained from G by replacing each edge uv of G by a pair of edges uv and vu in D. Two vertex sequences form internally disjoint x, y-paths in G if and only if they form internally disjoint x, y-paths in D, and a set S is an x, y-separating set in G if and only if it is an x, y-separating set in D. Hence $\kappa_G(x, y) = \kappa_D(x, y)$ (if $xy \notin E(G)$) and $\lambda_G(x, y) = \lambda_D(x, y)$, and we apply Theorem 4.2.17. $\square$

The global version for k-connected graphs, observed first by Whitney [1932], is commonly known as Menger's Theorem. The global versions for edges and digraphs were observed by Ford and Fulkerson [1956].

4.2.19. Corollary. ("Menger's Theorem") The connectivity of G equals the maximum k such that $\lambda(x, y) \geq k$ for all $x, y \in V(G)$. The edge-connectivity of G equals the maximum k such that $\lambda'(x, y) \geq k$ for all $x, y \in V(G)$. Both statements hold for graphs and for digraphs.

Proof. Theorem 4.2.16 immediately yields $\kappa'(G) = \lambda'(G)$. For $\kappa(G)$, we have $\kappa(x, y) = \lambda(x, y)$ when $xy \notin E(G)$, and $\kappa(G)$ is the minimum of these values. We need only show that $\lambda(x, y)$ cannot be smaller than $\kappa(G)$ when $xy \in E(G)$. Using Menger's Theorem for $G - xy$ and the effect of edge-deletion on $\kappa(G)$, we have

$$\lambda_G(x, y) = 1 + \lambda_{G-xy}(x, y) = 1 + \kappa_{G-xy}(x, y) \geq 1 + \kappa(G - xy) \geq \kappa(G). \qquad \square$$

APPLICATIONS OF MENGER'S THEOREM

Dirac extended Menger's Theorem to other families of paths. An x, U-*fan* is a set of x, U-paths such that any two of them share only the vertex x. If $x \in U$, an x, U-fan may contain a path of length 0.

4.2.20. Theorem. (Fan Lemma, Dirac [1960]). A graph is k-connected if and only if it has at least $k + 1$ vertices and, for every choice of x, U with $|U| \geq k$, it has an x, U-fan of size k.

Proof. Suppose G is k-connected, and construct G' from G by adding a new vertex y adjacent to all of U. The Expansion Lemma (4.2.2) implies that G' also is k-connected, and then Menger's Theorem yields k pairwise internally-disjoint x, y-paths in G'. Deleting y from these paths produces an x, U-fan of size k in G.

Conversely, suppose G satisfies the fan condition. For each $v \in V(G)$, there is a v, U-fan of size k with $U = V(G) - v$, so $\delta(G) \geq k$. Given

$x, y \in V(G)$, let $U = N(y)$. We can extend the k paths of an x, U-fan by edges from U to y to obtain $\lambda(x, y) \geq k$. Hence G is k-connected. $\square$

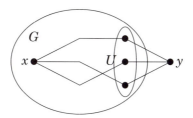

Suppose G is k-connected. Using the Fan Lemma, one can show that every k vertices in G lie on a cycle (Exercise 20). The Fan Lemma generalizes considerably; if X, Y are disjoint sets of vertices in G, then we can specify integer supplies on X and demands on Y (summing to k in each set) and obtain k pairwise internally-disjoint X, Y-paths with the specified multiplicities at endpoints (Exercise 23).

Applications of Menger's Theorem involve modeling a problem so that the desired objects correspond to paths in a graph or digraph, often by graph transformation arguments. For example, given a collection of sets $\mathbf{A} = A_1, \ldots, A_m$ with union X, a *system of distinct representatives* (SDR) is a set $\{x_1, \ldots, x_m\}$ of distinct elements such that $x_i \in A_i$. A necessary and sufficient condition for the existence of an SDR is that $|\cup_{i \in I} A_i| \geq |I|$ for all $I \subseteq [m]$. It is easy to prove this by modeling $\mathbf{A}$ with an appropriate bipartite graph and applying Hall's Theorem (Exercise 3.1.9). Hall's theorem was originally proved in the language of SDR's and is equivalent to Menger's Theorem (Exercise 18).

Ford and Fulkerson considered a more difficult problem. Let $\mathbf{A} = A_1, \ldots, A_m$ and $\mathbf{B} = B_1, \ldots, B_m$ be two systems of sets. We may ask when there exists a *common* system of distinct representatives (CSDR), meaning a set of m elements that is an SDR for $\mathbf{A}$ and also for $\mathbf{B}$. They found a necessary and sufficient condition.

4.2.21. Theorem. (Ford-Fulkerson [1958]) Families $\mathbf{A} = \{A_1, \ldots, A_m\}$ and $\mathbf{B} = \{B_1, \ldots, B_m\}$ have a common system of distinct representatives (CSDR) if and only if
$$|(\cup_{i \in I} A_i) \cap (\cup_{j \in J} B_j)| \geq |I| + |J| - m \quad \text{for each pair } I, J \subseteq [m].$$

Proof. We create a digraph G with distinguished vertices s and t and vertices a_i for $A_i \in \mathbf{A}$, b_j for $B_j \in \mathbf{B}$, and x for each element x in the sets. The edges are

$$\{sa_i: A_i \in \mathbf{A}\} \quad \{a_i x: x \in A_i\}$$
$$\{b_j t: B_j \in \mathbf{B}\} \quad \{x b_j: x \in B_j\}$$

Each s, t-path selects a member of the intersection of some A_i and some B_j. There is a CSDR if and only if there is a set of m pairwise internally-disjoint s, t-paths. By Menger's Theorem, it suffices to show that the stated condition is equivalent to having no s, t-cut of size less than m. Suppose $R \subseteq V(G) - \{s, t\}$, and let $I = \{a_i\} - R$ and $J = \{b_j\} - R$. We conclude that R is an s, t-cut if and only if $(\cup_{i \in I} A_i) \cap (\cup_{j \in J} B_j) \subseteq R$. Hence for an s, t-cut we have

$$|R| \geq |(\cup_{i \in I} A_i) \cap (\cup_{j \in J} B_j)| + (m - |I|) + (m - |J|).$$

This lower bound is at least m for every s, t-cut if and only if the stated condition holds. □

4.2.22. Example. *Digraph for CSDR.* In the example below, the elements are $\{1, 2, 3, 4\}$, $\mathbf{A} = \{12, 23, 31\}$, and $\mathbf{B} = \{14, 24, 1234\}$. Suppose $R \cap \{a_i\} = \{a_1, a_2\}$ and $R \cap \{b_j\} = \{b_1, b_2\}$. In the argument, we set $I = \{a_3\}$ and $J = \{b_3\}$, and we observe that R is an s, t-cut if and only if it also includes 1 and 3. □

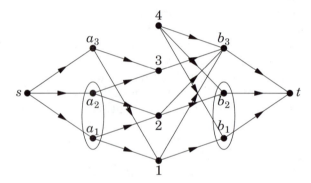

The notion of local edge-connectivity is intimately involved in generalizing Robbins' Theorem of Example 4.2.13. When G has a k-edge-connected orientation, Remark 4.2.11 implies that G must be $2k$-edge-connected. Nash-Williams [1960] proved that this obvious necessary condition is also sufficient. This is easy when G is Eulerian (Exercise 16), but the general case is quite difficult. Mader [1978] gave another proof of this using the following theorem (see Exercises 30-33). A thorough discussion of this and other orientation theorems appears in Frank [1993].

4.2.23. Theorem. (Mader [1978]) If z is a vertex of a graph G such that $d_G(z) \notin \{0, 1, 3\}$ and z is incident to no cut-edge, then z has neighbors x and y such that $\kappa_{G-xz-yz+xy}(u, v) = \kappa_G(u, v)$ for all $u, v \in V(G) - \{z\}$.

EXERCISES

4.2.1. (–) Determine $\kappa(u,v)$ and $\kappa'(u,v)$ in the graph drawn below.

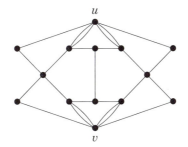

4.2.2. (–) Let G be the digraph with vertex set [12] in which $i \to j$ if and only if i divides j. Determine $\kappa(1,12)$ and $\kappa'(1,12)$.

4.2.3. (–) Prove or disprove: If P is a u,v-path in a 2-connected graph G, then there is a u,v-path Q that is internally-disjoint from P.

4.2.4. (–) Let G be a simple graph, and let $H(G)$ be the graph with vertex set $V(G)$ such that $uv \in E(H)$ if and only if u,v appear on a common cycle in G. Characterize the graphs G such that H is a clique.

4.2.5. Prove that a simple graph G is 2-connected if and only if for every ordered triple (x,y,z) of vertices, G has an x,z-path through y. (Chein [1968])

4.2.6. Prove that a graph G with at least four vertices is 2-connected if and only if for every pair X,Y of disjoint vertex subsets with $|X|,|Y| \geq 2$, there exist two completely disjoint paths P_1, P_2 in G such that each has an endpoint in X and an endpoint in Y and no internal vertex in X or Y.

4.2.7. Prove that a simple graph G is 2-connected if and only if G can be obtained from C_3 by a sequence of edge additions and edge subdivisions.

4.2.8. (+) Suppose v is a vertex of a 2-connected graph G. Prove that v has a neighbor u such that $G - u - v$ is connected. (Chartrand-Lesniak [1986, p51])

4.2.9. Suppose G is a 2-connected graph. Prove that if T_1, T_2 are two spanning trees of G, then T_1 can be transformed into T_2 by a sequence of operations in which a leaf is removed and reattached using another edge of G.

4.2.10. (!) *Membership in common cycles.*
    a) Prove that two distinct edges lie in the same block of a graph if and only if they belong to a common cycle.
    b) Suppose $e,f,g \in E(G)$, and suppose G has a cycle through e and f and a cycle through f and g. Prove that G also has a cycle through e and g. (Comment: this problem implies that "membership in a common cycle" is an equivalence relation whose equivalence classes define the blocks of G.)

4.2.11. (!) For a connected graph G with at least three vertices, prove that the following are equivalent.

A) G is 2-edge-connected.
B) Every edge of G appears in a cycle.
C) G has a closed trail containing any specified pair of edges.
D) G has a closed trail containing any specified pair of vertices.

4.2.12. (!) Suppose G is a 2-edge-connected graph. Define a relation R on $E(G)$ by $(e, f) \in R$ if $e = f$ or if $G - e - f$ is disconnected. (Lovász [1979, p277])
   a) Prove that $(e, f) \in R$ if and only if e, f belong to the same cycles.
   b) Prove that R is an equivalence relation on $E(G)$.
   c) For each equivalence class F, prove that F is contained in a cycle.
   d) For each equivalence class F, prove that $G - F$ has no cut-edge.

4.2.13. (!) A u, v-*necklace* is a sequence of cycles $C_1, \ldots, C_k$ such that $u \in C_1$, $v \in C_k$, consecutive cycles in the sequence share one vertex, and cycles that are not consecutive in the list share no vertices (see illustration). Use induction on $d(u, v)$ to prove that a graph G is 2-edge-connected if and only if for every pair $u, v \in V(G)$ there is a u, v-*necklace* in G. (Comment: guaranteeing a u, v-necklace is stronger than guaranteeing edge-disjoint u, v-paths.)

4.2.14. Determine the smallest graph with connectivity 3 having a pair of non-adjacent vertices linked by four pairwise internally-disjoint paths.

4.2.15. (!) Suppose G has no isolated vertices. Prove that if G has no even cycles, then every block of G is an edge or an odd cycle.

4.2.16. Suppose G is $2k$-edge-connected and has an Eulerian trail. Prove that G has a k-edge-connected orientation. (Nash-Williams [1960])

4.2.17. (!) Prove that every k-connected graph with diameter d has at least $k(d-1)+2$ vertices. Prove that this inequality is best possible for all k when $d \geq 2$.

4.2.18. (!) Use Menger's Theorem ($\kappa(x, y) = \lambda(x, y)$) to prove the König-Egerváry Theorem ($\alpha'(G) = \beta(G)$ when G is bipartite).

4.2.19. (!) Suppose that G is a k-connected graph and S, T are disjoint subsets of $V(G)$ with size at least k. Prove that G has k pairwise disjoint S, T-paths.

4.2.20. (!) Prove that if S is a set of k vertices in a k-connected graph G with $k \geq 2$, then G has a cycle containing S. (Dirac [1960]) (Hint: use induction and the Fan Lemma.)

4.2.21. For $k \geq 2$, prove that a graph with at least $k+1$ vertices is k-connected if and only if for every $T \subseteq S \subseteq V(G)$ with $|S| = k$ and $|T| = 2$, there is a cycle in G that contains T and avoids $S - T$. (Lick [1973])

4.2.22. A *vertex k-split* of a graph G is a graph H obtained from G by replacing one vertex $x \in V(G)$ by two adjacent vertices x_1, x_2 such that $d_H(x_i) \geq k$ and that $N_H(x_1) \cup N_H(x_2) = N_G(x) \cup \{x_1, x_2\}$.
   a) Prove that if G is k-connected graph and H is a vertex k-split of G, then H is k-connected.
   b) Conclude that any graph obtainable from a "wheel" $W_n = K_1 \vee C_{n-1}$ by a sequence of edge additions and vertex 3-splits on vertices of degree at least 4 is

3-connected. (Comment: Tutte proved also that every 3-connected graph can be constructed in this way, but this is much harder than the sufficiency proof in this problem. The characterization does not extend easily for $k > 3$.)

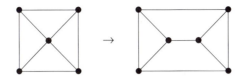

4.2.23. (!) Let X, Y be disjoint sets of vertices in G, and suppose $\kappa(x, y) \geq k$ for all $x \in X$, $y \in Y$. Suppose w assigns to each $v \in X \cup Y$ a nonnegative integer weight such that $\Sigma_{x \in X} w(x) = \Sigma_{y \in Y} w(y) = k$. Prove that there is a collection of k X, Y-paths intersecting pairwise only at endpoints such that the number of paths ending at each vertex equals the weight assigned to that vertex. (Hint: use part (a) of the preceding problem and Menger's Theorem).

4.2.24. Prove directly that $\kappa'(x, y) = \lambda'(x, y)$ for digraphs implies $\kappa'(x, y) = \lambda'(x, y)$ for graphs. (Hint: use a graph transformation argument.)

4.2.25. (!) Suppose that P and Q are paths of maximum length in a connected graph G. Prove that P and Q have a common vertex.

4.2.26. A graph may have more than one longest cycle; for example, $K_{2,4}$ has many 4-cycles, but no cycle of length more than 4. For $k = 2$ and $k = 3$, prove that two cycles of maximum length in a k-connected graph must share at least k vertices. (Hint: If they don't, construct a longer cycle.) For each $k \geq 2$, construct a k-connected graph such that distinct longest cycles do not intersect in more than k vertices.

4.2.27. *Graph splices.* Suppose G_1 and G_2 are disjoint k-connected graphs with $k \geq 2$. Choose $v_1 \in V(G_1)$ and $v_2 \in V(G_2)$. Let B be a bipartite graph with partite sets $N_{G_1}(v_1)$ and $N_{G_2}(v_2)$ that has no isolated vertex and has a matching of size at least k. Prove that $(G_1 - v_1) \cup (G_2 - v_2) \cup B$ is k-connected.

4.2.28. A 2-connected graph G is *critically 2-connected* if for every $xy \in E(G)$ the graph $G - xy$ is not 2-connected. Prove that if G is 2-connected, then $G - xy$ is 2-connected if and only if x and y lie on a cycle in $G - xy$. Conclude that a 2-connected graph is critically 2-connected if and only if no cycle has a chord. (Dirac [1967], Plummer [1968])

4.2.29. *More on critically 2-connected graphs.*
a) Prove that a minimally 2-connected graph contains a vertex of degree 2. (Hint: use the ear decomposition.) (Note: Halin [1969] proved that every minimally k-connected graph has a vertex of degree k.)
b) Prove that a critically 2-connected graph on $n \geq 4$ vertices has at most $2n - 4$ edges, with equality only for $K_{2,n-2}$. (Dirac [1967])

4.2.30. Given a graph G and nonempty set $S \subseteq V(G)$, let $d(S) = |[S, \bar{S}]|$. Suppose that X and Y are nonempty proper vertex subsets of G. Prove that $d(X \cap Y) + d(X \cup Y) \leq d(X) + d(Y)$.

4.2.31. (+) A k-edge-connected graph G is *critically k-edge-connected* if for every $xy \in E(G)$ the graph $G - xy$ is not 2-edge-connected. Prove that every critically k-edge-connected graph has a vertex of degree k. (Hint: consider a minimal vertex set X such that $\|[X]\| = k$. If $|X| \neq 1$, use $G - xx'$ for some $xx' \in E(G[X])$ to obtain another set Y such that $\|[Y]\| = k$, and such that X, Y contradict Exercise 30. Mader [1971], see also Lovász [1979, p285])

4.2.32. (−) Explain why Mader's Theorem (4.2.23) excludes vertices of degree 3.

4.2.33. (+) Use Mader's Theorem (4.2.23) and Exercise 31 to prove Nash-Williams' Orientation Theorem: every $2k$-edge-connected graph has a k-edge-connected orientation. (Comment: a weaker version of Mader's Theorem, presented in Lovász [1979, p286-288], also yields Nash-Williams' Theorem in the same way.)

4.3 Network Flow Problems

Consider a network of pipes having valves allowing flow in only one direction. Each pipe has a capacity per unit time. We can model this with a vertex for each junction and a (directed) edge for each pipe, weighted by the capacity. We also assume that flow cannot accumulate at an intermediate vertex. Given two locations s, t in the network, we may ask "what is the maximum flow (per unit time) from s to t?"

This question arises with any kind of transmission. The network can represent roads with traffic capacities, or links in a computer network with data transmission capacities, or currents in an electrical network. The problem has applications in industrial settings and to combinatorial min-max theorems. The seminal book on the subject is Ford-Fulkerson [1962]. More recently, Ahuja-Magnanti-Orlin [1993] presents a thorough treatment of many aspects of network flow problems.

MAXIMUM NETWORK FLOW

4.3.1. Definition. A *network* is a digraph with a nonnegative *capacity* $c(e)$ on each edge e and a distinguished *source vertex* s and *sink vertex* t. Vertices are also called *nodes*. A *flow* f assigns a value $f(e)$ to each edge e. We write $f^+(v)$ or $f^+(S)$ for the total flow on edges exiting a node v or a set of nodes S; the total flow on entering edges is $f^-(v)$ or $f^-(S)$. A *feasible flow* satisfies the *capacity constraints* $0 \le f(e) \le c(e)$ for each edge and the *conservation constraints* $f^+(v) = f^-(v)$ for each node $v \notin \{s, t\}$. The *value val(f)* of a flow f is the net flow $f^-(t) - f^+(t)$ into the sink. A *maximum flow* is a feasible flow of maximum value.

4.3.2. Example. *A feasible flow.* The *zero flow* asssigns flow 0 to each edge; this is feasible. In the network below with each edge having capacity 1, suppose $f(sx) = f(vt) = 0$ and $f(e) = 1$ for every other edge e. (We write flow values in parentheses.) This is a feasible flow of value 1.

The "greedy" approach seeks a maximum flow by augmenting the flow along source-sink paths with excess capacity. In this example, no path remains with excess capacity, but the flow f' with $f'(vx) = 0$ and $f'(e) = 1$ for $e \neq vx$ has value 2. The flow f is "maximal" in that no other feasible flow can be found by increasing the flow on some edges, but f is not a maximum flow. $\square$

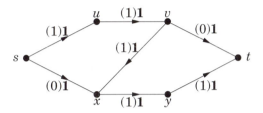

A more general notion of augmenting path allows us to find maximum flows. Suppose P takes us from s to t if we ignore directions on the edges. Following P may traverse some edges in the forward direction (with the arrows) and some edges in the backward direction (against the arrows). Let f be a feasible flow. If every forward edge on P has excess capacity (meaning $f(e) < c(e)$) and every backward edge on P has nonzero flow (meaning $f(e) > 0$), then we call P an f-*augmenting path* (even though it may not actually be a path), because we can change the values of f along P to obtain another feasible flow with higher value. In Example 4.3.2, s, x, v, t is an f-augmenting path.

Let the *leeway*[†] of P be the minimum of the excess capacity $c(e) - f(e)$ on all forward edges of P and the nonzero flow $f(e)$ on all backward edges of P. By the definition of f-augmenting path, the leeway of such a path is strictly positive.

4.3.3. Lemma. If P is an f-augmenting path with leeway ε, then increasing flow by ε along forward edges of P and decreasing flow by ε along backward edges of P produces a feasible flow f' with $val(f') = val(f) + \varepsilon$.

Proof. By the definition of leeway, we have $0 \leq f'(e) \leq c(e)$ for every edge e, so the capacity constraints hold. For the conservation constraints we need only consider vertices of P; flow elsewhere has not changed. The two edges forming a visit by P to an internal vertex v can

---

[†]"Leeway" is not a standard term. We use it here because the difference between the two sides of a *single* inequality constraint is called a "slack", but here we are talking about the minimum slack over a set of constraints.

be arranged in one of the four ways shown below. In each case, the change to the flow out of v is the same as the change to the flow into v, so the net flow out of v remains 0. Finally, the net flow into t is ε larger in f' than in f. $\square$

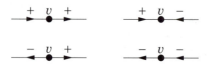

The flow on the backward edges did not disappear; it was redirected. The augmentation in Example 4.3.2 in effect cuts the flow path and extends each half to become a flow path. We will describe an algorithm to find augmenting paths. First, how do we determine whether what we have found is a maximum flow? We want to find something in the network that PROVES that the flow value can be no larger.

4.3.4. Definition. A *source/sink cut* partitions the nodes of a network into a *source set* S containing s and a *sink set* T containing t. As in Definition 4.1.6, the elements of the cut $[S,T]$ are the edges from S to T; every s,t-path uses at least one edge of $[S,T]$. The *capacity* of the cut $[S,T]$, written $cap(S,T)$, is the total of the capacities on the edges of $[S,T]$.

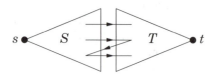

Given a cut $[S,T]$, intuition suggests that any flow from s to t must cross from S to T, so the capacity of edges from S to T should limit the value of a feasible flow. This rests on the intuition that net flow out of the source must reach the sink. In particular, it crosses every source/sink cut.

4.3.5. Lemma. If U is a set of nodes in a network, then the net flow out of U is the sum of the net flow out of the nodes in U. In particular, if f is a feasible flow and S,T is a source/sink cut, then the net flow out of S and net flow into T equal $val(f)$.

Proof. We consider the contribution of the flow $f(xy)$ on an edge xy to $\alpha = f^+(U) - f^-(U)$ and to $\beta = \Sigma_{v \in U} f^+(v) - f^-(v)$. If $x, y \in U$, then $f(xy)$ contributes 0 to α, and it contributes positively $(f^+(x))$ and negatively $(f^-(y))$ to β. If $x, y \notin U$, it contributes to neither sum. If $xy \in [U, \overline{U}]$, it contributes positively to each sum. If $xy \in [\overline{U}, U]$, it contributes negatively to each sum. Summing over all edges, we obtain $\alpha = \beta$.

If S, T is a source/sink cut and f is a feasible flow, then summing the net flow out of nodes of S yields $f^+(s) - f^-(s)$, and summing the net flow out of nodes of T yields $f^+(t) - f^-(t)$, which equals $-val(f)$. Hence the net flow across any source/sink cut equals both the net flow out of s and the net flow into t. $\square$

4.3.6. Corollary. (Weak duality) If f is a feasible flow and $[S, T]$ is a source/sink cut, then $val(f) \le cap(S, T)$.

Proof. By the lemma, the value of f equals the net flow out of S. Thus $val(f) = f^+(S) - f^-(S) \le f^+(S)$, since the flow into S is no less than 0. Since the capacity constraints require $f^+(S) \le cap(S, T)$, we obtain $val(f) \le cap(S, T)$. $\square$

The source/sink cut with the minimum capacity yields the best bound on the maximum value of a flow. This defines the *minimum cut* optimization problem. The max flow and min cut problems on a network are dual optimization problems.[†] Weak duality provides an easy way to PROVE that a solution is optimal. If we have a flow with value α, and we present a cut with value α, then by the weak duality inequality the cut is a minimum cut and the flow is a maximum flow.

If solutions to the max problem and the min problem having the same value always exist ("strong duality"), then a short proof of optimality always exists. This does not occur for all problems with weak duality (recall matching and covering in general graphs), but it does for the max flow and min cut problems. The Ford-Fulkerson algorithm seeks an augmenting path to increase the flow value. If it does not find such a path, then it finds a cut with the same value as this flow; by the weak duality inequality, both are optimal. If no infinite sequence of augmentations is possible, then the iteration leads to equality between the maximum flow value and minimum cut capacity.

4.3.7. Algorithm. *Ford-Fulkerson labeling algorithm.*
Input: A feasible flow f in a network.
Idea: Return an f-augmenting path or a cut with capacity $val(f)$. Seek an augmenting path by growing a set R of nodes reachable from s by paths having positive leeway. Reaching t completes an f-augmenting path. During the search, R is the set of nodes we have Reached, and S is the subset of these from which we have Searched to extend R.
Initialization: $R = \{s\}$, $S = \varnothing$.

[†]The precise notion of "dual problem" comes from linear programming. The reader may view a dual pair of optimization problems as a max problem and a min problem such that $a \le b$ whenever a and b are the values of feasible solutions to the max problem and min problem, respectively. See Section 8.1 for further discussion.

Iteration. Choose $v \in R - S$. Consider each edge vw leaving v and each edge uv entering v. If vw has excess capacity $(f(vw) < c(vw))$, include w in R. If uv has nonzero flow $(f(uv) > 0)$, include u in R. Record v as the vertex from which the new vertex of R was reached. After exploring all edges involving v, add v to S.

If the sink t has been put in R, trace back the path of inclusions leading to $t \in R$ to return an f-augmenting path. If $R = S$, terminate and return the cut $[S, \bar{S}]$. Otherwise, iterate. $\square$

4.3.8. Example. *Ford-Fulkerson labeling algorithm.* Consider the network below with capacities in bold. Suppose that $f(sx) = f(vt) = 0$ and $f(e) = 1$ for the other edges. When we run the labeling algorithm (so called because we "label" the vertices of R as "reached"), we first search from s and find excess capacity to u and x, labeling them. Then there is no excess capacity on uv or xy, but there is nonzero flow on vx. This enables us to label v from x. Now v is the only element of $R - S$, and we search from v to label t. We labeled t from v, v from x, and x from s, so we have found the augmenting path s, x, v, t.

After the augmentation (leeway $\varepsilon = 1$), every edge has unit flow except $f'(vx) = 0$. If we run the labeling algorithm again, we have excess capacity on $\{su, sx\}$ and can label $\{u, x\}$, but from these nodes we can label no others. We terminate with $R = S = \{s, u, x\}$. The capacity of the resulting cut $[S, \bar{S}]$ is 2, which equals $val(f')$ and proves that f' is a maximum flow. $\square$

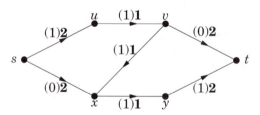

4.3.9. Theorem. (Max-flow Min-cut Theorem, Ford-Fulkerson [1956]). In every network, the maximum value of a feasible flow equals the minimum capacity of a source/sink cut.

Proof. In the max-flow problem, $f(e) = 0$ for all e is a feasible flow. Given a feasible flow, we apply the labeling algorithm. The labeling algorithm iteratively adds vertices to S (at most $n(D)$ times) and terminates with $t \in R$ ("breakthrough") or $S = R$.

In the breakthrough case, we have an augmenting path and increase the flow value. Then we repeat the labeling algorithm. When the capacities are rational, each augmentation increases the flow by a multiple of $1/a$, where a is the least common multiple of the denominators, so after finitely many augmentations the capacity of some cut is reached and we must terminate with $S = R$.

In this case, we claim that $[S, T]$ is a cut with capacity $val(f)$, where $T = \bar{S}$ and f is the present flow. First, $[S, T]$ is a source/sink cut, because $s \in S$ and $t \notin R = S$. Because no node of T entered R in the labeling algorithm, no edge from S to T has excess capacity, and no edge from T to S has nonzero flow in f. Hence $f^-(S) = 0$ and the total capacity on all edges from S to T equals $f^+(S) - f^-(S)$. Since the net flow out of any set containing the source but not the sink is $val(f)$, we have proved that $val(f) = cap(S, T)$. $\square$

The proof of $\kappa'(x, y) = \lambda'(x, y)$ in Theorem 4.2.16 is a special case of this argument without the terminology of flows, cuts, and labeling. The "legal sequence" produced an augmentation or a cut with the desired value. The maximum flow problem in general is merely a weighted version of the edge-disjoint path problem.

We have discussed only rational capacities. When the capacities can be irrational, the algorithm might not terminate! Ford and Fulkerson provided an example of this with only ten vertices; it appears in Papadimitriou-Steiglitz [1982, p126-128]. Edmonds and Karp [1972] modified the labeling algorithm to uses at most $(n^3 - n)/4$ augmentation iterations in an n-vertex network and work for arbitrary real capacities. As in the bipartite matching problem, the implementation that searches always for shortest augmenting paths accomplishes this. Faster algorithms are now known; again we cite Ahuja-Magnanti-Orlin [1993] for a thorough discussion.

INTEGRAL FLOWS

In combinatorial applications, we typically have integer capacities and want a solution in which the flow on each edge is an integer.

4.3.10. Corollary. (Integrality Theorem). If all capacities in a network are integers, then there is a maximum flow assigning integral flow to each edge. Furthermore, some maximum flow can be partitioned into flows of unit value along paths from source to sink.

Proof. In the labeling algorithm of Ford and Fulkerson, the amount by which flow values change when an augmenting path is found is always a flow value or the difference between a flow value and a capacity. If the present flow values and the capacities are integers, then this difference is a integer. Starting with the zero flow, there is thus no first time when a noninteger flow appears.

The algorithm therefore produces a maximum flow in which all edges have integer flow. At each internal node, we match units of entering flow to units of exiting flow. This forms s, t-paths and perhaps directed cycles (the conservation conditions guarantee that paths end only at s or t). If a cycle arises, then we can decrease flow around it by 1 and eliminate it without changing the flow value. This leaves precisely $val(f)$ paths from s to t, each corresponding to a unit of flow. □

The integrality theorem yields paths of unit flow. In applications, we give meaning to these units of flow. We can, for example, obtain the various versions of Menger's Theorem as corollaries (Exercises 6,7).

4.3.11. Application. *A well-balanced committee* (see Hall [1956]). The administration at a large university is creating an important committee. One professor will be chosen to represent each department. Many professors have joint appointments in two or more departments, but none can be the designated representative of more than one department. The administration also want the committee to have equal numbers of assistant professors, associate professors, and full professors. How can we find such a committee?

We design a maximum flow problem to answer this. Internal nodes represent the departments, the individual faculty members, and the three ranks. Edges of capacity one are *unit edges*. The source node s sends a unit edge to each department. Each department sends a unit edge to each of its professors. Each professor belongs to only one rank and sends a unit edge there. Finally, there is an edge of capacity k from each rank to the sink t, given that the university has $3k$ departments.

Units of flow in this network correspond to professors on the committee. The edges from the source to the departments guarantee that each department is represented at most once. Since capacity one leaves each professor, the professor can represent only one department. Finally, the capacity on the edges into the sink guarantees the desired balance across ranks. The desired committee exists if and only if this network has a flow of value $3k$. □

Menger's Theorem yields a similar test. Form a digraph D by replacing each edge of capacity k in the network of Example 4.3.11 with k parallel copies of one edge. The committee exists if and only if $\lambda'(s, t) = 3k$ in D.

More generally, Menger's Theorem directly implies Max-flow = Min-cut in every network having rational numbers as capacities. After scaling the capacities up to integers, we obtain a digraph D by splitting each edge of capacity j into j parallel edges. The $\lambda'_D(s, t)$ pairwise edge-disjoint s,t-paths in D map onto units of flow in the corresponding network N. Concerning cuts, we observed in proving Theorem 4.2.17 that a minimal s,t-disconnecting set containing one edge of a set of parallel edges contains them all; hence $\kappa'_D(s,t)$ counts full capacities of edges in N and is the minimum capacity of a source/sink cut in N.

When the focus is on combinatorial applications with unit capacities, Menger's Theorem may give a simpler proof of an application of network flow (compare Theorem 4.2.21 and Exercise 11, for example). For computation or for applications with non-unit capacities, network flow and the Ford-Fulkerson labeling algorithm are more appropriate.

4.3.12. Application. *The Baseball Elimination Problem* (Schwartz [1966]). After many games of the season have been played, we may wonder whether team T can still win the championship. In other words, can winners be assigned for the remaining games so that no team ends with more victories than T? If such an assignment exists, then one exists with T winning all its remaining games, reaching W wins. We want to know whether winners can be chosen for the games among the other teams so that no team obtains more than W wins.

Let $T_1,\ldots,T_n$ be the other teams. Let w_i be the current number of wins for T_i. Let a_{ij} be the remaining number of games to be played between T_i and T_j. Establish a network with n nodes $x_1,\ldots,x_n$ corresponding to the teams and nodes y_{ij} corresponding to the $\binom{n}{2}$ pairs of teams. Include a source s and sink t, with an edge from s to each team node and an edge from each pair node to t. Each pair node y_{ij} is entered by edges from x_i and x_j.

The capacities model the constraints. The capacity on edge $y_{ij}t$ is a_{ij}, the number of remaining games between T_i and T_j. The capacity on edge sx_i is $W-w_i$, the maximum number of additional wins we allow for T_i to keep T in contention. The capacity on edges x_iy_{ij} or x_jy_{ij} is ∞.

By the integrality theorem, a maximum flow breaks into flow units. Each unit corresponds to one game; the first edge specifies the winner, and the last edge specifies the pair. The network has a flow of value $\Sigma_{ij}a_{ij}$ if and only if *all* remaining games can be played with no team exceeding W wins, which is equivalent to T being in contention. □

These combinatorial applications have a common flavor: create a network such that the desired configuration exists if and only if the network has a sufficiently large flow. Often the Max-flow Min-cut Theorem then yields a necessary and sufficient condition for the desired configuration. Exercise 17 does this for the baseball elimination problem. Other examples include Exercise 6,7,9,11,16, Theorems 4.3.13 and 4.3.14, etc.

SUPPLIES AND DEMANDS (optional)

Next we consider a more general network model, allowing multiple sources $X = \{x_i\}$ and sinks $Y = \{y_j\}$. We also associate with each source x_i a *supply* $\sigma(x_i)$ and with each sink y_j a *demand* $\partial(y_j)$. To the capacity constraints for edges and conservation constraints for internal nodes, we add *transportation constraints* for the sources and sinks:

$$f^+(x_i) - f^-(x_i) \le \sigma(x_i) \text{ for } x_i \in X$$
$$f^-(y_j) - f^+(y_j) \ge \partial(y_j) \text{ for } y_j \in Y.$$

With positive values for $\{\partial(y_j)\}$, the zero is not feasible; we must find a feasible flow satisfying these additional constraints if one exists. The "supply/demand" terminology suggest the constraints; we must satisfy the demands at the sinks without exceeding the available supply at any source. The model applies to a company with multiple distribution centers (sources) and retail outlets (sinks).

We write $\sigma(A) = \Sigma_{v \in A}\sigma(v)$ and $\partial(B) = \Sigma_{v \in B}\partial(v)$ for the total supply or demand at a set A of sources or B of sinks, and we write $c(F)$ for $\Sigma_{e \in F}c(e)$. Given a set T of vertices, there is a *net demand* $\partial(Y \cap T) - \sigma(X \cap T)$ that must be satisfied by flow from the remaining vertices. Hence it is necessary that $c([\overline{T}, T])$ be at least this large. Satisfying this for every set T is also sufficient for a feasible flow.

4.3.13. Theorem. (Gale [1957]) In a network N with sources X, sinks Y, and transportation constraints, a feasible flow exists if and only if $c([S,T]) \ge \partial(Y \cap T) - \sigma(X \cap T)$ for every partition of the vertices of N into sets S and T.

Proof. We have already observed the necessity of the condition. For sufficiency, construct a new network N' by adding a supersource s and a supersink t, with an edge of capacity $\sigma(x_i)$ from s to each $x_i \in X$ and an edge of capacity $\partial(y_j)$ from each $y_j \in Y$ to t. Then N has a feasible flow if and only if N' has a flow saturating each edge to t; i.e., if and only if N' has a flow of value $\partial(Y)$.

By the Ford-Fulkerson Theorem, we know that N' has a flow of this value if and only if $cap(S \cup s, T \cup t) \ge \partial(Y)$ for each partition S, T of $V(N)$. The cut $[S \cup s, T \cup t]$ in N' consists of $[S, T]$ from N, plus edges from s to T and edges from S to t in N'. Hence $cap(S \cup s, T \cup t) = c(S, T) + \sigma(T \cap X) + \partial(S \cap Y)$. We now have $cap(S \cup s, T \cup t) \ge \partial(Y)$ if and only if $c(S, T) + \sigma(X \cap T) \ge \partial(Y) - \partial(Y \cap S) = \partial(Y \cap T)$, which is the condition assumed. $\square$

For specific instances, the construction of N' is the key point, because we produce a feasible flow in N (or prove impossibility) by running the Ford-Fulkerson algorithm on the network N'. When costs (per unit flow) are attached to the edges, we have a generalization of the Transportation Problem (see Examples 2.4.16 and 3.2.10). We leave the Min-cost Flow Problem for later study.

Gale's condition is useful for theoretical applications. Suppose we want to test whether $p = (p_1, \ldots, p_m)$ and $q = (q_1, \ldots, q_n)$ are realizable as the vertex degrees of a simple bipartite graph G with bipartition X, Y, with $d(x_i) = p_i$ for all $x_i \in X$ and $d(y_j) = q_j$ for all $y_j \in Y$. Clearly $\Sigma p_i = \Sigma q_j$ is necessary, but this is not sufficient. We can view each edge in the desired graph as a unit of flow. We can test realizability by creating a network N from $K_{m,n}$ by orienting each edge from X to Y, giving it capacity 1 (to prevent multiple edges), and letting $\sigma(x_i) = p_i$ and $\partial(y_j) = q_j$. Then (p, q) is realizable if and only if this network with transportation constraints has a feasible flow.

4.3.14. Theorem. (Gale [1957], Ryser [1957]) Suppose $p_1 \geq \cdots \geq p_m$ and $q_1 \geq \cdots \geq q_n$. Then (p, q) is realizable as the degree sequences of a simple bipartite graph if and only if $\Sigma_{i=1}^m \min\{p_i, k\} \geq \Sigma_{j=1}^k q_j$ for $1 \leq k \leq n$.

Proof. *Necessity.* Suppose (p, q) is realized by G, and consider the other endpoints of the edges incident to a set of k vertices in Y. Because G is simple, each $x_i \in X$ is incident to at most k of these edges, and also x_i is incident to at most p_i of these edges. Hence $\Sigma_{i=1}^m \min\{p_i, k\}$ is an upper bound on the number of edges incident to any k vertices of X, such as those with degrees $q_1, \ldots, q_k$.

Sufficiency. Form the network N described above. It suffices to show that the condition stated here implies Gale's condition for N. Given a set $S \subseteq V(N)$, define $I(S) = \{i : x_i \in S\}$ and $J(S) = \{j : y_j \in S\}$. For any partition S, T of $V(N)$, we then have $\sigma(X \cap T) = \Sigma_{i \in I(T)} p_i$ and $\partial(Y \cap T) = \Sigma_{j \in J(T)} q_j$, and we have $c([S, T]) = |I(S)| \cdot |J(T)|$.

Letting $k = |J(T)|$, this last quantity becomes $c([S, T]) = |I(S)|k = \Sigma_{i \in I(S)} k \geq \Sigma_{i \in I(S)} \min\{p_i, k\}$. Also $\Sigma_{i \in I(T)} p_i \geq \Sigma_{i \in I(T)} \min\{p_i, k\}$, and $\Sigma_{j \in J(T)} q_j \leq \Sigma_{j=1}^k q_j$. When we put all this together, the condition $\Sigma_{i=1}^m \min\{p_i, k\} \geq \Sigma_{j=1}^k q_j$ implies $c([S, T]) \geq \partial(Y \cap T) - \sigma(X \cap T)$. Since we considered an arbitrary partition S, T, the network has a feasible flow, which translates into the desired bipartite graph. □

Theorem 4.3.14 is the natural bipartite analogue of the Erdős-Gallai condition for graphic sequences (Exercise 3.3.16). The quantity $\Sigma_{i=1}^m \min\{p_i, k\}$ has a simple interpretation using partitions of integers, observed by Ryser [1957]. Having indexed $\{p_i\}$ in nonincreasing order, form a matrix of dots with p_i dots in the ith row, starting each time in the first column. This is called the *Ferrers diagram* of the partition of $t = \Sigma p_i$ into the numbers $p_1, \ldots, p_m$. The *conjugate partition* p^* is formed by counting the dots in each column, with p_j^* being the number of dots in the jth column. The quantity $\Sigma_{i=1}^{m} \min\{p_i, k\}$ counts the dots in the first k columns. Hence the condition for realizability can be expressed instead as

$$\Sigma_{j=1}^{k} p_j^* \geq \Sigma_{j=1}^{k} q_j \quad \text{for} \quad 1 \leq k \leq n.$$

A natural extension of the maximum flow problem incorporates lower bounds. Instead of merely imposing a capacity on each edge, we may have an upper bound and also a nonnegative lower bound on the permitted flow in each edge, such as $l(e) \leq f(e) \leq u(e)$. We still impose conservation constraints on the internal nodes. Given a feasible flow, we can easily modify the Ford-Fulkerson labeling algorithm to seek a maximum (or minimum) feasible flow (Exercise 4). The problem is finding an initial feasible flow. First we present an application.

4.3.15. Application. *Matrix rounding* (Bacharach [1966]). We may want to round the entries of a data matrix up or down to integers. If we are also presenting the row-sums and column sums, we should round those up or down to integers to be the column-sums and row-sums of the rounded matrix. This is a *consistent rounding*.

We can represent the consistent rounding problem as a feasible flow problem. Establish vertices $x_1, \ldots, x_n$ for the rows and vertices $y_1, \ldots, y_n$ for the columns of the matrix. Add a source s and a sink t. Add edges $sx_i, x_i y_j, y_j t$ for all values of i and j. If the matrix has entries a_{ij} with row-sums $r_1, \ldots, r_n$ and column-sums $s_1, \ldots, s_n$, set

$$l(sx_i) = \lfloor r_i \rfloor \quad l(x_i y_j) = \lfloor a_{ij} \rfloor \quad l(y_j t) = \lfloor c_j \rfloor$$
$$u(sx_i) = \lceil r_i \rceil \quad u(x_i y_j) = \lceil a_{ij} \rceil \quad u(y_j t) = \lceil c_j \rceil$$

The maximum flow problem generated to test for a feasible flow in a network with upper and lower bounds has integer capacities if all the $l(e)$'s and $u(e)$'s are integers. Hence the integrality theorem applies, and the matrix has a consistent rounding if and only if the network we constructed from it has a feasible flow. □

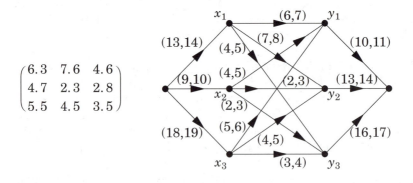

$$\begin{pmatrix} 6.3 & 7.6 & 4.6 \\ 4.7 & 2.3 & 2.8 \\ 5.5 & 4.5 & 3.5 \end{pmatrix}$$

4.3.16. Application. *Circulations and flows with lower bounds.* The first step in solving a flow problem with upper and lower bounds is to add an edge of infinite capacity from the sink to the source. The resulting network has a feasible flow with conservation at *every* node (called a *circulation*) if and only if the original network has a feasible flow.

We can convert a feasible circulation problem C into a maximum flow problem N by introducing supplies or demands at the nodes and then a new supersource s' and supersink t' to satisfy these. Given the flow constraints $l(e) \le f(e) \le u(e)$, let $c(e) = u(e) - l(e)$ for each edge e. Also, let $b(v) = l^-(v) - l^+(v)$ for each vertex v, where $l^-(v) = \Sigma_{e \in [V(C)-v,v]} l(e)$ and $l^+(v) = \Sigma_{e \in [v,V(C)-v]} l(e)$. Since every $l(uv)$ contributes equally to $l^+(u)$ and $l^-(v)$, we have $\Sigma b(v) = 0$. A flow f is a feasible circulation in C if and only if the flow f' defined by $f'(e) = f(e) - l(e)$ satisfies $0 \le f'(e) \le c(e)$ and $f'^+(v) - f'^-(v) = b(v)$.

If $b(v) \ge 0$, then v supplies flow $|b(v)|$ to the network; otherwise v demands $|b(v)|$. To restore conservation constraints, we add source s with an edge of capacity $b(v)$ to each v with $b(v) \ge 0$ and a sink t with an edge of capacity $-b(v)$ from each v with $b(v) < 0$. This completes the construction of N. Let α be the total capacity on the edges leaving s; since $\Sigma b(v) = 0$, the edges entering t also have total capacity α. Now C has a feasible circulation f if and only if N has a flow of value α (saturating all edges leaving s or entering t. $\square$

4.3.17. Corollary. A network D with conservation constraints at every node has a feasible circulation if and only if $\Sigma_{e \in [S,\bar{S}]} l(e) \le \Sigma_{e \in [\bar{S},S]} u(e)$ for every $S \subseteq V(D)$.

Proof. We can stop before the last step in the discussion of Application 4.3.16 and interpret our problem with supplies and demands in the model of Theorem 4.3.13. Since $\Sigma b(v) = 0$, the only way to satisfy all the demands is to use up all the supply. Hence there is a circulation if and only if the supply/demand problem with supplies $\sigma(v) = b(v)$ for $\{v \in V(D): b(v) \ge 0\}$ and demands $\partial(v) = -b(v)$ for $\{v \in V(D): b(v) < 0\}$ has a solution. Theorem 4.3.13 tells when this problem has a solution. Translated back into the lower and upper bounds on flow in the original problem (Exercise 21), the criterion of Theorem 4.3.13 becomes $\Sigma_{e \in [S,\bar{S}]} l(e) \le \Sigma_{e \in [\bar{S},S]} u(e)$ for every $S \subseteq V(D)$. $\square$

EXERCISES

4.3.1. (–) For the network below, list all integer-valued feasible flows (be careful not to overlook any). From the list, select a flow of maximum value. This illustrates the power of duality in comparison to exhaustive analysis. Prove that the selected flow is a maximum flow by exhibiting a cut with the same value. Determine the number of source/sink cuts.

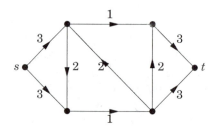

4.3.2. (–) Below is a network with edge capacities as indicated. Find the maximum value of a flow from s to t. Prove that your answer is optimal by using the dual problem, and explain why this is a proof of optimality.

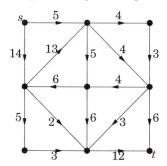

4.3.3. (–) The kitchen sink draws water from two tanks according to the network of pipes with capacities per unit time indicated below. Find the maximum flow per unit time. Prove that your answer is optimal by using the dual problem, and explain why this is a proof of optimality.

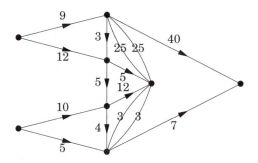

4.3.4. (–) Suppose N is a network with edge capacity and node conservation constraints, plus lower bound constraints $l(e)$ on the flow in edges, meaning that $f(e) \geq l(e)$ is required. If an initial feasible flow is given, how can the Ford-Fulkerson labeling algorithm be modified to search for a maximum feasible flow in this network?

4.3.5. (–) Suppose G is a directed graph and $x, y \in V(G)$. Instead of capacities on the edges of G, suppose capacities are specified on the vertices (other than

x, y); this means that for each vertex there is a fixed limit on the total flow in (or out). There is no restriction on flows in edges. Show how to use network flow theory to determine the maximum value of a feasible flow from x to y.

4.3.6. (!) Use the Ford-Fulkerson Theorem to prove Menger's Theorem for edges in digraphs: $\kappa'(x, y) = \lambda'(x, y)$.

4.3.7. (!) Use the Ford-Fulkerson Theorem to prove Menger's Theorem for non-adjacent vertices in digraphs: $\kappa(x, y) = \lambda(x, y)$.

4.3.8. Use network flows to prove that an undirected graph G is connected if and only if for every partition of $V(G)$ into two nonempty sets S, T, there is an edge with one endpoint in S and one endpoint in T. (Comment: Chapter 1 contains an easy direct proof of the conclusion, so this is an example of "using a sledgehammer to squash a bug".)

4.3.9. (!) Use network flows to prove the König-Egerváry Theorem ($\alpha'(G) = \beta(G)$ if G is bipartite).

4.3.10. Show that the Augmenting Path Algorithm for bipartite graphs (Algorithm 3.2.1) is a special case of the Ford-Fulkerson Labeling Algorithm.

4.3.11. (!) Suppose the families $\mathbf{A} = \{A_1, \ldots, A_m\}$ and $\mathbf{B} = \{B_1, \ldots, B_m\}$ are collections of subsets of an n-element set S. A *common system of distinct representatives* (CSDR) for $\mathbf{A}$ and $\mathbf{B}$ is a set U of m elements of S that is a system of distinct representatives (SDR) for $\mathbf{A}$ and for $\mathbf{B}$. For example, if $\mathbf{A} = \{123, 124, 34, 15\}$ and $\mathbf{B} = \{45, 23, 24, 34\}$, then $\{2, 3, 4, 5\}$ is a CSDR. Use the Ford-Fulkerson Theorem to prove that $\mathbf{A}$ and $\mathbf{B}$ have a CSDR if and only if, for every pair I, J of subsets of the index set $[m]$,

$$|(\cup_{i \in I} A_i) \cap (\cup_{j \in J} B_j)| \geq |I| + |J| - m.$$

(Hint: for sufficiency, use $\mathbf{A}, \mathbf{B}$ to construct a network that has a flow of value m if and only if $\mathbf{A}, \mathbf{B}$ have a CSDR, and prove that the condition given above implies that every source/sink cut in the network has capacity at least m.)

4.3.12. Prove that if $[S, \bar{S}]$ and $[T, \bar{T}]$ are minimum source/sink cuts in a network N, then $[S \cup T, \overline{S \cup T}]$ and $[S \cap T, \overline{S \cap T}]$ are also minimum cuts in N. (Hint: consider various types of contributions to the cuts; a picture may help guide the argument.)

4.3.13. Prove that if $[S, \bar{S}]$ and $[T, \bar{T}]$ are source/sink cuts in a network N, then $cap(S \cup T, \overline{S \cup T}) + cap(S \cap T, \overline{S \cap T}) \leq cap([S, \bar{S}]) + cap(T, \bar{T}) + [S \cap T, \overline{S \cap T}]$.

4.3.14. ($-$) Prove that there is no simple bipartite graph for which the vertices in one partite set have degrees (5,4,4,2,1) and the vertices in the other partite set also have degrees (5,4,4,2,1).

4.3.15. ($-$) Given two sequences $r = (r_1, \ldots, r_n)$ and $s = (s_1, \ldots, s_n)$, obtain necessary and sufficient conditions for the existence of a digraph D with vertices $v_1, \ldots, v_n$ such that each ordered pair occurs at most once as an edge and $d^+(v_i) = r_i$ and $d^-(v_i) = s_i$ for all i.

4.3.16. Several companies send representatives to a conference; the ith company sends m_i representatives. The organizers of the conference conduct simultaneous networking groups; the jth group can accommodate up to n_j participants. The organizers want to schedule all the participants into groups, but no two participants from the same company should be in the same group. The groups need not all be filled; the condition $\Sigma n_j \geq \Sigma m_i$ is necessary. Obtain necessary and sufficient conditions on $\{m_i\}$ and $\{n_j\}$ that characterize when there exists an assignment of participants to groups that satisfies all the constraints.

4.3.17. In the Baseball Elimination Problem (Application 4.3.12), prove that team T remains in contention if and only if for every set $Z \subseteq [n]$, the inequality $\Sigma_{i \in Z}(W - w_i) \geq \Sigma_{i,j \notin T} a_{ij}$ holds. (Hint: Use the Max-flow Min-cut Theorem.) (Schwartz [1966])

4.3.18. (−) Find a consistent rounding of the data in the matrix below.

$$\begin{pmatrix} .55 & .6 & .6 \\ .55 & .65 & .7 \\ .6 & .65 & .7 \end{pmatrix}$$

4.3.19. Prove that every two-by-two matrix can be consistently rounded.

4.3.20. Suppose that every entry in an n-by-n matrix is strictly between $1/n$ and $1/(n-1)$. Describe all consistent roundings.

4.3.21. Complete the details of proving Corollary 4.3.17, proving the necessary and sufficient condition for a circulation in a network with lower and upper bounds.

4.3.22. *Minimax theorem on weighted spanning trees.* The *capacity* of a spanning tree in a weighted (undirected) graph G is the minimum weight of its edges. The *limit* from a cut $[S, \bar{S}]$ is the maximum weight of its edges. Prove that the maximum capacity among spanning trees of G equals the minimum limit among cuts in G. (Ahuja-Magnanti-Orlin [1993, p538])

4.3.23. An $(m+n)$-regular graph G is (m, n)-*orientable* if it can be oriented so that each in-degree is m or n.

a) + Prove that G is (m, n)-orientable if and only if there is a partition X, Y of $V(G)$ such that for every $S \subseteq V(G)$,

$$\left| (m - n)|X \cap S| - |Y \cap S| \right| \leq |[S|.$$

b) Conclude that if G is (m, n)-orientable and $m > n$, then G is also $(m - 1, n + 1)$-orientable. (Bondy-Murty [1976, p210-211])

Chapter 5

Graph Coloring

5.1 Vertex Colorings and Upper Bounds

Coloring problems arise in many contexts. The committee-scheduling example of Section 1.1 has other settings: if $V(G)$ is a set of university courses, with edges between courses with common students, then the chromatic number is the minimum number of time periods needed to schedule examinations without conflicts. The problem of 4-coloring the regions of a map so regions with common boundaries receive different colors is another example; we return to it in Chapter 7.

DEFINITIONS AND EXAMPLES

5.1.1. Definition. A *k-coloring* of G is a labeling $f: V(G) \to \{1, \dots, k\}$. It is a *proper k-coloring* if $x \leftrightarrow y$ implies $f(x) \neq f(y)$. A graph G is *k-colorable* if it has a proper k-coloring. The *chromatic number* $\chi(G)$ is the minimum k such that G is k-colorable. If $\chi(G) = k$, then G is *k-chromatic*. If $\chi(G) = k$, but $\chi(H) < k$ for every proper subgraph H of G, then G is *color-critical* or *k-critical*.

5.1.2. Remark. The vertices having a given color in a proper coloring must form an independent set, so $\chi(G)$ equals the minimum number of independent sets needed to cover $V(G)$. Hence G is k-colorable if and only if G is k-partite. Multiple edges do not affect chromatic number, so in this chapter we consider only simple graphs. Although we define k-coloring using numbers from $\{1, \dots, k\}$ as labels, the numerical values are usually unimportant, and we may use any set of size k as labels. □

5.1.3. Example. *Elementary examples.* $\chi(K_n) = n$. Every bipartite graph is 2-colorable, since we can use color i on the ith partite set. Below we illustrate optimal colorings of the 5-cycle and the Petersen graph, which have chromatic number 3. □

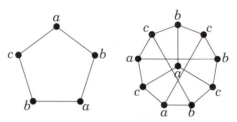

5.1.4. Example. *k-critical graphs for small k.* The k-colorable graphs are the k-partite graphs. Hence K_1 and K_2 are the only 1-critical and 2-critical graphs. Because the bipartite graphs are those without odd cycles, the 3-critical graphs are the odd cycles. The simplicity of chromatic number ends there; we have no characterization of 4-critical graphs and no good algorithm to test 3-colorability (see Section 6.3). We can test 2-colorability by using breadth-first search to compute distances from a vertex x (in each component), because a connected graph G is bipartite if and only if $G[X]$ and $G[Y]$ are independent sets, where $X = \{u \in V(G): d(u, x) \text{ is even}\}$ and $Y = \{u \in V(G): d(u, x) \text{ is odd}\}$. □

Recall that $\alpha(G)$ denotes the size of the largest independent set in G; we use $\omega(G)$ to denote the size of the largest set inducing a clique.[†] Since the vertices of a clique need different colors, $\chi(G) \geq \omega(G)$. Since each label appears only on an independent set, $\chi(G) \geq n(G)/\alpha(G)$.

5.1.5. Example. $\chi(G)$ *may exceed* $\omega(G)$. Let $G = C_{2r+1} \vee K_s$ be the join of a clique and an odd cycle. Since C_{2r+1} has no triangle, $\omega(G) = s + 2$. To color the vertices of the induced $2r + 1$-cycle C, we need at least three colors. The remaining vertices induce a clique and need s colors. Since every vertex not on C is adjacent to every vertex of C, these s colors must differ from the first three, and a proper coloring of G requires at least $s + 3$ colors. We conclude that $\chi(G) > \omega(G)$. □

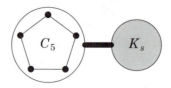

[†]Just as cliques and independent sets are "opposites", so α and ω are the opposite ends of the Greek alphabet. We can view independent sets and cliques as the beginning and end of the "evolution" of a graph (see Section 8.5).

Chromatic number is known for many special families of graphs (Exercises 3-8, for example). We also study the behavior of chromatic number (or other parameters) under various graph operations. For example, $\chi(G + H) = \max\{\chi(G), \chi(H)\}$, and $\chi(G \vee H) = \chi(G) + \chi(H)$. We also study a natural graph product.

5.1.6. Definition. The *Cartesian product* of graphs G and H, written $G \square H$, is the graph with vertex set $V(G) \times V(H)$ specified by putting (u, v) adjacent to (u', v') if and only if (1) $u = u'$ and $vv' \in E(H)$, or (2) $v = v'$ and $uu' \in E(G)$.

5.1.7. Example. *Cartesian products.* The operation is symmetric; $G \square H \cong H \square G$. For example, $C_3 \square K_2$ appears below, and $Q_k = Q_{k-1} \square K_2$ if $k \geq 1$. In general, the edges of $G \square H$ can be partitioned into a copy of H for each vertex of G and a copy of G for each vertex of H (Exercise 10). We use $\square$ instead of $\times$ to avoid confusion with other product operations, reserving $\times$ for the Cartesian product of vertex sets. The symbol $\square$ visually represents the observation that $K_2 \square K_2 = C_4$.

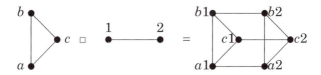

Cartesian product enables us to compute the chromatic number of G by computing the independence number of a product graph. A graph G is m-colorable if and only if the Cartesian product graph $G \square K_m$ has an independent set of size $n(G)$ (Exercise 11). A more elementary observation is $\chi(G \square H) = \max\{\chi(G), \chi(H)\}$ (Exercise 10).

UPPER BOUNDS

Easy bounds on $\chi(G)$ include $\chi(G) \leq n(G)$, $\chi(G) \geq \omega(G)$, and $\chi(G) \geq n(G)/\alpha(G)$. These are best possible, since all hold with equality for cliques. We seek refinements of the bounds that are best possible for more graphs. There are many graphs with $\chi(G) = \omega(G)$, but the cliques are the only graphs with $\chi(G) = n(G)$, so we refine the upper bound.

Most upper bounds on $\chi(G)$ come from coloring algorithms. The bound $\chi(G) \leq n(G)$ uses nothing about the structure of G. We can improve the bound by coloring vertices successively using the "least" available color.

5.1.8. Algorithm. (Greedy coloring). The *greedy coloring* with respect to a vertex ordering $v_1, \ldots, v_n$ of $V(G)$ is obtained by coloring vertices in the order $v_1, \ldots, v_n$, assigning to v_i the smallest-indexed color not already used on its lower-indexed neighbors.

5.1.9. Proposition. $\chi(G) \leq \Delta(G) + 1$.

Proof. Because each vertex has at most $\Delta(G)$ neighbors, a greedy coloring can never be forced to use more than $\Delta(G) + 1$ colors (for any ordering). This proves constructively that $\chi(G) \leq \Delta(G) + 1$. □

In proving $\chi(G) \leq \Delta(G) + 1$ by greedy coloring, we can use any vertex ordering; choosing the ordering carefully may improve the bound. Indeed, every graph G has a vertex ordering on which the greedy algorithm uses only $\chi(G)$ colors (Exercise 13a). Usually it is hard to find this ordering, but it is easy in some special classes.

5.1.10. Example. *Register allocation and interval graphs.* A computer program stores the values of its variables in memory. For arithmetic computations, the values must be entered in "registers". Registers are expensive, so we want to use them efficiently. If two variables are not used at the same time, we can allocate them to the same register. For each variable, we compute the first and last time when it is used. A variable is *active* during the interval between these times. We define a graph with the variables as vertices, in which two vertices are adjacent if they are active at a common time. The number of registers needed is the chromatic number of this graph.

Given any family of intervals, we can define a graph whose vertices are the intervals, with vertices adjacent when the intervals intersect. A graph formed in this way is an *interval graph*, and the family of intervals is an *interval representation* of the graph. In the illustration below, if we color the vertices in the order a, b, c, d, e, f, g, then the greedy algorithm assigns them colors 1,2,3, 1,1,2,3, respectively, which is optimal. □

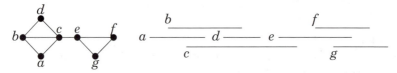

5.1.11. Proposition. If G is an interval graph, then $\chi(G) = \omega(G)$.

Proof. Put the vertices in increasing order by their left endpoints and apply greedy coloring. Suppose x receives the color k of maximum index. Since x cannot receive a smaller color, the left endpoint a of its interval belongs also to intervals that already have colors 1 through

$k - 1$. These intervals all share the point a, so we have a k-clique consisting of x and neighbors of x with colors 1 through $k - 1$. Hence $\omega(G) \geq k \geq \chi(G)$. Since $\chi(G) \geq \omega(G)$ always, this coloring is optimal. □

We can further strengthen $\chi(G) \leq 1 + \Delta(G)$ by replacing $\Delta(G)$ with something that never exceeds $\Delta(G)$. First we consider greedy coloring with carefully chosen vertex orderings. Vertices of high degree cause trouble; we can dispose of them first by putting the vertices in decreasing order of degree. (Exercise 15 considers another ordering.)

5.1.12. Proposition. (Welsh-Powell [1967]) If a graph G has degree sequence $d_1 \geq \cdots \geq d_n$, then $\chi(G) \leq 1 + \max_i \min\{d_i, i - 1\}$.

Proof. We apply greedy coloring with the vertices in nonincreasing order of degree. When we color the ith vertex, at most $\min\{d_i, i - 1\}$ of its neighbors have already been colored, so its color is at most $1 + \min\{d_i, i - 1\}$. Maximizing this over i yields an upper bound on $\chi(G)$. □

The greedy coloring algorithm runs rapidly. It is "on-line" in the sense that it produces a proper coloring even if it sees only one new vertex at each step and must color it with no option to change earlier colors. For a random vertex ordering in a random graph (see Section 8.5), greedy coloring almost always uses only about twice as many colors as the minimum, although with a bad ordering it may use many colors on a tree (Exercise 13b).

Other bounds follow from the properties of k-critical graphs, but k-critical subgraphs don't provide proper colorings: although every k-chromatic graph *has* a k-critical subgraph, we have no good algorithm for *finding* one. We presented greedy coloring first to underscore the algorithmic nature of upper bounds on chromatic number.

5.1.13. Lemma. If H is a k-critical graph, then $\delta(H) \geq k - 1$.

Proof. Suppose x is a vertex of H. Because H is k-critical, $H - x$ is $k - 1$-colorable. If $d_H(x) < k - 1$, then the $k - 1$ colors used on $H - x$ do not all appear on $N(x)$, and we can assign a missing one to x to extend the coloring to H. This contradicts our hypothesis that H has no proper $k - 1$-coloring. Hence every vertex of H has degree at least $k - 1$. □

5.1.14. Corollary. (Szekeres-Wilf [1968]) If G is a graph, then $\chi(G) \leq 1 + \max_{H \subseteq G} \delta(H)$.

Proof. Let $k = \chi(G)$, and let H' be a k-critical subgraph of G. By Lemma 5.1.13, we have $\chi(G) - 1 \leq \delta(H') \leq \max_{H \subseteq G} \delta(H)$, as desired. □

The next bound involves orientations (see also Exercises 23-26).

5.1.15. Theorem. (Gallai [1968], Roy [1967], Vitaver [1962]) If D is an orientation of G with longest path length $l(D)$, then $\chi(G) \leq 1 + l(D)$. Furthermore, equality holds for some orientation of G.

Proof. Suppose D is an orientation of G. Let D' be a maximal acyclic subdigraph of D; D' is an orientation of a spanning subgraph H of G. Color $V(G)$ by letting $f(v)$ be one more than the length of the longest path in D' that ends at v. Because D' has no cycle, each path ending at a vertex of a path in D' can be extended along the path. Therefore, f strictly increases along each path in D'.

The coloring f uses colors 1 through $1 + l(D')$ on $V(D) = V(G)$. We claim that f is a proper coloring of G. For each $uv \in E(D)$, there is a path in D' between its endpoints (since uv is an edge of D' or its addition to D' creates a cycle). This implies $f(u) \neq f(v)$, since f increases along paths of D'.

To prove the second statement, we construct an orientation D^* such that $l(D^*) \leq \chi(G) - 1$. Let f be an optimal coloring of G. For each edge uv in G, orient the edge as $u \to v$ in D^* if and only if $f(u) < f(v)$. Since f is a proper coloring, this defines an orientation. Since the labels used by f increase along each path in D^*, and there are only $\chi(G)$ labels in f, we have $l(D^*) \leq \chi(G) - 1$. □

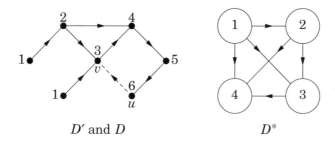

D' and D D^*

BROOKS' THEOREM

The bound $\chi(G) \leq 1 + \Delta(G)$ is tight for cliques and odd cycles. By choosing the vertex ordering more carefully, we can show that these are essentially the only graphs with $\chi(G) > \Delta(G)$. This implies, for example, that the Petersen graph is 3-colorable. To avoid unimportant complications, we phrase the statement only for connected graphs. It extends to all graphs because the chromatic number of a graph is the maximum chromatic number of its components. Many proofs are known; we present a modification of the proof by Lovász [1975].

5.1.16. Theorem. (Brooks [1941]) If G is a connected graph other than a clique or an odd cycle, then $\chi(G) \le \Delta(G)$.

Proof. Suppose G is connected but is not a clique or an odd cycle, and let $k = \Delta(G)$. We may assume that $k \ge 3$, since G is a clique when $k = 1$, and G is an odd cycle or is bipartite when $k = 2$.

If G is not k-regular, choose v_n so that $d(v_n) < k$. Since G is connected, we can grow a spanning tree of G from v_n, assigning indices in decreasing order as we reach vertices. Each vertex other than v_n in the resulting ordering $v_1, \ldots, v_n$ has a higher-indexed neighbor along the path to v_n in the tree. Hence each vertex has at most $k - 1$ lower-indexed neighbors, and the greedy coloring uses at most k colors.

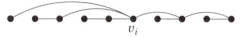

v_i

In the remaining case, G is k-regular. If G has a cut-vertex x, let G' be a component of $G - x$ together with its edges to x. The degree of x in G' is less than k, and we obtain a proper k-coloring of G' as above. By permuting the names of colors in each such subgraph, we can make the colorings agree on x to complete a proper k-coloring of G.

If G has a vertex with two neighbors whose deletion leaves a connected subgraph, then we number these v_1, v_2 and let their common neighbor be v_n. Because $G - \{v_1, v_2\}$ is connected, it has a spanning tree, and we number this tree using $3, \ldots, n$ such that labels increase along paths to the root v_n. As before, each vertex before n has at most $k - 1$ lower indexed neighbors. The greedy coloring also uses at most $k - 1$ colors on neighbors of v_n, since v_1 and v_2 receive the same color.

Hence it suffices to show that every 2-connected k-regular graph with $k \ge 3$ has such vertices. Choose a vertex x. If $\kappa(G - x) \ge 2$, let v_1 be x and let v_2 be a vertex with distance two from x, which exists because G is regular and not a clique. If $\kappa(G - x) = 1$, then x has a neighbor in every leaf block of $G - x$ (since G has no cut-vertex). Neighbors v_1, v_2 of x in two such blocks are nonadjacent. Furthermore, $G - \{x, v_1, v_2\}$ is connected, since blocks have no cut-vertices. Now $k \ge 3$ implies that $G - \{v_1, v_2\}$ also is connected. $\square$

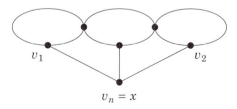

v_1 v_2

$v_n = x$

Brooks' Theorem implies that the cliques and odd cycles are the only $k - 1$-regular k-critical graphs (Exercise 27). The bound $\chi(G) \le$

$\Delta(G)$ can be improved if G has no large clique (Exercise 28). We can also extend Brooks' Theorem by considering *generalized coloring parameters*. Given a property P, we want to partition $V(G)$ into the smallest number of color classes such that the subgraph induced by each color class has property P. When P is the property "independent set", the minimum number of classes is the ordinary chromatic number. The theorem below becomes Brooks' Theorem when $j = 1$, because a graph with no 1-edge-connected subgraph is an independent set.

5.1.17. Theorem. (Matula [1973]) If G is a connected graph, then G has a $\lceil \Delta(G)/j \rceil$-coloring such that the subgraph induced by each color class has no j-edge-connected subgraph, unless G is 1) a j-regular j-edge-connected graph, 2) an n-clique with $n \equiv 1 \bmod j$, or 3) an odd cycle when $j = 1$. In these cases, $\lceil (\Delta(G) + 1)/j \rceil$ colors are necessary and sufficient. □

EXERCISES

5.1.1. (–) For each G below, compute $\chi(G)$ and give a $\chi(G)$-critical subgraph.

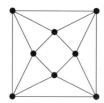 

5.1.2. (–) Suppose the blocks of G are $G_1, \ldots, G_k$. Prove that $\chi(G) = \max_i \chi(G_i)$.

5.1.3. (!) Place n points on a circle, where $n \geq k(k+1)$. Let $G_{n,k}$ be the $2k$-regular graph obtained by joining each point to the k nearest points in each direction on the circle. For example, $G_{n,1} = C_n$, and $G_{7,2}$ appears below. Prove that $\chi(G_{n,k}) = k + 1$ if $k + 1$ divides n and $\chi(G_{n,k}) = k + 2$ if $k + 1$ does not divide n. Prove that lower bound on n cannot be weakened, by proving that $\chi(G_{k(k+1)-1,k}) > k + 2$ if $k \geq 2$.

5.1.4. (!) Suppose the odd cycles in G are pairwise intersecting, meaning that every two odd cycles in G have a common vertex. Prove that $\chi(G) \leq 5$.

5.1.5. Suppose that every edge of G appears in at most one cycle. Prove that every block of G is an edge or a cycle, and use this to prove $\chi(G) \leq 3$.

5.1.6. (!) Let G be the *unit-distance graph* in the plane; the vertices are all points in the plane, with vertices joined by an edge if the Euclidean distance between them is exactly 1. Prove that G is 7-colorable but not 3-colorable. (Hint: for the upper bound, present an explicit coloring by regions, with attention to their boundaries.)

5.1.7. (!) Given a set of lines in the plane with no three meeting at a point, form a graph G whose vertices are the intersections of the lines, with two vertices adjacent if they appear consecutively on one of the lines. Prove that $\chi(G) \leq 3$. (Comment: this can fail when three lines may meet at a point.) (H. Sachs)

5.1.8. (+) Let $S = \binom{[n]}{2}$ denote the collection of 2-sets of the n-element set $[n]$. Define the graph G_n by $V(G_n) = S$ and $E(G_n) = \{ij, jk: i < j < k\}$ (disjoint pairs, for example, are nonadjacent). Prove that $\chi(G_n) = \lceil \lg n \rceil$. (Hint: Prove that G_n is k-colorable if and only if $[k]$ has at least n distinct subsets. Comment: G_n is called the *shift graph* of K_n.) (attributed to A. Hajnal)

5.1.9. (−) Draw $K_{1,3} \square P_3$ and exhibit an optimal coloring of the graph.

5.1.10. Prove that the Cartesian product $G \square H$ can be expressed as $aH \cup bG$, where G has a vertices and H has b vertices. Prove that $\chi(G \square H) = \max\{\chi(G), \chi(H)\}$.

5.1.11. (!) Prove that G is m-colorable if and only if the Cartesian product $G \square K_m$ has an independent set of size $n(G)$. (Berge [1973, p379-380])

5.1.12. (−) Prove or disprove: Always $\chi(G) \leq 1 + a(G)$, where $a(G) = 2e(G)/n(G)$ is the average vertex degree.

5.1.13. (!) *Greedy coloring.*
    a) Prove that every graph G has a vertex ordering with respect to which greedy coloring uses $\chi(G)$ colors.
    b) For all $k \in \mathbb{N}$, inductively construct a tree T_k with maximum degree k and an ordering σ of $V(T_k)$ such that greedy coloring with respect to σ uses $k + 1$ colors. (Comment: This shows that the performance ratio of greedy coloring may be as bad as $(\Delta(G) + 1)/2$.) (Bean [1976])

5.1.14. Suppose G has no induced subgraph isomorphic to P_4. Prove that for every vertex ordering, greedy coloring produces an optimal coloring of G. (Hint: Suppose the algorithm uses k colors for the ordering $v_1, \ldots, v_n$, and let i be the smallest integer such that G has a clique consisting of vertices assigned colors i through k in this coloring.)

5.1.15. Given an ordering σ of $V(G)$, let $G_i = G[\{v_1, \ldots, v_i\}]$, and let $f(\sigma) = 1 + \max_i d_{G_i}(x_i)$. The greedy coloring with respect to σ establishes $\chi(G) \le f(\sigma)$. Define σ^* as follows: let v_n be a vertex of minimum degree in G, and for $i < n$ let v_i be a vertex of minimum degree in $G - \{v_{i+1}, \ldots, v_n\}$. Prove that σ^* minimizes $f(\sigma)$ and that $f(\sigma^*) = 1 + \max_{H \subseteq G} \delta(H)$. (Halin [1967], Matula [1968], Finck-Sachs [1969], Lick-White [1970])

5.1.16. Prove that $V(G)$ can be partitioned into $1 + \max_{H \subseteq G} \delta(H)/r$ classes such that the subgraph induced by each class has a vertex of degree less than r. (Hint: consider ordering σ^* of Exercise 15. Comment: this generalizes the Szekeres-Wilf Theorem (Corollary 5.1.14).)

5.1.17. (!) Prove that $\chi(\overline{H}) = \omega(\overline{H})$ when H is bipartite and has no isolated vertices. (Hint: Phrase the conclusion in terms of H and apply results about bipartite graphs.)

5.1.18. (!) Prove that every k-chromatic graph has at least $\binom{k}{2}$ edges. Use this to prove that if G is the union of m cliques of order m, then $\chi(G) < 1 + m\sqrt{m-1}$. (Hint: $\sqrt{x+1} < \sqrt{x} + 1$ if $x > 0$. Comment: this bound is tight, but the Erdős-Faber-Lovász Conjecture (see Erdös [1981]) asserts that $\chi(G) = m$ when G the cliques are pairwise-edge-disjoint.)

5.1.19. Suppose G is a simple n-vertex graph. Use Turán's Theorem (Theorem 1.3.20) to prove that if $\omega(G) \le r$, then $e(G) \le (1 - 1/r)n^2/2$. Use this to prove that $\chi(G) \ge n^2/(n^2 - 2e(G))$. (Myers-Liu [1972])

5.1.20. *Chromatic number of G and $\overline{G}$.* Prove that $\chi(G) \cdot \chi(\overline{G}) \ge n(G)$, use this to prove that $\chi(G) + \chi(\overline{G}) \ge 2\sqrt{n(G)}$, and provide a construction achieving these bounds whenever $\sqrt{n(G)}$ is an integer. (Nordhaus-Gaddum [1956], Finck [1968])

5.1.21. (!) Prove that $\chi(G) + \chi(\overline{G}) \le n(G) + 1$. (Hint: use induction on $n(G)$.) (Nordhaus-Gaddum [1956])

5.1.22. (!) *Looseness of $\chi(G) \ge n(G)/\alpha(G)$.* Suppose G has n vertices, and let $b = (n + 1)/\alpha(G)$. Use Exercise 22 to prove that $\chi(G) \cdot \chi(\overline{G}) \le (n + 1)^2/4$, and use this to prove that $\chi(G) \le b(n + 1)/4$. For each odd n, construct a graph such that $\chi(G) = b(n + 1)/4$. (Nordhaus-Gaddum [1956], Finck [1968])

5.1.23. (–) Use the Gallai-Roy Theorem to prove that every tournament has a spanning path.

5.1.24. (!) *Paths and chromatic number in digraphs.*
  a) Suppose $G = F \cup H$. Prove that $\chi(G) \le \chi(F)\chi(H)$.
  b) Suppose D is an orientation of G and $\chi(G) > rs$. Suppose each $v \in V(D)$ is assigned a real number $f(v)$. Use (a) and the Gallai-Roy Theorem to prove that D has a path $u_0 \to \cdots \to u_r$ with $f(u_0) \le \cdots \le f(u_r)$ or a path $v_0 \to \cdots \to v_s$ with $f(v_0) > \cdots > f(v_s)$.
  c) Use (b) to prove that every sequence of $rs + 1$ distinct real numbers has an increasing subsequence of size $r + 1$ or a decreasing subsequence of size $s + 1$. (Erdős-Szekeres [1935])

5.1.25. a) Prove that $\chi(G \cup H) \le \chi(G)\chi(H)$.

b) A sequence $a_0, \ldots, a_k$ of points in the plane is ε-*linear* if the angle between the segments $a_{i-1}a_i$ and $a_i a_{i+1}$ is between $(1 - \varepsilon)\pi$ and π for all i. Prove that every set of $k^n + 1$ points in the plane contains a $1/n$-linear sequence of $k + 1$ points. (Hint: Apply part (a) to an appropriate partition of the edges of a directed graph defined on the points.)

5.1.26. Prove the Gallai-Roy Theorem using Minty's Theorem [1962], which states that G is k-colorable if and only if G has an orientation that on each cycle C points at least $n(C)/k$ edges in each direction.

5.1.27. (!) Prove that Brooks' Theorem is equivalent to the following statement: every $k - 1$-regular k-critical graph is a clique or an odd cycle.

5.1.28. *Improvement of Brooks' Theorem.*

(+) a) Given values $\Delta_1, \ldots, \Delta_t$ such that $\Sigma\Delta_i \ge \Delta(G) - t + 1$, prove that $V(G)$ can be partitioned into sets $V_1, \ldots, V_t$ inducing subgraphs $G_1, \ldots, G_t$ so that $\Delta(G_i) \le \Delta_i$ for each i. (Hint: use induction on t.) (Lovász [1966])

b) Given $4 \le r \le \Delta(G) + 1$, use (a) to prove that $K_r \not\subseteq G$ implies $\chi(G) \le \lceil \frac{r-1}{r}(\Delta(G) + 1) \rceil$. (Borodin-Kostochka [1977], Catlin [1978], Lawrence [1978])

5.1.29. (+) Prove that the 4-regular triangle-free graphs below are 4-chromatic. (Hint: consider the maximum independent sets. Comment: The graph on the left is the smallest triangle-free 4-regular 4-chromatic graph. Chvátal [1970])

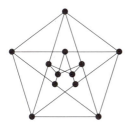 

5.1.30. Suppose G is a simple graph and $t = \lceil(\Delta(G) + 1)/j\rceil$. Prove that G can be t-colored such that for each i the subgraph G_i induced by the ith color class has no j-edge-connected subgraph. Prove that no smaller number of classes suffices if G is an j-regular j-edge-connected graph or an n-clique with $n \equiv 1 \bmod j$ (or an odd cycle when $j = 1$). (Matula [1973])

5.1.31. (+) *Chromatic number is bounded by one plus longest odd cycle length.*

a) Suppose G is a 2-connected nonbipartite graph containing an even cycle C. Prove that there exist vertices x, y on C and an x, y-path P internally disjoint from C such that $d_C(x, y) \ne d_P(x, y) \bmod 2$.

b) Suppose $\delta(G) \ge 2k$ and G has no odd cycle longer than $2k - 1$. Prove that G has a cycle of length at least $4k$. (Hint: consider the neighbors of an endpoint of a maximal path.)

c) Suppose G is 2-connected and has no odd cycle longer than $2k - 1$. Use part (a) and part (b) to prove that $\chi(G) \le 2k$. (Erdős-Hajnal [1966])

5.2 Structure of k-chromatic Graphs

We have observed that $\chi(H) \geq \omega(H)$ for all H. If equality holds in this bound for G and all its induced subgraphs, then we say G is *perfect*. We will present some examples of perfect graphs in Section 5.3 and study perfect graphs in more detail in Section 8.1. Our concern with the bound $\chi(G) \geq \omega(G)$ in this section is how *bad* it can be. Almost always $\chi(G)$ is much larger than $\omega(G)$.[†]

GRAPHS WITH LARGE CHROMATIC NUMBER

The bound $\chi(G) \geq \omega(G)$ can be tight, but it can also be arbitrarily bad. There have been many constructions of graphs having arbitrarily large chromatic number even though they do not contain K_3. In addition to the construction below, we present two earlier constructions in Exercises 5 and 6.

5.2.1. Example. *Mycielski's construction.* Mycielski [1955] found a construction that builds from any k-chromatic triangle-free graph G a $k+1$-chromatic triangle-free supergraph G'. Given G with vertex set $V = \{v_1, \ldots, v_n\}$, add vertices $U = \{u_1, \ldots, u_n\}$ and one more vertex w. Beginning with $G'[V] = G$, add edges to make u_i adjacent to all of $N_G(v_i)$, and then make $N(w) = U$. Note that U is an independent set in G'. From the 2-chromatic graph K_2, one iteration of Mycielski's construction yields the 3-chromatic C_5, and a second iteration yields the 4-chromatic *Grötzsch graph* drawn below. These graphs are the triangle-free k-chromatic graphs with fewest vertices for $k = 2, 3, 4$. $\square$

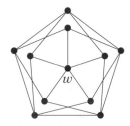

5.2.2. Theorem. Mycielski's construction produces a $k+1$-chromatic triangle-free graph from a k-chromatic triangle-free graph.

[†]The average values of $\omega(G)$, $\alpha(G)$, and $\chi(G)$ over all labeled n-vertex graphs are very close to $2\lg n$, $2\lg n$, and $n/(2\lg n)$ (see Section 8.5). Hence $\omega(G)$ is generally a bad lower bound on $\chi(G)$, while $n/\alpha(G)$ is generally a good lower bound on $\chi(G)$.

Proof. Suppose $V(G) = \{v_1, \ldots, v_n\}$ and $V(G') = \{v_i\} \cup \{u_i\} \cup \{w\}$, as described above. Since $\{u_i\}$ is independent in G', the other vertices of any triangle containing u_i belong to $V(G)$ and are neighbors of v_i, which completes a triangle in G. Therefore, if G is triangle-free, then G' is also triangle-free.

A proper k-coloring f of G extends to a proper $k+1$-coloring of G' by setting $f(u_i) = f(v_i)$ and $f(w) = k+1$. Now suppose G' has a proper k-coloring g. We may assume $g(w) = k$, which restricts g to $[k-1]$ on $\{u_i\}$. Let $A = \{v_i: g(v_i) = k\}$. We will change the coloring on A to obtain a proper $k-1$-coloring of G and thereby prove that $\chi(G) < \chi(G')$. For each $v_i \in A$, we change the color of v_i to $g(u_i)$. Because g properly colors G', A is an independent set in G, so we need only check edges of the form $v_i v'$ with $v' \in V(G) - A$. If $v' \leftrightarrow v_i$, then we have constructed G' so that $v' \leftrightarrow u_i$, which implies $g(u_i) \neq g(v')$. Hence our alteration does not violate edges within G. We ignore possible conflicts between v_i and u_j, because we now delete $U \cup \{w\}$ and have a proper $k-1$-coloring of G. $\square$

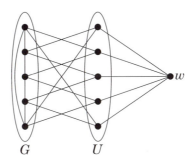

If G is color-critical, then the new graph G' generated by Mycielski's construction is also color-critical (Exercise 4). Although the graphs generated by Mycielski's construction starting with $G_2 = K_2$ are the smallest triangle-free k-chromatic graphs for $k = 2, 3, 4$, the graphs grow rapidly: $n(G_k) = 2n(G_{k-1}) + 1$. With $n(G_2) = 2$, this yields $n(G_k) = 3 \cdot 2^{k-2} - 1$ (exponential growth). Using probabilistic (non-constructive) methods, Erdős proved that there is a triangle-free k-chromatic graph with at most $ck^{2+\varepsilon}$ vertices, where ε is any positive constant and c depends on ε but not on k.

Blanches Descartes[†] [1947, 1954] constructed color-critical graphs that have no 3-cycles, 4-cycles or 5-cycles (Exercise 6). The *girth* of a graph is the length of its shortest cycle, or is infinite if there is no cycle. The search for graphs with large chromatic number and large girth helped motivate the study of random graphs. Erdős [1959] showed that almost every graph has a subgraph with chromatic number at least k and girth at least g (Theorem 8.5.9). By considering a more general

---

[†]This is a pseudonym used by W.T. Tutte.

question, Lovász [1968] found an explicit construction, later simplified by Nešetřil and Rödl [1979].

CRITICAL GRAPHS

Every k-chromatic graph has a k-critical subgraph. Describing the k-critical graphs could lead to an algorithm for chromatic number. We already know (Lemma 5.1.13) that $\delta(G) \geq k - 1$ if G is k-critical graph. The argument applies also to *vertex-color-critical* graphs, which are the graphs where deletion of any vertex decreases the chromatic number. Every color-critical graph is vertex-color-critical, but the converse is not true when $\chi(G) > 3$ (Exercises 10 and 11). We restrict our attention to color-critical graphs.

5.2.3. Remark. A graph H is color-critical if and only if 1) H has no isolated vertex, and 2) $\chi(H - e) < \chi(H)$ for every $e \in E(H)$. Hence when we prove that a connected graph is color-critical, we need only consider subgraphs obtained by deleting a single edge. □

5.2.4. Proposition. If $v \in V(G)$ and $\chi(G - v) < \chi(G) = k$, then G has a proper k-coloring in which $k - 1$ colors appear on $N(v)$ and the color on v appears nowhere else. If $e \in E(G)$ and $\chi(G - e) < \chi(G) = k$, then in every proper $k - 1$-coloring of $G - e$, the endpoints of e have the same color.

Proof. Since $G - v$ is $k - 1$-colorable, we can complete a k-coloring of G by using color k on v alone. The other colors must all appear on $N(v)$, else we would obtain a $k - 1$-coloring of G. If a proper $k - 1$-coloring of $G - e$ gives distinct colors to the endpoints of e, then we can add e to obtain a proper $k - 1$-coloring of G. □

For k-critical graphs, we can improve the requirement of $\delta(G) \geq k - 1$ to $\kappa'(G) \geq k - 1$ by using the König-Egerváry Theorem.

5.2.5. Lemma. (Kainen) Suppose G is a graph with $\chi(G) > k$, and X, Y is a partition of $V(G)$. If $G[X]$ and $G[Y]$ are k-colorable, then the edge cut $[X, Y]$ has at least k edges.

Proof. Let $X_1, \ldots, X_k$ and $Y_1, \ldots, Y_k$ be the partitions of X and Y formed by the color classes in the k-colorings of $G[X]$ and $G[Y]$. Form a bipartite graph H with vertices $X_1, \ldots, X_k$ and $Y_1, \ldots, Y_k$, putting $X_i Y_j \in E(H)$ if in G there is *no edge* between the set X_i and the set Y_j. If there are fewer than k edges between the sets X and Y in G, then H has more than $k(k - 1)$ edges. Since m vertices can cover at most km edges in a subgraph of $K_{k,k}$, $E(H)$ cannot be covered by fewer than k

vertices. By the König-Egerváry Theorem, H has a matching of size k, which is a complete matching. If $X_i \leftrightarrow Y_j$ in this matching, give color i to the vertices of X_i and Y_j in G. This completes a k-coloring of G. This contradicts the hypothesis that $\chi(G) > k$, so $|[X,Y]| \geq k$. $\square$

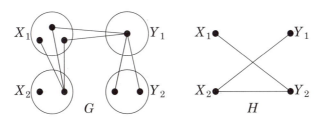

5.2.6. Theorem. (Dirac [1953]) Every k-critical graph is $k - 1$-edge-connected.

Proof. Suppose G is k-critical, and let $[X,Y]$ be a minimum edge cut. Since G is k-critical, $G[X]$ and $G[Y]$ are $k - 1$-colorable. Apply using $k - 1$ as the parameter, Lemma 5.2.5 then states that $|[X,Y]| \geq k - 1$. $\square$

High edge-connectivity does not prevent small vertex cuts, but we can restrict the behavior of small vertex cutsets in k-critical graphs.

5.2.7. Definition. Suppose S is a set of vertices in a graph G. An S-*component* of G is an induced subgraph of G whose vertex set consists of S and the vertices of a component of $G - S$.

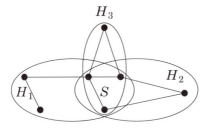

5.2.8. Proposition. If G is k-critical, then G has no cutset of vertices inducing a clique. In particular, if G has a cutset $S = \{x, y\}$, then $x \nleftrightarrow y$ and G has an S-component H such that $\chi(H + xy) \geq k$.

Proof. Suppose G is k-critical and S is a vertex cut. Let $H_1, \ldots, H_t$ be the S-components of G. Since each H_i is a proper subgraph of G and G is k-critical, each H_i is $k - 1$-colorable. If each H_i has a $k - 1$-coloring that assigns distinct colors to the vertices of S, then the names of the colors in the $k - 1$-colorings of $G_1, \ldots, G_t$ can be permuted to agree on S. In this case, the colorings can be combined to obtain a $k - 1$-coloring of

G, which is impossible.

Hence for some S-component H every $k-1$-coloring assigns the same color to some pair of vertices in S. In particular, S is not a clique. If $|S| = \{x, y\}$, then every $k-1$-coloring of H assigns the same color to x and y, which means that $H + xy$ is not $k-1$-colorable. $\square$

Exercise 18 strengthens this for $|S| = 2$. We also know more about the overall structure of k-critical graphs. Brooks' Theorem implies that the only $k-1$-regular k-critical graphs are the cliques and odd cycles (Exercise 5.1.27). Gallai [1963] strengthened this by proving that in the subgraph of a k-critical graph induced by the vertices of degree $k-1$, every block is a clique or an odd cycle (Exercise 25).

FORCED SUBDIVISIONS (optional)

We need not have a k-clique to have chromatic number k, but perhaps we must have some weakened form of a k-clique. Hajós [1961] conjectured that every k-chromatic graph contains a subdivision of K_k (a graph obtained from K_k by a sequence of edge subdivisions). This is trivial for $k \le 3$, since for $k = 3$ it states that every 3-chromatic graph contains a cycle. Dirac [1952a] proved the conjecture for $k = 4$. If F is a subdivision of H, then we call F an H-subdivision.

5.2.9. Theorem. (Dirac [1952a]) Every graph with chromatic number at least 4 contains a K_4-subdivision.

Proof. We use induction on $n(G)$. When $n(G) = 4$, the graph G is K_4 itself. Suppose $\chi(G) \ge 4$ and $n(G) > 4$, and let H be a a 4-critical subgraph of G. By Proposition 5.2.8, H has no cut-vertex. If $\kappa(H) = 2$ and $S = \{x, y\}$ is a 2-cut, then $x \not\leftrightarrow y$ and Proposition 5.2.8 yields an S-component H' of H such that $\chi(H' + xy) \ge 4$. Since $n(H') < n(G)$, we can apply the induction hypothesis to obtain a K_4-subdivision in H'. This subgraph F appears also in G unless it contains xy. In that case, we modify F to obtain a K_4-subdivision in G by replacing the edge xy with an x, y-path through another S-component of H. Such a path exists because, when S is a minimal cutset, each vertex of S has a neighbor in each component of $H - S$.

Hence we may assume that H is 3-connected, which implies $\delta(H) \ge 3$. Minimum degree at least 3 forces a cycle of length at least 4 (Lemma 1.2.18); let C be such a cycle in H. If C induces a clique, we are finished, so assume u, v are nonadjacent vertices on C. Since $H - u - v$ is connected, there is a shortest path P between the two portions of C in $H - u - v$. Suppose x, y are the endpoints of P. Again $H - x - y$ is connected, and there is a shortest path Q between the two portions of C in $H - x - y$. Let z, w be the endpoints of Q; the endpoints of P and Q appear around C in the order x, z, y, w. If P, Q do not intersect, then $C \cup P \cup Q$ is a K_4-subdivision. If P, Q do intersect, let Q' be the portion of Q from w to its first intersection with P. Now $C \cup P \cup Q'$ is a K_4-subdivision. $\square$

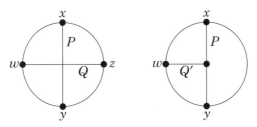

Catlin [1979] proved that Hajos' conjecture fails for $k \geq 7$ (see Exercise 23 for $k = 7$ and $k = 8$). Hadwiger [1943] proposed a weaker conjecture, that every k-chromatic graph contains a subgraph contractible to K_k, meaning a subgraph that becomes K_k after a sequence of edge contractions. This is a weaker conjecture because a K_k-subdivision is a special type of subgraph contractible to K_k. Hadwiger's conjecture is still open. For $k = 4$, it is implied by the truth of Hajós' Conjecture. For $k = 5$, it is equivalent to the famous Four Color Problem (see Chapter 7). For $k = 6$, it was proved using the Four Color Theorem by Robertson, Seymour, and Thomas [1993].

Some results about k-critical graphs extend to graphs with $\delta(G) \geq k - 1$. In Theorem 5.2.9, we proved that every 4-critical graph contains a K_4-subdivision; also every graph with $\delta(G) \geq 3$ contains a K_4-subdivision (Exercise 19). Dirac [1965] and Jung [1965] proved a weakened form of Hajós' Conjecture: G must contain a K_k-subdivision if $\chi(G)$ is sufficiently large. This extends in two ways: for every graph F, sufficiently large $\delta(G) \geq 2^{e(F)}$ forces an F-subdivision.

5.2.10. Lemma. (Mader [1967], see Thomassen [1988]) If $\delta(G) \geq 2k$, then G contains disjoint subgraphs G' and H such that 1) $\delta(G') \geq k$, 2) each vertex of G has a neighbor in H, and 3) H is connected.

Proof. We may assume that G is connected. When we contract the edges of a connected subgraph H', the set $V(H')$ becomes a single vertex, and we call the resulting graph $G \cdot H'$. Consider all connected subgraphs H' of G such that $G \cdot H'$ has at least $k(n(G) - n(H') + 1)$ edges . Since $\delta(G) \geq 2k$, every 1-vertex subgraph H' of G has this property. Choose H to be a maximal subgraph having this property. Let G' be the subgraph induced by the vertices of $G - V(H)$ that have neighbors in H.

We claim that $\delta(G') \geq k$. Suppose $d_{G'}(x) < k$ for some $x \in V(G')$. Choose $y \in N(x) \cap V(H)$. In $G \cdot H$ there are $d_{G'}(x)$ edges from x within

$V(G')$. In $G \cdot (H \cup xy)$ these edges collapse onto edges from $V(G')$ to H that appear in $G \cdot H$. The edge xy also contracts, but all other edges of $G \cdot H$ remain in $G \cdot (H \cup xy)$. Hence $e(G \cdot (H \cup xy)) = e(G \cdot H) - d_{G'}(x) - 1 \geq e(G \cdot H) - k$, which contradicts the maximality of H. $\square$

5.2.11. Theorem. (Mader [1967], see Thomassen [1988]) If $e(F) = m$ and $\delta(F) \geq 1$, then $\delta(G) \geq 2^m$ implies that G has a subdivision of F.

Proof. Proof by induction on m; the statement is trivial for $m \leq 1$. Consider $m \geq 2$. By the lemma, we may choose disjoint H and G' in G such that $\delta(G') \geq 2^{m-1}$, H is connected, and every vertex of G' has a neighbor in H. If e is an edge of F such that $F - e$ has no isolated vertices, then by induction G' contains a subdivision J of $F - e$, and a path through H can be added between the vertices of J that represent the endpoints of e to serve as a subdivision of e. If every edge of F is incident to a vertex of degree 1, then F is a forest of stars, and the minimum degree bound allows us to find F itself in G. $\square$

Let $f(k)$ be the smallest value of $\delta(G)$ that forces a K_k-subdivision in G. Since $K_{m,m-1}$ has no K_{2k}-subdivision when $m = k(k+1)/2$ (Exercise 22), we have $f(k) > k^2/8$. Szemerédi proved that $f(k) < ck^2 \log k$ for some constant c, which greatly improves the bound $f(k) \leq 2^{\binom{k}{2}}$ from Theorem 5.2.11. Dirac proved $f(4) = 3$, and Thomassen [1974] proved $f(5) \leq 8$. Because the graph of the icosahedron is 5-regular and has no K_5-subdivision, $f(5) \geq 6$. We could conclude $f(5) = 6$ from a proof of Dirac's striking conjecture [1964] that every n-vertex graph with at least $3n - 5$ edges contains a subdivision of K_5.

EXERCISES

5.2.1. Determine the minimum number of edges in a connected n-vertex graph with chromatic number k. (Eršov-Kožuhin [1962] - see Bhasker-Samad-West [1994] for higher connectivity.)

5.2.2. Suppose f is a proper k-coloring of a k-chromatic graph G. Prove that for each color there is a vertex having that color under f that is adjacent to vertices of all other colors.

5.2.3. Use properties of color-critical graphs to prove Prop. 5.1.12 again: $\chi(G) \leq 1 + \max_i \min \{d_i, i - 1\}$, where $d_1 \geq \cdots \geq d_n$ are the vertex degrees in G.

5.2.4. (!) Prove that if G is a color-critical graph, then the graph G' generated from it by applying Mycielski's construction is also color-critical.

5.2.5. Let $G_1 = K_1$. For $k > 1$, construct G_k from $G_1, \ldots, G_{k-1}$ by adding to the disjoint union $G_1 + \cdots + G_{k-1}$ a set T of $\Pi_{i=1}^{k-1} n(G_i)$ additional vertices. For each $v_1, \ldots, v_{k-1} \in V(G_1) \times \cdots \times V(G_{k-1})$, add one vertex with neighborhood

$v_1, \ldots, v_{k-1}$. The construction of G_4 is sketched below.

    a) Prove that $\omega(G_k) = 2$ and $\chi(G_k) = k$. (Zykov [1949])

    b) Prove that G_k is k-critical. (Schäuble [1969])

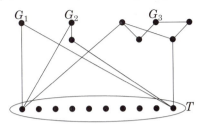

5.2.6. (+) *Construction of k-chromatic graph G_k with girth six.* Start with $G_2 = C_6$. Suppose k-chromatic G_k with n vertices is given. To construct G_{k+1}, let T be an independent set of kn new vertices. Take $\binom{kn}{n}$ pairwise disjoint copies of G_k, one for each way to choose an n-set $S \subset T$. Add a matching between each copy of G_k and the associated n-set S. Prove that the resulting graph G_{k+1} has chromatic number $k + 1$ and girth 6. (Blanche Descartes [1947, 1954])

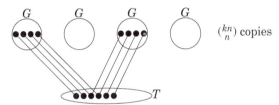

5.2.7. *Doubly-critical graph.* Suppose that $\chi(G - x - y) = \chi(G) - 2$ for all pairs x, y of distinct vertices. Prove that G is a clique. (Comment: Lovász conjectures that the same conclusion holds when $\chi(G - x - y) = \chi(G) - 2$ is required only for pairs of adjacent vertices.)

5.2.8. Prove that if G has no induced $2K_2$, then $\chi(G) \le \binom{\omega(G)+1}{2}$. (Hint: Use a maximum clique to define a collection of $\binom{\omega(G)}{2} + \omega(G)$ independent sets that cover the vertices. Comment: Mycielski's construction shows that forbidding a fixed clique cannot place a bound on $\chi(G)$ in terms of $\omega(G)$; forbidding an induced $2K_2$ does.) (Wagon [1980])

5.2.9. (+) Prove that every proper k-coloring of a k-chromatic graph contains each labeled k-vertex tree as a labeled subgraph. In particular, suppose $V(T) = \{w_1, \ldots, w_k\}$, and prove that there is an adjacency-preserving map $\phi: V(T) \to V(G)$ such that $f(\phi(w_i)) = i$ for all i. (Comment: ϕ need not preserve nonadjacency; for example, the result is trivial for $G = K_k$.)

5.2.10. (+) Prove that if G has a proper coloring g in which every color class has at least two vertices, then G has an optimal coloring f (minimum number of colors) in which every color class has at least two vertices. (Hint: If f has a color class with only one vertex, use g to make an alteration in f. The proof can be given algorithmically or by induction on $\chi(G)$.) (Gallai [1964])

5.2.11. Prove that the color-critical graphs with chromatic number 3 are the same as the vertex-color-critical graphs with chromatic number 3.

5.2.12. Prove that the graph below is vertex-color-critical but not color-critical.

5.2.13. (–) *The smallest k-critical graphs.*

a) Suppose x, y are vertices in a k-critical graph G. Prove that $N(x) \subseteq N(y)$ is impossible. Conclude that no k-critical graph has $k + 1$ vertices.

b) Prove that $\chi(G \vee H) = \chi(G) + \chi(H)$, and that $G \vee H$ is color-critical if and only if both G and H are color-critical. Conclude that $C_5 \vee K_{k-3}$, with $k + 2$ vertices, is k-critical.

5.2.14. *The Hajós construction.* Suppose G, H are k-critical graphs sharing only vertex v, with $vu \in E(G)$ and $vw \in E(H)$. Prove that $(G - vu) \cup (H - vw) \cup uw$ is k-critical. Use this to construct 4-critical graphs with n vertices for all odd $n \geq 7$. Give a separate construction of 4-critical graphs with n vertices for all even $n \geq 4$. (Hajós [1961])

5.2.15. (+) *Alternative proof that k-critical graphs are $k - 1$-edge-connected.*

a) Suppose G is k-critical with $k \geq 3$. Prove that for every $e, f \in E(G)$ there is a $k - 1$-critical subgraph of G containing e but not f. (Toft [1974])

b) Use part (a) and induction on k to prove Dirac's Theorem that every k-critical graph is $k - 1$-edge-connected. (Toft [1974])

5.2.16. (+) Prove that if G is k-critical and every $k - 1$-critical subgraph of G is isomorphic to K_{k-1}, then $G = K_k$ (if $k \geq 4$) (Hint: use Toft's critical graph lemma -- part (a) of Exercise 15.) (Stiebitz [1985])

5.2.17. (–) Find a subdivision of K_4 in the Grötzsch graph (Example 5.2.1).

5.2.18. Suppose G is k-critical and has a separating set $S = \{x, y\}$. Prove that 1) $x \not\leftrightarrow y$, 2) $G - S$ has exactly two S-components, and 3) the S-components of G can be named G_1, G_2 such that $G_1 + xy$ is a k-critical graph and $G_2 \cdot xy$ (add xy and then contract it) is a k-critical graph. Use these facts to shorten the proof that every 4-chromatic graph contains a subdivision of K_4.

5.2.19. (!) Prove that every simple graph with minimum degree at least 3 contains a K_4-subdivision. (Hint: prove the stronger result that every nontrivial graph with at most one vertex of degree less than 3 contains a K_4-subdivision. Use the result from the proof of Theorem 5.2.9 that every 3-connected graph contains a K_4-subdivision.) (Dirac [1952a])

5.2.20. (!) Given that $\delta(G) \geq 3$ forces a K_4-subdivision, prove that the maximum number of edges in a simple n-vertex graph with no K_4-subdivision is $2n - 3$.

5.2.21. Suppose $\chi(G) = k$ and G has girth at least 5. Prove that G contains every k-vertex tree as an induced subgraph. (Gyárfás-Szemeredi-Tuza [1980])

5.2.22. Let $m = k(k + 1)/2$. Prove that $K_{m,m-1}$ has no K_{2k}-subdivision.

5.2.23. Heavy edges below indicate that every vertex in one circle is adjacent to every vertex in the other. Prove that $\chi(G) = 7$ but G has no K_7-subdivision. Prove that $\chi(H) = 8$ but H has no K_8-subdivision. (Catlin [1979])

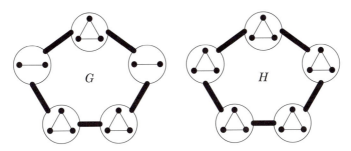

5.2.24. Suppose G is connected and k-chromatic and is not a complete graph or a cycle of length congruent to 3 modulo 6. Prove that every proper k-coloring of G has two vertices of the same color with a common neighbor. (Tomescu)

5.2.25. (+) *Minimum-degree vertices in k-critical graphs.*
    a) Prove that if every vertex of every even cycle of a graph is incident to a chord of that even cycle, then every block of the graph is a clique or an odd cycle.
    b) Suppose G is k-critical, v is a vertex of degree $k - 1$ in G, and f is a proper k-coloring of G using color k only on v. Prove that the two colors on any edge incident to v can be interchanged to obtain another proper k-coloring of G.
    c) Suppose G is k-critical and H is the subgraph of G induced by the vertices of degree $k - 1$. Prove that every block of H is a clique or an odd cycle. (Hint: using part (a), find an even cycle in H with a vertex v incident to no chord, and consider a k-coloring of G in which v is the only vertex of color k. Using part (b) repeatedly, produce a k-coloring of G that omits some lower color from the neighborhood of the sole vertex of color k.) (Gallai [1963])

5.3 Enumerative Aspects

Sometimes we can shed light on a difficult problem by considering a more general problem. We know no good algorithm to compute the minimum k such that G has a proper k-coloring (see Section 6.3), but we can define $\chi(G; k)$ to be the number of proper k-colorings of G. Knowing $\chi(G; k)$ for all k would permit finding the minimum k where the value is positive, which is the number $\chi(G)$. Birkhoff [1912] introduced this function as a possible way to attack the Four Color Problem.

In this section, we will explore properties of this counting function and classes where it is easy to compute. This leads us in one direction to the study of perfect graphs and in another direction to a relationship between colorings and acyclic orientations.

COUNTING PROPER COLORINGS

5.3.1. Definition. The function $\chi(G; k)$ counts the mappings $f: V(G) \to [k]$ that properly color G from the set $[k] = \{1, \dots, k\}$. In this definition, the k colors need not all be used, and permuting the colors used produces a different coloring.

5.3.2. Example. *Elementary examples.* When coloring the vertices of an independent set, we can independently choose one of the k colors at each vertex. Each of the k^n functions to $[k]$ is a proper coloring, and hence $\chi(\overline{K}_n; k) = k^n$.

Although K_3 has only one partition into three independent sets and none into four, we have $\chi(K_3; 3) = 6$ and $\chi(K_3; 4) = 24$. If we color $V(K_n)$ in some order, the colors chosen earlier cannot be used on the ith vertex, but there remain $k - i + 1$ choices available for the ith vertex no matter how the earlier colors were chosen. Hence $\chi(K_n; k) = k(k-1)\cdots(k-n+1)$. We obtain the same count by choosing n distinct colors and then multiplying by $n!$ to count the ways each such choice can be assigned to the vertices. The value of the formula is 0 if $k < n$, as it should be since K_n is k-chromatic.

If we choose some vertex of a tree as a root, we can color it in k ways. If we grow the tree from the root, along with a coloring, at every stage only the color of the parent is forbidden, and we have $k - 1$ choices for the color of the new vertex. Furthermore, by deleting a leaf, we can see inductively that every proper k-coloring arises in this way. Hence $\chi(T; k) = k(k-1)^{n-1}$ for every n-vertex tree. □

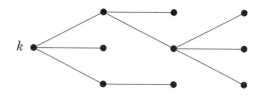

The answers are polynomials in k of degree $n(G)$. This holds for every graph, and hence $\chi(G; k)$ is called the *chromatic polynomial* of G.

5.3.3. Proposition. Let $x_{(r)} = x(x-1)\cdots(x-r+1)$. If $p_r(G)$ denotes the number of partitions of $V(G)$ into exactly r independent sets, then $\chi(G; k) = \sum_{r=1}^{n(G)} p_r(G) k_{(r)}$, which is a polynomial of degree $n(G)$.

Proof. When r colors are actually used in a proper coloring, they partition $V(G)$ into exactly r independent sets. With k colors available, the number of ways to assign colors to a partition when exactly r colors are used is $k_{(r)}$. This accounts for all colorings, so the formula for $\chi(G; k)$ holds. Since $k_{(r)}$ is a polynomial in k and $p_r(G)$ is a constant for each r, this formula implies that $\chi(G; k)$ is a polynomial function of k. Suppose G has n vertices. There is exactly one partition of G into n independent sets and no partition using more sets, so the leading term is k^n. $\square$

Listing partitions into independent sets is no easier than finding the smallest such partition (chromatic number), so computing the chromatic polynomial in this way is not feasible. There is also a recurrence, which looks much like that used in Chapter 2 to count spanning trees. Again $G \cdot e$ denotes the graph obtained by contracting the edge e in G, but since we are counting colorings we discard multiple copies of edges that arise from the contraction.

5.3.4. Theorem. (Chromatic recurrence) If G is a simple graph and $e \in E(G)$, then $\chi(G; k) = \chi(G - e; k) - \chi(G \cdot e; k)$.

Proof. Every proper k-coloring of G is a proper k-coloring of $G - e$. A proper k-coloring of $G - e$ is a proper k-coloring of G if and only if it gives distinct colors to the endpoints u, v of e. Hence we can count the proper k-colorings of G by subtracting from $\chi(G - e; k)$ the proper k-colorings of $G - e$ that assign u and v the same color. We claim that the number of these colorings equals $\chi(G \cdot e; k)$. This follows by establishing a bijection from the proper k-colorings of $G \cdot e$ to the proper k-colorings of $G - e$ that assign u and v the same color. Given a proper k-coloring of $G \cdot e$, we copy that coloring to $G - e$, except that u and v both receive the color that was assign to the contracted vertex. The inverse map gives the contracted vertex the color assigned to both u and v. $\square$

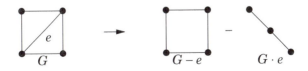

5.3.5. Example. *Proper k-colorings of C_4.* Deleting an edge of C_4 produces P_4, while contracting an edge produces K_3. Since P_4 is a tree and K_3 is a clique, we have $\chi(P_4; k) = k(k-1)^3$ and $\chi(K_3; k) = k(k-1)(k-2)$. Using the chromatic recurrence, we obtain $\chi(C_4; k) = \chi(P_4; k) - \chi(K_3; k) = k(k-1)(k^2 - 3k + 3)$. $\square$

Because both $G - e$ and $G \cdot e$ have fewer edges than G, we can use the chromatic recurrence inductively to compute $\chi(G; k)$. We need initial conditions for graphs with no edges, which we have already computed: $\chi(\overline{K}_n; k) = k^n$.

5.3.6. Theorem. $\chi(G; k)$ is a polynomial in k of degree $n(G)$, with integer coefficients alternating in sign and beginning $1, -e(G), \cdots$.

Proof. Proof by induction on $e(G)$. These claims hold trivially when $e(G) = 0$, where $\chi(\overline{K}_n; k) = k^n$. Suppose G is an n-vertex graph with $e(G) \geq 1$. Each of $G - e$ and $G \cdot e$ has fewer edges than G, and $G \cdot e$ has $n - 1$ vertices. By the induction hypothesis, there are nonnegative integers $\{a_i\}$ and $\{b_i\}$ such that $\chi(G - e; k) = \Sigma_{i=0}^{n}(-1)^i a_i k^{n-i}$ and $\chi(G \cdot e; k) = \Sigma_{i=0}^{n-1}(-1)^i b_i k^{n-1-i}$. By the chromatic recurrence,

$$
\begin{array}{llll}
\chi(G - e; k): & k^n - [e(G) - 1]k^{n-1} + a_2 k^{n-2} - \cdots & & +(-1)^i a_i k^{n-i} \cdots \\
-\chi(G \cdot e; k): & -(\quad\quad\quad k^{n-1} - b_1 k^{n-2} + \cdots & & +(-1)^{i-1} b_{i-1} k^{n-i}) \cdots \\
\hline
= \chi(G; k): & k^n - e(G)k^{n-1} + (a_2 + b_1)k^{n-2} - \cdots & +(-1)^i(a_i + b_{i-1})k^{n-i} \cdots
\end{array}
$$

Hence $\chi(G; k)$ is a polynomial with leading coefficient $a_0 = 1$ and next coefficient is $-(a_1 + b_0) = -e(G)$, and its coefficients alternate in sign. $\square$

5.3.7. Example. *Near-complete graphs.* If a graph has many edges, we may want to work up to cliques instead of down to their complements. Instead of $\chi(G; k) = \chi(G - e; k) - \chi(G \cdot e; k)$, we can write $\chi(G - e; k) = \chi(G; k) + \chi(G \cdot e; k)$. For example, to compute $\chi(K_n - e; k)$, we can let G be K_n in this alternative formula to obtain $\chi(K_n - e; k) = \chi(K_n; k) + \chi(K_{n-1}; k) = (k - n + 2)^2 \Pi_{i=0}^{n-3}(k - i)$. $\square$

The result of iterating the chromatic recurrence to the very end can be described directly. The formula has theoretical but not practical interest, because the summation has exponentially many terms. The formula immediately expresses $\chi(G; k)$ as a polynomial with leading terms $k^{n(G)} - e(G)k^{n(G)-1} \cdots$, but it doesn't immediately guarantee alternating coefficients.

Readers unfamiliar with the inclusion-exclusion principle should skip this theorem. The principle states that the number of items in a universe U that lie outside sets $A_1, \ldots, A_n$ is $\Sigma_S(-1)^{|S|}|\cap_{i \in S} A_i|$. The proof is that the sum counts an element positively for each even collection of sets it lies in and negatively for each odd collection, so the net count is 0 for elements belonging to any sets and 1 for elements belonging to none.

5.3.8. Theorem. Let $c(G)$ denote the number of components of a graph G. Given a set $S \subseteq E(G)$ of edges in G, let $G(S)$ denote the spanning subgraph of G with edge set S. Then the number $\chi(G; k)$ of

proper k-colorings of G is given by

$$\chi(G; k) = \Sigma_{S \subseteq E(G)}(-1)^{|S|} k^{c(G(S))}.$$

Proof. From the universe of $k^{n(G)}$ colorings, we want only those that don't assign the same color to the endpoints of any edge. Hence we define $e(G)$ sets; A_i is the set of k-colorings of $V(G)$ that violate edge e_i by assigning its endpoints the same color. By the inclusion-exclusion principle, the number of proper colorings is $\Sigma_S(-1)^{|S|}|\cap_{i \in S}A_i|$, where the summation runs over all subsets of the indices on the sets A_i. These indices correspond to the edges of G.

The set $\cap_{i \in S}A_i$ consists of the k-colorings violating each edge of S. In such a coloring, each vertex we can reach from x using edges in S must have the same color as x. Hence vertices in the same component of $G(S)$ must have the same color, which we can pick in k ways. The choice made for one component does not affect the choices for any other component, so there are $k^{c(G(S))} = |\cap_{i \in S}A_i|$ ways to make the choices for all the vertices of G to violate the spanning subgraph $G(S)$ (additional edges may also be violated). In the illustration below, G has 14 edges, the solid edges form a set S of size 6, and $c(G(S)) = 5$ (counting the two isolated vertices); there are k^5 ways to choose a map from $V(G)$ to $[k]$ that violates this set of 14 edges. $\square$

5.3.9. Example. *A chromatic polynomial.* When $G = K_4 - e$, every spanning subgraph with one edge has three components, and every spanning subgraph with two edges has two components. When $|S| = 3$, the number of components is 2 if and only if the three edges form a triangle. There are two such sets of three edges, and the other $\binom{5}{3} - 2$ sets of three edges yield spanning subgraphs with one component. All spanning subgraphs with four or five edges have only one component. Hence the inclusion-exclusion computation is $\chi(G; k) = k^4 - 5k^3 + 10k^2 - (2k^2 + 8k^1) + 5k - k = k^4 - 5k^3 + 8k^2 - 4k$. We can check this by computing directly that $\chi(G; k) = k(k-1)(k-2)(k-2)$, but such a direct computation is seldom available. $\square$

CHORDAL GRAPHS

Counting colorings is easy for cliques and trees because each such graph can be grown from K_1 by successively adding a vertex joined to a clique. The chromatic polynomial of such a graph is a product of linear factors. We next study the graphs where this phenomenon holds.

5.3.10. Definition. A vertex of G is *simplicial* if its neighborhood in G induces a clique. A *simplicial elimination ordering* is an order $v_n, \ldots, v_1$ in which vertices can be deleted so that each vertex v_i is a simplicial vertex of the remaining graph induced by $\{v_1, \ldots, v_i\}$. (Historically, these have been called *perfect elimination orderings*.)

5.3.11. Example. *Chromatic polynomials from simplicial elimination orderings.* In a tree, a simplicial elimination ordering is a successive deletion of leaves. We have observed that $\chi(G; k) = k(k-1)^{n-1}$ when G is an n-vertex tree. In general, when $v_n, \ldots, v_1$ is a simplicial elimination ordering for G, we can apply the product rule of elementary combinatorics to the reverse order $v_1, \ldots, v_n$ to count proper k-colorings of G. When we add v_i in this reverse order, there are $k - d(i)$ ways to color v_i, where $d(i) = |N(v_i) \cap \{v_1, \ldots, v_{i-1}\}|$. The factor $k - d(i)$ is the same regardless of how previous color choices were made, because the neighbors of v_i that have been colored form a clique of size $d(i)$ and have distinct colors. Furthermore, by deleting the simplicial vertex that starts the simplicial elimination ordering, we see inductively that every proper k-coloring of G arises in this way. The resulting formula factors the chromatic polynomial and shows that the chromatic polynomial of such a graph has only nonnegative integer roots. For example, the chromatic polynomial of the graph below is $k(k-1)^2(k-2)^2(k-3)$. $\square$

Trees, cliques, near-complete graphs $(K_n - e)$, and interval graphs (Exercise 19) all have simplicial elimination orderings. If G is a cycle of length more than 3, then G cannot have a simplicial elimination ordering, because a cycle has no simplicial vertex to start the elimination. As in Example 5.3.5, the chromatic polynomial of a graph with an induced cycle of length at least 4 cannot be expressed as a product of linear factors. We can use induced cycles to characterize the graphs having simplicial elimination orderings. We prove equivalence via a convenient third property.

5.3.12. Definition. A *chordless cycle* in G is an induced subgraph of G isomorphic to C_t for some $t \geq 4$. A graph G is *chordal* if it has no chordless cycle (every cycle of length at least 4 has a "chord"). If $x \nleftrightarrow y$, then an x, y-*separator* is a set $S \subseteq V(G)$ such that x and y lie in distinct components of $G - S$. A *minimal vertex separator* is a minimal x, y-separator for some nonadjacent pair x, y.

Although every minimal separating set of G is a minimal vertex separator, a minimal vertex separator need not be a minimal separating set of G (Exercise 15).

5.3.13. Theorem. For a simple graph G, the following properties are equivalent (and characterize the chordal graphs).
A) G has a simplicial elimination ordering.
B) G has no chordless cycle.
C) Every minimal vertex separator of G induces a clique.

Proof. A $\Rightarrow$ B. Let C be a cycle in G of length at least 4. When the simplicial elimination ordering first deletes a vertex of C, its remaining neighbors must induce a clique. Since this includes its neighbors on C, we obtain a chord of C.

B $\Rightarrow$ C. Let S be a minimal x, y-separator in G, and suppose u, v are distinct vertices of S. By the minimality of the separator, each of u, v has an edge to the components A, B of $G - S$ containing x and y. The union of shortest u, v-paths through A and through B is a cycle of length at least 4. By the choice of the paths and the absence of edges from A to B, this cycle has no chord other than uv. Since G has no chordless cycle, we conclude that $u \leftrightarrow v$. Since $u, v \in S$ were chosen arbitrarily, S induces a clique.

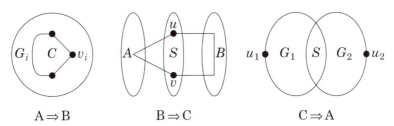

A $\Rightarrow$ B   B $\Rightarrow$ C   C $\Rightarrow$ A

C $\Rightarrow$ A. We show first that if H is an induced subgraph of G, then every minimal x, y-separator of H is contained in a minimal x, y-separator of G; this implies that C holds for every induced subgraph of G. If S is a minimal x, y-separator in H, then $S \cup (V(G) - V(H))$ separates x and y in G. Hence $S \cup (V(G) - V(H))$ contains a minimal x, y-separator of G. Such a set T must contain S, since otherwise $G - T$ contains an x, y-path within H.

With condition C holding for every induced subgraph of G, it suffices to prove that G itself has a simplicial vertex to start the elimination ordering. Using induction on $n(G)$, we prove the stronger result that if G satisfies C and is not a clique, then G has a pair of nonadjacent simplicial vertices. When G is a clique, every vertex is simplicial. When G is not a clique, let x_1, x_2 be a nonadjacent pair of vertices in G, let S be a minimal x_1, x_2-separator, and let G_i be the S-component of G containing x_i (see Definition 5.2.7). Since C holds for induced subgraphs, it holds for G_i. By the induction hypothesis, G_i has a simplicial vertex $u_i \notin S$ (whether or not G_i is a clique). Since there are no edges between $V(X)$ and $V(Y)$, these vertices u_1, u_2 are also simplicial in G, and they are nonadjacent in G. $\square$

A HINT OF PERFECT GRAPHS

For bipartite graphs and interval graphs, $\chi(G) = \omega(G)$. This also holds for every induced subgraph of such graphs.

5.3.14. Definition. A graph G is *perfect* if $\chi(H) = \omega(H)$ for every induced subgraph $H \subseteq G$. Equivalently, $\chi(G[A]) = \omega(G[A])$ for all $A \subseteq V(G)$. A family of graphs **G** is *hereditary* if every induced subgraph of a graph in **G** is also a graph in **G**. The *clique covering number* $\theta(G)$ of a graph G is the minimum number of cliques in G needed to cover $V(G)$; note that $\theta(G) = \chi(\overline{G})$.

Since cliques and independent sets exchange roles under complementation, the statement of perfection for $\overline{G}$ is "$\alpha(H) = \theta(H)$ for all induced subgraphs H of G". Lovász proved the *Perfect Graph Theorem (PGT)*: G is perfect if and only if its complement $\overline{G}$ is perfect. We will prove this in Section 8.1. Here we merely illustrate perfect graphs by exploring our earlier results about interval graphs and bipartite graphs. In order to verify that every graph in a hereditary class is perfect, it suffices to verify that $\chi(G) = \omega(G)$ for every graph G in the class (because the same argument applies to its induced subgraphs).

5.3.15. Example. *Bipartite graphs and their line graphs.* Bipartite graphs form a hereditary class, and $\chi(G) = 2 = \omega(G)$ for every bipartite graph having an edge; hence bipartite graphs are perfect. If H is bipartite, then the statement of perfection for $\overline{H}$ is Exercise 5.1.18 and follows from $\alpha(H) = \beta'(H)$ (Corollary 3.1.15). We could obtain $\alpha(G) = \theta(G) = \beta'(G)$ immediately from the trivial $\chi(G) = \omega(G)$ by the PGT.

In Section 6.1, we will study line graphs; here we introduce them briefly for illustration. The *line graph* $L(G)$ of a graph G has a vertex for each edge of G, with vertices of $L(G)$ adjacent if as edges in G they have a common endpoint. Each independent set in $L(G)$ corresponds to a matching in G, and a clique in $L(G)$ consists of edges in G sharing a single vertex or forming a triangle. Hence the statement that $\alpha(L(G)) = \theta(L(G))$ when G is bipartite is the König-Egerváry Theorem [1931] ($\alpha'(G) = \beta(G)$) about matchings in bipartite graphs.

Since complements of line graphs of bipartite graphs form a hereditary class, we conclude that they are perfect. From this the PGT yields $\chi(L(G)) = \omega(L(G))$. A proper coloring of $L(G)$ is a partition of $E(G)$ into matchings, and $\omega(L(G)) = \Delta(G)$ when G is bipartite. Hence $\chi(L(G)) = \omega(L(G))$ says that $E(G)$ can be partitioned into $\Delta(G)$ matchings when G is bipartite (König [1916]). We will prove this directly in Section 6.1. $\square$

Perfection of interval graphs is a special case of perfection of chordal graphs, since every interval graph is a chordal graph (Exercise 19). We will explore other characterizations of interval graphs and chordal graphs in Section 8.1.

5.3.16. Theorem. (Berge [1960]) Chordal graphs are perfect.

Proof. Deleting vertices cannot create chordless cycles, so the chordal graphs form a hereditary family, and we need only prove $\chi(G) = \omega(G)$. We use induction on $n(G)$. When $n(G) = 1$, the only example is K_1.

For the induction step, suppose $n(G) > 1$. In Theorem 5.3.11, we proved that every chordal graph has a simplicial elimination ordering, which begins with a simplicial vertex x. Since $N(x)$ induces a clique, we have $d_G(x) \le \omega(G - x)$. Since $G - x$ is chordal, the induction hypothesis yields $\chi(G - x) = \omega(G - x) = k$. We can color x from among these k colors unless $d_G(x) = \omega(G - x)$, in which case $\omega(G) = k + 1$ and we introduce a new color for x to complete a proper $\omega(G)$-coloring. This argument shows that a chordal graph is optimally colored by the greedy coloring with respect to the reverse of a simplicial elimination ordering. $\square$

Another fundamental class of perfect graphs generalizes the family of bipartite graphs:

5.3.17. Example. *Comparability graphs (Berge [1960]).* A simple graph G is a *comparability graph* if it has a *transitive orientation*, which is an orientation such that if $x \to y$ and $y \to z$, then also $x \to z$. If G is bipartite with bipartition X, Y, then the orientation obtained by directing every edge from X to Y is transitive, so every bipartite graph is a comparability graph. The name "comparability" arises from order relations; $x \to y$ could mean "x must happen before y".

Every induced subdigraph of a transitive digraph is transitive, so the class of comparability graphs is hereditary, and we need only show $\omega(G)$-colorability for each comparability graph G. Let Q be a maximal clique in G, and let F be a transitive orientation of G. On Q, F is a tournament. Since a transitive orientation is acyclic, some vertex v of Q has no predecessor in Q. If v has a predecessor u outside Q, then transitivity yields edges from u to the rest of Q, contradicting the maximality of Q in G. Hence every maximal clique of G contains a vertex with in-degree 0 in F. These form an independent set, and we use them as a color class. Deleting them decreases every maximal clique; hence the clique size decreases and we can proceed by induction. The sets obtained by iteratively stripping the set of vertices with in-degree 0 form an optimal coloring. The color assigned to x is 1 plus the length of the longest path ending at x, as in the argument for Theorem 5.1.15. $\square$

COUNTING ACYCLIC ORIENTATIONS (optional)

Surprisingly, $\chi(G; k)$ has meaning when evaluated at negative integers. In particular, setting $k = -1$ enables us to count the acyclic orientations of G.

5.3.18. Example. Since C_4 has 4 edges, it has 16 orientations. Of these, 14 are acyclic. In Example 5.3.5, we proved that $\chi(C_4; k) = k(k-1)(k^2 - 3k + 3)$. Evaluated at $k = -1$, this equals $(-1)(-2)(7) = 14$. $\square$

5.3.19. Theorem. (Stanley [1973]) The value of $\chi(G; k)$ at $k = -1$ is $(-1)^{n(G)}$ times the number of acyclic orientations of G.

Proof. Let $a(G)$ be the number of acyclic orientations of G. When G has no edges, $a(G) = 1$ and $\chi(G; -1) = (-1)^{n(G)}$, and the claim holds. We have a recurrence for $\chi(G; -1)$; if we provide a similar recurrence for $a(G)$, we can prove this claim by induction on $e(G)$. We will prove that $a(G) = a(G - e) + a(G \cdot e)$. If this holds for $e(G) > 0$, then the induction hypothesis yields $a(G) = a(G - e) + a(G \cdot e) = (-1)^{n(G)}\chi(G - e; -1) + (-1)^{n(G)-1}\chi(G \cdot e; -1) = (-1)^{n(G)}\chi(G; -1)$.

Now we prove that $a(G) = a(G - e) + a(G \cdot e)$ for $e(G) > 0$. Every acyclic orientation of G contains an acyclic orientation of $G - e$. How many acyclic orientations of G arise from a given acyclic orientation of $G - e$? Let D be an acyclic orientation of $G - e$, where $e = uv$. If D has no u, v-path, then we can orient e from v to u to complete an acyclic orientation of G. If D has no v, u-path, then we can orient e from u to v to complete an acyclic orientation of G. Since D is acyclic, D cannot have both a u, v-path and a v, u-path, so the possibilities cannot both be forbidden. Hence every such D extends to G in at least one way, and $a(G)$ equals $a(G - e)$ plus the number of orientations that have no u, v-path *and* no v, u-path. This latter number is $a(G \cdot e)$, because a u, v-path or a v, u-path in an orientation of $G - e$ becomes a cycle in $G \cdot e$. $\square$

This is an instance of the phenomenon of "combinatorial reciprocity". For example, the number of selections of n elements from k types of elements without repetition allowed is $\binom{k}{n}$, which is a polynomial in k of degree n. If we allow repetition, the answer is $\binom{k+n-1}{n}$, which turns out to be $(-1)^n$ times the value of the first polynomial at $-k$ instead of k. Stanley [1974] explores this more deeply; see also Exercise 23.

EXERCISES

Keep in mind that for each nonnegative integer k, $\chi(G;k)$ equals the number of proper k-colorings of G.

5.3.1. (–) Compute the chromatic polynomial of the graph drawn below.

5.3.2. (–) Use the chromatic recurrence to derive the chromatic polynomial of a tree with n vertices.

5.3.3. a) If G is a cycle on n points, prove that $\chi(G;k) = (k-1)^n + (-1)^n(k-1)$.
    b) If $H = G \vee K_1$, prove that $\chi(H;k) = k\chi(G;k-1)$. From this and (a), give the chromatic polynomial of the wheel $C_n \vee K_1$.

5.3.4. Let G_n be the graph on $2n$ vertices and $3n-2$ edges pictured below, for $n \geq 1$. Prove that the chromatic polynomial of G_n is $(k^2 - 3k + 3)^{n-1}k(k-1)$.

5.3.5. (–) Prove that the chromatic polynomial of an n-vertex graph has no real root larger than $n-1$. (Hint: use Proposition 5.3.3.)

5.3.6. Use Proposition 5.3.3 to give a non-inductive proof that the coefficient of $k^{n(G)-1}$ in $\chi(G;k)$ is $-e(G)$.

5.3.7. (!) Prove that the last nonzero term in the chromatic polynomial of G is the term whose exponent is the number of components of G. Conclude that if $p(k) = k^n - ak^{n-1} + \cdots \pm ck^r$ and $a > \binom{n-r+1}{2}$, then p is not a chromatic polynomial. (For example, $k^4 - 4k^3 + 3k^2$ is not a chromatic polynomial.)

5.3.8. (!) Prove that the sum of the coefficients of $\chi(G;k)$ is 0 unless G has no edges. (Hint: what computes the sum of the coefficients polynomial?)

5.3.9. (!) Prove that the number of proper k-colorings of a connected graph G is less than $k(k-1)^{n-1}$ if $k \geq 3$ and G is not a tree. What happens if $k = 2$?

5.3.10. Write the chromatic polynomial as $\chi(G; k) = \sum_{i=0}^{n-1}(-1)^i a_i k^{n-i}$. If G is a connected graph, prove that $a_i \geq \binom{n-1}{i}$ for $1 \leq i \leq n$. (Hint: Use the chromatic recurrence.)

5.3.11. Suppose that $F = G \cup H$ and that $G \cap H$ is a clique. Prove that $\chi(F; k) = \frac{\chi(G; k)\chi(H; k)}{\chi(G \cap H; k)}$. What happens if $G \cap H$ is not a clique?

5.3.12. (!) Let P be the Petersen graph. By Brooks' Theorem, the Petersen graph is 3-colorable, and hence by the pigeonhole principle it has an independent set S of size four.

a) Prove that $P - S = 3K_2$.

b) Using (a) and symmetry, determine the number of vertex partitions of P into three independent sets.

c) In general, how can the number of partitions into the minimum number of independent sets be obtained from the chromatic polynomial of G?

5.3.13. Prove that a graph with chromatic number k has at most k^{n-k} vertex partitions into k independent sets, with equality achieved only by $K_k + (n-k)K_1$ (a k-clique and $n-k$ isolated vertices. (Hint: use induction on n and consider the deletion of a single vertex.) (Tomescu [1971])

5.3.14. (!) Prove that $\chi(G; x + y) = \sum_{U \subseteq V(G)} \chi(G[U]; x)\chi(G[\bar{U}]; y)$.

5.3.15. (−) Construct an example of a connected chordal graph G having a minimal vertex separator that is not a minimal separating set of G.

5.3.16. (!) a) Without computing the chromatic polynomials, give a short proof that the two graphs below have the same chromatic polynomial.

b) Express this chromatic polynomial as the sum of the chromatic polynomials of two chordal graphs, and use this to give a short computation of it.

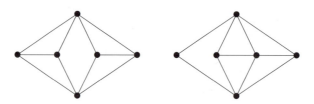

5.3.17. Among the statements below, B$\Rightarrow$C was proved in Theorem 5.3.6, and C$\Rightarrow$A holds because every minimal separating set is a minimal vertex separator. Prove A$\Rightarrow$B directly.

A) Every minimal separating set of G induces a clique.

B) G has no chordless cycle.

C) Each minimal vertex separator of G induces a clique.

5.3.18. Suppose G is a chordal graph. Use a simplicial elimination ordering of G to prove the following statements:

a) G has at most n maximal cliques, with equality if and only if G has no edges. (Fulkerson-Gross [1965])

b) Every maximal clique of G that contains no simplicial vertex of G is a separating set of G.

5.3.19. (!) Suppose G is an interval graph. Prove that $\overline{G}$ is a comparability graph and that G is a chordal graph. (Hint: establish a simplicial elimination ordering.)

5.3.20. Determine the smallest imperfect graph G such that $\chi(G) = \omega(G)$.

5.3.21. The number $a(G)$ of acyclic orientations of G satisfies the recurrence $a(G) = a(G - e) + a(G \cdot e)$ (Theorem 5.3.19). The number of spanning trees of G appears to satisfy the same recurrence; does the number of acyclic orientations of G always equal the number of spanning trees? Why or why not?

5.3.22. In an acyclic orientation of G, an edge is *dependent* if reversing it would create a cycle. Use the technique of the Gallai-Roy Theorem (Theorem 5.1.14) to prove that if $\chi(G)$ is less than length of the shortest cycle in G, then G has an orientation with no dependent edges.

5.3.23. Suppose D is an acyclic orientation of G and f is a coloring of $V(G)$ from the set $[k]$. We say that (D, f) is a *compatible pair* if $u \rightarrow v$ in D implies $f(u) \le f(v)$. Let $\eta(G; k)$ be the number of compatible pairs. Prove that $\eta(G; k) = (-1)^{n(G)} \chi(G; k)$. (Stanley [1973])

Chapter 6

Edges and Cycles

6.1 Line Graphs and Edge-coloring

Many questions about vertices have natural analogues involving edges. Independent sets have no adjacent pair of vertices; matchings have no intersecting pair of edges. Vertex colorings partition the vertices into independent sets; we can instead partition the edges into matchings. These pairs of problems are related via line graphs. The definition below applies also to digraphs, but in this chapter we discuss only undirected graphs. We use "line graph" instead of "edge graph" because $E(G)$ denotes the edge set.

6.1.1. Definition. The *line graph* of G, written $L(G)$, is the graph whose vertices are the edges of G, with $ef \in E(L(G))$ when $e = uv$ and $f = vw$ in G (i.e., when e and f share a vertex).

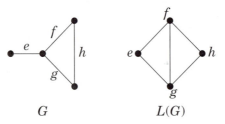

$$G \qquad\qquad L(G)$$

Some questions about edges in graphs can be phrased as questions about vertices in line graphs. When asked for general graphs, the vertex question becomes more general and perhaps more difficult. On the other hand, if we can answer the vertex question for general graphs, then we can answer the edge question by applying the vertex result to a line graph.

In Chapter 2, we discussed Eulerian graphs. An Eulerian circuit in G becomes a spanning cycle in $L(G)$. (Exercise 6.2.6 shows that the converse does not hold!) In Section 6.2, we study spanning cycles in general graphs. As discussed in Section 6.3, this problem is hard both for general graphs and for line graphs.

In Chapter 3, we studied matchings. A matching in G becomes an independent set in $L(G)$; we have $\alpha'(G) = \alpha(L(G))$. The study of $\alpha'(G)$ for general graphs is the study of α for line graphs. Computing α is much harder for general graphs than for line graphs. We discussed the bipartite case in Section 3.1 and discuss the general case briefly in Section 6.3.

In Chapter 4, we studied connectivity. Given Menger's Theorem for internally-disjoint paths in G, we can use line graphs to prove the corresponding result for edge-disjoint paths in G (Exercise 10). The desired equality between $\kappa'(G)$ and $\lambda'(G)$ for edge-disjoint paths becomes the known equality between $\kappa(L(G'))$ and $\lambda(L(G'))$ for a slight modification G' of G. Applying the vertex theorem to a line graph proves the edge theorem for general graphs.

In Chapter 5, we studied vertex colorings. We could also color edges so that edges sharing vertices have different colors; this is a vertex coloring of the line graph. Hence edge-coloring is a special case of vertex coloring, and we can hope for stronger results. In this section we present an algorithm that computes $\chi(H)$ within one when H is the line graph of a simple graph.

Thus line graphs lead us to the problems of edge-coloring and spanning cycles. There are other connections between these two problems, particularly for the class of planar graphs studied in Chapter 7. Meanwhile, if we have an algorithm that works for line graphs, and we want to apply it to G, we need to know whether G is a line graph. There are good algorithms to check this; they depend on characterizations of line graphs, which we postpone to the end of this section.

EDGE-COLORINGS

6.1.2. Example. *Edge-coloring of K_{2n}.* In a league with $2n$ teams, we may want to schedule games so that each pair of teams plays each other, but each team plays at most once a week. Since each team must play $2n - 1$ others, the season lasts at least $2n - 1$ weeks. The games of each week must form a matching. We can schedule the season in $2n - 1$ weeks if and only if we can partition $E(K_{2n})$ into $2n - 1$ disjoint matchings. Since K_{2n} is $2n - 1$-regular, these must be complete matchings.

The figure below describes the solution. Arrange $2n-1$ vertices cyclically, and let the *length* of an edge between two of these be the number of steps between them along the circle. This creates $2n-1$ edges of each length $1, \ldots, n-1$. In the figure, the solid matching has one edge of each length, plus an edge from the central vertex to the left-over vertex on the circle. Rotating the picture as indicated by the dashed matching yields new edges, again one of each class. The $2n-1$ rotations of the figure yield the desired matchings. □

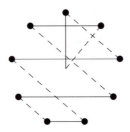

6.1.3. Definition. A *k-edge-coloring* of G is a labeling $f: E(G) \to [k]$; the labels are "colors". A *proper k-edge-coloring* is a k-edge-coloring such that edges sharing a vertex receive different colors; equivalently, each color is used on a matching. A graph G is *k-edge-colorable* if it has a proper k-edge-coloring. The *edge-chromatic number* or *chromatic index* $\chi'(G)$ is the minimum k such that G is k-edge-colorable.

Because the edges incident to one vertex need different colors, $\chi'(G) \geq \Delta(G)$. Vizing and Gupta independently proved that $\Delta(G)+1$ colors suffice when G is simple; this theorem is the main objective of this section. We can interpret it as saying that line graphs of simple graphs are "almost perfect". A clique in $L(G)$ corresponds to a set of pairwise-intersecting edges; these can only be the edges of a star or a triangle (Exercise 2). Therefore, for line graphs of simple graphs, which form a hereditary class, Vizing's Theorem says that the chromatic number is at most one more than the clique number.

6.1.4. Example. *Multigraphs and the "fat triangle".* Since we are coloring edges, we can extend the problem to multigraphs, for which $\chi'(G)$ may exceed $\Delta(G)+1$. Shannon [1949] proved that the maximum value of $\chi'(G)$ in terms of $\Delta(G)$ alone is $3\Delta(G)/2$ (Exercise 18). Vizing and Gupta proved that $\chi'(G) \leq \Delta(G) + \mu(G)$, where $\mu(G)$ is the maximum edge multiplicity. In the graph below, the edges are pairwise intersecting, so they must receive distinct colors, and $\chi'(G) = 3\Delta(G)/2 = \Delta(G) + \mu(G)$. □

To prove that $\chi'(G) \le 2\Delta(G) - 1$ for all multigraphs, we color the edges in some order, always assigning the current edge the least-indexed color different from those already appearing on edges incident to it. Since no edge is incident to more than $2(\Delta(G) - 1)$ other edges, we never use more than $2\Delta(G) - 1$ colors. This is precisely a greedy coloring for vertices of $L(G)$: $\chi'(G) = \chi(L(G)) \le \Delta(L(G)) + 1 \le 2\Delta(G) - 1$. For bipartite graphs, we can use the results of Chapter 3 to do much better, achieving the trivial lower bound.

6.1.5. Theorem. (König [1916]) If G is a bipartite multigraph, then $\chi'(G) = \Delta(G)$.

Proof. Corollary 3.1.8 states that every regular bipartite multigraph H has a 1-factor. By induction on $\Delta(H)$, this yields a proper $\Delta(H)$-edge-coloring. It therefore suffices to show that every bipartite multigraph G with maximum degree k has a k-regular bipartite supergraph H.

We construct such a supergraph. If G does not have the same number of vertices in each partite set, add vertices to the smaller set to equalize the sizes. If the resulting G' is not regular, then each partite set has a vertex with degree less than $\Delta(G') = \Delta(G)$. Add an edge consisting of this pair. Continue adding such edges until the graph becomes regular. $\square$

A regular graph G has a $\Delta(G)$-edge-coloring if and only if decomposes into 1-factors. Such subgraphs form a *1-factorization* of G, and we then say G is *1-factorable*. For an odd cycle, $\chi'(G) = 3 > \Delta(G)$. The Petersen graph also requires an extra color, but only one.

6.1.6. Example. *The Petersen graph is 4-edge-chromatic.* Consider the usual drawing of the Petersen graph, consisting of an outer 5-cycle, an inner (twisted) 5-cycle, and a matching between them (the *cross edges*). Since C_5 is 3-edge-colorable, giving the five cross edges the same color produces a 4-edge-coloring. To prove that the Petersen graph is not 3-edge-colorable, we need only show that every 2-factor of the Petersen graph is isomorphic to $2C_5$.

Since a 2-factor is a union of disjoint cycles, a 2-factor H of the Petersen graph has an even number m of cross edges. If $m = 0$, then $H = 2C_5$. If $m = 2$, then the two cross edges have nonadjacent endpoints on the inner cycle or the outer cycle. On the cycle where their endpoints are nonadjacent, the remaining vertices force all five edges of that cycle into H (see illustration), which is impossible. Finally, if $m = 4$, then the cycle edges forced into H by the unused cross edge yield $2P_5$ in H in such a way that the only completion to a 2-factor is again $2C_5$. $\square$

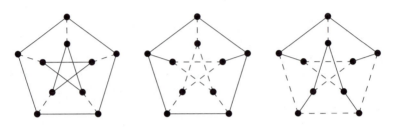

Now we consider simple graphs in general. We make $\Delta(G) + 1$ colors available and build a proper edge-coloring, incorporating edges one by one until we have a proper $\Delta(G) + 1$-edge-coloring of G. The algorithm runs surprisingly quickly, considering that it is hard to tell whether a graph has a proper $\Delta(G)$-edge-coloring (Section 6.3).

6.1.7. Theorem. (Vizing [1964, 1965], Gupta [1966]) Every simple graph with maximum degree Δ has a proper $\Delta + 1$-edge-coloring.

Proof. Suppose uv is an edge left uncolored by a proper $\Delta(G) + 1$-edge-coloring f of a proper subgraph G' of G. After possibly recoloring some edges, we extend the coloring to include uv; call this an *augmentation*. After $e(G)$ augmentations, we obtain a proper $\Delta(G) + 1$-coloring of G.

Since the number of colors exceeds $\Delta(G)$, every vertex has some color *not* appearing on its incident edges. Let a_0 be a color missing at u, and let a_1 be a color missing at v. We may assume that a_1 appears at u, or we could use a_1 on uv. Suppose v_1 is the neighbor of u along the edge of color a_1. At v_1 some color a_2 is missing. We may assume that a_2 appears at u, or we could recolor uv_1 from a_1 to a_2 and then use a_1 on uv to extend the coloring.

For $i \geq 2$, we continue this process. Having selected a new color a_i that appears at u, let v_i be the neighbor of u along the edge of color a_i. Let a_{i+1} be a color missing at v_i. If a_{i+1} is missing at u, then we shift color a_j from uv_j to uv_{j-1} for $1 \leq j \leq i$ (where $v_0 = v$) to complete the augmentation. We call this shifting of colors *downshifting from i* (see illustrations). We are finished unless a_{i+1} appears at u, in which case the process continues.

Since we have only $\Delta(G)+1$ colors to choose from, the iterative selection of a_{i+1} eventually repeats a color. Let l be the smallest index such that a color a_{l+1} missing at v_l is in the list $a_1,\ldots,a_l$. Suppose $a_{l+1}=a_k$; this color is missing at v_{k-1} and appears on uv_k. If a_0 does not appear at v_l, then we downshift from v_l and use color a_0 on uv_l to complete the augmentation.

Hence we may assume that a_0 appears at v_l and that a_k does not. Let P be the maximal alternating path of edges colored a_0 and a_k that begins at v_l. There is only one such path, because each vertex has at most one incident edge in each color (we ignore edges not yet colored). *Switching on P* means interchanging colors a_0 and a_k on the edges of P. Depending on the location of the other end of P, we describe a recoloring that completes the augmentation.

If P reaches v_k, then it reaches v_k along an edge with color a_0, continues along $v_k u$ in color a_k, and stops at u, which lacks color a_0. In this case, we downshift from v_k and switch on P (leftmost picture below). Similarly, if P reaches v_{k-1}, then it reaches v_{k-1} on color a_0 and stops there, because a_k does not appear at v_{k-1}. In this case, we downshift from v_{k-1}, give color a_0 to uv_{k-1}, and switch on P (middle picture below). Finally, suppose P does not reach v_k or v_{k-1}, so it ends at some vertex outside $\{u,v_l,v_k,v_{k-1}\}$. In this case, we downshift from v_l, give color a_0 to uv_l, and switch on P (rightmost picture below). $\square$

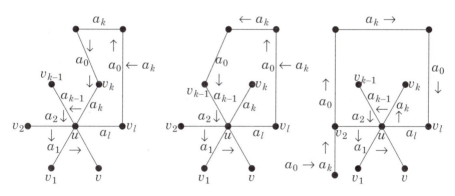

For simple graphs, we now have only the possibilities $\chi'(G)=\Delta(G)$ (*Class 1*) and $\chi'(G)=\Delta(G)+1$ (*Class 2*). Holyer [1981] proved that determining whether a simple graph is Class 1 or Class 2 is NP-complete (see Section 6.3). Interest remains high in finding conditions that forbid or guarantee $\Delta(G)$-edge-colorability, especially for regular graphs. For example, regular graphs with cut-vertices are Class 2 (Exercise 24). Although the Petersen graph is Class 2, related graphs are Class 1 (Exercise 13). Finally, $G\square H$ is Class 1 if G or H is Class 1 (Exercise 19) or if G and H both have 1-factors (Exercise 20).

CHARACTERIZATION OF LINE GRAPHS (optional)

Characterizations of line graphs lead to good algorithms for recognizing line graphs and for retrieving the original graph.

6.1.8. Example. To illustrate the ideas, we prove that the rightmost graph below is not a line graph. The graph $G = K_4 - e$ consisting of two triangles with a common edge is the line graph of $H = K_{1,3} + e$. By case analysis, we find that H is the only graph whose line graph is G, and the edges becoming the 2-valent vertices in G must be the dashed edges. We cannot add another vertex to G with only these two vertices as neighbors, because there is no way to add another edge to H that intersects both dashed edges without intersecting a solid edge. □

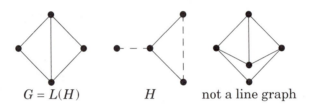

$$G = L(H) \qquad H \qquad \text{not a line graph}$$

Our first characterization encodes the process of taking the line graph. If $G = L(H)$ and H is simple, then each $v \in V(H)$ with $d(v) \geq 2$ generates a clique $Q(v)$ in G corresponding to edges incident to v, and these cliques partition $E(G)$. Furthermore, each vertex $e \in V(G)$ belongs only to the cliques generated by endpoints of $e \in E(H)$. For example, when $G = K_4 - e$, we can partition $E(G)$ into three cliques (a triangle plus two edges), each vertex covered at most twice. These three cliques correspond to the vertices of degree at least 2 in $K_{1,3} + e$. The rightmost graph above does not have such a partition.

6.1.9. Theorem. (Krausz [1943]). A simple graph G is the line graph of some simple graph if and only if $E(G)$ has a partition into cliques using each vertex of G at most twice.

Proof. We argued above that the condition is necessary. Note that when $G = L(H)$, the only vertices of G that do not belong to two of the cliques we have defined are those corresponding to edges of H that are incident to leaves.

For sufficiency, suppose $E(G)$ has such an partition, using cliques with vertex sets $\{S_1, \ldots, S_k\}$. We construct H such that $G = L(H)$. We may assume that G has no isolated vertices, since they correspond to isolated edges in H. Let $v_1, \ldots, v_l$ be the vertices of G (if any) that appear in exactly one of $\{S_i\}$. Let H have one vertex for each set in the list $\mathbf{A} = S_1, \ldots, S_k, \{v_1\}, \ldots, \{v_l\}\}$, and let vertices of H be adjacent if the corresponding sets intersect. Each vertex of G appears in exactly two sets in $\mathbf{A}$, and no two vertices appear in the same pair of sets. Hence H is a simple graph with one edge for each vertex of G. If vertices are adjacent in G, then they appear together in some S_i, and the corresponding edges of H share the vertex corresponding to S_i. Hence $G = L(H)$. □

Krausz's characterization does not directly yield an efficient test for line graphs, because it characterizes them by the existence of a special edge partition, and there are too many partitions to test. A later characterization by van Rooij and Wilf tests substructures of fixed size and therefore yields a good algorithm. (Lehot [1974] presents a linear time algorithm for testing whether G is a line graph and retrieving the unique H such that $G = L(H)$.) For the van Rooij-Wilf characterization, we define a triangle T in G to be *odd* if $|V(T) \cap N(v)|$ is odd for *some* $v \in V(G)$. Otherwise, $|V(T) \cap N(v)|$ is even for *every* $v \in V(G)$, and then we say T is *even*. An induced copy of $K_4 - e$ is a *double triangle*; it consists of two triangles sharing an edge.

6.1.10. Theorem. (van Rooij and Wilf [1965]). A simple graph G is the line graph of a simple graph if and only if $K_{1,3}$ is not an induced subgraph of G and no double triangle of G has two odd triangles.

Proof. If $G = L(H)$, the necessity of forbidding an induced $K_{1,3}$ is clear; an edge of H has two endpoints, so if it is incident to three other edges it must meet two of them at the same endpoint, and then they also are incident. For the other condition, we saw earlier that the vertices of a double triangle in G must correspond to the edges of a $K_{1,3} + e$ in H. In particular, one of these triangles in G is generated by a triangle in H. A triangle in G generated by a triangle in H must be even, because an edge incident to a triangle in H intersects exactly two of its edges.

For sufficiency, suppose that G satisfies the specified conditions. We may assume that G is connected, else we apply the construction to each component. The case where G is $K_{1,3}$-free and has a double triangle with both triangles even is very special; there are only three such graphs (Exercise 27). Here we consider only the general case, in which every double triangle of G has exactly one odd triangle.

By Theorem 6.1.9, it suffices to partition $E(G)$ into cliques that cover each vertex at most twice. Let $S_1, \ldots, S_k$ be the maximal cliques of G that are not even triangles, and let $T_1, \ldots, T_l$ be the edges that belong to one even triangle and no odd triangle. We claim that $\mathbf{B} = \{S_i\} \cup \{T_j\}$ partitions $E(G)$ into cliques using each vertex at most twice.

Every edge appears in a maximal clique, but every triangle in a clique with more than three vertices is odd. Hence T_j is not in any clique S_i. Also S_i and $S_{i'}$ share no edge, because G has no double

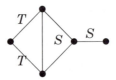

triangles with both triangles odd. Hence the cliques in **B** are edge-disjoint. If $e \in E(G)$, we have e in some S_i unless the only maximal clique containing e is an even triangle. In this case e is a T_j, since we have forbidden double triangles with both triangles even.

It remains to show that each $v \in G$ appears at most twice in **B**. Suppose v belongs to $A, B, C \in \mathbf{B}$. Edge-disjointness implies that v has neighbors x, y, z with each belonging to only one of $\{A, B, C\}$. Since G has no induced $K_{1,3}$, we may assume $x \leftrightarrow y$. By edge-disjointness, the triangle vxy cannot belong to a member of **B**. Hence it must be an even triangle. Therefore, z must have exactly one other edge to vxy, say $z \leftrightarrow x$ and $z \not\leftrightarrow y$. But now the same argument shows zvx is an even triangle, and we have a double triangle with both triangles even. □

This characterization is close to a forbidden subgraph characterization. By Theorem 6.1.10, the proof of Theorem 6.1.11 consists of showing that the eight graphs listed other than $K_{1,3}$ are the vertex-minimal $K_{1,3}$-free graphs containing a double triangle with both triangles odd. Each such graph has a double triangle and one or two additional vertices that make the triangles odd by having one or three neighbors in the triangles.

6.1.11. Theorem. (Beineke [1968]) A simple graph G is the line graph of some simple graph if and only if G does not contain any of the nine graphs below as an induced subgraph.

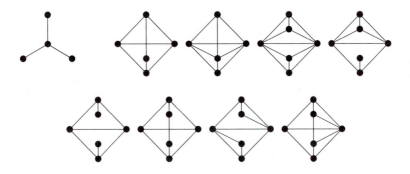

Proof. (Exercise 29). □

EXERCISES

6.1.1. (–) Determine whether $\bar{P}_5$ is a line graph. If so, determine H such that $\bar{P}_5 = L(H)$.

6.1.2. (–) Prove that a set of edges in a simple graph is pairwise intersecting if and only if the edges have one common endpoint or form a triangle.

6.1.3. (–) Prove that $L(K_{m,n}) = K_m \square K_n$.

6.1.4. (–) Prove that the Petersen graph is the complement of $L(K_5)$.

6.1.5. (–) Prove that if $L(G)$ is connected and regular, then either G is regular or G is a bipartite graph in which vertices of the same partite set have the same degree. (Ray-Chaudhuri [1967])

6.1.6. (!) Suppose G is a simple graph and $L(G)$ is the line graph of G.
    a) Prove that the number of edges in $L(G)$ is $\Sigma_{v \in V(G)} \binom{d(v)}{2}$.
    b) Prove that G is isomorphic to $L(G)$ if and only if G is 2-regular.

6.1.7. Suppose G is connected. Use part (a) of Exercise 6 to determine when $e(L(G)) < e(G)$.

6.1.8. Suppose G is k-edge-connected. Prove that $L(G)$ is k-connected and is $2k - 2$-edge-connected.

6.1.9. (!) *Matchings in line graphs.* Use Tutte's 1-factor Theorem to prove that every connected line graph of even order has a complete matching. Conclude from this that the edges of a connected simple graph of even size can be partitioned into paths of length 2.

6.1.10. Given vertices x, y in a graph or digraph G, modify G to obtain G' by adding two new vertices s, t and two new edges sx and yt. Use the line graph or line digraph $L(G')$ to prove that $\kappa(x, y) = \lambda(x, y)$ implies $\kappa'(x, y) = \lambda'(x, y)$.

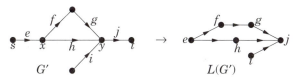

6.1.11. (–) Give explicit edge-colorings to prove that $\chi'(G) = \Delta(G)$ when G is a k-dimensional cube Q_k or a complete bipartite graph $K_{m,n}$.

6.1.12. (–) Compute the edge-chromatic number for the two graphs below and draw their line graphs.

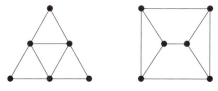

6.1.13. (!) The *generalized Petersen graph* $P(n, k)$ is the graph with vertices $\{u_1, \ldots, u_n\}$ and $\{v_1, \ldots, v_n\}$ and edges $\{u_i u_{i+1}\}$, $\{u_i v_i\}$, and $\{v_i v_{i+k}\}$, where addition is modulo n. The usual Petersen graph is $P(5, 2)$ with $\chi' = 4$ (Example 6.1.6).

a) Prove that the subgraph of $P(n, 2)$ induced by k consecutive pairs $\{u_i, v_i\}$ has a spanning cycle if $k \equiv 1 \bmod 3$ and $k \geq 4$.

b) Use part (a) to prove that $\chi'(P(n, 2)) = 3$ if $n \geq 6$.

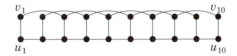

6.1.14. Exhibit six perfect matchings in the Petersen graph such that each edge belongs to exactly two of the matchings. (Hint: consider the drawing of the Petersen graph having a 9-cycle on the "outside".)

6.1.15. Use Brooks' Theorem to give a short proof of a special case of Vizing's Theorem: if G is loopless and $\Delta(G) = 3$, then G is 4-edge-colorable.

6.1.16. (!) Prove that every simple bipartite graph G has a $\Delta(G)$-regular simple bipartite supergraph.

6.1.17. (!) Suppose D is a digraph (loops allowed) such that each in-degree and out-degree is at most d. Prove that D has an edge-coloring with d colors such that the edges entering each vertex have distinct colors and the edges exiting each vertex have distinct colors. (Hint: use a graph transformation.)

6.1.18. (!) *Shannon's bound on* $\chi'(G)$.

a) Prove that a loopless multigraph G has a loopless $\Delta(G)$-regular supergraph. (Comment: vertices and/or edges may be added.)

b) Use (a) and Petersen's Theorem (every $2k$-regular multigraph has a 2-factor - Theorem 3.3.5) to prove that $\chi'(G) \leq 3\Delta(G)/2$ if $\Delta(G)$ is even and G is a loopless multigraph. (Shannon [1949])

6.1.19. (!) Suppose G and H are simple graphs. Use Vizing's Theorem to prove that $\chi'(H) = \Delta(H)$ implies $\chi'(G \square H) = \Delta(G \square H)$. (Hint: It may help to consider first the special case $H = K_2$, though this need not be proved separately.)

6.1.20. *Kotzig's Theorem for Cartesian product of simple graphs.*

a) Use Vizing's Theorem to prove that $\chi'(G \square K_2) = \Delta(G \square K_2)$.

b) Suppose G_1, G_2 are edge-disjoint graphs with vertex set V and H_1, H_2 are edge-disjoint graphs with vertex set W. Prove that $(G_1 \cup G_2) \square (H_1 \cup H_2) = (G_1 \square H_2) \cup (G_2 \square H_1)$.

c) Use parts (a) and (b) to prove that $\chi'(G \square H) = \Delta(G \square H)$ if both G and H have 1-factors. (Kotzig [1979], J. George [1991]) (Comment: As a result, the Cartesian product of the Petersen graph with itself in Class 1, which does not follow from Exercise 19. Here neither factor need be Class 1, there G need not have a 1-factor.)

6.1.21. *Optimal edge-colorings.* Given an edge-coloring of a multigraph G, let $c(v)$ denote the number of distinct colors appearing on edges incident to v. Among all k-edge-colorings of G, a coloring is *optimal* if it maximizes $\Sigma_{v \in V(G)} c(v)$.

a) Suppose no component of G is an odd cycle. Prove that G has a 2-edge-coloring in which both colors appear at each vertex of degree at least 2. (Hint: Use Eulerian circuits.)

b) Suppose f is an optimal k-edge-coloring of G, in which color a appears at least twice at $u \in V(G)$ and color b does not appear at u. Let H be the subgraph of G consisting of edges colored a or b. Prove that the component of H containing u is an odd cycle. (Fournier [1973])

c) Suppose G is a bipartite multigraph. Prove that G is $\Delta(G)$-edge-colorable and also that G has a $\delta(G)$-edge-coloring in which each color appears at every vertex. (Gupta [1966])

6.1.22. Suppose G is a bipartite multigraph with minimum degree δ. Prove that G has a δ-edge-coloring in which at each vertex v, each color appears $\lceil d(v)/\delta \rceil$ or $\lfloor d(v)/\delta \rfloor$ times. (Hint: use a graph transformation.)

6.1.23. Use Vizing's Theorem to prove that every graph with maximum degree Δ has an "equitable" $\Delta + 1$-edge-coloring: an edge coloring with each color used $\lceil e(G)/\Delta \rceil$ or $\lfloor e(G)/\Delta \rfloor$ times. (de Werra [1971], McDiarmid [1972])

6.1.24. (!) Prove that if G is regular and has a cut-vertex, then $\chi'(G) > \Delta(G)$.

6.1.25. *Density conditions for $\chi'(G) > \Delta(G)$.*

a) Prove that if G has $2m + 1$ vertices and has more than $m \cdot \Delta(G)$ edges, then $\chi'(G) > \Delta(G)$.

b) Prove that if G is obtained from a k-regular graph with $2m + 1$ vertices by deleting fewer than $k/2$ edges, then $\chi'(G) > \Delta(G)$.

c) Prove that if G is obtained by subdividing an edge of a regular graph with $2m$ vertices, then $\chi'(G) > \Delta(G)$.

6.1.26. Bounds on the edge-chromatic number of multigraphs were proved by Shannon [1949], Vizing [1964, 1965], Gupta [1966], Ore [1967a], Andersen [1977], and Goldberg [1977,84]. Let P denote the set of 3-vertex paths in G, expressed as edges xy and yz, and let $\mu(e)$ denote the multiplicity of edge e.

Shannon: $\chi'(G) \le \lfloor 3\Delta(G)/2 \rfloor$.

Vizing, Gupta: $\chi'(G) \le \Delta(G) + \mu(G)$.

Ore: $\chi'(G) \le \max \{\Delta(G), \max_P \lfloor \frac{1}{2} d(x) + d(y) + d(z) \rfloor\}$.

Andersen, Goldberg: $\chi'(G) \le \max \{\Delta(G), \max_P \lfloor \frac{1}{2}(d(x) + \mu(xy) + \mu(yz) + d(z)) \rfloor\}$.

Prove that the Andersen-Goldberg bound implies the earlier bounds.

6.1.27. Suppose that G is connected and simple, contains no induced $K_{1,3}$, and has a double triangle H with each triangle even. Prove that G is one of the three graphs below, and conclude that G is a line graph. (Comment: this completes the proof of Theorem 6.1.10).

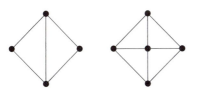

6.1.28. A *Krausz decomposition* of a simple graph H is a partition of $E(H)$ into cliques such that each vertex of H appears in at most two of the cliques.

a) Prove that two Krausz decompositions of a connected graph that have a common clique are identical.

b) Find distinct Krausz decompositions for the graphs in Exercise 26.

c) Prove that no connected graph with more than six vertices has two distinct Krausz decompositions (use Exercise 26 and the proof of Theorem 6.1.10?).

d) Conclude that $K_{1,3}, K_3$ is the only pair of nonisomorphic simple graphs with isomorphic line graphs. (Whitney [1932])

6.1.29. Complete the proof of Theorem 6.1.11 by proving that a graph with no induced $K_{1,3}$ contains a double triangle with both triangles odd if and only if it contains an induced subgraph among the other eight graphs shown there.

6.1.30. (+) *Line graphs of cliques.* If $n \neq 8$, prove that $G = L(K_n)$ if and only if G is a $2n - 4$-regular graph with $\binom{n}{2}$ vertices in which nonadjacent vertices have four common neighbors and adjacent vertices have $n - 2$ common neighbors. (When $n = 8$, there are three exceptional graphs satisfying the conditions.) (Chang [1959], Hoffman [1960])

6.1.31. (+) *Line graphs of complete bipartite graphs.* Unless $n = m = 4$, prove that $G = L(K_{m,n})$ if and only if G is an $n + m - 2$-regular mn-vertex graph in which every pair of nonadjacent points have two common neighbors, $n\binom{m}{2}$ pairs of adjacent points have $m - 2$ common neighbors, and $m\binom{n}{2}$ pairs of adjacent points have $n - 2$ common neighbors. (Moon [1963], Hoffman [1964]) (Comment: for $n = m = 4$, there is one exceptional graph - Shrikande [1959].)

6.2 Hamiltonian Cycles

A *Hamiltonian cycle* is a spanning cycle in a graph (a cycle through every vertex). The *circumference* of a graph is the length of its longest cycle. Introduced by Kirkman in 1855, Hamiltonian cycles are named for Sir William Hamilton, who described a game on the graph of the dodecahedron in which one player would specify a 5-vertex path and the other would have to extend it to a spanning cycle. The game was

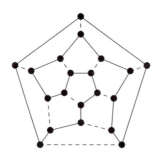

marketed as the "Traveller's Dodecahedron", a wooden version in which the vertices were named for 20 important cities. It was not commercially successful.

Until the 1970s, interest in Hamiltonian cycles centered on their relationship to the Four Color Problem (Section 7.3). More recently, the study of Hamiltonian cycles in general graphs has been fueled by practical applications and by the issue of complexity (Section 6.3).

A *Hamiltonian path* is a spanning path. Every graph with a spanning cycle has a spanning path, but P_n shows that the converse is not true. A graph with a spanning cycle is a *Hamiltonian graph*. No easily testable characterization is known for Hamiltonian graphs; we will study necessary conditions and sufficient conditions.

Loops and multiple edges are irrelevant to the existence of a spanning cycle; a graph is Hamiltonian if and only if the simple graph obtained by keeping one copy of each edge is Hamiltonian. Therefore, in this section we confine our attention to simple graphs, which becomes important when we discuss conditions on vertex degrees.

NECESSARY CONDITIONS

Every Hamiltonian graph is 2-connected, because deleting a vertex leaves a subgraph with a spanning path. Bipartite graphs suggest a way to strengthen this necessary condition.

6.2.1. Example. *Bipartite graphs.* A spanning cycle in a bipartite graph visits the two partite sets alternately, so there can be no such cycle unless the parts have the same size. Hence $K_{m,n}$ is Hamiltonian only if $m = n$. Alternatively, we can argue that the cycle returns to different vertices of one part after each visit to the other part. □

6.2.2. Theorem. If G has a Hamiltonian cycle, then for any set $S \subseteq V$, the graph $G - S$ has at most $|S|$ components.

Proof. When leaving a component of $G - S$, a Hamiltonian cycle can go only to S, and the arrivals in S must visit different vertices of S. Hence S must have at least as many vertices as $G - S$ has components. □

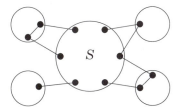

We use $c(H)$ to denote the number of components of H, writing the necessary condition as $c(G - S) \le |S|$ for all $\varnothing \ne S \subseteq V$. This condition guarantees that G is 2-connected (deleting one vertex leaves at most one component), but it does not guarantee a Hamiltonian cycle.

6.2.3. Example. *Necessary but not sufficient.* The graph on the left below fails this necessary condition, even though it is bipartite with partite sets of equal size; hence it is not Hamiltonian. The graph is 3-regular but not 3-connected. Tutte [1971] conjectured that every 3-connected 3-regular bipartite graph is Hamiltonian, but Horton [1982] found a counterexample with 96 vertices, and the smallest known counterexample now has 50 vertices (Georges [1989]).

The graph on the right shows that the necessary condition is not sufficient. This graph satisfies the condition but has no spanning cycle. All edges incident to 2-valent vertices must be used, but in this graph that requires three edges incident to the central vertex.

The Petersen graph also satisfies the condition but is not Hamiltonian. A spanning cycle is a connected 2-factor, but we proved in Example 6.1.6 that $2C_5$ is the only 2-factor of the Petersen graph. □

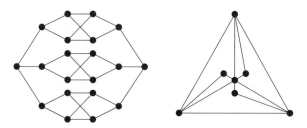

Appropriate additional conditions may turn a necessary condition into a sufficient condition. Perhaps requiring $|S| \ge 2c(G - S)$ for every cutset S will guarantee a spanning cycle. A graph G is *t-tough* if $|S| \ge tc(G - S)$ for every cutset $S \subset V$. The *toughness* of G is the maximum t such that G is t-tough. For example, the toughness of the Petersen graph is 4/3, because $c(G - S) = 2$ requires $|S| \ge 3$, $c(G - S) = 3$ requires $|S| \ge 4$, $c(G - S) = 4$ requires $|S| = 6$, and no larger value of $c(G - S)$ is possible (Exercise 13).

Theorem 6.2.2 says that toughness at least 1 is necessary. Chvátal conjectured that a high enough value of toughness is sufficient; it is now conjectured that this value is 2. No value of toughness larger than 1 is necessary, since C_n itself is only 1-tough. No smaller value than 2 is sufficient, because non-Hamiltonian graphs exist with toughness $2 - \varepsilon$ for each $\varepsilon > 0$ (Enomoto-Jackson-Katerinis-Saito [1985]). If Chvátal's conjecture holds, then toughness at least 1 is necessary and toughness at least 2 is sufficient.

SUFFICIENT CONDITIONS

The number of edges needed to force an n-vertex graph to be Hamiltonian is quite large (Exercise 21). Under conditions that "spread out" the edges, we can reduce the number of edges while still guaranteeing Hamiltonian cycles. The simplest such condition is a lower bound on the minimum degree; $\delta(G) \geq n(G)/2$ suffices. We first note that no smaller minimum degree is sufficient.

6.2.4. Example. *Best-possible degree bound.* The graph consisting of cliques of orders $\lfloor (n+1)/2 \rfloor$ and $\lceil (n+1)/2 \rceil$ sharing a vertex has minimum degree $\lfloor (n-1)/2 \rfloor$ but is not Hamiltonian (not even 2-connected). □

6.2.5. Theorem. (Dirac [1952b]). If G is a simple graph with at least three vertices and $\delta(G) \geq n(G)/2$, then G is Hamiltonian.

Proof. The condition $n(G) \geq 3$ is annoying but must be included, since K_2 is not Hamiltonian but satisfies $\delta(K_2) = n(K_2)/2$.

The proof combines the techniques of contradiction and extremality. If there is a non-Hamiltonian graph satisfying the hypotheses, then adding edges cannot reduce the minimum degree, so we may restrict our attention to *maximal* non-Hamiltonian graphs with minimum degree at least $n/2$. By "maximal", we mean that no proper supergraph is also non-Hamiltonian, so $G + uv$ is Hamiltonian whenever $u \not\leftrightarrow v$. (We can also describe this as reducing the theorem to a special case.)

The maximality of G implies that G has a spanning path from $u = v_1$ to $v = v_n$, because every spanning cycle in $G + uv$ contains the new edge uv. We use most of this path $v_1, \ldots, v_n$, with a small switch, to obtain a spanning cycle in G. If some neighbor of u immediately follows a neighbor of v on the path, say $u \leftrightarrow v_{i+1}$ and $v \leftrightarrow v_i$, then G has the spanning cycle $(u, v_{i+1}, v_{i+2}, \ldots, v, v_i, v_{i-1}, \ldots, v_2)$ shown below.

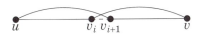

$$\begin{array}{cccc} \bullet & \bullet & \bullet & \bullet \\ u & v_i & v_{i+1} & v \end{array}$$

To prove that such a cycle exists, we show that there is a common index in the sets S and T defined by $S = \{i : u \leftrightarrow v_{i+1}\}$ and $T = \{i : v \leftrightarrow v_i\}$. Summing the sizes of these sets yields

$$|S \cup T| + |S \cap T| = |S| + |T| = d(u) + d(v) \geq n.$$

Neither S nor T contains the index n. This implies that $|S \cup T| < n$, and hence $|S \cap T| \geq 1$. We have established a contradiction by finding a spanning cycle in G; hence there is no (maximal) non-Hamiltonian graph satisfying the hypotheses. □

Ore observed that this argument uses $\delta(G) \geq n(G)/2$ only to show $d(u) + d(v) \geq n$. Therefore, we can weaken the requirement of minimum degree $n/2$ to require only that $d(u) + d(v) \geq n$ whenever $u \not\leftrightarrow v$. We also did not need that G was a maximal non-Hamiltonian graph, only that $G + uv$ was Hamiltonian and thereby provided a spanning u,v-path.

6.2.6. Lemma. (Ore [1960]) If G is a simple graph and u, v are distinct nonadjacent vertices of G with $d(u) + d(v) \geq n(G)$, then G is Hamiltonian if and only if $G + uv$ is Hamiltonian.

Proof. One direction is trivial, and the proof of the other direction is the same as for Theorem 6.2.5. □

Bondy and Chvátal [1976] phrased the essence of Ore's argument in a much more general form that yields sufficient conditions for cycles of length l and other subgraphs. We discuss only the case of spanning cycles. Using the lemma above to add edges, we can test whether G is Hamiltonian by testing whether the larger graph is Hamiltonian.

6.2.7. Definition. The *(Hamiltonian) closure* of a graph G, denoted $C(G)$, is the supergraph of G on $V(G)$ obtained by iteratively adding edges between pairs of nonadjacent vertices whose degree sum is at least n, until no such pair remains.

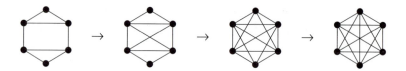

Ore's lemma yields the following theorem:

6.2.8. Theorem. (Bondy-Chvátal [1976]). A simple n-vertex graph is Hamiltonian if and only if its closure is Hamiltonian. □

Fortunately, the closure does not depend on the order in which we choose to add edges when more than one is available.

6.2.9. Lemma. The closure of G is well-defined.

Proof. Suppose adding $e_1, \ldots, e_r$ in sequence yields G_1 and adding $f_1, \ldots, f_s$ in sequence yields G_2, each addition joining vertices with degrees summing to at least $n(G)$. If in either sequence a pair (u, v) becomes addable, then it must eventually be added before the sequence ends. Hence f_1, being initially addable to G, must belong to G_1. Similarly, if $f_1, \ldots, f_{i-1} \in E(G_1)$, then f_i becomes addable to G_1 and therefore belongs to G_1. Hence neither sequence contains a first edge omitted by the other sequence, and we have $G_1 \subseteq G_2$ and $G_2 \subseteq G_1$. $\square$

We now have a necessary and sufficient condition to test for Hamiltonian graphs. It doesn't help much, because it requires us to test whether another graph is Hamiltonian! Nevertheless, it does furnish a method for proving sufficient conditions. A condition that forces $C(G)$ to be Hamiltonian is sufficient for Hamiltonian graphs. For example, the condition may imply $C(G) = K_n$. Chvátal used this method to prove the best possible sufficient condition on vertex degrees.

6.2.10. Theorem. (Chvátal [1972]) Suppose G has vertex degrees $d_1 \leq \cdots \leq d_n$. If $i < n/2$ implies that $d_i > i$ or $d_{n-i} \geq n - i$, then G is Hamiltonian.

Proof. Because adding edges to form the closure reduces no entry in the degree sequence, and a graph is Hamiltonian if and only if its closure is Hamiltonian, it suffices to consider the special case where G is closed. In this case, we prove that the condition implies $G = K_n$. We prove the contrapositive; if $G = C(G) \neq K_n$, then we construct a value of i less than $n/2$ for which Chvátal's condition is violated. Violation means that at least i vertices have degree at most i and at least $n - i$ vertices have degree less than $n - i$.

If $G \neq K_n$, we choose among the pairs of nonadjacent vertices a pair u, v with maximum degree sum. Because G is closed, $u \not\leftrightarrow v$ implies $d(u) + d(v) < n$, and we may choose the labels on u, v so that $d(u) \leq d(v)$. Since $d(u) + d(v) < n$, this implies $d(u) < n/2$; let $i = d(u)$.

Because we chose a nonadjacent pair with maximum degree sum, every vertex of $V - \{v\}$ that is not adjacent to v has degree at most $d(u) = i$; furthermore, there are at least $n - 1 - d(v) \geq d(u) = i$ of these vertices. Similarly, every vertex of $V - \{u\}$ that is not adjacent to u has degree at most $d(v) < n - d(u) = n - i$, and there are $n - 1 - d(u)$ of these. Since $d(u) \leq d(v)$, we can also add u to the set of vertices with degree at most $d(v)$, so we obtain $n - i$ vertices with degree less than $n - i$. Hence we have proved $d_i \leq i$ and $d_{n-i} < n - i$ for this specially chosen i, which contradicts the hypothesis. $\square$

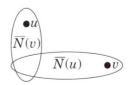

6.2.11. Example. *Non-Hamiltonian graphs with "large" vertex degrees.* Chvátal's theorem characterizes the degree sequences that force

Hamiltonian cycles. Suppose G fails Chvátal's condition "just barely". If it fails at i, then the largest we can make the terms in $d_1, \ldots, d_n$ is $d_j = i$ for $j \leq i$, $d_j = n - i - 1$ for $i + 1 \leq j \leq n - i$, and $d_j = n - 1$ for $j > n - i$. There is a unique graph realizing this degree sequence (Exercise 20), and it is not Hamiltonian. The graph is $(\overline{K}_i + K_{n-2i}) \vee K_i$. It is not Hamiltonian because deleting the i vertices of degree $n - 1$ leaves a subgraph with $i + 1$ components. If a graph H is non-Hamiltonian and has vertex degrees $d_1' \leq \cdots \leq d_n'$, then Chvátal's result implies that for some i the graph $(\overline{K}_i + K_{n-2i}) \vee K_i$ with vertex degrees $d_1 \leq \cdots \leq d_n$ satisfies $d_j \geq d_j'$ for all i. $\square$

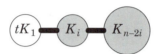

We could make arguments like those above to prove sufficient conditions for Hamiltonian paths, but it is easier to use our previous work and prove the new theorem by invoking a theorem about cycles. To do this, we use a standard transformation: G has a spanning path if and only if $G \vee K_1$ has a spanning cycle.

6.2.12. Theorem. Suppose G has vertex degrees $d_1 \leq \cdots \leq d_n$. If $d_i \geq i$ or $d_{n+1-i} \geq n - i$ whenever $i < (n + 1)/2$, then G has a spanning path.

Proof. Let $G' = G \vee K_1$, let $n' = n + 1$, and let $d_1' \leq d_2' \leq \cdots d_{n'}'$ be the degree sequence of G'. Since a spanning cycle in $G \vee K_1$ becomes a spanning path in G when the extra vertex is deleted, it suffices to show that G' satisfies Chvátal's sufficient condition for Hamiltonian cycles. Since the new vertex is adjacent to every vertex of G, we have $d_{n'}' = n$ and $d_j' = d_j + 1$ for $j < n'$. If $i < n'/2 = (n + 1)/2$, then the hypothesis on G tells us that $d_i' = d_i + 1 \geq i + 1 > i$ or $d_{n'-i}' = d_{n-i} + 1 \geq n - i + 1 = n' - i$. This is precisely Chvátal's sufficient condition, so G' has a spanning cycle, and deleting the extra vertex leaves a spanning path in G. $\square$

The degree requirements can be weakened when other "uniformity" conditions hold. A lower bound on toughness has this effect. Also, every k-regular graph G with at most $3k$ vertices is Hamiltonian (Jackson [1980]), meaning that regularity drops the minimum-degree requirement from $n(G)/2$ to $n(G)/3$. Zhu-Liu-Yu [1985] proved that only the Petersen graph prevents improvement to $(n(G) - 1)/3$, and Bondy-Kouider [1988] simplified the proof, but it remains difficult. Our last sufficient condition for Hamiltonian cycles is not a degree sequence condition; it involves connectivity and independence number.

6.2.13. Theorem. (Chvátal-Erdős [1972]) If $\kappa(G) \geq \alpha(G)$, then G has a Hamiltonian cycle (unless $G = K_2$).

Proof. With $G \neq K_2$, the conditions require $\kappa(G) > 1$. Suppose that $k = \kappa(G) \geq \alpha(G)$, and let C be a longest cycle in G. Let H be a component of $G - C$. Since $\delta(G) \geq \kappa(G)$ and every graph with $\delta(G) \geq 2$ has a cycle of length at least $\delta + 1$ (Lemma 1.2.18), C has at least $k + 1$ vertices. Also C has at least k vertices with edges to H, else the vertices of C with edges to H contradict $\kappa(G) = k$. Let $u_1, \ldots, u_k$ be k vertices of C with edges to H, in clockwise order. For $i = 1, \ldots, k$, let a_i be the vertex immediately following u_i on C. If any two of these vertices are adjacent, say $a_i \leftrightarrow a_j$, then we construct a longer cycle by using $a_i a_j$, the portions of C from a_i to u_j and a_j to u_i, and a u_i, u_j-path through H (see illustration). This argument includes the case $a_i = u_{i+1}$, so we also conclude that no a_i has a neighbor in H. Hence $\{a_1, \ldots, a_k\}$ plus a vertex of H forms an independent set of size $k + 1$. This contradiction implies that C is a Hamiltonian cycle. $\square$

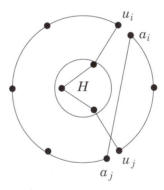

This theorem can be strengthened by using a more detailed extremal structure. A longest cycle is selected, and from the remaining graph a longest cycle is selected, and so on until the remaining graph is a tree or is empty. This approach leads to a sufficient condition that applies more generally than the Chvátal-Erdős condition.

6.2.14. Theorem. (Lu [1994]) For $\varnothing \neq S \subset V(G)$, let $t(S) = \frac{|\bar{S} \cap N(S)|}{|\bar{S}|}$, and let $\theta(G) = \min t(S)$. If $\theta(G)n(G) \geq \alpha(G)$, then G is Hamiltonian. $\square$

To see that this strengthens the Chvátal-Erdős result, let $k = \kappa(G)$. If $|S| \geq n - k$, then $t(S) = 1$. If $|S| < n - k$, then $t(S) \geq k/(n - |S|)$. Since $n > n - |S|$, this implies $t(S)n(G) > k = \kappa(G)$. Hence $\kappa(G) \geq \alpha(G)$ implies $\theta(G)n(G) \geq \alpha(G)$.

Most sufficient conditions for Hamiltonian cycles have *long-cycle* versions. A weakened form of the sufficient condition may force a long cycle if not a spanning cycle. Dirac [1952b] proved the first long-cycle result: a 2-connected graph with minimum degree k has circumference at least min $\{n, 2k\}$. This improves Lemma 1.2.18, which guarantees a cycle of length at least $k + 1$. Long-cycle results are more general and more difficult than the corresponding sufficient conditions for Hamiltonian cycles; we consider these in Section 8.4.

CYCLES IN DIRECTED GRAPHS (optional)

The theory of cycles in digraphs is similar to that of cycles in graphs. For a digraph G, let $\delta^-(G) = \min d^-(v)$ and $\delta^+(G) = \min d^+(v)$. The arguments of Chapter 1 using maximal paths guarantee paths of length k and cycles of length $k + 1$, where $k = \max \{\delta^-(G), \delta^+(G)\}$.

Here we consider spanning cycles in digraphs. Although cliques are trivially Hamiltonian, the question becomes interesting for tournaments. The necessary condition of 2-connectedness becomes a necessary condition of strong connectedness; for tournaments, this is also sufficient (Exercise 38).

For arbitrary digraphs, we consider degree conditions. The basic result is analogous to Dirac's theorem. By applying it to the digraph obtained from a graph G by replacing every edge by a pair of opposed edges with the same endpoints, we obtain Dirac's theorem as a special case. A digraph is *strict* if it has no loops and has at most one copy of each ordered pair as an edge. Meyniel [1973] subtantially strengthened the theorem by weakening the hypothesis - see Theorem 8.4.38.

6.2.15. Theorem. (Ghouila-Houri [1960]) If D is a strict digraph, and min $\{\delta^+(D), \delta^-(D)\} \geq n(D)/2$, then D is Hamiltonian.

Proof. Again contradiction and extremality. If D is an n-vertex counterexample, we consider a longest cycle C in D, with length $l < n$. As mentioned above, $l > \max \{\delta^+, \delta^-\} \geq n/2$. Let P be a longest path in $D - V(C)$, beginning at u, ending at w, and having length $m \geq 0$. Now $l > n/2$ and $n \geq l + m + 1$ imply $m < n/2$.

Let S be the set of predecessors of u on C, and let T be the set of successors of w on C. Since the maximality of P forces all predecessors of u and successors of w into $V(C) \cup V(P)$, each of S, T has size at least min $\{\delta^+, \delta^-\} - m \geq n/2 - m > 0$. The maximality of C guarantees that the distance along C from a vertex $u' \in S$ to a vertex $w' \in T$ must exceed $m + 1$. Otherwise, traveling along P instead of C from u' to w' yields a longer cycle. Hence we may assume that every vertex of S is followed on C by more than m vertices not in T.

If the distance between successive vertices of S along C is always at most $m + 1$, then there is no legal place to put a vertex of T. Since

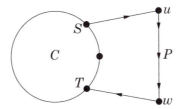

both S and T are nonempty, we may thus assume there is a vertex of S followed on C by at least $m + 1$ vertices not in S. These are forbidden from T, as is the immediate successor on C of all the other vertices of S. We thus have at least $|S| - 1 + m + 1 \geq n/2$ vertices of C that cannot be vertices of T. Together with the actual vertices of T, this forces $|V(C)| \geq n - m$, which contradicts $l \leq n - m - 1$. □

EXERCISES

6.2.1. (–) For which values of r is $K_{r,r}$ a Hamiltonian graph?

6.2.2. (–) Is the Grötzsch graph (Example 5.2.1) a Hamiltonian graph?

6.2.3. (–) Prove that G has a Hamiltonian path only if for every $S \subseteq V(G)$, the number of components of $G - S$ is at most $|S| + 1$.

6.2.4. (–) Prove that $K_{n,n}$ has $(n - 1)!n!/2$ Hamiltonian cycles.

6.2.5. Prove that every 5-vertex path in the dodecahedron graph extends to a Hamiltonian cycle.

6.2.6. (!) Prove that the line graph of G is Hamiltonian if and only if G has a closed trail that contains at least one endpoint of each edge. Exhibit a pair of graphs G, H such that $H = L(G)$, H is Hamiltonian, and G is not Eulerian. (Harary and Nash-Williams [1965])

6.2.7. (!) Suppose G is a Hamiltonian bipartite graph and $x, y \in V(G)$. Prove that $G - x - y$ has a complete matching if and only if x and y are on opposite sides of the bipartition of G. Conclude that deleting two unit squares from an 8 by 8 chessboard leaves a board that can be partitioned into 1 by 2 rectangles if and only if the two missing squares have opposite colors.

6.2.8. A mouse eats its way through a $3 \times 3 \times 3$ cube of cheese by eating all the $1 \times 1 \times 1$ subcubes. If it starts at a corner subcube and always moves on to an adjacent subcube (sharing a face of area 1), can it do this and eat the center subcube last? Give a method or prove impossible. (Ignore gravity.)

6.2.9. (+) On a chessboard, a *knight's move* is a move from one square to another that differs from it by 1 in one coordinate and by 2 in the other coordinate (such moves are shown below). Prove that no $4 \times n$ chessboard has a knight's tour, meaning a traversal visiting each square exactly once and returning to the starting square using knight's moves. (Hint: find an appropriate set of vertices in the corresponding graph to violate the necessary condition.)

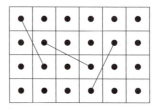

6.2.10. G is *uniquely* k-edge-colorable if all proper k-edge-colorings of G induce the same partition of the edges. Prove that every uniquely 3-edge-colorable 3-regular graph is Hamiltonian. (Greenwell-Kronk [1973])

6.2.11. Place n points around a circle. Let G_n be the 4-regular graph obtained by joining each point to the nearest two points in each direction. If $n \geq 5$, prove that G_n is the union of two Hamiltonian cycles.

6.2.12. For $k \geq 3$, let G_k be the graph obtained from two disjoint copies of $K_{k,k-2}$ by adding a matching between the two "partite sets" of size k. Determine all values of k for which G_k is Hamiltonian.

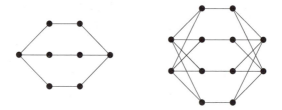

6.2.13. Prove that the Petersen graph has toughness 4/3.

6.2.14. (!) Prove that the toughness of a $K_{1,3}$-free graph is half its connectivity. (Matthews-Sumner [1984])

6.2.15. (−) Prove that the Cartesian product of two Hamiltonian graphs is Hamiltonian. Conclude that the k-dimensional cube Q_k is Hamiltonian for $k \geq 2$.

6.2.16. (+) Suppose that k is odd. Construct a $k-1$-connected k-regular bipartite graph that is not Hamiltonian.

6.2.17. (!) The kth *power* of a simple graph G is the simple graph G^k with vertex set $V(G)$ and edge set $\{uv: d_G(u,v) \leq k\}$.

a) Suppose $G - x$ has at least three nontrivial components in which x has exactly one neighbor. Prove that G^2 is not Hamiltonian. (Hint: consider the non-Hamiltonian graph in Example 6.2.2.)

b) Prove that the cube of each connected graph (with at least three vertices) is Hamiltonian. (Hint: reduce this to the special case of trees, and prove it for trees by proving the stronger result that if xy is an edge of the tree T, then T^3 has a Hamiltonian cycle using the edge xy. Comment: Fleischner [1974] proved that the square of each 2-connected graph is Hamiltonian.)

6.2.18. Suppose that $n = k(2l + 1)$. Construct a non-Hamiltonian complete k-partite graph with n vertices and minimum degree $\frac{n}{2} \frac{k-1}{k} \frac{2l}{2l+1}$. (Snevily)

6.2.19. Let $\mathbf{G}(k, t)$ be the class of connected k-partite graphs in which each partite set has size t and the edges between each pair of parts form a matching of size t. For $k \geq 4$ and $t \geq 4$, construct a graph in $\mathbf{G}(k, t)$ that is not Hamiltonian. (Hint: there is a graph in $\mathbf{G}(4, 4)$ with a 3-set whose deletion leaves four components; generalize this example. Comment: $\mathbf{G}(3, t) = \{C_{3t}\}$, and also every graph in $\mathbf{G}(k, 3)$ is Hamiltonian.) (Ayel [1982])

6.2.20. (–) Prove that Chvátal's graph $C_{n,k} = (\overline{K}_i + K_{n-2i}) \vee K_i$ is the only (unlabeled) simple graph realizing its degree sequence.

6.2.21. (!) Prove that if G fails Chvátal's condition, then $\overline{G}$ has at least $n - 2$ edges. Use this to conclude that the maximum number of edges in a non-Hamiltonian n-vertex graph is $\binom{n-1}{2} + 1$. (Ore [1961], Bondy [1972])

6.2.22. *Generalization of the edge bound.*
　　a) Let $f(i) = 2i^2 - i + (n - i)(n - i - 1)$, and suppose that $n \geq 6k$. Prove that on the interval $k \leq i \leq n/2$, the maximum value of $f(i)$ is $f(k)$.
　　b) Let G be a simple graph with minimum degree k. Use part (a) and Chvátal's condition to prove that if G has at least $6k$ vertices and has more than $\binom{n(G)-k}{2} + k^2$ edges, then G is Hamiltonian. (Erdős [1962])

6.2.23. (!) Suppose G has n vertices and vertex degrees $d_1 \leq d_2 \leq \cdots \leq d_n$, and let $d'_1 \leq d'_2 \leq \cdots \leq d'_n$ be the vertex degrees in $\overline{G}$. Prove that if $d_i \geq d'_i$ for all $i \leq n/2$, then G has a Hamiltonian path. Conclude that every graph isomorphic to its complement has a Hamiltonian path. (Clapham [1974])

6.2.24. (!) Suppose that n is even and G is a bipartite graph with partite sets X, Y each of size $n/2$. Suppose the vertex degrees of G are $d_1 \leq \cdots \leq d_n$. Let G' be the supergraph of G obtained by added edges to make Y induce a clique.
　　a) Prove that G is Hamiltonian if and only if G' is Hamiltonian, and describe the relationship between d and the degree sequence of G'.
　　b) Suppose that $d_k > k$ or $d_{n/2} > n/2 - k$ whenever $k \leq n/4$. Prove that G is Hamiltonian. (Hint: suppose the degree sequence of G' fails Chvátal's condition for some $i < n/2$, and obtain a contradiction.) (Chvátal [1972])

6.2.25. (!) A graph is *Hamiltonian-connected* if for every pair of vertices u, v there is a Hamiltonian path from u to v. Prove that $e(G) \geq \binom{n(G)-1}{2} + 2$ implies G is Hamiltonian and that $e(G) \geq \binom{n(G)-1}{2} + 3$ implies G is Hamiltonian-connected. (Proving the two statements together permits a simpler proof.) (Ore [1963])

6.2.26. *Hamiltonian-connected graphs - necessary condition.* (Moon [1965])
　　a) Prove that a Hamiltonian-connected graph G with at least four vertices must have at least $\lceil 3n(G)/2 \rceil$ edges.
　　b) Prove that the bound in part (a) is best possible by showing that $C_m \square K_2$ is Hamiltonian-connected if m is odd.

6.2.27. (!) *Hamiltonian-connected graphs - sufficient condition.* (Ore [1963])
　　a) Prove that G is Hamiltonian-connected if $d(x) + d(y) > n(G)$ whenever

$x \not\leftrightarrow y$. (Hint: construct graphs related to G.)

b) Prove that the condition in part (a) is best possible by constructing an n-vertex graph with minimum degree $n/2$ that is not Hamiltonian-connected, for each even n at least 4.

6.2.28. (–) Prove or disprove: If G is a simple graph with at least three vertices, and G has at least $\alpha(G)$ vertices of degree $n(G) - 1$, then G is Hamiltonian.

6.2.29. *Las Vergnas' condition.* Suppose G has a vertex ordering $v_1, \ldots, v_n$ for which there is no nonadjacent pair v_i, v_j such that $i < j$, $d(v_i) \le i$, $d(v_j) < j$, $d(v_i) + d(v_j) < n$, and $i + j \ge n$. Las Vergnas [1971] proved that $C(G)$ is complete (and hence G is Hamiltonian).

a) Prove that Chvátal's condition for Hamiltonian cycles implies Las Vergnas' condition, which means that Las Vergnas' theorem strengthens Chvátal's theorem.

b) Prove that each of the graphs below fails Chvátal's condition but has a complete graph as its Hamiltonian closure. Prove that the smaller graph satisfies Las Vergnas' condition but the larger one does not.

6.2.30. (!) *Long paths and cycles.* Suppose G is a connected simple graph with $\delta(G) = k \ge 2$ and $n(G) > 2k$.

a) Let P be a maximal path in G (not a subgraph of any longer path). If $n(P) \le 2k$, prove that the induced subgraph $G[V(P)]$ has a spanning cycle (this cycle need not have its vertices in the same order as P).

b) Use (a) to prove that G has a path with at least $2k + 1$ vertices. Give an example for each odd value of n to show that G need not have a cycle with more than $k + 1$ vertices.

6.2.31. Prove that if a simple graph G has degree sequence $d_1 \le \cdots \le d_n$ and $d_1 + d_2 < n$, then G has a path of length at least $d_1 + d_2 + 1$ unless G is the join of $n - (d_1 + 1)$ isolated vertices with a graph on $d_1 + 1$ vertices or $G = pK_{d_1} \vee K_1$ for some $p \ge 3$. (Ore [1967b])

6.2.32. (!) Dirac proved that every 2-connected graph G has a cycle of length at least $\min\{n(G), 2\delta(G)\}$. Use this to prove that every $2k$-regular graph with $4k + 1$ vertices is Hamiltonian. (Nash-Williams)

6.2.33. Scott Smith conjectured that any two longest cycles in a k-connected graph have at least k common vertices. The approach below works for small k.

a) Suppose that G is a 4-regular multigraph with n vertices that is the union of two edge-disjoint Hamiltonian cycles. Let G' be the 4-regular multigraph on $n + 2$ vertices obtained from G by subdividing two edges and adding a double edge between the two new vertices. Show that G' is also the union of two edge-disjoint spanning cycles if $n \le 5$.

b) Use (a) to conclude that any pair of longest cycles in a k-connected graph intersect in at least k points if $k \leq 6$. (Smith, Burr)

6.2.34. Prove that the Cartesian product of two graphs with Hamiltonian paths fails to have a Hamiltonian cycle if and only if both graphs are bipartite and have odd order, in which case the product has a Hamiltonian path.

6.2.35. (+) Suppose G is Eulerian. Let V' be the set of Eulerian circuits of G, considering a circuit and its reversal to be the same. Let G' be the graph with vertex set V' such that two circuits are adjacent if and only if one arises from the other by reversing the edge order on a proper closed subcircuit. Prove that G' is Hamiltonian if $\Delta(G) \leq 4$. (Hint: use induction on the number of 4-valent vertices, proving that there is a Hamiltonian cycle through every edge of G'. Comment: The conclusion also holds without restriction on $\Delta(G)$.) (Xia [1982], Zhang-Guo [1986])

6.2.36. Prove that the Eulerian circuit graph G' of the preceding problem is regular, and derive a formula for its vertex degree. Compare $\delta(G')$ and $n(G')$ when $n(G) = 2$ to show that the preceding problem cannot be solved by applying general results on Hamiltonicity of regular graphs with specified degree.

6.2.37. Prove that every tournament has a Hamiltonian path (a spanning directed path). (Hint: use extremality). (Rédei [1934]).

6.2.38. Suppose T is a strong tournament. For each $u \in V(T)$ and each k satisfying $3 \leq k \leq n$, prove that u belongs to a cycle of length k in T. (Hint: use induction on k.) (Moon [1966])

6.2.39. (+) Prove that every tournament has a Hamiltonian path with the edge between beginning and end directed from beginning to end, except the cyclic tournament on three vertices and the tournament T_5 on five vertices drawn below. (Grünbaum, in Harary [1969, p211])

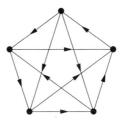

(Hint: this can be proved by induction, which requires a bit of care for invoking the induction hypothesis to prove the claim for six vertices. In all cases, find the desired configuration or $G = T_5$.)

6.2.40. Prove that Ghouila-Houri's Theorem is best possible by showing that the strictness condition on the digraph cannot be weakened to allow loops. In particular, construct for each even n a n-vertex digraph D that is not Hamiltonian even though it satisfies "at most one copy of each ordered pair is an edge" and $\min \{\delta^-(D), \delta^+(D)\} \geq n/2$.

6.3 Complexity (optional)

A salesman plans a trip to visit $n-1$ other cities and return home. The natural objective is to minimize the total travel time. If we assign each edge of K_n a weight equal to the travel time between the corresponding cities, then we seek the spanning cycle of minimum total weight. This is the famous *Traveling Salesman Problem (TSP)*. Seemingly analogous to the Minimum Spanning Tree problem, the TSP as yet has no good algorithm.

Similarly, although we have a good algorithm for finding maximum matchings, we have none for finding the maximum size of an independent set of vertices. Since the former is the special case of the latter for line graphs, it is not too surprising that it is easier to solve.

INTRACTABILITY

We defined a good algorithm (Definition 2.3.1) as one that runs (correctly) in time bounded by a polynomial function of the input size. One algorithm for the TSP considers all the possible spanning cycles and selects the cheapest one. This is not a good algorithm, because K_n has $(n-1)!/2$ spanning cycles, and this has unbounded ratio to every polynomial function of n. The computation takes too long for graphs of any substantial size, and practical applications require solving TSPs on graphs with hundreds or thousands of vertices.

No one has found a good algorithm, and no one has proved that none exists. The TSP belongs to a large class of problems having the property that a good algorithm for any one of them will yield a good algorithm for every one of them. A good algorithm for B yields a good algorithm for A if we can "reduce" problem A to problem B.

As an easy example of this, we can use a good algorithm for the TSP (problem B) to recognize Hamiltonian graphs (problem A). From a graph G form an instance of the TSP on vertex set $V(G)$ by assigning weight 0 to vertex pairs that are edges of G and weight 1 to pairs that are not. The graph G has a Hamiltonian cycle if and only if the optimal solution to this TSP has cost 0. The time for the transformation is polynomial in $n(G)$, so a good algorithm for the TSP produces a good algorithm to test for spanning cycles. We conclude that the TSP is at least as hard as the Hamiltonian cycle problem.

In the formal discussion, we consider only *decision problems*, where the answer is YES or NO. This makes sense for recognizing Hamiltonian graphs, but the TSP is an optimization problem. When formulated as a decision problem (called MINIMUM SPANNING CYCLE), the input for the TSP is a weighted graph G and a number k, and the problem is to test whether G has a spanning cycle with weight at most k. Repeated applications of this decision problem (at most a polynomial number of applications) can be used to find the minimum weight of a spanning cycle. Similarly, MAXIMUM INDEPENDENT SET takes a graph G and an integer k as input and tests $\alpha(G) \geq k$.

We judge a graph algorithm by its worst-case running time, over all possible inputs on n vertices, as a function of n. The *complexity* of a decision problem is the minimum worst-case running time over all algorithms that solve it, again as a function of the size of the problem.[†] In measuring growth rates, the set of functions whose magnitude is bounded above by a constant multiple of f (for large arguments) is called $O(f)$. Similarly, $\Omega(f)$ denotes those functions with magnitude bounded below by a constant multiple of f (for large arguments). If $c_1|f(n)| \leq |g(n)| \leq c_2|f(n)|$ for $n > a$, then $g \in \Theta(f) = O(f) \cap \Omega(f)$. To describe functions that grow more slowly or quickly than f, we use $o(f)$ and $\omega(f)$ to mean the sets of functions g such that $|g(n)/f(n)|$ approaches 0 or ∞ (respectively).

The class of problems with complexity bounded by a polynomial in the size of the input (solvable by a good algorithm) is called "P". Here we consider deterministic algorithms, meaning that each input gives rise to exactly one polynomial-time computation.

We also discuss nondeterministic algorithms. For many decision problems with no known good algorithm, short proofs exist for YES answers. For example, if we guess the right sequence of vertices in the HAMILTONIAN CYCLE problem (specified by a sequence of $O(n \log n)$ bits), then we can verify rapidly that the sequence is a spanning cycle. A *nondeterministic polynomial-time algorithm* tries all possible values of a polynomial-length sequence of bits simultaneously, applying polynomial-time computation to each guess in parallel. (Again, polynomials in the length of the input.) If any of the guesses demonstrates a YES answer to the decision problem, then the algorithm says YES. Otherwise, the answer is NO. This amounts to saying that when the answer is YES, there is a polynomial-time proof of this. The nondeterminism lies not in the answer but in the choice of the computation path.

The class of problems solvable by nondeterministic polynomial-time algorithms is called "NP". A machine that has the power to follow many computation paths in parallel can also follow one; hence $P \subseteq NP$. It is commonly believed that $P \neq NP$. This has not been proved, so NP cannot be taken to mean "non-polynomial". Instead, we use the

---

[†] Technically, the *size* of the problem instance refers to its length in bits in some encoding of the problem, but measuring the size of a graph problem by the number of vertices will suffice for our purposes. A polynomial in n is also a polynomial in n^2 or n^3, so the distinction is unimportant unless the problem involves edge weights larger than exponential in the number of vertices.

informal term *intractable* for the problems in NP that are essentially as hard as all the problems in NP.

A problem is *NP-hard* if a polynomial-time algorithm for it could be used to construct a polynomial-time algorithm for each problem in NP. It is *NP-complete* if it belongs to NP and is NP-hard. If some NP-complete problem belongs to P, then P=NP. Researchers have found no polynomial-time algorithm for any of the many NP-complete problems; this supports the prevailing belief that P≠NP. Garey-Johnson [1979] presents a thorough introduction to the theory of NP-completeness.

Given one NP-complete problem, NP-completeness of other problems is shown by reduction arguments, as suggested earlier. We present the details of several such arguments at the end of this section. Here we list complexities for some basic problems of graph theory discussed in this book.

Standard style in computer science uses upper-case names for decision problems. When the problem stated is an optimization problem, we mean the decision problem of testing whether the value is as extreme as a number that is given as part of the input. Parameters that appear in the name, however, are fixed as part of the statement of the problem. This is an important distinction. For example, k-INDEPENDENT SET for fixed k is in P, since the number of k-sets of vertices that are candidates for independent sets is a polynomial in n of degree k when k is fixed, and we can simply test them all. On the other hand, MAXIMUM INDEPENDENT SET is NP-complete; this is the problem of testing whether G has an independent set of size at least k where k is part of the input (and can grow with n).

Problems in P	NP-complete Problems
k-INDEPENDENT SET	MAXIMUM INDEPENDENT SET
GIRTH (SHORTEST CYCLE)	CIRCUMFERENCE (LONGEST CYCLE)
EULERIAN CIRCUIT	HAMILTONIAN CYCLE
DIAMETER	LONGEST PATH, HAMILTONIAN PATH
CONNECTIVITY	
2-COLORABILITY	k-COLORABILITY (for any fixed $k \geq 3$)
MAXIMUM MATCHING	$\Delta(G)$-EDGE-COLORABILITY
PLANARITY	GENUS

NP-completeness is closely related to the lack of an easily testable necessary and sufficient condition for YES answers. A *good characterization* is a characterization by a condition that we can check in polynomial time. The characterization for Eulerian graphs is good, and GIRTH, DIAMETER, and 2-COLORABILITY can all be solved in polynomial time using breadth-first search algorithms. For CONNECTIVITY, it is not immediately obvious that the characterization by the existence of internally-disjoint paths can be checked in polynomial time, but min-max relations like Menger's Theorem generally lead to polynomial-time algorithms to determine optimum values, often based on the Ford-Fulkerson labeling algorithm for maximum flow.

HEURISTICS AND BOUNDS

Our poor traveling salesman still awaits instruction. NP-completeness of a problem does not end the need for an answer. We seek heuristic algorithms that find solutions close to optimal. Perhaps we can prove a guarantee about how far from the optimum the result may be. For example, we may be content to have a solution whose cost is at most twice the optimum, if we have an algorithm that can quickly generate such a solution. An *approximation algorithm* always generates a solution whose ratio to the optimum is bounded by a constant.[†]

Greediness is a simple heuristic. For the minimum spanning tree problem, the result is optimal. In general, greedy algorithms may perform very badly. For example, consider MAXIMUM INDEPENDENT SET. Suppose we generate an independent set iteratively by picking a vertex and deleting its closed neighborhood. The question is how to pick the next vertex; if we always choose right, the result is a maximum independent set. A greedy heuristic may pick a vertex of minimum degree in what remains, since this leaves the largest set of potential vertices for the next step. The result may be arbitrarily bad.

6.3.1. Example. *Defeating the greedy algorithm for maximum independent set.* Consider the graph $(K_1 + K_m) \vee \overline{K}_m$. This graph has one vertex of degree m, m vertices of degree $m + 1$, and m vertices of degree $2m - 1$. The greedy heuristic picks the vertex of minimum degree and deletes it and its neighbors, leaving a clique. Hence the greedy algorithm finds an independent set of size 2, when in fact $\alpha(G) = m$. □

Nevertheless, the greedy algorithm works well on large graphs generated randomly. In this model (see Section 8.5), it is almost certain to find a stable set of size at least half the maximum. Exercise 11 presents two heuristics for MINIMUM VERTEX COVER; one fails as in Example 6.3.1, but the other yields an approximation algorithm.

Next we consider simple heuristics for the TSP, where $\{v_1, \ldots, v_n\}$ are the vertices and w_{ij} denotes the weight (cost) of edge $v_i v_j$. From an arbitrary starting vertex, it seems reasonable to move to a new vertex

---

[†]Better yet: an *approximation scheme* is a family of algorithms indexed by a parameter ε, say $\varepsilon = 1/k$ for the kth algorithm, such that the kth algorithm has a performance ratio bounded by $1 + \varepsilon$. The complexities of the algorithms are polynomial functions, but the degree of the kth algorithm is a growing function of k; we are willing to spend more time if we need to guarantee a better performance ratio.

via the least-cost incident edge, and continue always visiting the closest unvisited neighbor of the current vertex. This is a "greedy" algorithm, and it runs very quickly. This is the *nearest-neighbor* heuristic.

6.3.2. Example. *Nearest-neighbor is not an approximation algorithm for the TSP.* Suppose the edge weight is 0 on a Hamiltonian path P, is n^2 on the other edges incident to the endpoints of P, and is 1 on all remaining edges. In this example, there are many spanning cycles of weight n, but the nearest-neighbor heuristic yields a cycle of weight at least n^2 from any starting vertex. The cost of the cycle produced by the algorithm is not bounded by a constant multiple of the optimal cost. □

There are many similar heuristics. We could try to grow a cycle one vertex at a time, greedily absorbing the vertex whose insertion in the cycle causes the least increase in cost. This *nearest-insertion* heuristic has a better chance than nearest-neighbor, because at stage i of the nearest-neighbor heuristic we make a choice among $n - i$ alternatives, whereas at stage i in nearest-insertion we choose among $(n - i)i$ alternatives (which to add and where to insert it). Even so, this also is not an approximation algorithm (Exercise 5).

Another approach is to start with a candidate spanning cycle and try to improve it. Maintaining a feasible solution (an actual cycle) and considering small changes to improve it is called *local search*. Allowing changes takes us beyond greedy algorithms and may perform better.

To improve the current cycle, we consider changing a pair of edges. If $(v_1, \ldots, v_n)$ is our cycle, we could substitute the edges $v_i v_j$ and $v_{i+1} v_{j+1}$ for $v_i v_{i+1}$ and $v_j v_{j+1}$ and obtain a new cycle (the other possible switch leads to two disjoint cycles instead of one cycle). The switch is beneficial if $w_{i,j} + w_{i+1,j+1} < w_{i,i+1} + w_{j,j+1}$. The current cycle has $\binom{n}{2} - n = \binom{n-1}{2}$ pairs of nonincident edges to consider switching. The algorithm of Lin-Kernighan [1973], which has proved remarkably difficult to improve in practice, considers switches among three edges at a time.

The following theorem seems to doom efforts to find an approximation algorithm for the general TSP.

6.3.3. Theorem. (Sahni-Gonzalez [1976]) If there is a constant $c \geq 1$ and a polynomial-time algorithm A such that A produces for each instance of the TSP a spanning cycle whose cost is at most c times the optimum, then P=NP.

Proof. We show that such an algorithm A could be used to construct a polynomial-time algorithm for HAMILTONIAN CYCLE, which is NP-complete (Corollary 6.3.11). Given an n-vertex graph G, we construct an n-vertex instance of TSP such that any cycle with cost at most cn has cost exactly n and corresponds to a spanning cycle of G. Since A

produces a solution with cost at most c times the optimum, it produces one of cost n if and only if G has a spanning cycle. The construction is very simple: let $w_{ij} = 1$ if $v_i v_j \in E(G)$, and $w_{ij} = cn$ otherwise. □

Approximation algorithms exist for some special classes of TSP problems. We will need a lower bound on the optimal solution. Let M be the cost of a minimum spanning tree in the weighted graph G. If we delete an edge from the optimal cycle in G, we obtain a spanning path. Since this is a spanning tree, its cost is at least M. The cost of the optimal cycle is at least M plus the minimum cost of an edge not in some tree with cost M. We can run Kruskal's algorithm for minimum spanning trees to compute this bound.

6.3.4. Theorem. On the class of Traveling Salesman Problems where the input satisfies the triangle inequality, there is an approximation algorithm that finds a spanning cycle with cost at most twice the optimum.

Proof. Satisfying the triangle inequality means that the matrix of costs satisfies $w_{i,j} + w_{j,k} \geq w_{i,k}$ for all i, j, k. We know that the cost of the optimal cycle is at least M, where M is the cost of the minimum spanning tree. We use the triangle inequality and the minimum spanning tree to obtain a spanning cycle with cost at most $2M$. Replace each edge in a minimum spanning tree by a pair of oppositely-oriented directed edges, as on the left below. Since the in-degree and out-degree at each node are equal, this digraph has an Eulerian circuit; it has $2n$ edges and total cost $2M$. We successively reduce the number of edges without increasing the cost until only n edges remain. If we also maintain the property that the circuit visits all vertices, then the circuit at the end corresponds to a spanning cycle with cost at most $2M$.

The proof that we can do this is quite simple. If the circuit has more than n edges, then it visits some vertex more than once, say via edges $v_i \to v_j \to v_k$ and $v_r \to v_j \to v_s$. Replace the edges $v_i v_j$ and $v_j v_k$ by $v_i v_k$. The smaller graph is still an Eulerian circuit visiting all the original vertices. Furthermore, the triangle inequality guarantees that the total cost of the edges is no larger than before. □

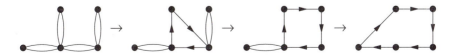

This algorithm using the minimum spanning tree was discovered many times independently. Christofides [1976] improved the performance ratio to 3/2 when the triangle inequality holds. After finding a minimum spanning tree, we needn't double all the edges to obtain a

directed Eulerian circuit. It suffices to add edges to pair up vertices
that have odd degree in the tree. The resulting graph has an Eulerian
circuit, and applying the same algorithm as before to reduce the number
of edges produces a spanning cycle whose cost is at most M plus the cost
of the matching. It suffices to show that there is such a matching with
cost at most half the cost of the optimal cycle (Exercise 6). The idea is
useful because we can find a minimum weight complete matching
among these vertices in polynomial time.

NP-COMPLETENESS PROOFS

We now consider details of NP-completeness proofs. A *transformation* of problem A to problem B is a procedure that converts instances of
problem A into instances of problem B so that the answer to A on the
original instance is determined by the answer to B on the transformed
instance. If we have an efficient (polynomial-time) transformation of
problem A to problem B and an efficient algorithm for problem B, then
we have an efficient algorithm for problem A. We say that A *reduces to*
or *transforms to* B.

If A is NP-hard, and A reduces to B by a polynomial-time transformation, then B is also NP-hard (a polynomial algorithm for B yields a
polynomial algorithm for A, which in turn yields a polynomial algorithm
for every problem in NP). If we also know that B is in NP, then we say
that B is NP-complete by *reduction from* or *transformation from* A.

The direction of the reduction is crucial. For example, we can
reduce EULERIAN CIRCUIT to HAMILTONIAN CYCLE. Given an
input graph G, replace each edge with a path of length four through
three new vertices, then add a clique on the neighbors of each original
vertex, and delete $V(G)$. The graph G is Eulerian if and only if the
resulting graph G' is Hamiltonian, so we could apply an algorithm for
HAMILTONIAN CYCLE to G' to determine whether G is Eulerian.
This tells use that HAMILTONIAN CYCLE is as hard as EULERIAN
CIRCUIT, up to a polynomial factor. Since EULERIAN CIRCUIT is
easy (in P), this tells us nothing of use about the complexity of HAMIL-
TONIAN CYCLE.

The reduction technique requires an initial NP-complete problem.
Cook [1971] gave us SATISFIABILITY to satisfy this need. Consider a
logical formula expressed as a list of clauses; each clause is a collection
of literals (variables or their negations), and a clause is considered true
if at least one of its literals is true. A formula is *satisfiable* if there is an
assignment of truth values to the variables such that every clause is
true. SATISFIABILITY takes such a formula as input and asks
whether it is satisfiable. Cook proved that for every problem A in NP,

an instance of SATISFIABILITY can be produced in polynomial time from an instance of A such that the answer to the SATISFIABILITY instance is the same as the answer to the instance of A.

This effort need not be repeated for each NP-complete problem. To prove that B is NP-hard, we can reduce SATISFIABILITY to B. With each additional problem proved NP-complete, we obtain another problem that can be used in this way for NP-completeness proofs. In principle the proofs become easier to find, but in practice there are a few fundamental NP-complete problems that serve as the known NP-complete problem in most NP-completeness proofs.

Starting from SATISFIABILITY, Karp [1972] provided 21 such problems. These include many fundamental problems of graph theory, including HAMILTONIAN CYCLE and MAXIMUM INDEPENDENT SET. It helps to have as restrictive a version of an NP-complete problem as possible while still remaining NP-complete. Since a restricted version has less flexibility, it is easier to transform it to the problem we are trying to prove NP-complete.

For example, SATISFIABILTY remains NP-complete when we restrict it by requiring that every clause have three literals. The restricted problem is called 3-SATISFIABILITY or 3-SAT. The NP-completeness of 3-SAT is proved by considering an arbitrary instance of SATISFIABILITY and replacing each clause by an equivalent collection of clauses with three literals, at the cost of introducing some additional variables. 3-SAT is sufficiently restrictive that many, many NP-completeness proofs procede by reduction from 3-SAT. It is easier yet to reduce the more restrictive 2-SAT, but logical formulas with two literals per clause are so restrictive that 2-SAT is solvable in polynomial time.

We take 3-SAT as our starting point, because the NP-completeness of SATISFIABILITY and the reduction of it to 3-SAT do not involve graph theory. Our reductions to 3-COLORABILITY and DIRECTED HAMILTONIAN PATH follow the presentation of Gibbons [1985]. We begin with the coloring problems.

6.3.5. Definition. 3-SATISFIABILITY (3-SAT)
   Instance: A set of logical *variables* $U = \{u_j\}$ and a set $\mathbf{C} = \{C_i\}$ of *clauses*, where each clause consists of three literals, a *literal* being a variable u_i or its negation $\bar{u}_i$.
   Question: Can the value of each variable be set to True or False so that each clause is "satisfied" (contains at least one true literal)?

6.3.6. Theorem. (Karp [1972]) 3-SAT is NP-complete. □

6.3.7. Theorem. (Karp [1972]) 3-COLORABILITY is NP-complete.

Proof. In this problem, we are given a graph and asked whether is it 3-colorable. If the answer is YES, then there exists a proper 3-coloring, and we can verify in quadratic time that the coloring is proper. Hence 3-COLORABILITY is in NP. To prove that it is NP-hard, we reduce 3-SAT to 3-COLORABILITY.

Consider an instance of 3-SAT with variables $U = \{u_j\}$ and clauses $\mathbf{C} = \{C_i\}$. We transform this into a graph G that is 3-colorable if and only if the instance of 3-SAT is satisfiable. We use the auxiliary graph H drawn below, calling $\{u'_1, u'_2, u'_3\}$ the *inputs* and v the *output*. When we use H in the transformation, we will attach it to a larger graph at the inputs, as suggested on the right.

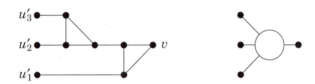

We consider 3-colorings using the color set $\{0, 1, 2\}$. Every proper 3-coloring of H in which the inputs all have color 0 also assigns color 0 to v. On the other hand, if the inputs receive colors that are not all 0, then this coloring extends to a proper 3-coloring of H in which v does not have color 0.

From our instance of 3-SAT, we construct a graph G having vertices u_j and $\bar{u}_j$ for each variable in U, a copy H_i of H for each clause C_i, and two special vertices a, b. For each j, the vertices $a, u_j, \bar{u}_j$ form a triangle. For each clause C_i, the subgraph H_i attaches to the graph formed thus far at the vertices for the literals in C_i. The vertex for the jth literal in C_i plays the role of u'_j in H_i. Except for these attachment vertices, the subgraphs H_i are pairwise disjoint. Finally, the vertex b is adjacent to a and to the output vertex v_i for each H_i. Below we draw the graph G resulting from an instance of 3-SAT having four variables and three clauses.

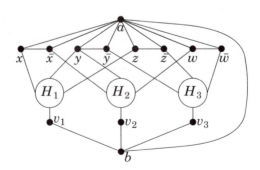

A satisfying truth assignment for $\{c_i\}$ leads to a proper 3-coloring f of G. If u_i is true in the assignment, then we let $f(u_i) = 1$ and $f(\overline{u}_i) = 0$; otherwise, we set $f(u_i) = 0$ and $f(\overline{u}_i) = 1$. For each clause, some literal is true; hence for each H_i at least one of $\{u_1', u_2', u_3'\}$ has color 1. By our observation about H, we can extend f so that each v_i has a color other than 0. Now we can complete the proper 3-coloring by setting $f(b) = 0$ and $f(a) = 2$.

Conversely, suppose G has a proper 3-coloring f. By renaming colors if necessary, we may assume that $f(a) = 2$ and $f(b) = 0$. Since $f(a) = 2$, for each variable we have one literal colored 0 and one colored 1. Consider the truth assignment in which variable u_j is true or false when $f(u_j)$ is 1 or 0, respectively. We claim that this is a satisfying truth assignment. Since $f(b) = 0$, every output vertex v_i has nonzero color. By our observation about H, the vertices corresponding to the inputs in H_i cannot all have color 0. Therefore each clause contains at least one true literal. □

Exercise 1 extends this to k-COLORABILITY for each fixed $k \geq 3$. For each k, this is a special case of CHROMATIC NUMBER, which thus also is NP-complete.

6.3.8. Theorem. INDEPENDENT SET, CLIQUE, and VERTEX COVER are NP-complete.

Proof. A YES answer to the question of whether an input graph G has an independent set as large as an input integer k can be verified by exhibiting the set and checking (in quadratic time) that its vertices are independent. Hence INDEPENDENT SET is in NP.

Exercise 5.1.11 states that G is m-colorable if and only if $G \square K_m$ has an independent set of size $n(G)$. This transformation reduces CHROMATIC NUMBER to INDEPENDENT SET. The construction of $G \square K_m$ is quadratic in $n(G)$, since CHROMATIC NUMBER is trivial if $m > n$. We conclude that INDEPENDENT SET is NP-hard.

Since cliques in G are independent sets in $\overline{G}$, CLIQUE and INDE-PENDENT SET are polynomially equivalent. Since $\alpha(G) + \beta(G) = n(G)$ (Lemma 3.1.13), INDEPENDENT SET and VERTEX COVER are polynomially equivalent. □

We next consider problems of traversing graphs via spanning paths and cycles. Digraph problems are more general than graph problems, since we can model graphs by using symmetric digraphs. Thus, it may be easiest to prove a digraph version of a problem NP-hard and then obtain the graph version by a simple restriction.

6.3.9. Definition. Given vertices x, y in a digraph D, the DIRECTED HAMILTONIAN PATH problem asks whether G has a spanning x, y-path.

6.3.10. Theorem. DIRECTED HAMILTONIAN PATH is NP-complete.

Proof. A spanning x, y-path in a digraph D can be verified in linear time. Thus DIRECTED HAMILTONIAN PATH is in NP. To show that it is NP-hard, we reduce VERTEX COVER to DIRECTED HAMILTO-NIAN PATH.

Consider an instance of VERTEX COVER, consisting of a graph G and an integer k; we want to know whether G has a vertex cover of size (at most) k. We construct a digraph D such that D has a Hamiltonian path if and only if G has a vertex cover of size at most k. We index the edges incident to each vertex, arbitrarily. When the edge $e = uv$ is the ith edge incident to u and the jth edge incident to v, we write $e_i(u) = e = e_j(v)$.

To build D, we start with $k + 1$ special vertices $z_0, \ldots, z_k$. For each $v \in V(G)$, we add a path $P(v) = v_1, \ldots, v_{2r}$ to D, where $r = d_G(v)$. We add edges from each of $z_0, \ldots, z_{k-1}$ to the start of each $P(v)$ and from the end of each $P(v)$ to each of $z_1, \ldots, z_k$. Also, for each edge $e = e_i(u) = e_j(v)$ in $E(G)$, we create the edges $u_{2i-1}v_{2j-1}$, $v_{2j-1}u_{2i-1}$, $u_{2i}v_{2j}$, and $v_{2j}u_{2j}$.

Suppose that G has a vertex cover of size k, consisting of vertices $v^1, \ldots, v^k$. We form a z_0, z_k-path in D by traversing $z_0, P(v^1), z1, P(v^2), z_2, \ldots, P(v^k), z_k$. This path omits all of $P(u)$ for each uncovered vertex u. We absorb these vertices in pairs, absorbing each pair $u_{2i-1}u_{2i}$ by making a detour from the path $P(v)$ for the vertex that covers the edge $uv = e_i(u) = e_j(v)$. The detour is $v_{2j-1}u_{2j-1}u_{2j}v_{2j}$, as shown below. Because the vertices v_{2j-1}, v_{2j} are associated with only one edge, each such detour is requested at most once. After implementing all the detours, we have a Hamiltonian z_0, z_k-path in D.

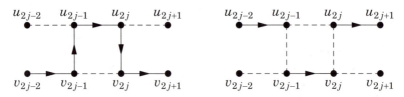

Conversely, suppose there is such a path Q. Note that every vertex of $P(u)$ except that first and last has in-degree and out-degree 2, for each $u \in V(G)$. We show first for $i \geq 1$ and $u \in V(G)$ that $u_{2i}u_{2i+1} \in E(Q)$ if and only if $u_{2i-2}u_{2i-1} \in E(Q)$, where for $i = 1$ we take u_0 to mean some element of $\{z_r\}$. Similarly, when $i = d(u)$ we tak u_{2i+1} to mean some element of $\{z_r\}$ Since the ith edge incident to u is well-defined, we can let the vertex v and index j be such that $e_i(u) = e_j(v)$.

If $u_{2i-2}u_{2i-1} \notin E(Q)$, then Q must enter u_{2i-1} from v_{2j-1}. This implies that Q can only leave u_{2i-1} on $u_{2i-1}u_{2i}$ and can only enter v_{2j} on $u_{2i}v_{2j}$. This in turn implies $u_{2i}u_{2i+1} \notin E(Q)$.

If $u_{2i-2}u_{2i-1} \in E(P)$, then P cannot leave v_{2j-1} on $v_{2j-1}u_{2i-1}$ and must leave v_{2j-1} on $v_{2j-1}v_{2j}$. This implies that Q enters v_{2j} on $v_{2j-1}v_{2j}$ and not on $u_{2i}v_{2j}$. Hence Q does not leave u_{2i} on $u_{2i}v_{2j}$ and must leave Q on $u_{2i}u_{2i+1}$. (In this case, Q may include $\{u_{2i-1}u_{2i}, v_{2i-2}v_{2i-1}, v_{2i}v_{2i+1}\}$ or $\{u_{2i-1}v_{2i-1}, v_{2i}u_{2i}\}$.)

Now let $S = \{v \in V(G): z_i v_1 \in Q$; these are the k vertices in G whose initial copies are entered from $z_0, \ldots, z_{k-1}$ by Q. Our argument above shows for each edge uv that $u \notin S$ implies $v \in S$. Hence S is a vertex cover, and we have the desired reduction of VERTEX COVER to DIRECTED HAMILTONIAN PATH. $\square$

6.3.11. Corollary. DIRECTED HAMILTONIAN CYCLE, HAMILTONIAN PATH, and HAMILTONIAN CYCLE are NP-complete.

Proof. All these problems are in NP. To reduce DIRECTED HAMILTONIAN PATH to DIRECTED HAMILTONIAN CYCLE, add one vertex z and edges vz and zu to an instance requesting a spanning u, v-path in G. The reduction of HAMILTONIAN PATH (with specified endpoints) to HAMILTONIAN CYCLE is the same. To reduced DIRECTED HAMILTONIAN PATH to HAMILTONIAN PATH, consider an instance requesting a u, v-path in D. Form G by splitting each vertex x into a path x^-, x^0, x^+, with x^- inheriting all edges with heads at x and x^+ inheriting all edges with tails at x. A spanning u, v-path in D becomes a spanning u^-, v^+-path in G by replacing each vertex x by the sequence x^-, x^0, x^+. Conversely, since a spanning u^-, v^+-path in G must visit each x^0, it must visit traverse all sequences of the form x^-, x^0, x^+, forwards or backwards. Since no vertices of the same sign are adjacent, these traversals must all be in the same direction, and then they collapse to the desired u, v-path in D. $\square$

There are two main reasons to explore the boundary between P and NP-complete. The more important refinement is finding polynomial-time algorithms for large classes of inputs. We have also mentioned that tighter restrictions of the input for which a problem remains NP-complete make it easier to use that problem to prove other NP-completeness results. We illustrate the approach by proving NP-completeness of 3-COLORABILITY for planar graphs. (Readers unfamiliar with the notion of a planar graph can find the relevant definitions in Section 7.1.)

6.3.12. Theorem. (Garey-Johnson-Stockmeyer [1976]) PLANAR 3-COLORABILITY is NP-complete.

Proof. As usual, it is easy to verify that a 3-coloring is proper. We reduce 3-COLORABILITY to PLANAR 3-COLORABILITY. Given an arbitrary graph G, we construct a planar graph G' such that G' is 3-colorable if and only if G is 3-colorable. Consider a drawing of G in the plane. We replace each crossing by a planar "gadget" that has the effect of propagating color across the crossing in a 3-coloring. Due to its triangles, the graph H below has only one partition into three independent sets. In particular, each 3-coloring of H gives the same color to the top and bottom vertex, and it gives the same color to the left and right vertices. As indicated by the partial edges, these are the terminals where H will hook up to the rest of the graph.

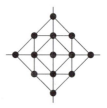

Given a drawing of G, each edge involved in k crossings is cut into $k + 1$ segments by the crossings. On each segment, add a new vertex. Replace each crossing by a copy of H attached by its terminals to the new vertices on the four segments incident to the crossing. Finally, for each original edge u, v, choose one endpoint and contract the edge between it and the vertex on the segment of uv incident to it. In particular, an edge involved in no crossings returns to its original state.

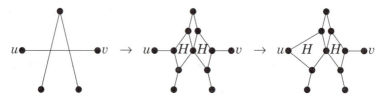

If the new graph G' has a proper 3-coloring, the propagation of color across the gadgets requires the endpoints of each original edge to have different colors. Hence restricting this coloring to the original vertices yields a 3-coloring of G. Conversely, given a coloring of G, we can start along each edge from the endpoint involved in a copy of H and propagate the colors to obtain a coloring of G. $\square$

HAMILTONIAN CYCLE also remains NP-complete for planar graphs. Indeed, it remains NP-complete for graphs that are planar, 3-regular, 3-connected, and have no face of length less than five (Garey-Johnson-Tarjan [1976]), or for bipartite graphs.

The HAMILTONIAN CYCLE and 3-COLORABILITY problems

also remain NP-complete for line graphs. Exercise 14 covers part of this reduction for Hamiltonian cycles.

Finally, 3-EDGE-COLORABILITY is NP-complete for 3-regular graphs, by reduction from 3-SAT.

EXERCISES

6.3.1. (!) Theorem 6.3.7 states that 3-COLORABILITY is NP-complete. Given this, prove that k-COLORABILITY is NP-complete for each fixed value of k that is at least 3.

6.3.2. Give a polynomial-time algorithm for 2-COLORABILITY.

6.3.3. Prove that HAMILTONIAN CYCLE and HAMILTONIAN PATH are polynomial-time equivalent. (Show that a polynomial-time algorithm for either one can be used to obtain a polynomial-time algorithm for the other.)

6.3.4. Testing for a cycle of fixed length k in an input graph with n vertices can be done in time bounded by a multiple of $k!n^k$: look at each of the $\binom{n}{k}$ vertex subsets of size k in turn and test all possible orderings. Since k is a constant, this is polynomial time. For a 4-cycle, this runs in time $O(n^4)$. Devise an algorithm that will test for the presence of a 4-cycle in time $O(n^2)$. (Richards-Liestman [1985])

6.3.5. Construct a family of examples to prove that the performance ratio of the nearest-insertion heuristic for the TSP is not bounded by any constant.

6.3.6. Suppose the costs in an instance of the TSP satisfy the triangle inequality. Prove that there is a pairing up of the vertices of odd degree in the minimum spanning tree whose cost is at most half the cost of the minimum spanning cycle. Use this and the fact that the optimal tour costs at least M to prove that Christofides' algorithm produces a spanning cycle with performance ratio bounded by 3/2.

6.3.7. Prove that in order to solve the TSP exactly, it suffices to have an algorithm that solves the TSP in the special case where the edge weights satisfy the triangle inequality. (Hint: given an arbitrary instance of the TSP, produce in polynomial time an instance of the TSP in which the edge weights satisfy the triangle inequality $(w(ij) + w(jk) \ge w(ik))$ and the set of optimal tours is the same as in the original instance.)

6.3.8. Prove that 2-SAT belongs to P.

6.3.9. Each city has one snowplow. It must plow its narrow streets, which the plow can plow completely in one traversal. There are also county roads that the plow can traverse to get from one place to another, though these don't need to be plowed. In other words, we have a weighted graph with edges of two types; those of type 1 must be traversed, those of type 2 need not be. The state wants an algorithm that will find a minimum-length tour traversing the type 1 edges in such a graph. Prove that this problem is NP-hard, by reduction from the

HAMILTONIAN CYCLE problem.

6.3.10. (–) Using algorithms developed earlier in this text, describe a good algorithm to compute $\alpha(G)$ when G is bipartite.

6.3.11. (!) *Heuristic algorithms for vertex covering.* Algorithm 1: include a vertex of maximum remaining degree, delete, iterate until the remaining graph is a stable set. Algorithm 2: choose an arbitrary edge, include both endpoints, delete them, iterate until the remaining graph is a stable set. The heuristic in Algorithm 1 may seem more powerful, but Algorithm 2 has a better performance guarantee!

　　a) Prove that algorithm 2 always produces a vertex cover with size at most twice the minimum.

　　b) Prove that algorithm 1 may produce a vertex cover of size about $\log n$ times the minimum. (Hint: construct a bipartite graph G with $\beta(G) = t$ for which Algorithm 1 chooses about t/i vertices of degree i for each $1 \le i \le t$.)

6.3.12. (+) A graph G is α-*critical* if $\alpha(G - e) > \alpha(G)$ for every $e \in E(G)$. Prove that a connected α-critical graph has no cut-vertex. (Hint: if e_1, e_2 are edges incident to a cut-vertex x, use the maximum independent sets in $G - e_1$ and $G - e_2$ to build an independent set in G with more than $\alpha(G)$ vertices.)

6.3.13. SATISFIABILITY differs from 3-SAT in that clauses may have arbitrary size. Prove that a clause containing more than three literals can be replaced by clauses with 3 literals (and possible the addition of a few "dummy" variables) so that the original clause is satisfiable if and only if the new instance of 3-SAT is satisfiable. Conclude that SATISFIABILITY reduces to 3-SAT. (Karp [1972])

6.3.14. Given that HAMILTONIAN CYCLE is NP-complete for 3-regular graphs, prove that COVERING CIRCUIT is NP-complete. This is the question of whether the input graph G contains a closed trail that includes at least one endpoint of every edge. Prove that the line graph of G is Hamiltonian if and only if G has a covering circuit. Conclude that HAMILTONIAN CYCLE is NP-complete for line graphs.

Chapter 7

Planar Graphs

7.1 Embeddings and Euler's Formula

Topological graph theory, broadly conceived, is the study of graph layouts. Initial motivation for this involved the famous Four Color Problem: can the regions of any map on the globe can be colored with four colors so that regions sharing a nontrivial boundary have different colors? More recent motivation comes from the study of circuit layouts on silicon chips. Crossings cause problems in a layout, so we want to know which circuits have layouts without crossings.

DRAWINGS IN THE PLANE

The following brain teaser appeared as early as Dudeney [1917].

7.1.1. Example. *Gas-water-electricity.* Three sworn enemies A, B, C live in houses in the woods. We must cut paths so that each has a path to each of three utilities, which by tradition are gas, water, and electricity. In order to avoid confrontations, we don't want any of the paths to cross. Can this be done? This asks whether $K_{3,3}$ can be drawn in the plane without edge crossings; we will give two proofs that it cannot. □

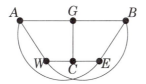

We need a precise notion of drawing. We have drawn graphs in this book using continuous curves, but these present technical difficulties in proofs. To avoid such difficulties, we consider only curves formed from finitely many line segments. These can approximate any continuous curve well enough that the eye cannot tell the difference.

7.1.2. Definition. A *polygonal path* or *polygonal curve* in the plane is the union of finitely many line segments such that each segment starts at the end of the previous one and no point appears in more than one segment except for common endpoints of consecutive segments. In a *polygonal u, v-path*, the beginning of the first segment is u and the end of the last segment is v.

A *drawing* of a graph G is a function that maps each vertex $v \in V(G)$ to a point $f(v)$ in the plane and each edge uv to a polygonal $f(u), f(v)$-path in the plane. The images of vertices are distinct. A point in $f(e) \cap f(e')$ other than a common end is a *crossing*. A graph is *planar* if it has a drawing without crossings. Such a drawing is a *planar embedding* of G. A *plane graph* is a particular drawing of a planar graph in the plane with no crossings.

We commonly abuse terminology by using "G" when discussing a particular drawing of G and by referring to the points and curves in a drawing as the vertices and edges of the graph. By slight perturbations, we may assume that no three edges meet at a point, that an edge contains no vertex except its endpoints, and that no pair of edges is tangent. If two edges cross more than once, then we can cut and paste them as shown below to reduce the number of crossings; hence we may assume that edges cross at most once. We consider only drawings with all these properties.

Defining the basics terms used to discuss planar graphs requires elementary notions about segments and distances.

7.1.3. Definition. An *open set* in the plane is a set $U \in \mathbb{R}^2$ such that for every $p \in U$, all points within some small distance from p belong to U. A *region* is an open set U that contains a polygonal u, v-path for every pair $u, v \in U$. The *faces* of a plane graph are the maximal regions of the plane that are disjoint from the drawing.

By the definition of region, a finite plane graph G has one un-bounded face. If we view all the points *not* in the drawing as vertices of a graph H, adjacent if the segment between them does not intersect the drawing of G, then the components of H are the faces of G. In particular, the faces of G are pairwise disjoint. Points p, q on no edge of a plane graph G are in the same face if and only if there is a polygonal p, q-path that crosses no edge of G.

A curve in the plane is *closed* if its first and last points are the same, and it is *simple* if it does not otherwise intersect itself. Computation with faces uses the notion that a simple closed curve cuts the plane into two regions. In topology this is a deep result; in graph theory we take it almost as an axiom. The full topological details are difficult, but the case of polygonal curves is simpler. We present some detail of the polygonal case in order to explain how to compute whether a point is in the inside or the outside. This proof appears in Tverberg [1980].

7.1.4. Theorem. (Restricted Jordan Curve Theorem). A simple closed polygonal curve C consisting of finitely many segments partitions the plane into exactly two faces, each having C as boundary.

Proof. Because we have finitely many segments, nonintersecting segments cannot be arbitrarily close together, and hence we can leave a face only by crossing C. As we traverse C in one direction, the nearby points on our right are all in the same face, and similarly for the points on the left. (There is a precise algebraic notion of the meaning of left and right here.) If $x \notin C$ and $y \in C$, the segment xy first meets C somewhere, approaching it from the right or the left. Hence every point not in the plane lies in the same face with at least one of the two sets we have described.

Intuitively, the sets are distinct because in the plane we cannot traverse a closed curve and switch the meaning of left and right. We can also distinguish the inside from the outside without defining an orientation for the curve. Consider a ray that starts at a given point p. A direction for the ray is "bad" if the resulting ray contains an endpoint of a segment of C. Since C has finitely many segments, there are finitely many bad directions, and in each good direction the ray crosses C finitely often. As the ray rotates, the number of crossings can change only when the ray passes through a bad direction, but before and after those directions the parity of the number of crossings is the same. Call this the *parity* of p.

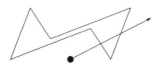

Suppose x, y are points in the same face of C, and let P be a polygonal x, y-path that avoids C. Because C has finitely many segments, the ends of the segments of P can be adjusted slightly (if necessary) so that the rays along segments on P are in good directions for their endpoints. A segment of P belongs to a ray from one end that contains the other. Since the segment does not intersect C, the parity of the two points is the same. Hence any pair of points in the same face have the same parity. Because the endpoints of a short segment intersecting C exactly once have opposite parity, there must be two distinct faces. The even points and odd points comprise the outside face and inside face, respectively. □

7.1.5. Example. K_5 *and* $K_{3,3}$ *are nonplanar.* Consider a drawing of K_5 or $K_{3,3}$ in the plane using polygonal curves. Let C be a spanning cycle in the graph; C also is drawn as a closed polygonal curve. If the drawing is an embedding, chords of C must be drawn inside or outside this curve. Two chords *conflict* if their endpoints on C occur in alternating order. Conflicting chords must embed in opposite faces of C. Considering a 6-cycle from $K_{3,3}$, we are left with three pairwise conflicting chords and can put at most one inside and one outside. Considering a 5-cycle from K_5, at most two chords can go inside or outside. Hence neither of these graphs is planar. □

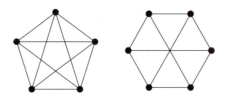

DUAL GRAPHS

We can view a geographic map on the plane or the sphere as a plane graph in which the faces are the territories of the map, the vertices are places where several boundaries meet, and the edges are the portions of the boundaries that join two vertices. We allow the full generality of loops and multiple edges. From any plane graph G, we can form another plane graph called its "dual".

7.1.6. Definition. Suppose G is a plane graph. The *dual graph* G^* of G is a plane graph having a vertex for each region in G. The edges of G^* correspond to the edges of G as follows: if e is an edge of G that has region X on one side and region Y on the other side, then the corresponding dual edge $e^* \in E(G^*)$ is an edge joining the vertices x, y of G^* that correspond to the faces X, Y of G.

7.1.7. Example. *A simple plane graph and its dual.* Below we have drawn a plane graph G with dashed edges and its dual G^* with solid edges. Since G has four vertices, four edges, and two faces, G^* has four faces, four edges, and two vertices. As in this example, a simple plane graph may have loops and multiple edges in its dual. A cut-edge of G becomes a loop in G^*, because the faces on both sides of it are the same. Multiple edges arise in the dual when distinct regions of G have more than one common boundary edge. □

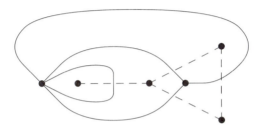

7.1.8. Remark. *Geometry of the dual.* Some arguments require a careful geometric description of the placement of vertices and edges in the dual. For each face F of G, we place the dual vertex for F in the interior of F; hence each face of G contains one vertex of G^*. For each edge in the boundary of F, we place a "half-edge" emanating from the dual vertex for F to a point on the edge. These half-edges do not cross each other, and each meets another half-edge at the boundary to form a dual edge. No other edges enter F. Hence G^* is a plane graph, and each edge of G^* in this layout crosses exactly one edge of G. Such arguments lead to a proof that $(G^*)^*$ is isomorphic to G if and only if G is connected (Exercise 3).

7.1.9. Example. *Variability in the dual.* Different embeddings of a planar graph may have nonisomorphic duals. For example, the plane graph on the left below has faces of lengths 6,4,3,3, while the plane graph on the right has faces of lengths 5,5,3,3. Although these two graphs are isomorphic, their duals have degree sequences 6,4,3,3 and 5,5,3,3, respectively, and hence cannot be isomorphic. Every 3-connected planar graph has essentially one embedding; Whitney's 2-Isomorphism Theorem [1933b] gives a procedure for generating all the duals of a planar graph, and it implies that a 3-connected planar graph has only one dual. □

7.1.10. Remark. *The plane vs. the sphere.* We have not distinguished between maps on the sphere and graphs in the plane because a graph embeds in the plane if and only if it embeds on the sphere. Given an embedding on the sphere, we can puncture the sphere within any face and project from there onto a plane tangent to the antipodal point to obtain a planar embedding. The punctured face on the sphere becomes the unbounded face in the plane, and the process is reversible. □

A statement about a connected plane graph becomes a statement about the dual graph when we interchange the roles of vertices and faces. Edges incident to a vertex become edges bounding a face, and vice versa. What does the degree-sum formula say about edges and faces instead of edges and vertices? The *length* of a face in a plane graph G is the length of the walk in G that bounds it. A cut-edge belongs to the boundary of only one face, and it contributes twice to the boundary of that face as we traverse the boundary. The graph below has three faces, with lengths 3,6,7. The sum of the lengths is 16, which is twice the number of edges.

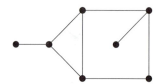

7.1.11. Proposition. If $l(F_i)$ denotes the length of face F_i in a plane graph G, then $2e(G) = \Sigma l(F_i)$.

Proof. Since bounding edges for a face X correspond to dual edges incident to the dual vertex x, and $e(G) = e(G^*)$, the statement $2e(G) = \Sigma f_i$ is the same as the degree-sum formula $2e(G^*) = \Sigma d_{G^*}(x)$ for G^*. Alternatively, adding up the face lengths counts each edge twice. □

Similarly, we can interpret the problem of coloring G^*. The edges of G^* represent shared boundaries between regions of G. Hence the chromatic number of G^* equals the number of colors needed to properly color the regions of G. Since the dual of the dual of a connected plane graph is the original graph, this means that the maximum number of colors needed to color regions of planar maps is the same as the maximum chromatic number of planar graphs.

The Jordan Curve Theorem states that a simple closed curve cuts its interior from its exterior. In plane graphs, this duality between curve and cut becomes a duality between cycles and bonds.

7.1.12. Theorem. Edges in a plane graph G form a cycle in G if and only if the corresponding dual edges form a bond in G^*.

Proof. Suppose $D \subseteq E(G)$. It suffices to prove that D contains a cycle if and only if the set D^* of dual edges contains a bond of G^*. If D contains a cycle C, then by the Jordan Curve Theorem some face of G lies inside C and some face lies outside C. These faces correspond to vertices v^*, w^* in G^*, one drawn inside C and one outside C. A v^*, w^*-path in G^* crosses C and hence uses an edge of G^* that is dual to an edge of C. Thus D^* disconnects v^* from w^*, and hence D^* contains a bond.

Conversely, if D does not contain a cycle, then D encloses no region (see Exercise 9a). It remains possible to reach each face of G from every other without crossing D. Hence $G^* - D^*$ is connected, and D^* contains no bond. $\square$

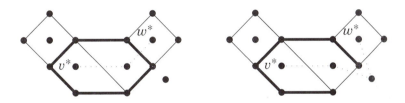

Face boundaries allow us to characterize bipartite planar graphs. The characterization can also be proved by induction (Exercise 6).

7.1.13. Theorem. The following statements are equivalent for a plane graph G.
A) G is bipartite.
B) Every face of G has even length.
C) The dual graph G^* is Eulerian.

Proof. We first prove A$\Leftrightarrow$B. A face of G is a closed walk, and an odd closed walk contains an odd cycle, so a bipartite plane graph has no face of odd length. Conversely, suppose G has an odd cycle C. Since G has no crossings in the plane, C is laid out as a simple closed curve; let F be the region enclosed by C. Every region of G is wholly within F or wholly outside F. If we sum the face lengths for the regions inside F, we obtain an even number, since each face length is even. This sum counts each edge of C once. It also counts each edge inside F twice, since each such edge belongs twice to faces in F. Hence the parity of the length of C is the same as the parity of the full sum, which is even. The equivalence of B and C follows because the dual graph is connected, and its vertex degrees are the face lengths of G. □

Many of the questions we study for general planar graphs can be answered more easily for a special class of planar graphs.

7.1.14. Definition. A graph is *outerplanar* if it has an embedding in the plane such that every vertex lies on the unbounded face. An *outerplane* graph is a planar embedding with every vertex on the unbounded face. The *weak dual* of a plane graph G is the graph obtained from the dual G^* by deleting the vertex corresponding to the unbounded face of G. A *maximal* outerplanar graph is a simple outerplanar graph that is not a spanning subgraph of any other simple outerplanar graph.

The graph in Example 7.1.9 is outerplanar, and the drawing of it on the left is an outerplanar embedding.

7.1.15. Proposition. The weak dual of an outerplane graph is a forest. The boundary of the unbounded face in a maximal outerplane graph is a cycle. Every simple outerplanar graph has a vertex of degree at most 2.

Proof. A cycle in the dual graph G^* passes through faces that surround a vertex of G. If every vertex of G lies on the unbounded face, then every cycle of G^* passes through the vertex v^* of G^* that corresponds to the unbounded face in G, and hence $G^* - v^*$ is a forest.

If an outerplane graph is disconnected, then we can add an edge joining any pair of vertices and still have an outerplane graph. If the boundary of the outer face in a connected outerplane graph is not a cycle, then it is a walk that visits some vertex v more than once. If $\cdots u, v, w \cdots$ is one such visit, then we can add the edge uw and still have every vertex on the outer face. Hence the outer boundary of a maximal outerplane graph must be a cycle.

Adding edges to a graph does not reduce vertex degrees, so we have $\delta(G) \le 2$ for all outerplanar graphs if we prove it for outerplane graphs in which the outer face is bounded by a cycle. For such an outerplane graph G, the weak dual is connected, i.e. a tree. A leaf of this tree corresponds to a bounded face of G that has exactly one internal edge e. Since G is simple, there are at least two other edges on this face. The vertices of this face not belonging to e have degree 2. □

The existence of vertices of degree at most two in an outerplanar graph can also be proved by induction (Exercise 8).

EULER'S FORMULA

Euler's Formula is the basic computational tool for planar graphs.

7.1.16. Theorem. (Euler [1758]): If a connected plane graph G has n vertices, e edges, and f faces, then $n - e + f = 2$.

Proof. Proof by induction on n. If $n = 1$, then G is a "bouquet" of loops, each a closed curve in the embedding. If $e = 0$, then we have one face, and the formula holds. Each added loop passes through a region and partitions it into two regions (by the Jordan Curve Theorem), so the formula holds for $n = 1$ and any $e \geq 0$.

Now suppose $n(G) > 1$. Since G is connected, we can find an edge that is not a loop. When we contract such an edge, we obtain a plane graph G' with n' vertices, e' edges, and f' faces. The contraction does not change the number of faces (we merely shortened boundaries), but it reduces the number of edges and vertices by one. Applying the induction hypothesis, we find $n - e + f = n' + 1 - (e' + 1) + f = 2$. $\square$

7.1.17. Remark. 1) Euler's formula implies that all planar embeddings of a connected graph G have the same number of faces. Thus, although the dual may depend on the embedding chosen for G, the number of vertices in the dual does not.

2) Contracting an edge of G has the effect of deleting an edge in G^*. Similarly, deleting an edge of G has the effect of contracting an edge in G^*, as two faces of G merge into a single face (see Exercise 4 and Section 8.2).

3) Euler's formula as stated fails for disconnected graphs. If a plane graph G has k components, then we can adjust Euler's formula by observing that adding $k - 1$ edges to G will yield a connected plane graph without changing the number of faces. Hence the general formula for a plane graph with k components is $n - e + f = k + 1$ (for example, consider a graph with n vertices and no edges).

Euler's formula has many applications. An n-vertex planar multigraph may have infinitely many edges, but a simple graph cannot.

7.1.18. Theorem. A simple planar graph G has at most $3n(G) - 6$ edges. A simple planar triangle-free graph G has at most $2n(G) - 4$ edges.

Proof. It suffices to consider connected graphs, since otherwise we could add edges. We can use Euler's formula to relate $n(G)$ and $e(G)$ if we can dispose of f. Proposition 7.1.8 provides an inequality between e and f. Every face boundary in a simple graph contains at least three edges (if $n(G) \geq 3$). If $\{f_i\}$ is the sequence of face-lengths, this yields $2e = \Sigma f_i \geq 3f$. Substituting into $n - e + f = 2$ yields $e \leq 3n - 6$. If G is triangle-free, then the faces have length at least four. In this case $2e = \Sigma f_i \geq 4f$, and we obtain $e \leq 2n - 4$. $\square$

7.1.19. Example. K_5 *and* $K_{3,3}$. Euler's formula incorporates the earlier geometric reasoning we used to show that K_5 and $K_{3,3}$ are nonplanar. The nonplanarity follows immediately from the edge bound. For K_5, we have $e = 10 > 9 = 3n - 6$, and for the triangle-free $K_{3,3}$ we have $e = 9 > 8 = 2n - 4$. $\square$

7.1.20. Remark. *Maximal planar graphs / triangulations.* The proof of Theorem 7.1.17 shows that having $3n - 6$ edges in a simple n-vertex planar graph requires $2e = 3f$, meaning that every face is a triangle. If G has some face that is not a triangle, then we can add an edge between nonadjacent vertices on the boundary of this face to obtain a larger plane graph. Hence the simple plane graphs with $3n - 6$ edges, the triangulations, and the *maximal* plane graphs are all the same family.

7.1.21. Example. *Regular Polyhedra.* Informally, we think of a regular polyhedron as a solid whose boundary consists of regular polygons of the same length, with the same number of faces meeting at each vertex. When we lay the surface out in the plane, we obtain a regular planar graph with faces of the same length. Hence the dual is also regular. We can prove that there are only five regular polyhedra by proving that there are only five simple regular planar graphs whose duals are also simple and regular.

Suppose G is a plane graph with n vertices, e edges, and f faces. Suppose also that G is regular of degree k and G^* is regular of degree l (i.e., G has faces of length l). By the degree-sum formula for G and G^*, we have $kn = 2e = lf$. Substituting for n and f into Euler's Formula, we have $e(\frac{2}{k} - 1 + \frac{2}{l}) = 2$. Since e and 2 are positive, the other factor must also be positive, which yields $(2/k) + (2/l) > 1$, and hence $2l + 2k > kl$. This inequality is equivalent to $(k - 2)(l - 2) < 4$. Because the dual of a 2-regular graph is not simple, we conclude that $k, l \geq 3$. We also have $k, l \leq 5$, since Theorem 7.1.18 and the degree-sum formula imply that every planar graph has a vertex of degree at most 5. This yields only five solution pairs in integers for (k, l): (3,3), (3,4), (3,5), (4,3), (5,3). Once we specify k and l, there is only one way to lay out the plane graph when we start with any face. Hence there are no more than the five known Platonic solids. □

k	l	$(k-2)(l-2)$	e	n	f	name
3	3	1	6	4	4	tetrahedron
3	4	2	12	8	6	cube
4	3	2	12	6	8	octahedron
3	5	3	30	20	12	dodecahedron
5	3	3	30	12	20	icosahedron

EXERCISES

7.1.1. (–) Determine the number of isomorphism classes of planar graphs that can be obtained as planar duals of the graph below

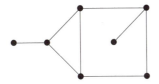

7.1.2. If G is a 2-connected simple plane graph with minimum degree 3, does it follow that the dual graph G^* is simple? Give a proof or a counterexample.

7.1.3. Suppose G is a plane graph, and the dual G^* is drawn so each dual edge intersects only the edge corresponding to it in G. Prove the following.
 a) G^* is connected.
 b) If G is connected, then each face of G^* contains exactly one vertex of G.
 c) $(G^*)^* = G$ if and only if G is connected.

7.1.4. Prove that contracting an edge of a plane graph corresponds to deleting the dual edge from the dual; i.e., prove that $(G \cdot e)^* = G^* - e^*$. Use this to give an inductive proof of Theorem 7.1.12: a set $D \subseteq E(G)$ is a cycle in G if and only if the corresponding set $D^* \subseteq E(G^*)$ is a bond in G^*.

7.1.5. (–) Suppose G is a maximal simple planar graph. Prove that G^* is 2-edge-connected and 3-regular.

7.1.6. Prove by induction on $e(G)$ that a plane graph is bipartite if and only if every face has even length.

7.1.7. (!) Prove that a set of edges in a connected plane graph G forms a spanning tree of G if and only if the duals of the remaining edges form a spanning tree of G^*.

7.1.8. Prove by induction that every simple outerplanar graph with at least four vertices has two nonadjacent vertices with degree at most 2. Use this to determine the maximum number of edges in a simple n-vertex outerplanar graph.

7.1.9. (!) *Alternative proof of Euler's Formula.*
   a) Use polygonal paths (not Euler's Formula) to prove by induction on n that every planar embedding of a tree has one face.
   b) Prove Euler's Formula by induction on $e(G) - n(G) + 1$, using the operation of edge-deletion.

7.1.10. (!) Prove that every n-vertex plane graph isomorphic to its dual has $2n - 2$ edges. For each $n \geq 4$, construct a simple n-vertex plane graph isomorphic to its dual.

7.1.11. Use Euler's formula to determine the maximum number of edges in a simple n-vertex outerplanar graph.

7.1.12. (!) Suppose G is a 3-regular plane graph in which every vertex is incident to one face of length 4, one face of length 6, and one face of length 8. Without drawing G, determine the number of faces of G. (Hint: use Euler's formula and an expression for the number of faces of each length in terms of $n(G)$.)

7.1.13. *Complements of planar graphs.*
   a) Prove that the complement of a simple planar graph with at least 11 vertices is nonplanar.
   b) Construct a self-complementary simple planar graph with eight vertices.

7.1.14. (!) Suppose G is planar, 2-edge-connected, and has no cycle of length less than k. Prove that $e(G) \leq (n(G) - 2)k/(k - 2)$, and use this to prove that the Petersen graph is nonplanar.

7.1.15. Let F be a figure that can be drawn in the plane by a continuous (closed) movement of a pencil point, without retracing any portion (this can be considered an Eulerian graph). Prove that F can be drawn continuously so that in addition the pencil point never crosses what has previously been drawn. For example, the figure below has two traversals; one crosses itself and the other doesn't.

7.1.16. Prove that there is no bipartite planar graph with minimum degree at least 4. For each natural number n, construct a planar graph with minimum degree 5 that has more than n vertices (a sequence of arbitrarily large graphs must be constructed, but it is not necessary to construct one of each order).

7.1.17. Prove that every planar graph with at least four vertices has at least four vertices with degree less than 6. For each even value of n with $n \geq 8$, construct an n-vertex planar graph G that has exactly four vertices with degree less than 6. (Grünbaum-Motzkin [1963])

7.1.18. Let S be a set of n points in the plane such that the distance in the plane between every two points in S is at least 1. Prove that there are at most $3n - 6$ pairs for which the distance is exactly 1.

7.1.19. *Directed plane graphs.* Suppose G is a plane graph, and D is an orientation of G. The *dual* D^* of D is an orientation of G^* such that when an edge of D is traversed from tail to head, the dual edge in D^* crosses it from right to left. For example, if the solid edges below are in D, then the dashed edges are in D^*.

Prove that if D is strongly connected, then D^* has no directed cycle, and $\delta^-(D^*) = \delta^+(D^*) = 0$. Conclude that if D is strongly connected, then D has a face on which the edges form a clockwise cycle and another face on which the edges form a counterclockwise cycle.

7.2 Characterization of Planar Graphs

Which graphs embed in the plane? We now know that K_5 and $K_{3,3}$ do not. In a natural sense, these are the critical graphs and yield a characterization of planarity. Before 1930, the most actively sought result in graph theory was the characterization of planar graphs, and the characterization using K_5 and $K_{3,3}$ is known as Kuratowski's Theorem. Kasimir Kuratowski once asked Harary who originated the notation for K_5 and $K_{3,3}$; Harary replied that "K_5 stands for Kasimir and $K_{3,3}$ stands for Kuratowski!"

Recall that *subdividing* an edge or performing an *elementary subdivision* means replacing the edge with a path of length 2. A *subdivision* of G is a graph obtained from G by a sequence of elementary subdivisions, turning edges into paths through new vertices of degree 2. If $\delta(G) \geq 3$ and H is a subdivision of G, then vertices of H having degree at least 3 are the *branch vertices*; these are the images of the original vertices.

Subdividing edges does not affect planarity, so we seek a characterization by finding the *topologically minimal* nonplanar graphs - those that are not subdivisions of other nonplanar graphs. We already know that a graph containing any subdivision of K_5 or $K_{3,3}$ is nonplanar. Kuratowski [1930] proved that G is planar if and only if G contains no subdivision of K_5 or $K_{3,3}$.

Wagner [1937] proved another characterization. Deletion and contraction of edges preserve planarity, so we can seek the minimal nonplanar graphs using these operations. Wagner proved that G is planar if and only if it has no subgraph contractible to K_5 or $K_{3,3}$ (Section 8.2).

We can also place additional requirements on a planar embedding. Wagner [1936], Fary [1948], and Stein [1951] showed that every finite planar graph has an embedding in which all edges are straight line segments; this is known as Fary's Theorem. If also each face boundary (including the unbounded face) is a convex polygon, the representation is called *convex*. Tutte [1960,1963] proved that every 3-connected planar graph has a convex representation. This is best possible in the sense that the 2-connected planar graph $K_{2,n}$ does not have a convex representation if $n \geq 4$. We follow Thomassen's approach to simultaneously proving Kuratowski's Theorem and Tutte's Theorem.

PREPARATION FOR KURATOWSKI'S THEOREM

Thomassen's contribution concerns the 3-connected graphs, where the stronger results hold. We first reduce the problem to the 3-connected case. For convenience, we call a subgraph of G that is a subdivision of K_5 or $K_{3,3}$ a *Kuratowski subgraph* of G. A *minimal nonplanar graph* is a nonplanar graph such that every proper subgraph is planar. We reduce the problem to the 3-connected case by proving that a minimal nonplanar graph having no Kuratowski subgraph must be 3-connected. Then a proof of planarity for 3-connected graphs with no Kuratowski subgraph completes the proof of Kuratowski's Theorem.

7.2.1. Lemma. If E is the edge set of a face in some planar embedding of G, then G has an embedding in which E is the edge set of the unbounded face.

Proof. Project the embedding onto the sphere, where the edge sets of regions remain the same and all regions are bounded, and then return to the plane by projecting from inside the face bounded by E. $\square$

7.2.2. Lemma. Every minimal nonplanar graph is 2-connected.

Proof. Suppose G is a minimal nonplanar graph. If G is disconnected, we can embed one component of G inside one face of an embedding of the rest of G. If G has a cut-vertex v, let $G_1, \ldots, G_k$ be the subgraphs of G induced by v together with a component of $G - v$. By the minimality of G, these subgraphs are planar. By Lemma 7.2.1, we can embed each with v on the outside face. We can squeeze each embedding to fit in an angle smaller than $360/k$ degrees at v, after which we can merge the embeddings at v to obtain an embedding of G. $\square$

7.2.3. Lemma. Suppose $S = \{x, y\}$ is a 2-cut of G and G_1, G_2 are subgraphs of G such that $G_1 \cup G_2 = G$ and $V(G_1) \cap V(G_2) = S$. Let $H_i = G_i \cup xy$. If G is nonplanar, then at least one of H_1, H_2 is nonplanar.

Proof. Suppose H_1 and H_2 are planar. Then the edge xy occurs on some face in a planar embedding of H_1. By Lemma 7.2.1, H_1 has a planar embedding with xy on the outside face. This allows H_1 to be attached to an embedding of H_2, embedded in a face with xy on the boundary. Deleting the edge xy if it does not appear in G, we have constructed a planar embedding of G. □

7.2.4. Lemma. Suppose G is a nonplanar graph with no Kuratowski subgraph, and G has the fewest edges among such graphs. Then G is 3-connected.

Proof. If we delete an edge of G, we cannot create a Kuratowski subgraph. Therefore, the hypotheses on G guarantee that deleting one edge produces a planar subgraph, and hence G is a minimal nonplanar graph. By Lemma 7.2.2, G is 2-connected. Suppose G has a 2-cut $S = \{x, y\}$. Since G is nonplanar, Lemma 7.2.3 guarantees that H_1 or H_2 is non-planar; suppose H_1 is nonplanar. Since H_1 has fewer edges than G, the choice of G forces H_1 to have a Kuratowski subgraph. All of H_1 appears in G, except possibly the edge xy. The role of xy in the Kuratowski subgraph of H_1 can be played by an x, y-path H_2 to obtain a Kuratowski subgraph of G. This contradicts the hypothesis that G has no Kuratowski subgraph, so G has no 2-cut. □

CONVEX EMBEDDINGS

To complete the proof of Kuratowski's Theorem, it suffices to prove that 3-connected graphs without Kuratowski subgraphs are planar. We prove the stronger statement that 3-connected graphs without Kuratowski subgraphs have convex embeddings (Tutte's Theorem).

7.2.5. Lemma. (Thomassen [1980]). A 3-connected graph with at least five vertices contains an edge whose contraction leaves a 3-connected graph.

Proof. Proof by contradiction; assume that for each $e \in E(G)$, the graph $G \cdot e$ is not 3-connected. Hence $G \cdot e$ has a separating 2-set, which must include the vertex obtained by shrinking $e = xy$. If z is the other vertex of this set, then $\{x, y, z\}$ is a separating set in G. Among all the edges of G, choose $e = xy$ and their guaranteed companion z such that the resulting disconnected graph $G - \{x, y, z\}$ has a component H with the largest possible number of vertices. Let H' be another component of this graph (see the figure below). Since $\{x, y, z\}$ is a minimal separating set, each of x, y, z has a neighbor in each of H, H'. Let u be a neighbor of z in H', and let v be the guaranteed companion of uz, so that $\{z, u, v\}$ separates G. The subgraph of G induced by $V(H) \cup \{x, y\}$ is connected. Deleting v from this subgraph, if it occurs there, cannot disconnect it, since then $G - \{z, v\}$ would be disconnected. Therefore $G_{V(H) \cup \{x,y\}} - v$ belongs to a component of $G - \{z, u, v\}$ that has more vertices than H, which contradicts the choice of x, y, z. $\square$

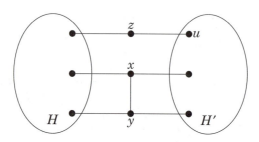

Our inductive proof of Tutte's Theorem involves edge contraction; we need to show that it preserves the absence of Kuratowski subgraphs.

7.2.6. Lemma. If $G \cdot e$ has a Kuratowski subgraph, then G also has a Kuratowski subgraph.

Proof. Let H be a Kuratowski subgraph of $G' = G \cdot e$, and let z be the vertex of G' obtained by contracting $e = xy$. If z is not a branch vertex of H, then G also contains a Kuratowski subgraph, obtaining by lengthening a path through z if necessary (i.e., expanding z back into the edge xy). If z is a branch vertex in H and at most one of the edges incident to z in H is incident to x in G, then z can be expanded into xy to lengthen that path, and y becomes the corresponding branch vertex for a Kuratowski subgraph in G.

The only remaining case (illustrated below) is when H is a subdivision of K_5, z is a branch vertex of H, and each of x, y is incident in G to two of the four edges incident to z in H. In this case, let u_1, u_2 be the branch vertices of H that are at the other ends of the paths leaving z on edges incident to x in G, and let v_1, v_2 be the branch vertices of H that are at the other ends of the paths leaving z on edges incident to y in G. By deleting the u_1, u_2-path and v_1, v_2-path from H, we obtain a subdivision of $K_{3,3}$ in G, in which y, u_1, u_2 are the branch vertices for one partite set and x, v_1, v_2 are the branch vertices of the other. $\square$

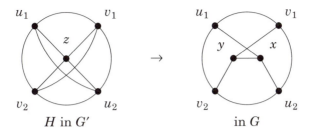

H in G' in G

Now we can prove Tutte's Theorem.

7.2.7. Theorem. (Tutte [1960,1963]) If G is a 3-connected graph with no subdivision of K_5 or $K_{3,3}$, then G has a convex embedding in the plane.

Proof. (Thomassen [1980,1981]) We use induction on $n(G)$. The only 3-connected graph with at most four vertices is K_4, which has a convex embedding, so suppose $n \geq 5$. Let $e = xy$ be an edge such that $G \cdot e$ is 3-connected, as guaranteed by the contraction lemma; let z the vertex obtained by contraction. By Lemma 7.2.6, $G \cdot e$ has no Kuratowski subgraph, so the induction hypothesis guarantees a convex embedding of $H = G \cdot e$. Consider such an embedding. The plane graph obtained by deleting the edges incident to z has a face containing z (this may be the unbounded face); let C be the cycle of $H - z$ bounding this face.

Since we started with a convex embedding of H, we have straight segments from z to all its neighbors. Let $x_1, \ldots, x_k$ be the neighbors of x in order on C. If all neighbors of y belong to a single segment from x_i to x_{i+1} on C, then we obtain a convex embedding of G by putting x at z in H and putting y at a point close to z in the wedge formed by xx_i and xx_{i+1}, as illustrated on the left below.

If this case does not occur, then either a) y shares three neighbors with x, in which case C together with these six edges involving x and y form a subdivision of K_5, or b) y has two u, v in C that are in different components of the subgraph of C obtained by deleting x_i and x_{i+1} (for some i), in which case C together with the paths uyv, $x_i x x_{i+1}$, and xy form a subdivision of $K_{3,3}$. These cases appear on the right below. $\square$

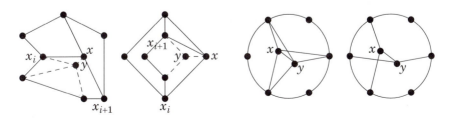

Now that we have Kuratowski's Theorem, Fary's Theorem can be obtained separately: if a graph has a planar embedding, then it has a straight-line planar embedding (Exercise 6). For applications in computer science, one would like more---a straight-line planar embedding in which the vertices are located at the integer points in a relatively small grid. Schnyder [1992] proved that every n-vertex planar graph has a straight-line embedding in which the vertices are located at integer points in the grid $[n-1] \times [n-1]$.

BRIDGES AND PLANARITY TESTING (optional)

Dirac and Schuster [1954] gave the first short proof of Kuratowski's Theorem. Appearing in the texts of Harary [1969, 109-112], Bondy-Murty [1976, p153-156], and Chartrand-Lesniak [1986, p96-98], it uses a particular type of subgraph:

7.2.8. Definition. If H is a subgraph of G, an H-*bridge* of G is either 1) an edge not in H whose endpoints are in H or 2) a component of $G - V(H)$ together with the edges (and vertices of attachment) that connect it to H. The H-bridges are the "pieces" that must be added to an embedding of H to obtain an embedding of G.

An H-bridge differs from a $V(H)$-component because the H-bridge omits the edges of H. After disposing of the non-3-connected case (along the lines suggested earlier), Dirac and Schuster considered a minimal nonplanar 3-connected graph G with no Kuratowski subgraph. Deleting an edge e yields a planar 2-connected graph. After choosing an appropriate cycle H through the endpoints of e, we can add e to the embedding unless we have H-bridges embedded inside and outside H that "conflict" with e. As in Thomassen's proof for the 3-connected case, this produces a Kuratowski subgraph of G. The idea of conflicting H-bridges was formalized by Tutte to obtain another characterization of planar graphs.

7.2.9. Definition. Given a cycle H in a graph G, two H-bridges A, B *conflict* if they have three common vertices of attachment or if there are four vertices v_1, v_2, v_3, v_4 in cyclic order on H such that v_1, v_3 are vertices of attachment of A and v_2, v_4 are vertices of attachment of B. In the latter case, we say the H-bridges *cross*. The *conflict graph* of H is a graph whose vertices are the H-bridges of G, with conflicting H-bridges adjacent.

Tutte [1958] proved that G is planar if and only if for every cycle C in G, the conflict graph of C is bipartite (Exercise 8). We used this idea in our first proof that K_5 and $K_{3,3}$ are nonplanar (Example 7.1.5); the conflict graph of a spanning cycle in $K_{3,3}$ is C_3, and the conflict graph of a spanning cycle in K_5 is C_5.

Nonplanar 3-connected graphs have Kuratowski subgraphs of a special type. Kelmans [1981] conjectured this extension of Kuratowski's Theorem, and it was proved independently by Kelmans [1984] and by Thomassen [1984]: Every 3-connected nonplanar graph with at least six vertices contains a cycle with three pairwise crossing chords.

There are linear-time planarity-testing algorithms due to Hopcroft and Tarjan [1974] and to Booth and Luecker [1976], but these are very complicated (Gould [1988, p177-185] discusses the ideas used in the Hopcroft-Tarjan algorithm). A much simpler earlier algorithm is not linear but runs in polynomial time. Due to Demoucron, Malgrange, and Pertuiset [1964], it uses H-bridges. The idea is that if a planar embedding of H can be extended to a planar embedding of G, then in that extension every H-bridge of G appears inside a single face of H. The algorithm builds increasingly larger plane subgraphs H of G that can be extended to an embedding of G if G is planar. We want to enlarge H by making small decisions that won't lead to trouble.

The basic enlargement step is the following. 1) Choose a face F that can accept an H-bridge B; a necessary condition for F to accept B is that its boundary contain all vertices of attachment of B. 2) Although we do not know the best way to embed B in F, each particular path in B between vertices of attachment by itself has only one way to be added across F, so we add a single such path. The details of choosing F appear below. This algorithm and those mentioned earlier are constructive, producing an embedding if G is planar.

7.2.10. Algorithm. *(Planarity Testing).*
Input: A 2-connected planar graph. (Since G is planar if and only if each block of G is planar, and Algorithm 4.1.19 computes blocks, we may assume that G is a block with at least three vertices.)
Idea: Successively add ears from current bridges. Maintain the vertex sets forming face boundaries in the subgraph already embedded.
Initialization: G_0 is an arbitary cycle in G embedded in the plane, with two face boundaries consisting of its vertices.
Iteration: Having determined G_i, find G_{i+1} as follows.

    1. Determine all G_i-bridges of the input block G.

    2. For each G_i-bridge B, determine all faces of G_i that contain all vertices of attachment of B; call this set $F(B)$.

    3. If $F(B)$ is empty for some B, return NONPLANAR. If $|F(B)| = 1$ for some B, select some such B. If $|F(B)| > 1$ for every B, select an arbitrary B.

    4. Choose an arbitrary path P between two vertices of attachment of the selected B. Embed P across a face in $F(B)$. Call the resulting

graph G_{i+1} and update the list of face boundaries.

    5. If G_{i+1} contains all edges of G, return PLANAR. Otherwise, augment i and return to Step 1.

7.2.11. Example. *Planarity Testing.* Consider the two graphs below (from Bondy/Murty [1976, p165-166]). The algorithm produces a planar embedding of the graph on the left, but it terminates in step 3 for the graph on the right. The cycle 12348765 has three crossing chords; 14,27,36. □

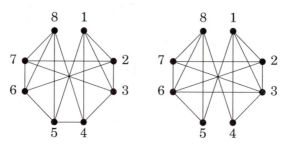

7.2.12. Theorem. (Demoucron-Malgrange-Pertuiset [1964]) The algorithm above produces a planar embedding if G is planar.

Proof. Since we may assume that G is 2-connected, every bridge of every subgraph has at least two vertices of attachment. If G is planar, every cycle of G appears as a simple closed curve in each planar embedding. Since we can flip the plane over, every embedding of a single cycle can be extended to an embedding of G. Hence G_0 is extendable to a planar embedding of G if G is planar. It suffices to show that if the plane graph G_i is extendable to a planar embedding of G and the algorithm produces a plane graph G_{i+1}, then G_{i+1} is also extendable to G.

    If some G_i-bridge B has $|F(B)| = 1$, then there is only one possible way to add P to G_i; since G_i is extendable, G_{i+1} must also be extendable. Problems can arise only if every B has $|F(B)| > 1$ and we select the wrong face in which to embed a path P from B. Suppose we embedded P in face $f \in F(B)$ and G_i can be extended to a planar embedding $\hat{G}$ of G in which P is inside face $f' \in F(B)$. We want to modify $\hat{G}$ to show that G_i can be extended to another embedding G' of G in which P is inside f. This shows that G_{i+1} is extendable.

    Let C be the set of vertices in the boundary of both f and f'. The vertices of attachment of B are contained in C. We draw G' by switching between f and f' all G_i-bridges that have vertices of attachment contained in C and in $\hat{G}$ appear in f or f', as illustrated on the left below, where the edges *not* present in G_i are dashed. The change switches B and produces the desired embedding G' unless some other unswitched G_i-bridge $\hat{B}$ "conflicts" with the switch. Since we are only making changes in the interiors of f and f' and the switch treats f and f' symmetrically, we may assume that $\hat{B}$ appears in f in $\hat{G}$. The occurrence of a conflict means that $\hat{G}$ has some B' in f', which we are trying to move to f, such that $\hat{B}$ and B' are adjacent in the conflict graph of f.

Let $\hat{A}, A'$ denote the vertex sets where $\hat{B}, B'$ attach to the boundary of f. Since $\hat{B} \leftrightarrow B'$ in the conflict graph of f, $\hat{A}, A'$ have three common vertices or four alternating vertices on the boundary of f. Since $A' \subseteq C$ but $\hat{A} \not\subseteq C$, the first possibility implies the second. Let x, v, y, u be the alternation, with $x, y \in A' \subseteq C$ and $u, v \in \hat{A}$. We may assume $v \notin C$, as illustrated on the right below; if there is no such alternation, then $\hat{B}, B'$ do not conflict or $\hat{B}$ can switch to f'.

As suggested in the illustration, $\hat{B}$ now fails to have its vertices of attachment in the boundary of at least two faces, which contradicts the hypothesis that $|F(B)| > 1$. The reason is that an x, y curve through f and an y, x-curve through f' complete a closed curve Q that encloses exactly one of $\{u, v\}$. Hence $\hat{A}$ is not contained in the boundary of any face entirely inside or entirely outside Q. The only faces that intersect both the interior and the exterior of Q are f and f', and we have assumed that $\hat{B}$ can embed in only one of those two. $\square$

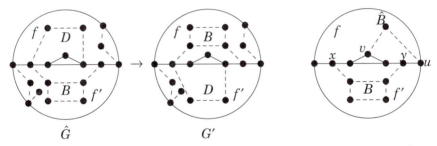

Since we can begin by checking that G has at most $3n - 6$ edges, can maintain appropriate lists for the face boundaries, and can perform the other operations via searches of linear size, it appears that this algorithm runs in quadratic time. The proof of Kuratowski's Theorem by Klotz [1989] also gives a quadratic algorithm to test planarity, and it finds a Kuratowski subgraph if G is not planar.

EXERCISES

7.2.1. (–) Prove that the complement of the 3-dimensional cube Q_3 is nonplanar.

7.2.2. (–) Give three proofs that the Petersen graph is nonplanar.
   a) Using Kuratowski's Theorem.
   b) Using Euler's formula and the fact that the Petersen graph has girth 5.
   c) Using the planarity-testing algorithm of Demoucron-Malgrange-Pertuiset.

7.2.3. (–) Find a convex embedding in the plane for the graph below.

7.2.4. (–) For each graph below, prove nonplanarity or give a convex embedding.

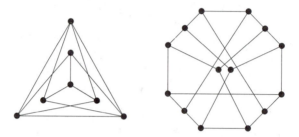

7.2.5. (!) Prove that every 3-connected graph with at least six vertices that contains a subdivision of K_5 also contains a subdivision of $K_{3,3}$.

7.2.6. (!) *Fary's Theorem.*

a) Let R be a region in the plane bounded by a simple polygon with at most five sides (*simple polygon* means the edges are line segments that do not cross). Prove there is a point x inside R that "sees" all of R, meaning that the segment from x to any point of R does not cross the boundary of R.

b) Use part (a) to prove inductively that every planar graph has a straight-line embedding.

7.2.7. (!) Use Kuratowski's Theorem to prove that G is outerplanar if and only if it has no subgraph that is a subdivision of K_4 or $K_{2,3}$. (Hint: to *apply* Kuratowski's Theorem, find an appropriate modification of G. This is *much* easier than trying to mimic a proof of Kuratowski's Theorem.)

7.2.8. Prove that a graph G is planar if and only if for every cycle C in G, the conflict graph for C is bipartite. (Tutte [1958])

7.2.9. Suppose x, y are vertices of a planar graph G. Prove that G has a planar embedding with x and y on the same face unless $G - x - y$ has a cycle C with x and y in conflicting C-bridges in G. (Hint: Use Kuratowski's Theorem.)

7.2.10. Let G be a 3-connected plane graph containing a cycle C. Prove that C is the boundary of a face in G if and only if G has exactly one C-bridge.

7.3 Parameters of Planarity

Every property and parameter we have studied for general graphs can be studied for planar graphs. The problem of greatest historical interest is the maximum chromatic number of planar graphs. We also consider conditions for Hamiltonian cycles in planar graphs. Since planarity is a severe restriction on the inputs to these NP-complete problems (almost all graphs have too many edges to be planar - Section 8.5), one might hope for polynomial-time algorithms for planar graphs. Nevertheless, 3-COLORING and HAMILTONIAN CYCLE remain NP-complete for planar graphs (Theorem 6.3.12). In this section, we discuss colorings and Hamiltonian cycles in planar graphs and measures of distance from planarity.

COLORING OF PLANAR GRAPHS

Because every simple planar graph with n vertices has at most $3n - 6$ edges, every simple planar graph has a vertex of degree at most 5. Hence by induction planar graphs are 6-colorable. Heawood improved this. The proof sounds like contradiction but is essentially induction.

7.3.1. Theorem. (Heawood [1890]) Every planar graph is 5-colorable.

Proof. This holds for K_1; suppose there is a minimal counterexample G. Let v be a vertex of degree at most 5 in G. The choice of G implies that $G - v$ is 5-colorable; let $f: V(G - v) \to [5]$ be a 5-coloring of $G - v$. Since G is not 5-colorable, each color appears at one of the neighbors of v (and hence $d(v) = 5$). We may label the colors to assume that the neighbors of v in a planar embedding of G are v_1, v_2, v_3, v_4, v_5 in clockwise order around v, with $f(v_i) = i$.

Let G_{ij} denote the subgraph of G induced by the vertices of colors i and j. We can exchange the two colors on any component of G_{ij} to obtain another 5-coloring of $G - v$. Hence the component of G_{ij} that contains v_i must also contain v_j, else we could make the interchange on the component of G_{ij} containing v_i to remove color i from $N(v)$, and then we could assign color i to v to extend f to be a 5-coloring of G. Let P_{ij} be a path in G_{ij} from v_i to v_j.

Consider the cycle C completed with $P_{1,3}$ by v; this separates v_2 from v_4. By the Jordan Curve Theorem, the path $P_{2,4}$ must cross C. Since G is planar, such a crossing can happen only at a shared vertex. This is impossible, because the vertices of $P_{1,3}$ all have color 1 or 3, and the vertices of $P_{2,4}$ all have color 2 or 4. $\square$

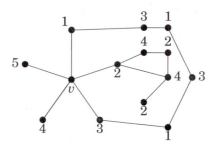

Is Heawood's bound best possible; i.e., is there a 5-chromatic planar graph? This infamous question was open for more than a hundred years. Its ease of statement and geometric subtleties invite fallacious proofs; some were published and remained unexposed for years. Some thought it would suffice to show that there cannot be five pairwise-adjacent connected countries, which is the statement that K_5 is nonplanar. Our brief discussion of the history follows that of Aigner [1984, 1987]; see also Ore [1967a], Saaty-Kainen [1977, 1986], and Appel-Haken [1989].

The Four Color Problem first appeared in a letter of October 23, 1852 to Sir William Hamilton from Augustus de Morgan at University College in London. The question was asked by de Morgan's student Frederick Guthrie, who later attributed it to his brother Francis Guthrie. The problem was phrased in terms of map coloring, in which the faces of a planar graph are to be colored.

Cayley announced the problem to the London Mathematical Society in 1878. Within a year, Kempe [1879] published a "solution," and soon he was elected a Fellow of the Royal Society. In 1890, Heawood published a refutation. Nevertheless, Kempe's idea was that of the alternating paths in the Five Color Theorem above, and it led eventually to a proof by Appel and Haken [1976, 1977, 1986] (portions with Koch). A path on which the colors alternate between two specified colors is a *Kempe chain*, or specifically an α, β-chain when the colors are α and β.

In proving the Five Color Theorem, we argued that any minimal counterexample contains a vertex of degree at most 5 and that a planar graph with a vertex of degree at most 5 cannot be a minimal counterexample. This aspects suggests an approach to the Four Color Problem; we need an *unavoidable* set of *reducible* configurations. "Unavoidable" means that any minimal counterexample must contain one of these configurations. "Reducible" means that a planar graph containing the configuration cannot be a minimal counterexample. We prove that a configuration is reducible by showing that if it appears in G, then we can delete or shrink it, apply induction to obtain a 4-coloring of the smaller graph, and then alter that coloring (if necessary) to obtain a 4-coloring of G. We also may restrict our attention to triangulations, since every planar graph is contained in a triangulation.

Since every planar graph has a vertex of degree at most 5, stars of degree at most 5 form an unavoidable set. As in Theorem 7.3.1, consider a vertex v of degree at most 5 in a minimal counterexample G. We can immediately extend a 4-coloring of $G - v$ unless all four colors appear on vertices adjacent to v. If $d(v) = 4$, the Kempe chain argument works as before. If $d(v) = 5$, the restriction to triangulations implies that the repeated color appears on nonconsecutive neighbors of v. Suppose $N(v)$ is $\{v_1, v_2, v_3, v_4, v_5\}$ in order, and in the 4-coloring f of $G - v$ we have $f(v_i) = i$, except $f(v_5) = 2$. Define G_{ij} as before.

We can eliminate color 1 from the neighborhood of v unless there is a 1,3-chain P from v_1 to v_3 and a 1,4-chain Q from v_1 to v_4. Now the component H of $G_{2,4}$ containing v_2 is separated from v_4 and v_5 by the cycle completed with P by v, and the component H' of $G_{2,3}$ containing v_4 is separated from v_2 and v_3 by the cycle completed with Q by v. The situation is illustrated on the left below. We can eliminate color 2 from the neighborhood of v by switching colors 2 and 4 in H and colors 2 and 3 in H'. Right? This was the final case in Kempe's proof.

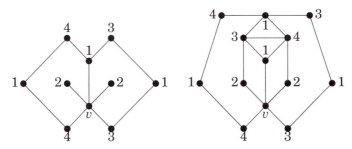

The problem is that the paths P and Q can intertwine, intersecting at a vertex with color 1 as indicated on the right above. We can make the change in H or the change in H' to remove color 2 from the neighborhood of v, but if we make them both, then we wind up with a pair of adjacent vertices having color 2.

Because of this difficulty, a vertex of degree 5 is not a reducible configuration, and we must consider larger configurations when our supposed minimal counterexample has minimum degree 5. Instead of seeking small clumps of vertices, we seek small separating cycles, potentially with many vertices inside. This was among the important contributions of Heesch [1969]. A *configuration* in a planar triangulation is a separating cycle C (called the *ring*) together with the portion of the graph inside C. A configuration is *reducible* if whenever it appears in a triangulation G, its interior can be shrunk or replaced to obtain a subgraph with fewer vertices such that any 4-coloring of the new graph G' can be manipulated to obtain a 4-coloring of G. We have proved that the configurations below are reducible.

It is not too hard to show that every configuration having ring size 3 or 4 is reducible (Exercise 4), which is equivalent to showing that no minimal 5-chromatic triangulation has a separating cycle of length at most 4. Birkhoff pushed this farther. First, he proved that any configuration with ring size 5 that has more than one vertex inside is reducible. He also proved that the configuration with ring size 6 on the left below, called the *Birkhoff diamond*, is reducible.

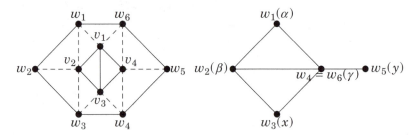

7.3.2. Theorem. (Birkhoff [1913]) The Birkhoff diamond is reducible.

Proof. Suppose G is a minimal 5-chromatic triangulation containing the Birkhoff diamond, labeled as above. Since $\delta(G) \geq 5$, the vertices w_4 and w_6 cannot be adjacent outside the ring. Hence we can form a smaller triangulation G' by contracting the edges v_4w_4 and v_4w_6 and replacing everything else inside the ring by an edge from w_2 to the combined vertex; the replacement is illustrated on the right above. By the minimality of G, G' is 4-colorable. We want to show that an any proper 4-coloring of G' can be used to obtain a proper 4-coloring of G.

Since w_1, w_2, w_4 induce a triangle in G', they receive distinct colors, which we may assume are α, β, γ, respectively. Let x, y denote the colors on w_3 and w_5, respectively; note that x must be α or the fourth color δ, and $y \neq \gamma$. If we copy this coloring onto G, with w_4, w_6 both getting color γ, then the outside of the ring is properly colored. Inside the ring, we want to choose colors for $\{v_i\}$ to complete a proper coloring of G. In the cases listed below for x, y, this works:

x	y	v_1	v_2	v_3	v_4
α or δ	β or δ	δ	γ	β	α
δ	α	δ	γ	α	β

The remaining possibility is $x = y = \alpha$; this coloring cannot be extended. Colors β and δ are forced onto $\{v_1, v_3\}$, but then v_4 is adjacent to all four colors. This essentially was where Kempe's argument failed, but here the restricted structure of the colors on w_1, w_2, w_3 allows a Kempe-chain argument to work.

We inherit from G' the coloring $\alpha, \beta, \alpha, \gamma, \alpha, \gamma$ for $w_1, \ldots, w_6$; we temporarily leave $\{v_i\}$ uncolored. If either of w_1, w_3 is not in the component H of $G_{\alpha, \delta}$ that contains w_5, we may assume by symmetry it is w_1. Then we can switch α, δ on H to obtain one of the cases $x, y = \alpha, \delta$ or $x, y = \delta, \delta$ considered earlier. Hence we may assume H contains a w_5, w_1-path and a w_5, w_3-path. With paths through $\{v_i\}$, these complete cycles that separate w_2, w_4, w_6 into disjoint regions. Hence w_2, w_4, w_6 lie in distinct components of $G_{\beta, \gamma}$. By switching β, γ on the component containing w_4, we obtain a proper 4-coloring with color β on w_4. We extend this to a proper 4-coloring of G by giving v_1, v_2, v_3, v_4 colors $\beta, \delta, \gamma, \delta$. □

The intricacy of the case analysis observed here suggests that we have barely touched the tip of the iceberg, and the detail remaining is enormous. From 1913 to 1950, additional configurations were shown to be reducible, enough to prove that all planar graphs with at most 36 vertices are 4-colorable. This was slow progress. In the 1960s, Heinrich Heesch focused attention on the size of the ring in the configuration, gave heuristics for finding reducible configurations, and developed useful ways to describe unavoidable sets. By the end of the decade, many reducible configurations were known, but they were far from an unavoidable set.

The search for an unavoidable set of reducible configurations was enormous. A ring of size 13 has 66430 different 4-colorings, each of which must be used to produce a 4-coloring of the full graph in order to prove reducibility. Given the Kempe chain arguments that might be needed to prove reducibility, it is not at all clear how to perform reducibility tests in a systematic manner. We are also ignoring altogether the methods for generating unavoidable sets.

Kenneth Appel and Wolfgang Haken, working also with John Koch, devised heuristics improving upon those suggested by Heesch and others to restrict the computer searches to "promising" configurations. Using 1000 hours of computer time on three computers in the first six months of 1976, they obtained an unavoidable set of 1936 reducible configurations, all with ring size at most 14. By 1983 this had been modified to an unavoidable set of 1258 reducible configurations. It is worth noting that advances in computer power have been substantial enough that the computation could now run in a day.

The proof of the Four Color Theorem was greeted with considerable uproar. Some mathematicians objected in principle to the use of a computer. Others merely objected to the length of the proof, complaining that it could not be verified by hand and worrying about computer error. Those of us who have attempted to check calculations by hand recognize that the probability of human error in a mathematical proof is considerably higher than the probability of computer error, especially if the algorithm has been proved correct. To be fair, a few errors were found in the original algorithms, but these were fixed. Robertson, Sanders, Seymour, and Thomas [1996] have completed a check of the proof by hand.

EDGE-COLORINGS AND HAMILTONIAN CYCLES

The Four Color Problem is equivalent to several other problems, including the case $k = 5$ of Hadwiger's Conjecture (Section 5.2). In 1878, Tait proved a theorem relating face-coloring of planar maps to proper edge-colorings of planar graphs. This was an early reason for interest in edge-coloring. We have seen that a planar graph is 4-colorable if and only if some dual of it is 4-face-colorable. We have also observed that we need only prove the 4-colorability of triangulations, which is the same as 4-face-coloring the duals of plane triangulations. The dual G^* of a plane triangulation G is 3-regular and 2-edge-connected (Exercise 7.1.5). Tait observed that a 3-edge-coloring of G^* could be used to obtain a 4-face-coloring of G^* and hence a 4-coloring of the original triangulation G. He used this in an approach to the Four Color Problem. Before discussing that, we prove his theorem.

7.3.3. Theorem. (Tait [1878]) A simple 2-edge-connected 3-regular planar graph is 3-edge-colorable if and only if it is 4-face-colorable.

Proof. Let G be such a graph. Suppose first that G is 4-face-colorable; we obtain a 3-edge-coloring. Let the four colors be denoted by binary ordered pairs: $c_0 = 00$, $c_1 = 01$, $c_2 = 10$, $c_3 = 11$. We obtain a proper 3-edge-coloring of G by assigning to the edge between faces with colors c_i and c_j the color obtained by adding c_i and c_j as vectors of length 2, using coordinatewise addition modulo 2. Because G is 2-edge-connected, each edge bounds two distinct faces, and hence the color 00 never occurs as a sum. It suffices to prove that the 3 edges at a vertex receive distinct colors. At vertex v the faces bordering the three incident edges are pairwise adjacent, so these three faces must have three distinct colors $\{c_i, c_j, c_k\}$, as illustrated below. If color 00 is not in this set, then the sum of any two of these is the third, and hence $\{c_i, c_j, c_k\}$ is the set of colors on the three edges. If $c_k = 00$, then c_i and c_j appear on two of the edges, and the third receives color $c_i + c_j$, which is the color not in $\{c_i, c_j, c_k\}$.

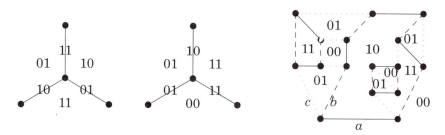

Now suppose G has a proper 3-edge-coloring using colors a, b, c on the subgraphs E_a, E_b, E_c; we construct a 4-face-coloring using the four colors defined above. Since G is 3-regular, each color appears at every vertex, and the union of any two of E_a, E_b, E_c is 2-regular, which makes it a union of disjoint cycles. Each face of this subgraph is a union of faces of the original graph. Let $H_1 = E_a \cup E_b$ and $H_2 = E_b \cup E_c$. To each face of G, assign the color whose ith coordinate ($i \in \{1, 2\}$) is the parity of the number of cycles in H_i that contain it (0 for even, 1 for odd). We claim this is a proper 4-face-coloring, as illustrated above. If two faces F, F' are separated by an edge e, they are distinct faces, since G is 2-edge-connected. This edge belongs to a cycle C in at least one of H_1, H_2 (in both if e has color b. By the Jordan Curve Theorem, one of F, F' is inside C and the other is outside. However, for every other cycle in H_1 or H_2, F, F' are on the same side. Hence if e has color a, c, or b, then the parity of the number of cycles containing F and F' is different in H_1, H_3 or *both*, respectively. This means F and F' receive different colors in the face-coloring we have constructed. $\square$

Due to this theorem, a proper 3-edge-coloring of a 3-regular graph is called a *Tait coloring*. The problem of showing that every 2-edge-connected 3-regular planar graph is 3-edge-colorable can be reduced to showing that every 3-connected 3-regular planar graph is 3-edge-colorable (Exercise 12). It is easy to show that every Hamiltonian 3-regular graph has a Tait coloring (Exercise 8). Together, Tait believed that this gave a proof of the Four Color Theorem, because he assumed that every 3-connected 3-regular planar graph is Hamiltonian. Although the gap was noticed earlier, it was not until 1946 that an explicit counterexample was found. Later, Grinberg [1968] discovered a simple necessary condition that led to many 3-regular 3-connected non-Hamiltonian planar graphs, including the Grinberg graph of Exercise 16.

7.3.4. Theorem. (Grinberg [1968]) If G is a loopless plane graph with a Hamiltonian cycle C, and G has f_i' faces of length i inside C and f_i'' faces of length i outside C, then $\Sigma_i (i-2)(f_i' - f_i'') = 0$.

Proof. If we separate what is happening inside and outside C, we want to show $\Sigma_i(i-2)f_i' = \Sigma_i(i-2)f_i''$. No changes inside or outside C affect the sum on the other side, and furthermore we can exchange the inside and outside by projecting the embedding onto a sphere and puncturing a face inside C instead of the outside face. Hence we need only show that the sum $\Sigma_i(i-2)f_i'$ is an invariant as we add edges inside a cycle of length n. If there are no edges, the sum is $n-2$. We proceed by induction to show that the sum is always $n-2$.

Suppose $\Sigma_i(i-2)f_i' = n-2$ for any graph with k edges inside C. We can obtain any graph with $k+1$ edge inside C by adding an edge to such a graph. The edge addition cuts a face of some length r into two faces of lengths s and t. We have $s+t = r+2$, because the new edge contributes to each face and each of the edges on the old face contributes to one of the new faces. All other contributions to the sum remain unchanged. From this equality we obtain $(s-2)+(t-2)=(r-2)$, so the total contribution from these faces is also the same as before. By the induction hypothesis, the sum is still $n-2$. $\square$

Being a necessary condition, Grinberg's condition can be used to show that graphs are *not* Hamiltonian. The arguments can often be simplified by using modular arithmetic. If two quantities are not congruent mod k, then they are not equal.

We apply such arguments to the first known non-Hamiltonian 3-connected 3-regular planar graph (Tutte [1946]). Tutte used an *ad hoc* argument to prove that this graph is not Hamiltonian, and for many years it was the only known example (see Exercise 17 for the smallest known example).

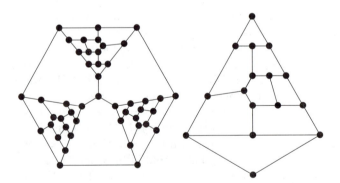

7.3.5. *Grinberg's condition and the Tutte graph.* The Tutte graph G appears on the left above. Let H denote each component of the subgraph obtained by deleting the central vertex and the three long edges. Since a Hamiltonian cycle must visit the central vertex of G, it must contain a Hamiltonian path in one copy of H between the two other entrances to that subgraph. To apply Grinberg's condition to show this impossible, we need a graph that has a Hamiltonian cycle if and only if H has the desired Hamiltonian path. Such a graph H' is obtained by adding a path of at least two edges between the desired endpoints of the Hamiltonian path in H.

If we add a path of two edges, as on the right above, we then have five 5-faces, three 4-faces, and one 9-face, and Grinberg's condition becomes $2a_4 + 3a_5 + 7a_9 = 0$, where $a_i = f_i' - f_i''$. Since the unbounded face is outside, the equation reduces mod 3 to $2a_4 \equiv 7 \bmod 3$. Since $f_3' + f_3'' = 3$, the possibilities for a_4 are $+3, +1, -1, -3$. The only choice that satisfies the condition is $+1$, which requires that two of the 4-faces lie outside the cycle. These two faces must include one of the two 4-faces with a 2-valent vertex; call it x and its neighbors y, z. This is impossible, because the y, z-path through x and the other y, z-path on the cycle separate this face from the outside face.

If we subdivide an edge incident to each 2-valent vertex, however, the existence of a Hamiltonian cycle does not change, and we have seven 5-faces, one 4-face, and one 11-face. In this case the impossibility is immediate, since the equation modulo 3 becomes $2 \cdot (\pm 1) \equiv 0 \bmod 3$. $\square$

On the positive side, it is conjectured that every planar 3-connected 3-regular bipartite graph is Hamiltonian; this is known as Barnette's Conjecture. Other conjectures about edge-coloring of 3-regular graphs would be stronger than the Four Color Theorem. Tutte conjectured that a 2-connected 3-regular graph is 3-edge-colorable if it has no subgraph contractible to the Petersen graph. Fulkerson conjectured that every 2-connected 3-regular graph has a list of 6 perfect matchings such that every edge is in exactly two of them (see Exercise 6.1.13).

CROSSING NUMBER

Many parameters measure a graph's deviation from planarity. Here we briefly discuss the measure suggested by the definition of planarity. The *crossing number* $v(G)$ of a graph G is the minimum number of crossings in a drawing of G in the plane. The lower bound below is useful for small graphs.

7.3.6. Example. *Crossing number of small graphs.* It may be possible to determine the crossing number of a small graph by considering maximal planar subgraphs. Consider a drawing of G in the plane. If H is a maximal plane subgraph of this drawing, then every edge of G not in H crosses some edge of H in this drawing, so the drawing has at least $e(G) - e(H)$ crossings. If G has n vertices, then $e(H) \leq 3n - 6$. If also G has no triangles, then $e(H) \leq 2n - 4$.

The graph K_6 has 15 edges, but planar 6-vertex graphs have at most 12 edges. Hence $\nu(K_6) \geq 3$, and the drawing on the left below proves that $\nu(K_6) \leq 3$. The graph $K_{3,2,2}$ has 16 edges, and planar graphs with seven vertices have at most 15 edges, so $\nu(K_{3,2,2}) \geq 1$. The best drawing we find has two crossings, as shown on the right below. To improve the lower bound, observe that $K_{3,2,2}$ has $K_{3,4}$ as a subgraph. Because $K_{3,4}$ is triangle-free, there are at most $2 \cdot 7 - 4 = 10$ edges in a planar subgraph of $K_{3,4}$, and hence $\nu(K_{3,4}) \geq 2$. Because every drawing of $K_{3,2,2}$ contains a drawing of $K_{3,4}$, we have $\nu(K_{3,2,2}) \geq \nu(K_{3,4}) \geq 2$. $\square$

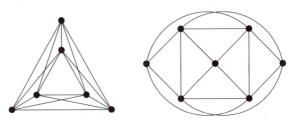

7.3.7. Proposition. Suppose G is an n-vertex graph with m edges. If k is the maximum number of edges in a planar subgraph of G, then
$$\nu(G) \geq \Sigma_{i=1}^{\lfloor m/k \rfloor}(m - ik).$$

Proof. Consider any planar drawing of G, and let H be a maximal subgraph of G whose edges have no crossings in this drawing. Every edge not in H crosses at least one edge in H, else it could be added to H. Since H has at most $k = 3n - 6$ edges, we have at least $e - k$ crossings between edges of H and edges of $G - E(H)$. We can discard $E(H)$ and apply the same argument for a lower bound on crossings in the remaining graph. With $l = \lfloor m/k \rfloor$, we obtain at least $\Sigma_{i=1}^{l} m - ik = ml - kl(l+1)/2$ crossings. $\square$

For graphs with many edges, this argument misses too many crossings. Consider the complete graph: here $m = (n^2 - n)/2$, $k = 3n - 6$, and l is a bit more than $n/6$. The formula above yields $\nu(K_n) \geq n^3/24 + O(n^2)$, but actually $\nu(K_n)$ grows like a multiple of n^4. The crossing number is never more than $\binom{n}{4}$, since we can choose a straight-line drawing, and in a straight-line drawing each set of four vertices contributes at most one crossing.

In drawing K_n, for example, we could place the vertices on the circumference of a circle and draw chords. This is the worst possible straight-line drawing of K_n. In any such drawing on K_n, four vertices determine a K_4 with or without a crossing, depending on whether any point is inside the triangle determined by the other three. In this convex drawing, each 4-tuple contributes a crossing. How many crossings can be saved by a better drawing?

7.3.8. Theorem. (R. Guy [1972]) $n^4/80 + O(n^3) \le \nu(K_n) \le n^4/64 + O(n^3)$.

Proof. The simplest lower bound is inductive. Any drawing of K_n has n copies of K_{n-1}. We can delete any vertex and count crossings in the subdrawing. The total count is at least $n\nu(K_{n-1})$, but each crossing in K_n is counted $(n-4)$ times this way, so $\nu(K_n) \ge n\nu(K_{n-1})/(n-4)$. The solution of the recurrence $f_n = \dfrac{n}{n-4} f_{n-1}$ is $f_n = A\binom{n}{4}$. Since K_5 has one crossing, this yields $\nu(K_n) \ge \dfrac{1}{5}\binom{n}{4}$. The denominator of the quartic term in the lower bound can be improved from 120 to 80 by considering copies of $K_{6,n-6}$, whose crossing number is known to be $6\lfloor (n-6)/2\rfloor\lfloor (n-7)/2\rfloor$ (Exercise 27b).

By using a more clever drawing, we lower the upper bound to $n^4/64 + O(n^3)$. Suppose $n = 2k$. Drawing K_n in the plane is equivalent to drawing it on a sphere or on the surface of a can. Place k vertices on the top rim of the can and k vertices on the bottom rim, drawing chords on the top and bottom for those k-cliques. Consider the k classes of vertical edges needed to complete the graph. The "class number" is the circular separation between the top and bottom endpoints; this ranges from $\lceil -(k-1)/2\rceil$ to $\lceil (k-1)/2\rceil$. It makes sense to wind an edge around the can as little as possible in passing from top to bottom, so edges in the same class don't cross. We have the same crossings if we twist the can so the class displacements run from 1 to k, which makes them easier to count. The 4-sets that yield crossings consist of 4 on the top, 4 on the bottom, and some sets with 2 on top and 2 on the bottom. If we take two top vertices x, y and two bottom vertices z, w, where xz has smaller positive displacement than xw, then we have a crossing for x, y, z, w if and only if $x + z > y$. That is, from a fixed x, we have a crossing if and only if the displacements to y, z, w are distinct positive values in increasing order. (In the illustration, x, y, z, a do not yield a crossing.) Hence there are $k\binom{k}{3}$ crossings on the side of the twisted can, and $\nu(K_n)$ $\le 2\binom{k}{4} + k\binom{k}{3} = n^4/64 + O(n^3)$. □

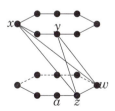

7.3.9. Example. $\nu(K_{m,n})$. The most naive drawing puts the vertices of one partite set on one side of a channel and the vertices of the other partite set on the other side, with all edges drawn straight across. This has $\binom{n}{2}\binom{m}{2}$ crossings, but it is easy to reduce this by a factor of 4. Place the nodes of the graph along two perpendicular axes. Put $\lceil n/2\rceil$ vertices along the positive y-axis and $\lfloor n/2 \rfloor$ along the negative y-axis; similarly split the m vertices along the positive and negative x-axis. Adding up the four types of crossings generated when we join every vertex on the x-axis to every vertex on the y-axis yields $v \le \lfloor \frac{m}{2} \rfloor \lfloor \frac{m-1}{2} \rfloor \lfloor \frac{n}{2} \rfloor \lfloor \frac{n-2}{2} \rfloor$.

The true value is not known. Zarankiewicz [1954] thought he had proved the conjectured formula, but flaws were found (see Guy [1969]). Kleitman [1970] proved the formula for $\min(n, m) \le 6$. The smallest unknown case is $K_{7,7}$, where the crossing number is 77, 79 or 81. Using Kleitman's result, Guy [1970] proved $v(K_{m,n}) \ge \lceil m(m-1)n(n-1)/20 \rceil$, which is not far from the upper bound (Exercise 27). □

SURFACES OF HIGHER GENUS (optional)

Instead of minimizing crossings in a planar drawing, we could change the surface to eliminate crossings. This is what highway engineers do by building overpasses and cloverleafs instead of installing traffic lights. Highways run on the surface of the earth, and for this discussion it is convenient to view our drawings on the sphere instead of in the plane; as discussed in Section 7.1, these settings are equivalent.

To avoid creating boundaries or cuts in the surface, we add the overpass by cutting two small holes in the sphere and joining the edges of the holes by a tube, as shown below. We have added a *handle*. By stretching the handle and squeezing the rest of the sphere, we obtain a doughnut, called the *torus*.

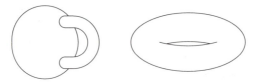

The torus is topologically the same as the sphere with one handle, in the sense that one surface can be continuously transformed into the other.[†]

A large graph may have many crossings and need more handles. For any graph, adding enough handles to a drawing on the sphere will eliminate all crossings and produce an embedding. The number of handles added is the *genus* of the resulting surface. The minimum number of handles that allows an embedding of G is the *genus* of G. The surfaces of genus 0, 1, 2 are the sphere, torus, and double torus (two handles), respectively, and the graphs embeddable on them are the *planar*, *toroidal*, and *double-toroidal* graphs. When we add some number of handles, it doesn't matter how we do it, because two surfaces obtained

---

[†]This is the source of the joke that a topologist is a person who can't tell the difference between a doughnut and a coffee cup.

by adding the same number of handles to a sphere can be continuously deformed into each other. The theory of planar graphs extends in various ways to graphs embeddable on higher surfaces; here we give only a brief discussion for cultural interest.

Drawings of large graphs on surfaces of large genus are hard to follow, even on a pretzel (S_3). Locally, the surface looks like a plane sheet of paper. To draw the graph we want to lay the entire surface out flat; to do this we must cut the surface. If we keep track of how the edges should be pasted back together to get the surface, we can describe the surface on a flat piece of paper. Consider first the torus.

7.3.10. Example. *Combinatorial description of the torus.*

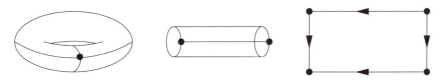

Cutting the closed tube once turns it into a cylinder, and then slitting the length of the cylinder allows us to lay it out flat as a rectangle. Labeling the edges of the rectangle lets us keep track of how to paste it back together; the two sides of a cut, labeled with the same letter, are "identified". This is particularly important because edges of an embedding may cross the imaginary cut on a surface. When the edge reaches one border of the rectangle, it is reaching one side of this imaginary cut. When it crosses the imaginary cut, it emerges from the identical point on the other copy of this border. Note also that the four "corners" of the rectangle correspond to the single point on the surface through which both cuts pass. These ideas lead to pleasant toroidal embeddings of K_5, $K_{3,3}$, and K_7. □

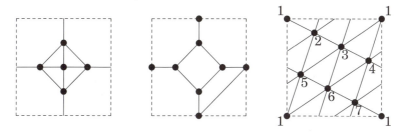

For surfaces of higher genus, there is some flexibility in how we make the cuts, but every way will cost two cuts per handle before we can lay the surface flat. The standard polygonal representation comes from viewing the handles as occuring on "lobes" of the surface, with the cuts having a common point on the hub.

7.3.11. Example. *Laying out the double torus.* Below we illustrate a polygonal representation for the double torus. Making the cuts is equivalent to adding loops to a single vertex on the surface until we have a one-face embedding of a bouquet of loops. In general, we can make 2γ cuts through a single point to lay S_γ flat. Keeping track of the borders arising from each cut, we represent S_γ by a 4γ-gon in which a clockwise traversal of the boundary can be described by reading out the cuts as we traverse them, writing a cut using inverted commands. we traverse it in the opposite order. Since we are actually following the boundary of a single face, with our left hand always on the wall, each edge will be followed once forward and once backward. For the example here, the traversal is $\alpha_1\beta_1\alpha_1^{-1}\beta_1^{-1}\alpha_2\beta_2\alpha_2^{-1}\beta_2^{-1}$. Each surface S_γ has a layout of the form $\alpha_1\beta_1\alpha_1^{-1}\beta_1^{-1}\cdots\alpha_\gamma\beta_\gamma\alpha_\gamma^{-1}\beta_\gamma^{-1}$. Other layouts result from other ways of making the cuts -- different ways of embedding a bouquet of 2γ loops. For example, the double torus can also be represented by an octagon with boundary $\alpha\beta\gamma\delta\alpha^{-1}\beta^{-1}\gamma^{-1}\delta^{-1}$. $\square$

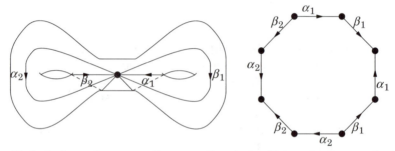

Euler's formula generalizes to S_γ. A *2-cell* is a region such that any closed curve in the interior can be continuously contracted to a point. A *2-cell embedding* is an embedding in which every region is a 2-cell, and the generalization of Euler's Formula says that $n - e + f = 2 - 2\gamma$ for any 2-cell embedding of a connected graph on S_γ (Exercise 32). For example, our embedding of K_7 on the torus ($\gamma = 1$) has 7 vertices, 21 edges, 14 faces, and $7 - 21 + 14 = 0$. The proof of Euler's formula in general is like the proof in the plane, except that we must take more care with the basis case of 1-vertex graphs, which requires showing that it takes 2γ cuts to lay the surface flat (i.e., to obtain an embedding of a graph with a single face that is a 2-cell).

Given Euler's formula for S_γ, we can follow the same argument as for planar graphs to obtain an upper bound on the number of edges in a graph embeddable on S_γ: $e \leq 3(n - 2 + 2\gamma)$ (Exercise 32). Note that K_7 satisfies this with equality, as every region in the toroidal embedding of K_7 is a 3-gon. Hence K_7 is a maximal toroidal graph. The bound $e \leq 3(n - 2 + 2\gamma)$ can be turned around to obtain a lower bound on the number of handles we must add to obtain a surface on which G is embeddable; this is $\gamma \geq 1 + (e - 3n)/6$.

EXERCISES

7.3.1. State a polynomial-time algorithm that takes an arbitrary planar graph as input and produces a proper 5-coloring of the graph.

7.3.2. (!) Prove that a plane graph is 2-face-colorable if and only if it is Eulerian.

7.3.3. Without using the Four Color Theorem, prove that every planar graph with at most 12 vertices is 4-colorable. Use this to prove that every planar graph with at most 32 edges is 4-colorable.

7.3.4. (+) Suppose there exists a minimal triangulation G requiring 5 colors. Without using anything that has not been proved in this section, prove that G has no separating cycle C of length three or four (i.e., any configuration with a ring of length at most four is reducible). (Hint: Consider the subgraphs A and B that are the C-components of G. Prove that A and B have 4-colorings that agree on C. By replacing the inside (or outside) with a single edge between opposite vertices of C (when C has length 4), one can force a 4-coloring of A or B to use distinct colors on two opposite vertices of the ring. Birkhoff [1913])

7.3.5. (!) Without using the Four Color Theorem, prove that every outerplanar graph is 3-colorable. Apply this to prove the Art Gallery Theorem: If an art gallery is laid out as a polygon with n segments, then it is possible to place $\lfloor n/3 \rfloor$ guards such that every point of the interior is visible to some guard. (Chvátal [1975], Fisk [1978])

7.3.6. (+) Prove that a maximal planar graph is 3-colorable if and only if it is Eulerian. (Hint: For the proof of sufficiency, use induction on n, choosing an appropriate pair of adjacent vertices to delete. This proof yields the additional observation that every Eulerian triangulation has even order, except for K_3.)

7.3.7. Grötzsch's Theorem [1959] (see Thomassen [1994a], Steinberg [1993]) states that every triangle-free planar graph G is 3-colorable and hence has an independent set of size at least $n(G)/3$. Tovey-Steinberg [1993] proved that $\alpha(G) > n(G)/3$ always. Prove that this is best possible by considering the family of graphs G_k defined as follows: G_1 is the 5-cycle, with vertices a, x_0, x_1, y_1, z_1 in order. For $k > 1$, G_k is obtained from G_{k-1} by adding the three vertices x_k, y_k, z_k and the five edges $x_{k-1}x_k, x_k y_k, y_k z_k, z_k y_{k-1}, z_k x_{k-2}$. The graph G_3 is shown below. (Fraughnaugh [1985])

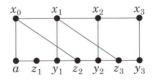

7.3.8. (−) Prove that every Hamiltonian 3-regular graph has a Tait coloring.

7.3.9. *Examples.* Exhibit 3-regular simple graphs with the following properties:
   a) planar but not 3-edge-colorable.
   b) 2-connected but not 3-edge-colorable.
   c) planar with connectivity 2, but not Hamiltonian.

7.3.10. Without using the Four Color Theorem, prove that every Hamiltonian plane graph is 4-face-colorable (nothing is assumed about the vertex degrees).

7.3.11. Use Tait's Theorem [1878] to prove that the graph below is 3-edge-colorable.

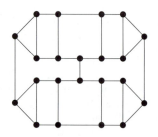

7.3.12. (!) *Reduction of Four Color Problem to Tait's Conjecture.* Let G be a simple cubic (i.e. 3-regular) graph with edge-connectivity 2.

a) Prove that there exist subgraphs $G_1, G_2 \subseteq G$ and vertices $u_1, v_1 \in G_1$ and $u_2, v_2 \in G_2$ such that $u_1 \not\leftrightarrow v_1$ and $u_2 \not\leftrightarrow v_2$ and G consists of G_1, G_2 and a "ladder" (of some length) joining them at u_1, v_1, u_2, v_2 as illustrated below.

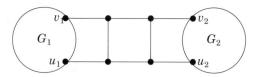

b) Prove that if $G_1 + u_1 v_1$ and $G_2 + u_2 v_2$ are 3-edge-colorable, then G is also 3-edge-colorable.

c) Tait [1878] proved that a simple cubic 2-edge-connected planar graph is 3-edge-colorable if and only if its dual (a maximal planar graph) is 4-colorable. Using this and parts (a) and (b) above, show that the Four Color Problem can be reduced to proving the following statement (Tait's conjecture): "every simple cubic 3-connected planar graph is 3-edge-colorable."

7.3.13. It has been conjectured that every planar triangulation has edge-chromatic number $\Delta(G)$, and this has been proved when $\Delta(G)$ is high enough. Show that $\chi'(G) = \Delta(G)$ for the graph of the icosahedron, illustrated below.

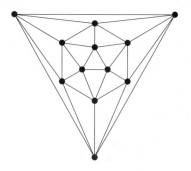

7.3.14. Determine the maximum of connectivity over the set of planar graphs.

7.3.15. (!) For each of the planar graphs below, present a Hamiltonian cycle or use planarity (Grinberg's condition) to prove that it is non-Hamiltonian.

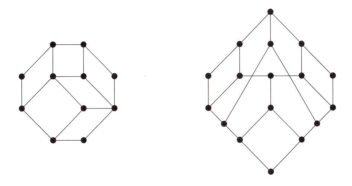

7.3.16. (!) Use Grinberg's condition to prove that the Grinberg graph (below) is not Hamiltonian.

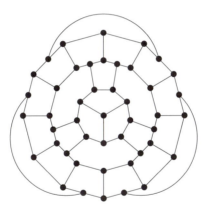

7.3.17. (!) The smallest known 3-regular planar graph that is not Hamiltonian has 38 vertices and appears below. Prove that this graph is not Hamiltonian. (Lederberg [1966], Bosák [1966], Barnette)

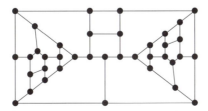

7.3.18. Let G be the grid graph $P_m \square P_n$. Suppose Q is a Hamiltonian path from the upper left corner vertex to the lower right corner vertex. As illustrated below, Q partitions the grid into regions, of which some open to the left or downward and others open to the right or upward. Prove that the total area of the up-right regions (B) equals the total area of the down-left regions (A). (Fisher-Collins-Krompart [1994])

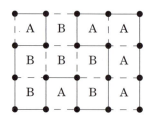

7.3.19. *Thickness.* The *thickness* of G is the minimum number of planar graphs whose union is G.

    a) Prove that the thickness of K_n is at least $\lfloor \frac{n+7}{6} \rfloor$. (Hint: $\lceil x/r \rceil = \lfloor (x + r - 1)/r \rfloor$).

    b) Find a self-complementary planar graph with 8 vertices. What is the thickness of K_8?

7.3.20. Suppose G is the union of two planar graphs (thickness 2). Prove that $\chi(G) \le 12$. Use the graph $C_5 \vee K_6$ to prove that $\chi(G)$ may be as large as 9 if G has thickness 2. (Sulanke)

7.3.21. Compute the crossing number for $K_{2,2,2,2}$, $K_{4,4}$, and the Petersen graph.

7.3.22. Prove that $K_{3,2,2}$ has no planar subgraph with 15 edges. Use this to give another proof that $\nu(K_{3,2,2}) \ge 2$.

7.3.23. Prove that the crossing number of the graph drawn below is at most 5.

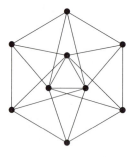

7.3.24. Let M_n be the graph obtained from the cycle C_n by adding chords between vertices that are opposite (if n is even) or nearly opposite (if n is odd). The graph is 3-regular if n is even, 4-regular if n is odd. Determine the crossing number of M_n. (Guy-Harary [1967])

7.3.25. The graph P_n^k is the graph with vertex set $[n]$ and edge set $\{ij: |i - j| \leq k\}$. Prove that P_n^3 is a maximal planar graph. Use a planar embedding of P_n^3 to prove that $\nu(P_n^4) = n - 4$. (Harary-Kainen [1993])

7.3.26. For every positive integer k, construct a graph that embeds on the torus but requires at least k crossings when drawn in the plane. (Hint: a single easily described toroidal family suffices. It is not necessary to provide a construction with crossing number exactly k.)

7.3.27. *Lower bounds on crossing number for special families.*
   a) Use Kleitman's computation of $\nu(K_{6,m})$ to show
$$\nu(K_{m,n}) \geq \lceil m(m - 1)n(n - 1)/20 \rceil. \text{ (Guy [1970])}$$
   b) Use $\nu(K_{6,m})$ to show $\nu(K_n) \geq n^4/80 + O(n^3)$.

7.3.28. *Crossing number of Cartesian products.* It is known that $\nu(C_m \Box C_n) = (m - 2)n$ if $m \leq \min\{5, n\}$, and also $\nu(K_4 \Box C_n) = 3n$. Find drawings in the plane to establish the upper bounds. (Comment: beyond $Q_4 = C_4 \Box C_4$, the crossing number of the k-dimensional cube is not known, but Madej [1991] has good bounds.)

7.3.29. (!) *Crossing number of complete tripartite graphs.* Let $f(n) = \nu(K_{n,n,n})$.
   a) Show that $3\nu(K_{n,n}) \leq f(n) \leq 3(\binom{n}{2})^2$.
   b) Show that $\nu(K_{3,2,2}) = 2$ and $\nu(K_{3,3,1}) = 3$. Show that $5 \leq \nu(K_{3,3,2}) \leq 7$ and $9 \leq \nu(K_{3,3,3}) \leq 15$.
   c) From Kleitman's computation of $K_{6,m}$, the lower bound in (a) at present is $(3/20)n^4 + O(n^3)$. Improve it by using a recurrence to show that $f(n) \geq n^3(n - 1)/6$.
   d) The upper bound in (a) is $(3/4)n^4 + O(n^3)$. Improve it by showing $f(n) \leq (9/16)n^4 + O(n^3)$. (Hint: one construction one embeds the graph on the surface of a tetrahedron and generalizes to a construction for $K_{l,m,n}$, another uses of K_n and generalizes to a construction for $K_{n,...,n}$.)

7.3.30. (!) Suppose n is at least 9 and n is not a prime or twice a prime. Construct a 6-regular toroidal graph with n vertices.

7.3.31. (!) A embedding of a graph on a surface is *regular* if its faces all have the same length. Construct regular embeddings of $K_{4,4}$, $K_{3,6}$, and $K_{3,3}$ on the torus (look for pretty pictures).

7.3.32. (!) Prove Euler's formula for genus γ: For every 2-cell embedding of a graph on a surface with genus γ, the numbers of vertices, edges, and faces satisfy $n - e + f = 2 - 2\gamma$. Conclude that an n-vertex graph embeddable on S_γ has at most $3(n - 2 + 2\gamma)$ edges.

7.3.33. Use Euler's formula for S_γ to prove that $\gamma(K_{3,3,n}) \geq n - 2$, and determine the value exactly for $n \leq 3$.

7.3.34. For every positive integer k, use Euler's formula for higher surfaces to prove that there exists a planar graph G such that $\gamma(G \Box K_2) \geq k$.

Chapter 8

Additional Topics

In this chapter we explore more advanced or specialized material. Each section gives a glimpse of a topic that deserves its own chapter (or book). Several sections treat more difficult material near the end.

8.1 Perfect Graphs

We defined perfect graphs in Section 5.3; a graph G is *perfect* if $\chi(G[A]) = \omega(G[A])$ for every $A \subseteq V(G)$. Our first goal is the *Perfect Graph Theorem (PGT)*, which states that a graph is perfect if and only if its complement is perfect. Later we will study some classes of perfect graphs and minimal imperfect graphs. Golumbic [1980] provides a thorough introduction to the subject. Berge-Chvátal [1984] collects and updates many of the classical papers.

Cliques in H become independent sets in $\overline{H}$, and vice versa. Coloring $\overline{H}$ is covering H by cliques; the *clique cover number* $\theta(H)$ is the minimum number of cliques in H needed to cover its vertices. Hence $\omega(\overline{H}) = \alpha(H)$ and $\chi(\overline{H}) = \theta(H)$. Since also $\overline{G}[A] = \overline{G[A]}$ for all $A \subseteq V(G)$, saying that $\overline{G}$ is perfect is saying that $\theta(G[A]) = \alpha(G[A])$ for all $A \subseteq V(G)$. Berge therefore defined two types of perfection: *γ-perfect* means $\chi(G[A]) = \omega(G[A])$ for all $A \subseteq V(G)$, and *α-perfect* means $\theta(G[A]) = \alpha(G[A])$ for all $A \subseteq V(G)$. We now use only one definition, because "G if perfect if and only $\overline{G}$ is perfect" (the PGT) has the same meaning as "G is γ-perfect if and only if G is α-perfect."

In this section, we use *clique* to mean a set of vertices inducing a complete subgraph, and we use *stable set* to mean an independent set of vertices. We say that a set of cliques or of stable sets *covers* G when it covers $V(G)$. As before, *maximum* means maximum-sized.

Since a clique and a stable set share at most one vertex, covering

G requires at least $\omega(G)$ stable sets and requires at least $\alpha(G)$ cliques. That is, $\chi(G) \geq \omega(G)$ and $\theta(G) \geq \alpha(G)$. A statement of perfection for a class of graphs or their complements is thus a min-max relation for the class. We saw in Section 5.3 that familiar min-max relations become statements that bipartite graphs, their line graphs, comparability graphs, and the complements of all such graphs are perfect.

8.1.1. Example. *Odd cycles and the Strong Perfect Graph Conjecture.* Not every graph is perfect. We have $\chi(G) > \omega(G)$ for odd cycles, and this inequality also holds for complements of odd cycles (Exercise 1), so these graphs are imperfect. The *Strong Perfect Graph Conjecture (SPGC)*, posed by Berge in 1960, asserts that G is perfect if and only if G does not have an odd cycle or its complement as an induced subgraph. This conjecture remains open. Since the condition in the conjecture is self-complementary, the SPGC would have the PGT as a corollary. □

THE PERFECT GRAPH THEOREM

Berge conjectured equivalence of γ-perfection and α-perfection in 1960. Using polyhedral methods, Fulkerson reduced the conjecture to a statement he thought was too strong to be true. When Berge told him that Lovász had proved the theorem, in a few hours he proved the missing lemma (Corollary 8.1.27), illustrating that a theorem becomes easier to prove when known to be true (Fulkerson [1971]).

Meanwhile, Lovász' proof of the conjecture announced his arrival in the world of combinatorics in dramatic fashion. He had resolved an important and well-known open question at the age of 22. He has subsequently produced a steady stream of elegant proofs, including many mentioned in this book.

We will prove the Perfect Graph Theorem using an operation that enlarges a graph w.ithout changing whether it is perfect.

8.1.2. Definition. *Duplicating a vertex x of G produces a supergraph $G \circ x$ by adding a vertex x' with $N(x') = N(x)$. More generally, the vertex multiplication of G by the nonnegative integer vector $h = (h_1, \ldots, h_n)$ is the graph $H = G \circ h$ having h_i copies of x_i, such that copies of x_i and x_j are adjacent in H if and only if $x_i \leftrightarrow x_j$ in G.*

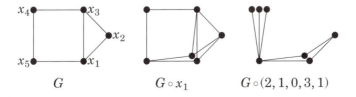

$$G \qquad\qquad G \circ x_1 \qquad\qquad G \circ (2, 1, 0, 3, 1)$$

8.1.3. Lemma. (Berge [1961]). Vertex multiplication preserves γ-perfection and α-perfection.

Proof. In considering $G \circ h$, let $A = \{i : h_i > 0\}$. We first observe that $G \circ h$ can be obtained from $G[A]$ by a sequence of vertex duplications. If every h_i is 0 or 1, then $G \circ h = G[A]$. Otherwise, start with $G[A]$ and duplicate vertices successively until there are h_i copies of each x_i. Each vertex duplication preserves the property that copies of x_i and x_j are adjacent if and only if $x_i x_j \in E(G)$, so the resulting graph is $G \circ h$.

If G is α-perfect but $G \circ h$ is not, then some step in the sequence of vertex duplications creating $G \circ h$ from $G[A]$ produces a graph that is not α-perfect from an α-perfect graph. It thus suffices to prove that vertex duplication preserves α-perfection. The same reduction holds for γ-perfection. Since every proper induced subgraph of $G \circ x$ is an induced subgraph of G or is a vertex duplication of a proper induced subgraph of G, we further reduce our claim to showing that $\chi(G \circ x) = \omega(G \circ x)$ when G is γ-perfect and that $\alpha(G \circ x) = \theta(G \circ x)$ when G is α-perfect.

When G is γ-perfect, we extend a proper coloring of G to a proper coloring of $G \circ x$ by giving x' the same color as x. No clique contains both x and x', so $\omega(G \circ x) = \omega(G)$. Hence $\chi(G \circ x) = \chi(G) = \omega(G) = \omega(G \circ x)$.

When G is α-perfect, we consider two cases. If x belongs to a maximum stable set in G, then adding x' to it yields $\alpha(G \circ x) = \alpha(G) + 1$. Since $\theta(G) = \alpha(G)$, we can obtain a clique covering of this size by adding x' as a 1-vertex clique to some set of $\theta(G)$ cliques covering G.

If x belongs to no maximum stable set in G, then $\alpha(G \circ x) = \alpha(G)$. Let Q be the clique containing x in a minimum clique cover of G. Since $\theta(G) = \alpha(G)$, Q intersects every maximum stable set in G. Since x belongs to no maximum stable set, $Q' = Q - x$ also intersects every maximum stable set. This yields $\alpha(G - Q') = \alpha(G) - 1$. Applying the α-perfection of G to the induced subgraph $G - Q'$ (which contains x) yields $\theta(G - Q') = \alpha(G - Q')$. To a set of $\alpha(G) - 1$ cliques covering $G - Q'$, we add the clique $Q' \cup x'$ to complete a set of $\alpha(G)$ cliques covering $G \circ x$. $\quad\square$

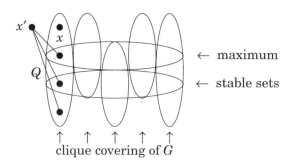

$\leftarrow$ maximum

$\leftarrow$ stable sets

clique covering of G

8.1.4. Theorem. (The Perfect Graph Theorem (PGT) - Lovász [1972a, 1972b]) A graph is perfect if and only if its complement is perfect.

Proof. It suffices to prove that α-perfection of G implies γ-perfection of G; the same statement for $\overline{G}$ yields the converse. Suppose G is a minimal counterexample. If G has a stable set S intersecting every maximal clique, then the minimality of G yields $\chi(G-S) = \omega(G-S) = \omega(G) - 1$, and S completes a proper $\omega(G)$-coloring of G. By the minimality of G, this would yield γ-perfection for G. Therefore, we may assume that every maximal stable set S in G misses some maximum clique $Q(S)$.

To show that this cannot happen for an α-perfect graph G, we will apply Lemma 8.1.3 to a special vertex multiplication of G. Let $\mathbf{S} = \{S_i\}$ be the collection of maximal stable sets of G. We weight each vertex x_j by its frequency in $\{Q(S_i)\}$, letting h_j be the number of stable sets $S_i \in \mathbf{S}$ such that $x_j \in Q(S_i)$. By Lemma 8.1.3, the vertex multiplication $H = G \circ h$ is α-perfect, yielding $\alpha(H) = \theta(H)$. We use counting arguments for $\alpha(H)$ and $\theta(H)$ to obtain a contradiction.

Let A be the 0,1-matrix of the incidence relation between $\{Q(S_i)\}$ and $V(G)$. The vertices $\{x_j\}$ index the columns, the maximal stable sets $\{S_i\}$ index the rows, and we put $a_{ij} = 1$ if and only if $x_j \in Q(S_i)$. By construction, h_j is the number of 1's in column j of A, and $n(H)$ is the total number of 1's in A. Since each row contributes $\omega(G)$ ones, we have $n(H) = \omega(G)|\mathbf{S}|$. Since vertex duplication cannot enlarge cliques, we have $\omega(H) = \omega(G)$. Therefore $\theta(H) \geq n(H)/\omega(H) = |\mathbf{S}|$.

We obtain a contradiction by proving that $\alpha(H) < |\mathbf{S}|$. Every stable set in H consists of copies of elements in some stable set of G, so a maximum stable set in H consists of all copies of all vertices in some maximal stable set of G. Hence $\alpha(H) = \max_{T \in \mathbf{S}} \sum_{i: x_i \in T} h_i$. The sum counts the 1's in A that appear in the columns indexed by T. If we count these 1's instead by rows, we obtain $\alpha(H) = \max_{T \in \mathbf{S}} \sum_{S \in \mathbf{S}} |T \cap Q(S)|$. Since T is a stable set, it has at most one vertex in each chosen clique $Q(S)$. Furthermore, T is disjoint from the chosen clique $Q(T)$. With $|T \cap Q(S)| \leq 1$ for all $S \in \mathbf{S}$, and $|T \cap Q(T)| = 0$, we have $\alpha(H) \leq |\mathbf{S}| - 1$. $\square$

$$V(G)$$

$Q(S_1)$	$\vdots$	$\vdots$	$\vdots$
$Q(T)$	0	0	0
$Q(S_n)$	$\vdots$	$\vdots$	$\vdots$
	$\uparrow$	$\uparrow$	$\uparrow$
	T	T	T

Clique-vertex incidence matrices also arise in expressing α and θ as integer optimization problems. A linear (maximization) program can be written in the form "maximize $c \cdot x$ over nonnegative vectors x such that $Ax \leq b$," where c is a vector of constants and each row of the matrix inequality is a linear constraint $a_i \cdot x \leq b_i$ on x. A vector x satisfying all the constraints is a *feasible solution*.

An *integer linear program* requires that each x_j also be an integer. Given a graph G, let A be the incidence matrix between maximal cliques $\{Q_i\}$ and vertices $\{v_j\}$; we set $a_{ij} = 1$ when $v_j \in Q_i$. By definition, $\alpha(G)$ is the solution to "max $\mathbf{1}_n \cdot x$ such that $Ax \leq \mathbf{1}_m$" when the variables are required to be nonnegative integers. In the solution, x_j is 1 or 0 depending on whether v_j belongs to the maximum stable set; the constraints prevent choosing two adjacent vertices. Similarly, when B is the incidence matrix between maximal stable sets and vertices, $\omega(G)$ is the solution to "max $\mathbf{1}_n \cdot x$ such that $Bx \leq \mathbf{1}_p$" with integer variables.

Every maximization program has a dual minimization program. When the max program is "max $c \cdot x$ such that $Ax \leq b$", the dual is "min $y \cdot b$ such that $y^T A \geq c$". This program has a variable y_i for each original constraint and a constraint for each original variable x_j, and it switches the roles of $c, \max, \leq$ with $b, \min, \geq$. When stated in this form, the variables in both programs must be nonnegative. The integer programs dual to ω and α ask for the minimum number of stable sets that cover the vertices and the minimum number of cliques that cover the vertices, respectively, which describe χ and θ.

Using the nonnegativity of the variables, the constraints yield $c \cdot x \leq y^T A x \leq y^T \cdot b$. The statement that $c \cdot x \leq y \cdot b$ whenever x, y are feasible solutions to dual programs is *weak duality*. The *strong duality* theorem states that when dual programs both have feasible solutions and integer solutions are not required, the programs have optimal solutions with the same value. The statements $\chi \geq \omega$ and $\theta \geq \alpha$ are statements of weak duality for dual pairs of linear programs. When we can guarantee strong duality using solutions that have only integer values, we obtain a combinatorial min-max relation. We have presented many such relations and remarked that they guarantee quick proofs of optimality. Optimal solutions in integers are not always guaranteed.

8.1.5. Example. *Fractional solutions for an imperfect graph.* For the 5-cycle, the linear programs for $\omega, \chi, \alpha, \theta$ all have optimal value 5/2. There are five maximal cliques and five maximal stable sets, each of size 2. Setting each $x_j = 1/2$ gives weight 1 to each clique and stable set, thereby satisfying the constraints for either maximization problem. Setting each $y_i = 1/2$ in the dual programs covers each vertex with a total weight of 1, so again the constraints are satisfied. These programs have no optimal solution in integers, and the integer programs have a "duality gap": $\chi = 3 > 2 = \omega$ and $\theta = 3 > 2 = \omega$. □

CHORDAL GRAPHS REVISITED

A class of graphs may have many characterizations. This holds for trees and also for the more general class of chordal graphs (Definition 5.3.12), which are the graphs having no *chordless cycle* (induced cycle of length at least 4). This definition is a **forbidden substructure characterization** of chordal graphs. A finite list of forbidden substructures such as induced subgraphs yields a fast algorithm for testing membership in the class, but for chordal graphs the list is infinite.

Alternatively, chordal graphs can be built from a single vertex by iteratively adding a vertex joined to a clique. A **construction procedure** produces the graphs in **G** (and no other graphs) by applying specified operations, starting with a special small class (in this case K_1). The reverse of a construction procedure is a **decomposition procedure**, which also may lead to fast algorithms for computations on graphs in **G**. For chordal graphs, the reverse of a construction ordering is a simplicial elimination ordering. Next we consider a way of using intersections of sets to produce the graphs in a given class.

8.1.6. Definition. An *intersection representation* of a graph G is a collection of sets $\{S_v : v \in V(G)\}$ such that $u \leftrightarrow v$ if and only if $S_u \cap S_v \neq \emptyset$. If $\{S_v\}$ is an *intersection representation* of G, then G is the *intersection graph* of $\{S_v\}$.

The interval graphs are the graphs having intersection representations in which every set assigned to a vertex is an interval on the real line. Line graphs also form an intersection class; the allowed sets are pairs of natural numbers, corresponding to vertices of the graph H such that $G = L(H)$. An intersection characterization for chordal graphs was discovered independently by Walter [1972, 1978], Gavril [1974], and Buneman [1974]. It requires a lemma.

8.1.7. Lemma. If $T_1, \ldots, T_k$ are pairwise intersecting subtrees of a tree, then there is a vertex belonging to all of $T_1, \ldots, T_k$.

Proof. We use induction on k. For $k = 2$, the conclusion is part of the hypothesis. For $k > 2$, the induction hypothesis yields vertices u, v, w such that $u \in T_1 \cap \cdots \cap T_{k-1}$, $v \in T_2 \cap \cdots \cap T_k$, and $w \in T_1 \cap T_k$. The host tree T has a unique u,v-path P, contained in all of $T_2, \ldots, T_{k-1}$. Let Q be the unique path in T from w to $V(P)$. Since $u, w \in T_1$ and $v, w \in T_k$, both T_1 and T_k contain Q. Hence the vertex z at which Q reaches P belongs to all of $T_1, \ldots, T_k$. □

8.1.8. Theorem. The following conditions are equivalent.

A) G has no chordless cycle.

B) G has a simplicial elimination ordering.

C) G has an intersection representation using subtrees of a tree (a *subtree representation*).

Proof. Theorem 5.3.13 establishes the equivalence of A and B. The equivalence of B and C is trivial for K_1; we proceed by induction on $n(G)$ for each implication. Suppose $n(G) > 1$.

B $\Rightarrow$ C. Suppose G has a simplicial elimination ordering $v_1, \ldots, v_n$. Since $v_2, \ldots, v_n$ is a simplicial elimination ordering for $G - v_1$, the induction hypothesis yields a subtree representation of $G - v_1$ in a host tree T. Since v_1 is simplicial in G, the set $S = N_G(v_1)$ induces a clique in $G - v_1$. The subtrees of T assigned to vertices of S are pairwise intersecting; by Lemma 8.1.7, they have a common vertex x. Extend T to a tree T' by adding a leaf y adjacent to x, and add the edge xy to the subtrees representing vertices of S. Represent v_1 by the subtree consisting only of y. This completes a subtree representation of G in T'.

C $\Rightarrow$ B. Let T be a smallest tree in which G has a subtree representation, with each $v \in V(G)$ represented by $T(v)$. If x, y are adjacent in T, then G must have a vertex u such that $T(u)$ contains x but not y; otherwise, contracting xy into y would produce a representation in a smaller tree. Suppose x is a leaf of T, and let u be a vertex such that $T(u)$ contains x but not its neighbor. The subtrees for all neighbors of u in G must contain x, and hence they are pairwise intersecting. We conclude that u is simplicial in G. Deleting $T(u)$ yields a subtree representation of $G - u$. By the induction hypothesis, $G - u$ has a simplicial elimination ordering, which completes the simplicial elimination ordering of G. $\square$

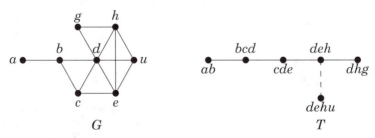

Because the class of chordal graphs is hereditary, a simplicial elimination ordering can start with any simplicial vertex. Naively, we could examine neighborhoods to find a simplicial vertex, delete it, and iterate; this tests for chordal graphs in time $O(n(G)^3)$. Rose-Tarjan-Lueker [1976] found a faster algorithm, which was simplified further by Tarjan in an unpublished manuscript. By using vertex labels, the algorithm produces a vertex numbering for the input graph, and this is a simplicial construction ordering (the reverse of a simplicial elimination ordering) if and only if the graph is chordal. The algorithm was published along with several applications in Tarjan-Yannakakis [1984]; we follow Golumbic [1984].

8.1.9. Algorithm. *Maximum Cardinality Search (MCS)*
Input: A graph G.
Idea: A construction ordering should quickly introduce dense clusters of vertices involving the start vertex. For each unnumbered vertex v, maintain a label $l(v)$ that equals its degree among the vertices already numbered. The output is a vertex numbering (a bijection f from $V(G)$ to $\{1, \ldots, n(G)\}$).
Initialization: Assign label 0 to every vertex. Set $i = 1$.
Iteration: Arbitrarily select an un-numbered vertex with maximum label. Number this vertex i and add one to the label of its neighbors. Augment i and iterate. □

8.1.10. Example. The first vertex chosen in the MCS order is arbitrary. An application of MCS to the graph G on page 294 could start by setting $f(c) = 1$ and hence $l(b) = l(d) = l(e) = 1$. Next we could select $f(e) = 2$ and update $l(d) = 2$, $l(h) = l(u) = 1$. Now d is the only vertex with label as large as 2, and hence $f(d) = 3$. We update $l(b) = l(f) = l(u) = 2$, $l(g) = 1$, $l(a) = 0$. Continuing the procedure can produce the order c, e, d, b, f, g, a, u in increasing order of f. This is a simplicial construction ordering, and u, a, g, f, b, d, e, c is a simplicial elimination ordering. □

8.1.11. Theorem. (Tarjan [1976]). A simple graph G is chordal if and only if the numbering $v_1, \ldots, v_n$ produced by the Maximum Cardinality Search algorithm is a simplicial construction ordering of G.

Proof. If MCS produces a simplicial construction ordering, then G is chordal. Conversely, suppose G is chordal, and let $f \colon V(G) \to [n]$ be the numbering produced by MCS. A *bridge* of f is a chordless path of length at least 2 whose lowest numbers occur at the endpoints. We prove first that f has no bridge. If f has a bridge, let $P = u, v_1, \ldots, v_k, w$ be a bridge that minimizes $\max\{f(u), f(w)\}$. By symmetry, we may assume $f(u) > f(w)$ (f is used as the vertical coordinate to position the vertices in the illustration below).

Since u is numbered in preference to v_k at time $f(u)$, and w is already numbered at that time, there exists a vertex $x \in N(u) - N(v_k)$ with $f(x) < f(u)$. Letting $v_0 = u$, set $r = \max\{j : x \leftrightarrow v_j\}$. The path $P' = x, v_r, \ldots, v_k, w$ is chordless, since $x \leftrightarrow w$ would complete a chordless cycle. Since both of $f(x), f(w)$ are less than $f(u)$, P' is a bridge that contradicts the choice of P. Hence f has no bridge.

With this claim, the proof follows by induction on $n(G)$. It suffices to show that v_n is simplicial, since the application of MCS to $G - v_n$ produces the same numbering $v_1, \ldots, v_{n-1}$ that leaves v_n at the end. If v_n is not simplicial, then v_n has nonadjacent neighbors u, w, in which case u, v_n, w is a bridge of f. $\square$

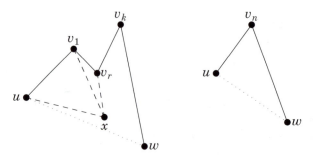

A careful implementation of this algorithm runs in time $O(n(G) + e(G))$. We maintain each set S_j consisting of vertices with label j as a doubly-linked list. For each vertex we also store its label and a pointer to its position in the lists. When v is numbered, in time $O(1 + d(v))$ we remove v from its list and move its neighbors into the next higher lists. To complete the test for G, we must also check whether the MCS order is a simplicial construction ordering (Exercise 7). Simplicial elimination or construction orderings quickly yield optimal colorings, cliques, stable sets, and clique coverings (Exercise 6).

Theorem 8.1.8 computes a subtree representation from a simplicial elimination ordering. When the list of maximal cliques in G is given, Kruskal's algorithm (Section 2.3) can be used to compute a subtree representation without knowing a simplicial elimination ordering.

8.1.12. Definition. A tree T is a *clique tree* of G if there is a bijection between $V(T)$ and the maximal cliques of G such that for each $v \in V(G)$ the cliques containing v induce a subtree of T.

8.1.13. Lemma. Every tree of minimum order in which G has a subtree representation is a clique tree of G.

Proof. Suppose T is a host tree of minimum order in which G has a subtree intersection representation. By Lemma 8.1.7, the vertices of a maximal clique Q in G occur at a common vertex q of T. If the vertices of G assigned to some $q' \in V(T)$ form a proper subclique Q' of Q, then the subtrees for these vertices contain the entire q', q-path in T. The first edge of T on that path can be contracted without changing the intersection graph, which yields a smaller host tree. $\square$

The *weighted intersection graph* of a collection **A** of finite sets is a weighted clique in which the elements of **A** are the vertices and the weight of each edge AA' is $|A \cap A'|$.

8.1.14. Theorem. (Acharya-Las Vergnas [1982]) Suppose $M(G)$ is the weighted intersection graph of the set of maximal cliques $\{Q_i\}$ of a simple graph G. If T is a spanning tree of $M(G)$, then $w(T) \le \Sigma n(Q_i) - n(G)$, with equality if and only T is a clique tree.

Proof. (McKee [1993]) Suppose T is a spanning tree of $M(G)$. Let T_v be the subgraph of T induced by $\{Q_i : v \in Q_i\}$. Each vertex $v \in V(G)$ contributes once to the weight of T for each edge of T_v; hence $w(T) = \Sigma_{v \in V(G)} e(T_v)$. Each T_v is a forest, so $e(T_v) \le n(T_v) - 1$, with equality if and only if T_v is a tree. The term $n(T_v)$ contributes one to the size of each clique containing v. Summing this yields $w(T) \le \Sigma n(Q_i) - n(G)$. Equality holds if and only if each T_v is a tree, which is true if and only if T is a clique tree. $\square$

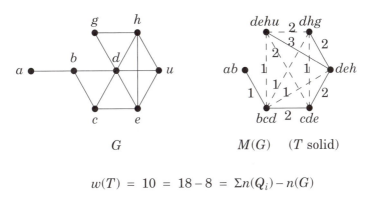

$$G \qquad\qquad M(G) \quad (T \text{ solid})$$

$$w(T) = 10 = 18 - 8 = \Sigma n(Q_i) - n(G)$$

As a consequence of Theorem 8.1.14, it is possible to tell whether G is a chordal graph by finding the maximum weight of a spanning tree in $M(G)$. Furthermore, when G is chordal the clique trees are precisely the maximum weight spanning trees of $M(G)$ (Bernstein-Goodman [1981], Shibata [1988] - see McKee [1993] for related material).

OTHER CLASSES OF PERFECT GRAPHS

Interval graphs are the intersection graphs of collections of intervals on a line; by Proposition 5.1.11, they are perfect. They arise in linear scheduling problems having constraints on concurrent events. Example 5.1.10 describes an application to register allocation in computers; we mention several others.

8.1.15. Example. *Classical applications of interval graphs.*

Analysis of DNA chains. Interval graphs were invented for the study of DNA. Benzer [1959] established the linearity of the chain for higher organisms. Each gene is encoded as an interval, except that the relevant interval may contain a dozen or more irrelevant junk pieces called "introns" among the relevant pieces called "exons". Under the hypothesis that mutations arise from alterations of connected segments, changes in traits of microorganisms can be studied to determine whether their determining amino-acid sets could intersect. This establishes a graph with traits as vertices and "common alterations" as edges. Under the hypotheses of linearity and contiguity, the graph is an interval graph, and this aids in locating genes along the DNA sequence.

Phasing of traffic lights. Given traffic streams at an intersection, a traffic engineer (or a person with common sense) can specify which pairs of streams are allowed to flow simultaneously. If the cycle has an "all-stop" moment, then each period of the cycle can be treated as an interval. In this case, the intersection graph of the green-light intervals for the various streams yields an interval graph that must be a subgraph of the graph of allowable pairs. These graphs can be examined to optimize some criterion such as average waiting time. Roberts [1978] contains further discussion.

Archeological seriation. Given pottery samples at an archeological dig, we seek a time-line of what styles were in use when. We assume that each style was used during a single time interval and that two styles appearing in the same grave were used concurrently. We form a graph with the styles as vertices, making two styles adjacent if they appear together in a grave. If this graph is an interval graph, then its interval representations are the possible time-lines. Otherwise, the information is incomplete, and the desired interval graph requires additional edges. □

We present two characterizations of interval graphs.

8.1.16. Definition. A matrix of 0's and 1's has the *consecutive 1's property* (for columns) if its rows can be permuted so that the ones in each column appear consecutively. The *clique-vertex incidence matrix* of a graph is the incidence matrix with rows indexed by the maximal cliques and columns indexed by the vertices.

8.1.17. Theorem. The following equivalent conditions on a graph G characterize the interval graphs.
A) G has an interval representation.
B) G is a chordal graph, and $\overline{G}$ is a comparability graph.
C) The clique-vertex incidence matrix of G has the consecutive 1's property.

Proof. We leave $A \Rightarrow B$ and $A \Leftrightarrow C$ to Exercise 21, proving $B \Rightarrow C$ here. Suppose G is a chordal graph and F is a transitive orientation of $\overline{G}$. We use F and the lack of chordless cycles in G to establish an ordering on the maximal cliques of G that exhibits the consecutive 1's property for the clique-vertex incidence matrix M.

Suppose Q_i and Q_j are maximal cliques in G. By maximality, each vertex of one clique has a non-neighbor in the other. Suppose that under F, some edge of $\overline{G}$ points from Q_i to Q_j and some edge of $\overline{G}$ points from Q_j to Q_i. If these edges have a common vertex, then the transitivity of F forces an edge of a clique in G to belong to $\overline{G}$. Hence the situation is as on the left below, with the (dashed) edges of F having four distinct vertices. If the two remaining pairs among these four vertices form edges in G, then G has an induced C_4. Hence at least one diagonal is in $\overline{G}$, but each possible orientation of it in F contradicts transitivity. We conclude that all the edges of $\overline{G}$ between vertex sets Q_i and Q_j point in the same direction in F.

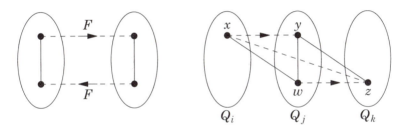

This allows us to define a tournament T with vertices corresponding to the maximal cliques of G. We put $Q_i \to Q_j$ in T when all edges of F between Q_i and Q_j point from Q_i to Q_j. By the preceding paragraph, T is an orientation of a complete graph. We claim that T is transitive. To prove this we need to show that $Q_i \to Q_j$ and $Q_j \to Q_k$ imply $Q_i \to Q_k$. Suppose $x \to y$ and $w \to z$ in F with $x \in Q_i$, $y, w \in Q_j$, and $z \in Q_k$. If $y = w$, transitivity of F immediately implies $x \to z$. Otherwise, consider the pair xz as in the figure on the right above. Joining x and z in G would form an induced C_4 in G, so $x \not\leftrightarrow z$. Hence this pair appears in F, and it must be directed from $x \to z$ to avoid violating transitivity. This implies $Q_i \to Q_k$ in T.

A transitive tournament specifies a unique linear ordering of the vertices consistent with the edges; use the transitive tournament T to order the rows of M as $Q_1 \to \cdots \to Q_m$. Suppose that under this ordering there is some column x where the 1's do not appear consecutively. Then we have Q_i, Q_j, Q_k such that $i < j < k$, $x \in Q_i, Q_k$, $x \notin Q_j$. Since $x \notin Q_j$, the clique Q_j must have some vertex y not adjacent to x, else Q_j could absorb x and would not be maximal. Now $x \in Q_i$ implies $x \to y$ in F, and $x \in Q_k$ implies $y \to x$ in F, which cannot both happen. $\square$

The interval graphs form a relatively small class of perfect graphs. We next discuss larger classes that maintain some of the nice properties of chordal graphs and comparability graphs.

8.1.18. Definition. *Classes of perfect graphs* (conditions on odd cycles apply only for length at least 5).
o-triangulated: every odd cycle has a non-crossing pair of chords.
parity: every odd cycle has a crossing pair of chords.
Meyniel: every odd cycle has at least two chords.
weakly chordal: no induced cycle of length at least 5 in G or $\overline{G}$.
strongly perfect: every induced subgraph has a stable set meeting all its maximal cliques.

Gallai [1962] proved that o-triangulated graphs are perfect; this class includes all bipartite graphs. Every chordal graph is o-triangulated (Exercise 28) and weakly chordal (Exercise 33), and it is immediate that o-triangulated graphs and parity graphs are Meyniel graphs. Meyniel graphs are perfect (Meyniel [1976], Lovász [1983]); indeed, they are also strongly perfect (Ravindra [1982]).

Parity graphs, shown to be perfect in Olaru [1969] and Sachs [1970], carry that name due to a later characterization by Burlet and Uhry [1984]: G is a parity graph if and only if, for every pair $x, y \in V(G)$, the chordless x, y-paths are all even or all odd (Exercise 29).

8.1.19. Example. The graphs below exhibit some differences among these classes. Here $\mathbf{T, C, O, P, M, W}$ respectively denote the classes of chordal (triangulated), comparability, o-triangulated, parity, Meyniel, and weakly chordal graphs. □

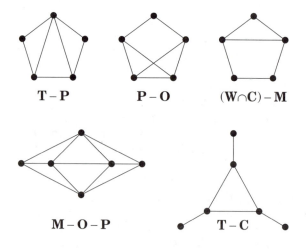

T – P   P – O   (W∩C) – M

M – O – P   T – C

Strongly perfect graphs appeared first in Berge-Duchet [1984]. The distinction between maximal cliques and maximum cliques is crucial. Changing maximal to maximum yields a weaker requirement that is equivalent to γ-perfection; a stable set meeting all maximum cliques can be used as the first color class in an $\omega(G)$-coloring constructed inductively. Thus strongly perfect graphs form a class of perfect graphs.

Chordal graphs are strongly perfect, but not every weakly chordal graph is strongly perfect (Exercise 33). The coloring algorithm for comparability graphs in Example 5.3.17 shows that comparability graphs are strongly perfect. Since every maximal path in a transitive orientation of G contains a source, the sources in the orientation form a stable set in G that meets all maximal cliques. The class is hereditary, so the statement holds for all induced subgraphs. Our next class contains all chordal graphs and comparability graphs and is properly contained in the class of strongly perfect graphs (Exercises 30,32).

8.1.20. Definition. A *perfect order* on a graph (Chvátal [1984]) is a vertex ordering such that applying the greedy coloring algorithm to the ordering inherited by any induced subgraph produces an optimal coloring of the subgraph. A *perfectly orderable graph* is a graph having a perfect order. Given an orientation of G, an *obstruction* is an induced 4-vertex path a, b, c, d whose first and last edges are oriented $a \leftarrow b$ and $d \leftarrow c$.

The proof of perfection for perfectly orderable graphs uses a forbidden substructure characterization of perfect orders. The orientation obtained from a perfect order L by putting $u \leftarrow v$ if $u < v$ in L has no obstruction, because the greedy coloring induced on such a P_4 would use three colors instead of two. Chvátal proved that a graph is perfectly orderable if and only if it has an orientation with no obstruction.

8.1.21. Example. *Perfectly orderable graphs.* Some Meyniel graphs are not perfectly orderable, and vice versa (Exercise 31). Nevertheless, Chvátal's characterization of perfectly orderable graphs implies that every comparability graph is perfectly orderable and every chordal graph is perfectly orderable. If we orient each edge of a chordal graph from earlier to later according to a perfect elimination ordering, then there is no induced $u \leftarrow v \rightarrow w$. A transitive orientation of a comparability graph has an no induced $u \rightarrow v \rightarrow w$. Every ordering with an obstruction has an induced $u \rightarrow v \rightarrow w$ and an induced $u \leftarrow v \rightarrow w$. Hence if G is a comparability graph or a chordal graph, then G has an ordering with no obstruction and is perfectly orderable. $\square$

Given a vertex ordering L of G, the orientation of G *associated with L* is obtained by putting $u \leftarrow v$ when $u \leftrightarrow v$ in G and $u < v$ in L. A vertex ordering L of G is *obstruction-free* if the orientation associated with L is obstruction-free.

8.1.22. Lemma. (Chvátal [1984]) Suppose that G has a clique Q and a stable set S disjoint from Q, and suppose that for each vertex $w \in Q$ there is a vertex $p(w) \in S$ such that $p(w) \leftrightarrow w$. If L is an obstruction-free ordering of G such that $p(w) < w$ for all $w \in Q$, then some $p(w) \in S$ is adjacent to all of Q.

Proof. We use induction on $n(G)$. The claim holds vacuously for $n(G) = 1$, so we may assume $n(G) > 1$. For each $w \in Q$, the graph $G - w$ satisfies the hypotheses using the clique $Q - w$ and the stable set $\{p(u): u \in Q - w\}$. By the induction hypothesis, there is a vertex $w^* \in Q - w$ such that $p(w^*) \leftrightarrow Q - w$. We obtain $p(w^*)$ adjacent to all of Q for some w unless $p(w^*) \nleftrightarrow w$ for every $w \in Q$. This assigns a unique w^* to every w, since $p(w^*)$ is nonadjacent only to w among Q, so setting $\sigma(w) = w^*$ defines a permutation σ on the vertices of Q.

This permutation leads to an obstruction, which proves that this case cannot occur and hence that $p(w^*) \leftrightarrow w$ for some w. Let v be the least vertex of Q in L. Let $b, c \in Q$ be the vertices such that $b^* = v$ and $c^* = b$. Let $a = p(b)$ and $d = p(v)$. Because $p(w^*) \nleftrightarrow w$, we have $d \nleftrightarrow b$ and $a \nleftrightarrow c$, which implies $a \neq d$ in the stable set S and yields the picture below, with each edge oriented toward the lesser vertex in L.

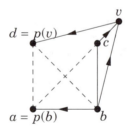

Because $d = p(b^*)$, the only vertex of Q nonadjacent to d is b; i.e. $c \leftrightarrow d$. Since $d = p(v) < v \leq c$ in L, this edge is oriented as $c \to d$, which implies that a, b, c, d induce an obstruction. $\square$

8.1.23. Theorem. (Chvátal [1984]) A vertex ordering of a simple graph G is a perfect order if and only if it is obstruction-free, and every graph with such an ordering is perfect.

Proof. Since the class of graphs having obstruction-free orderings is hereditary (the inherited ordering for an induced subgraph is obstruction-free), it suffices to show that an obstruction-free ordering L gives a greedy coloring of G that is optimal. If the greedy coloring uses k colors for the ordering L, then optimality can be established by showing that G has a k-clique, which also proves perfection.

Let $f: V(G) \to [k]$ be the coloring generated by the greedy algorithm with this ordering. Let i be the smallest integer such that G has a clique consisting of vertices $w_{i+1}, \ldots, w_k$ such that $f(w_j) = j$. Since f uses color k on some vertex, some such clique exists. If $i = 0$, then G has a k-clique. Suppose $i > 0$. For each w_j there is a vertex $p(w_j)$ such that $p(w_j) < w_j$ in L and $f(p(w_j)) = i$; otherwise the greedy coloring would use a lower color on w_j. Since the vertices in $S = \{p(w_{i+1}), \ldots, p(w_k)\}$ all have color i, S is a stable set. Hence the conditions of the lemma are satisfied, and there is a vertex of S that can be added to the clique and called w_i, which contradicts the minimality of i. □

Next we consider a different way of generating perfect graphs. An operation that preserves perfection can expand a class of perfect graphs into a larger class. Vertex multiplication, which expands each vertex into an independent set, is such a property. We generalize this. If $V(G) = \{v_1, \ldots, v_n\}$, and $H_1, \ldots, H_n$ are pairwise disjoint graphs, then the *composition* $G[H_1, \ldots, H_n]$ is the graph $H_1 + \cdots + H_n$ together with $\{xy: x \in V(H_i), y \in V(H_j), v_i v_j \in E(G)\}$. The special case $G[\overline{K}_{h_1}, \ldots, \overline{K}_{h_n}]$ is $G \circ h$. The example below uses $H_1 = 2K_1$, $H_2 = K_2 + K_1$, $H_3 = P_3$, $H_4 = K_2$, and $G = K_{1,3}$ with central vertex v_1.

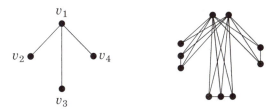

Lovász proved that composition preserves perfection. This is one corollary of Chvátal's Star-Cutset Lemma.

8.1.24. Definition. A *star-cutset* of G is a vertex cut S containing a vertex x adjacent to all of $S - \{x\}$. A *minimal imperfect graph* is an imperfect graph whose proper induced subgraphs are all perfect.

8.1.25. Lemma. (The Star-Cutset Lemma Lemma) If G has no stable set intersecting every maximum clique, and every proper induced subgraph of G is $\omega(G)$-colorable, then G has no star-cutset.

Proof. Suppose G has a star-cutset C, with w adjacent to all of $C - \{w\}$. Since $G - C$ is disconnected, we have a partition of $V(G - C)$ into nonempty sets V_1, V_2 such that no edge joins them. Let $G_i = G[V_i \cup C]$, and let f_i be a proper $\omega(G)$-coloring of G_i. Let S_i be the set of vertices in G_i with the same color in f_i as w; this includes w but no other vertex of C. Since there are no edges between V_1 and V_2, the union $S = S_1 \cup S_2$ is a stable set.

If Q is a clique in $G - S$, then Q is contained in $G_1 - S_1$ or in $G_2 - S_2$. Since f_i provides an $\omega(G) - 1$-coloring of $G_i - S_i$, we have $|Q| \leq \omega(G) - 1$. Since this applies to every clique Q in $G - S$, the stable set S meets every $\omega(G)$-clique of G, which contradicts the hypotheses. $\square$

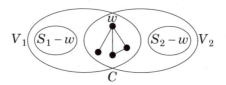

8.1.26. Theorem. (The Star-Cutset Lemma, Chvátal [1985]) No minimal imperfect graph has a star-cutset; equivalently, identification at star-subgraphs preserves perfection.

Proof. If G is a minimal imperfect graph, then $\chi(G) > \omega(G)$ and deletion of any stable set S leaves a perfect graph. Hence we have $1 + \omega(G) \leq \chi(G) \leq 1 + \chi(G - S) = 1 + \omega(G - S) \leq 1 + \omega(G)$. Hence $\omega(G - S) = \omega(G)$, which means that no stable set meets every maximum clique. Furthermore, since G is minimally imperfect, every proper induced subgraph G' satisfies $\chi(G') = \omega(G') \leq \omega(G)$, i.e. it is $\omega(G)$-colorable. By Lemma 8.1.25, we conclude that G has no star-cutset. $\square$

8.1.27. Corollary. (Replacement Lemma - Lovász [1972b]) Every composition of perfect graphs is perfect.

Proof. A composition can be constructed by a sequence of substitutions in which a single vertex v of G_1 is replaced by a graph G_2 and all edges added between $V(G_2)$ and $U = N_{G_1}(v)$ to form a graph G. Hence it suffices to prove that this operation preserves perfection. If the resulting graph G is not perfect, then it contains a minimal imperfect induced subgraph F. Such a subgraph cannot be contained in G_1 or G_2, which forces it to have at least two vertices of G_2 and at least one vertex of G_1.

If F has no vertex of G_1 outside U, then F is the join of two perfect graphs. Join preserves perfection, since $\chi(H_1 \vee H_2) = \chi(H_1) + \chi(H_2)$ and $\omega(H_1 \vee H_2) = \omega(H_1) + \omega(H_2)$ for every H_1, H_2. If F has a vertex of G_1 outside U, then $V(F) \cap U$ together with one vertex of G_2 in F is a star-cutset of F. Hence there is no minimal imperfect subgraph F. $\square$

The Star-Cutset Lemma also yields perfection of weakly chordal graphs. Hayward [1985] proved that G or $\overline{G}$ has a star-cutset when G is a weakly chordal graph that is not a clique or stable set. With the Star-Cutset Lemma and the Perfect Graph Theorem, this implies that no weakly chordal graph is a minimal imperfect graph. Since the class is hereditary, he concluded that every weakly chordal graph is perfect.

IMPERFECT GRAPHS

The *p-critical* graphs are the minimal imperfect graphs. The Strong Perfect Graph Conjecture (SPGC) is the statement that the only p-critical graphs are the odd cycles and their complements. With enough properties of p-critical graphs, perhaps we could prove that only odd cycles and their complements have all these properties; this would prove the SPGC. We begin with simple observations about p-critical graphs, some used earlier in discussing star-cutsets. (This presentation was originally modeled after Shmoys [1981].)

8.1.28. Lemma. If G is p-critical, then G is connected, $\overline{G}$ is p-critical, $\omega(G) \geq 2$, and $\alpha(G) \geq 2$. Furthermore, for every $x \in V(G)$, $\chi(G - x) = \omega(G)$ and $\theta(G - x) = \alpha(G)$.

Proof. G is perfect if and only if every component of G is perfect, and G is perfect if and only if $\overline{G}$ is perfect. Cliques and their complements are perfect. Finally, we observed when discussing star-cutsets that deleting a stable set from a p-critical critical graph cannot decrease the clique number. Since $G - x$ is perfect, we thus have $\chi(G - x) = \omega(G - x) = \omega(G)$. The condition $\theta(G - x) = \alpha(G)$ is this statement for $\overline{G}$. □

More subtle properties of p-critical graphs follow from Lovász's extension of the PGT.

8.1.29. Theorem. (Lovász [1972b]) A graph G is perfect if and only if $\omega(G[A])\alpha(G[A]) \geq |A|$ for all $A \subseteq V(G)$. □

The property "$\omega(G[A])\alpha(G[A]) \geq |A|$ for all $A \subseteq V(G)$" was suggested by Fulkerson; we call it "β-perfection". It follows immediately from α-perfection or γ-perfection; if we can color G with $\omega(G)$ stable sets, then some stable set has at least $n(G)/\omega(G)$ vertices. The proof of the converse involves counting arguments like those we gave for the PGT, but more delicate. Since β-perfection is unchanged under complementation, Theorem 8.1.29 immediately implies the PGT.

8.1.30. Theorem. If G is p-critical, then $n(G) = \alpha(G)\omega(G) + 1$. Furthermore, for every $x \in V(G)$, $G - x$ has a partition into $\omega(G)$ stable sets of size $\alpha(G)$ and a partition into $\alpha(G)$ cliques of size $\omega(G)$.

Proof. When G is p-critical, the condition for β-perfection fails only for the full vertex set $A = V(G)$. Hence $n(G) - 1 \leq \alpha(G - x)\omega(G - x) = \alpha(G)\omega(G) \leq n(G) - 1$ for each $x \in V(G)$, and equality holds throughout. Since $\chi(G - x) = \omega(G - x) = \omega(G)$, the vertices of $G - x$ can be covered by $\omega(G)$ stable sets in $G - x$. These stable sets have size at most $\alpha(G)$, and there are $\alpha(G)\omega(G)$ vertices to be covered, so the coloring partitions

$V(G - x)$ into the desired stable sets. Similarly, $\theta(G - x) = \alpha(G - x) = \alpha(G)$ yields the partition of $V(G - x)$ into $\alpha(G)$ cliques of size $\omega(G)$. $\square$

In studying p-critical graphs, it has been helpful to enlarge the class to include the graphs satisfying the properties in Theorem 8.1.30. Other structural properties have been proved for this class, and we can employ them when proving the SPGC for a particular class of graphs. Padberg [1974] began the study of these graphs. Other definitions were given to extend the class of p-critical graphs, but they turned out to be alternative characterizations of the same class. The definition here originates in Bland-Huang-Trotter [1979].

8.1.31. Definition. Given integers $a, w \geq 2$, a graph G is an a, w-*partitionable graph* if it has $n = aw + 1$ vertices and for each $x \in V(G)$ the subgraph $G - x$ has a partition into a cliques of size w and a partition into w stable sets of size a.

8.1.32. Theorem. (Buckingham-Golumbic [1983]) A graph G of order $n = aw + 1$ is an a, w-partitionable graph if and only $\chi(G - x) = w$ and $\theta(G - x) = a$ for every $x \in V(G)$. Furthermore, $\omega(G) = w$ and $\alpha(G) = a$ for such graphs, and the inequalities $\chi(G - x) \leq w$ and $\theta(G - x) \leq a$ are sufficient.

Proof. Suppose G is partitionable. Since $G - x$ has a clique of size w and a coloring of size w, we have $\chi(G - x) = w$. Since $a \geq 2$, G is not a clique. Choosing $x, y \in V(G)$ with $x \not\leftrightarrow y$, every clique containing x appears in $G - y$, so $\omega(G - y) = w$ prevents $\omega(G) > \omega(G - x)$. The same arguments for $\bar{G}$ yield the equalities involving a.

Conversely, suppose that $\chi(G - x) \leq w$ and $\theta(G - x) \leq a$ for every $x \in V(G)$. The minimal coloring of $G - x$ partitions it into at most w stable sets. If any of these stable sets has more than a vertices, then $\theta(G - x) \leq a$ is impossible. Hence covering the aw vertices of $G - x$ requires using w stable sets of size a. Similarly, the covering of $G - x$ by a cliques yields the desired clique partition. $\square$

Henceforth we use w interchangeably with $\omega(G)$ and a interchangeably with $\alpha(G)$. Theorems 8.1.30 and 8.1.32 imply that every p-critical graph is partitionable and that every partitionable graph is imperfect. Furthermore, G is a, w-partitionable if and only if $\bar{G}$ is w, a-partitionable.

8.1.33. Example. *Circulant graphs.* The circulant graph C_n^d is constructed by placing n vertices on a circle and making each vertex adjacent to the d nearest vertices in each direction on the circle. When $d = 1$, $C_n^d = C_n$. We view the vertices as the integers modulo n, in order.

The graph C_{10}^2, illustrated on the left below, is neither perfect nor p-critical, since the vertices $0, 2, 4, 6, 8$ induce C_5, but it is $3,3$-partitionable. When vertex i is removed, the unique partition of the remaining 9 vertices into three triangles is $\{(i + 1, i + 2, i + 3), \ (i + 4, i + 5, i + 6), (i + 7, i + 8, i + 9)\}$, and the unique partition into three stable sets is $\{(i + 1, i + 4, i + 7), (i + 2, i + 5, i + 8), (i + 3, i + 6, i + 9)\}$.

More generally, C_{aw+1}^{w-1} is a, w-partitionable, since each set of w consecutive vertices (i.e., in $G - x$) forms a clique, and each set of a vertices separated by jumps of length w forms a stable set. Chvátal proved that C_{aw+1}^{w-1} is p-critical if and only if $w = 2$ or $a = 2$ (Theorem 8.1.45) so the SPGC can be reduced to the statement that G is p-critical if and only if $G = C_{aw+1}^{w-1}$ for some a, w. $\square$

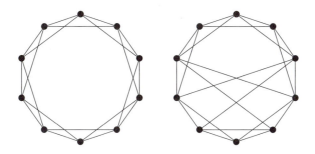

8.1.34. Example. *A non-circulant partitionable graph.* We can construct more partitionable graphs by adding unimportant edges to circulant graphs. In C_{10}^2, we can add any of the diagonals without changing the set of maximum cliques or set of maximum stable sets, so the resulting graph is still partitionable. The SPGC would still follow if all partitionable graphs were circulants plus such unimportant edges (Theorem 8.1.48), but there are others. Chvátal-Graham-Perold-Whitesides [1979] found a way to produce such graphs, such as the graph on the right above (also in Huang [1976]). Every edge in this graph belongs to a maximum clique, but it has two more edges than C_{10}^2. The partitions demonstrating that this is partitionable differ from those used for C_{10}^2 (Exercise 35). $\square$

8.1.35. Example. *Further properties.* Circulant graphs suggest additional properties. The graph C_{aw+1}^{w-1} has exactly n w-cliques, each consisting of w consecutive vertices on the cycle, and each vertex x belongs to w consecutive w-cliques. Vertices of stable sets must be separated by at least w, so in a maximum stable set there are $a - 1$ pairs of consecutive vertices separated by distance w and one pair separated by distance $w + 1$. There are n places for the larger gap and hence n such stable sets. For a maximum stable set containing x, there are a places for the larger gap, so x belongs to a maximum stable sets. Finally, the only way for a w-clique to miss an a-staset is to fit inside the gap of length $w + 1$, so there is a pairing $\{(S_i, Q_i)\}$ between the maximum stable sets and maximum cliques such that $S_i \cap Q_j = \varnothing$ if and only if $i = j$. $\square$

These properties comprise the characterization in the next theorem. The arguments are due originally to Padberg [1974], who used them to give a polyhedral characterization of perfect graphs. They can be viewed as "combinatorial linear algebra", in which combinatorial conclusions are drawn from linear algebraic properties of matrices. Other characterizations of partitionable graphs appeared in Bland-Huang-Trotter [1979], Golumbic [1980, p58-62], Tucker [1977], Chvátal-Graham-Perold-Whitesides [1979], and Buckingham [1980].

8.1.36. Theorem. A graph G with $n = aw + 1$ vertices is an a, w-partitionable graph if and only if the following conditions all hold:
  1) $\alpha(G) = a$ and $\omega(G) = w$, and each vertex of G belongs to a stable sets of size a and to w cliques of size w.
  2) G has exactly n maximum cliques and exactly n maximum stable sets, which can be indexed so that $S_i \cap Q_j = \varnothing$ if and only if $i = j$ (we say that S_i and Q_i are *mates*).

Proof. *Necessity.* We have proved that $\chi(G - x) = w = \omega(G)$ and that $\theta(G - x) = a = \alpha(G)$ for each $x \in V(G)$. Choose a maximum clique $Q \subset G$; it has w vertices. For each $x \in Q$, $G - x$ has a partition into a cliques of size w. From Q and the w partitions, we obtain a list of $n = aw + 1$ maximum cliques $Q_1, \ldots, Q_n$. Each vertex outside Q appears in one clique in each partition. Each vertex of Q appears in Q and once in $w - 1$ other partitions. Hence every vertex appears in exactly w cliques in the list.

For each Q_i, we obtain a maximum stable set S_i disjoint from Q_i. Given a vertex $x \in Q_i$, we have a proper w-coloring of $G - x$. The stable sets of the coloring can meet Q_i only in the $w - 1$ vertices other than x, so one of them misses Q_i; label it S_i. We still must show that $Q_1, \ldots, Q_n$ are distinct, that $S_1, \ldots, S_n$ are distinct, that G has no other maximum cliques or stable sets, and that $S_i \cap Q_j \neq \varnothing$ if $i \neq j$.

Let A be the clique-vertex incidence matrix with $a_{ij} = 1$ if $x_j \in Q_i$ and $a_{ij} = 0$ otherwise. We can prove that $Q_1, \ldots, Q_n$ are distinct by proving that the rows of A are distinct, which we can do by proving that A is non-singular. Similarly, let B be the incidence matrix for stable sets, with $b_{ij} = 1$ if $x_j \in S_i$ and $b_{ij} = 0$ otherwise. These are square matrices of order n, and if AB^T is non-singular then the product rule for determinants implies that A, B are both non-singular.

The ijth entry of AB^T is the dot product of row i of A with row j of B, which equals $|Q_i \cap S_j|$. By construction, this is 0 when $i = j$. Since cliques and stable sets intersect at most once, to show that all other entries equal 1 we need only show that each column of AB^T has sum $n - 1$. Multiplying by the row vector $\mathbf{1}_n^T$ on the left computes these sums. We constructed A so that each column has exactly w 1's and B so that each row has exactly a 1's. Therefore, $\mathbf{1}_n^T(AB^T) = (\mathbf{1}_n^T A)B^T = w\mathbf{1}_n^T B^T = wa\mathbf{1}_n = (n-1)\mathbf{1}_n^T$. Hence $AB^T = J - I$, where J is the matrix of all 1's. Since $J - I$ is non-singular, the cliques are distinct and the stable sets are distinct, and $Q_i \cap S_j \neq \varnothing$ if and only if $i \neq j$.

To prove that G has no other maximum cliques, we let c be the incidence vector of a maximum clique Q and show that c must be a row of A. Since A is nonsingular, its rows span $\mathbb{R}^n$ and hence c. Writing $c = tA$, we need only show that $t = (0 \ldots 010 \ldots 0)$. Because $A(\omega^{-1}J - B^T) = \omega^{-1}\omega J - (J - I) = I$, we can compute that $t = cA^{-1} = c(\omega^{-1}J - B^T) = \omega^{-1}cJ - cB^T = \mathbf{1}_n^T - cB^T$, using in the last step that c has ω 1's. Since each column of B^T is the incidence vector of a maximum stable set, coordinate i of cB^T equals $|Q \cap S_i|$. Hence t is a 0,1-vector; with $c = tA$, this means that c is a sum of rows of A. Since c has exactly ω 1's and no 2's, only one row can be used, so c is a row of A and $Q_1, \ldots, Q_n$ are the only maximum cliques.

The same argument applied to $\overline{G}$ shows that the distinct stable sets $S_1, \ldots, S_n$ are the only maximum stable sets in G, and also that each vertex appears in a of these.

Sufficiency. By Theorem 8.1.32, it suffices to prove that $\chi(G - x) \leq w$ and $\theta(G - x) \leq a$ for all $x \in V(G)$. Given the cliques $\{Q_i\}$ and $\{S_i\}$ as guaranteed by condition (2), define the incidence matrices A, B as above. Since each vertex appears in a stable sets among $\{S_i\}$, we have $JB = aJ = BJ$. The intersection requirements for Q_i and S_j imply $AB^T = J - I$. This is non-singular, so B is non-singular and $A^T B = B^{-1}BA^T B = B^{-1}(J - I)B = B^{-1}BJ - I = J - I$.

Combinatorially, the row of the product $A^T B = J - I$ corresponding to $x \in V(G)$ states that $V(G - x)$ is the disjoint union of the mates of the w maximum cliques containing x, and hence $\chi(G - x) \leq w$. Similarly, the column corresponding to x states that $V(G - x)$ is the disjoint union of the mates of the a maximum stable sets containing x, and hence $\theta(G - x) \leq a$. $\square$

8.1.37. Corollary. If G is a, w-partitionable and $w = 2$, then $G = C_{2a+1}$; if $a = 2$, then $G = \overline{C}_{2w+1}$. Hence the SPGC reduces to showing that p-critical graphs have $\omega = 2$ or $\alpha = 2$.

Proof. If $\omega = 2$, then every vertex belongs to exactly two cliques of size 2, so G is 2-regular. Furthermore, G is connected and has odd order $(2\alpha + 1)$, so G is an odd cycle. For $a = 2$, consider $\overline{G}$. $\square$

Since $\alpha(G) = a$ and $\omega(G) = w$ for a, w-partitionable graphs, we henceforth use α for a and ω for w.

8.1.38. Theorem. (Tucker [1977]). If G is partitionable and $x \in V$, then the following properties hold.

a) If a clique Q and stable set S are mates, then $x \in S$ if and only if Q appears in a minimum clique cover of $G - x$; similarly $x \in Q$ if and only if S appears in a minimum coloring of $G - x$.

b) The minimum clique covering and minimum coloring of $G - x$ are unique (denoted $\Theta(G - x)$ and $X(G - x)$), with $\Theta(G - x)$ consisting of the mates of the maximum stable sets containing x and $X(G - x)$ consisting of the maximum cliques containing x.

Proof. (a) Suppose $Q \cap S = \emptyset$. If $x \in S$, the $\alpha - 1$ vertices of $S - x$ can touch only $\alpha - 1$ cliques in a partition of $G - x$; the remaining clique must be the mate Q. Conversely, if Q is in a clique cover of $G - x$, then S contains at most one vertex each from the other $\alpha - 1$ cliques in the cover, and hence $|S| = \alpha$ implies $x \in S$. (a) states that every mate of a stable set containing x appears in every clique partition of $G - x$, and the uniqueness follows. $\square$

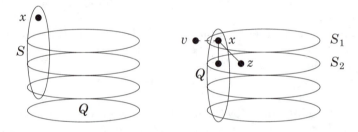

8.1.39. Theorem. (Buckingham-Golumbic [1983]). If x, v are vertices of a partitionable graph G, then

1) For $x \not\leftrightarrow v$, every maximum clique containing x consists of a vertex from each set in $X(G - v)$; for $x \leftrightarrow v$, every maximum stable set containing v consists of a vertex from each set in $\Theta(G - v)$.

2) $2\omega - 2 \le d(x) \le n - 2\alpha + 1$.

Proof. (1) Let Q be a maximum clique containing x. Since $v \notin Q$ and the ω stable sets of $X(G - v)$ cover Q, each stable set contains exactly one vertex of Q. The same argument applies to $\overline{G}$.

(2) Select a vertex $v \not\leftrightarrow x$. Let S_1 be the stable set of $X(G - v)$ containing x, and let S_2 be another stable set of $X(G - v)$. Choose $z \in N(x) \cap S_2$. In the clique cover $\Theta(G - z)$, some clique Q contains x. Since $v \not\leftrightarrow x$, (1) implies that Q has a vertex in each stable set of $X(G - v)$, including S_2. Since $Q \in \Theta(G - z)$ implies $z \notin Q$, we have $|S_2 \cap N(x)| \ge 2$. Since S_2 is arbitrary among the $\omega - 1$ stable sets of $X(G - v)$ other than S_1, we have $d(x) = |N(x)| \ge 2(\omega - 1)$. The same argument in $\overline{G}$ yields $n - 1 - d(x) = |N_{\overline{G}}(x)| \ge 2(\alpha - 1)$. $\square$

These degree bounds hold with equality for circulant graphs.

8.1.40. Definition. An edge of a graph is *critical* if deleting it increases the independence number. A pair of nonadjacent vertices is *co-critical* if adding it increases the clique number.

The characterization of critical edges in partitionable graphs is implicit in the work of several authors.

8.1.41. Theorem. For an edge xy in a partitionable graph G, the following statements are equivalent.
A) xy is a critical edge.
B) For some $S \subset V(G)$, $S \cup \{y\} \in X(G - x)$ and $S \cup \{x\} \in X(G - y)$.
C) xy belongs to $\omega - 1$ maximum cliques.

Proof. B $\Rightarrow$ A. $S \cup \{x, y\}$ is a stable set of size $\alpha + 1$ in $G - xy$.

A $\Rightarrow$ C. If xy is critical, then there is a set S such that $S \cup \{x\}$ and $S \cup \{y\}$ are maximum stable sets in G. Hence every maximum clique containing x but not y is disjoint from $S \cup \{y\}$. Since there are ω maximum cliques containing x and only one maximum clique disjoint from $S \cup \{y\}$, the remaining $\omega - 1$ maximum cliques containing x must also contain y.

C $\Rightarrow$ B. The stable sets in the unique coloring of $G - x$ are the mates of the cliques containing x. Since xy belongs to $\omega - 1$ maximum cliques, the mates of these $\omega - 1$ cliques belong to both $X(G - x)$ and $X(G - y)$. this leaves only $a + 1$ vertices in the graph, consisting of the vertices x, y and a stable set S such that $S \cup \{y\} \in X(G - x)$ and $S \cup \{x\} \in X(G - y)$. □

8.1.42. Corollary. Suppose G is a partitionable graph. If xy is an edge appearing in no maximum clique, then $G - xy$ is partitionable. If x, y is a nonadjacent pair appearing in no maximum stable set, then $G + xy$ is partitionable.

Proof. By complementation, we need only prove the first statement. If we delete an edge appearing in no maximum clique, then by Theorem 8.1.41 it is not a critical edge, and we have $\omega(G - xy) = \omega(G)$ and $\alpha(G - xy) = \alpha(G)$. Since we have not destroyed any maximum clique and have not created a bigger stable set, we can use the optimal coloring and clique partition of $G - u$ to conclude that $\chi(G - xy - u) \le \omega$ and $\theta(G - xy - u) \le \alpha$. Hence $G - xy$ is partitionable, by Theorem 8.1.32. □

The discussion in Example 8.1.34 suggests that edges appearing in no maximum clique are uninteresting "junk". Corollary 8.1.42 assures us that "junk is junk". The graphs in Examples 8.1.33 and 8.1.34 have no junk.

THE STRONG PERFECT GRAPH CONJECTURE

We have been proving properties of partitionable graphs in a "top down" approach to the SPGC, trying to find enough properties to eliminate all but odd cycles and their complements as p-critical graphs. The "bottom up" approach is to verify that the SPGC holds on larger and larger classes of graphs, until all are included.

8.1.43. Definition. Induced subgraphs of G isomorphic to odd cycles or their complements are *odd holes* or *odd antiholes* in G, respectively. A graph having no odd hole or antihole is a *Berge graph*.

Another way to prove that a class **G** satisfies the SPGC is to prove that every Berge graph in G is perfect. Alternatively, a hereditary class **G** satisfies the SPGC if the odd cycles and their complements are the only p-critical graphs in **G**. Properties of partitionable graphs lead to proofs of the SPGC for the three families of graphs defined below. It also holds for planar graphs (Tucker [1973]), for toroidal graphs (Grinstead [1981]), for graphs with $\Delta(G) \leq 6$ (Grinstead [1978]), for graphs with $\omega(G) \leq 3$ (Tucker [1977]), and for various classes defined by forbidding a fixed subgraph with four or five vertices (Meyniel [1976], Tucker [1977], Parthasarathy-Ravindra [1976, 1979], Chvátal-Sbihi [1988], Olariu [1989], Sun [1991]).

8.1.44. Definition. A *circular-arc graph* is the intersection graph of a family of arcs of a circle. A *circle graph* is the intersection graph of a family of chords of a circle. A $K_{1,3}$-*free* graph is a graph not having $K_{1,3}$ as an induced subgraph.

Every cycle is both a circle graph and a circular-arc graph, but neither of these classes contains the other (Exercise 39).

One way to prove the SPGC for a class **G** is to show that every partitionable graph in **G** belongs to another class **H** where the SPGC is known to hold. In this role we use the class of circulant graphs.

8.1.45. Theorem. (Chvátal [1976]) Circulant graphs satisfy the SPGC. In particular, a circulant graph C_{aw+1}^w is p-critical if and only if $w = 2$ or $a = 2$, in which case the graph is an odd hole or antihole.

Proof. Let the vertices be $\{v_0, \ldots, v_{aw}\}$. Consider the $2a$ vertices $\{v_{iw+1}, v_{(i+1)w}: 1 \leq i \leq a\}$. The subgraph induced by these vertices is a cycle, since consecutive vertices have distance 1 or $w - 1$ on the cycle (except that v_{aw} and v_1 are separated by 2), and non-consecutive vertices are at least distance w apart. To obtain an odd cycle of length $2a - 1$, replace $\{v_{(a-1)w}, v_{aw}, v_1, v_w\}$ with $\{v_{(a-1)w+1}, v_0, v_{w-1}\}$. This is an odd hole if $a, w > 2$. □

8.1.46. Theorem. (Tucker [1975]) The SPGC holds for circular-arc graphs.

Proof. We first observe that in a partitionable graph the closed neighborhood of one vertex cannot be contained in the closed neighborhood of another vertex. Consider the clique Q in $\Theta(G - x)$ that contains y. All vertices of this clique belong to $N[y]$. If $N[y]$ is contained in $N[x]$, then Q belongs to $N[x]$ but does not contain x, so x can be added to obtain a larger clique.

Now, if G is a partitionable circular-arc graph, it suffices to show that $G = C_n^{\omega(G)-1}$, because the SPGC holds for circulant graphs. Consider a circular-arc representation that assigns arc A_x to $x \in V$. Since $N[y]$ cannot contain $N[x]$ for distinct vertices in a partitionable graph, we cannot have A_x appear in another arc A_y of the representation. If no arc contains another, then every arc that intersects A_x contains exactly one of its endpoints. Since the vertices corresponding to arcs containing a particular point induce a clique, there are at most $\omega - 1$ other arcs containing each endpoint of A_x. On the other hand, $\delta(G) \geq 2\omega - 2$ for a partitionable graph, so every vertex of G has degree exactly $2\omega - 2$.

Pick a point on the circle in the representation, and let v_i be the vertex corresponding to the ith arc encountered in a clockwise direction from there. Since each arc meets exactly $\omega - 1$ others at each endpoint, v_i is adjacent to $v_{i+1}, \ldots, v_{i+\omega-1}$ (addition modulo n) for any i. Hence $G = C_n^{\omega-1}$. $\square$

The proof of the SPGC for $K_{1,3}$-free graphs also uses Chvátal's proof of the SPGC for circulant graphs, but the emergence of the circulant graphs is more subtle. The next theorem was essentially proved in Giles-Trotter-Tucker [1984], phrased as Corollary 8.1.49, the proof of which explains the equivalence. The proof we present for Theorem 8.1.47 is shorter.

8.1.47. Theorem. If a partitionable graph G has a cycle consisting of critical edges, then the subgraph G' obtained by deleting the edges belonging to no maximum clique is the partitionable circulant graph $C_n^{\omega-1}$.

Proof. (Hartman [1995b]) Deleting edges does not destroy any stable set. Deleting edges in no maximum clique does not destroy any maximum clique. Hence the colorings and clique coverings of vertex-deleted subgraphs of G also make G' partitionable (by Theorem 8.1.33). Also G' is connected.

We next prove that if G has a u, v-path of length k consisting of critical edges, then u and v belong to at least $\omega - k$ common maximum cliques. This follows from Theorem 8.1.41 by induction on k. The statement of Theorem 8.1.41 is the basis case $k = 1$. For $k > 1$, if w is the vertex before v on such a path, then the induction hypothesis puts u and w in $\omega - k + 1$ common maximum cliques. Since w belongs to exactly ω maximum cliques (by Theorem 8.1.36), and $\omega - 1$ of these contain v, at most one of the $\omega - k + 1$ cliques containing u and w can omit v.

Let C be a cycle of critical edges in G. Since critical edges belong to maximum cliques, C is contained in G'. By the preceding paragraph, every set of ω vertices forming a path in G' induces a maximum clique in G'. If the length of C exceeds ω, then this establishes ω successive maximum cliques containing a given vertex x of C. By Theorem 8.1.36, these are all the maximum cliques of G containing x, and hence they include all the edges of G' incident to x. Hence C is a component of G', but G' is connected, so C contains all vertices of G'. This explicitly expresses G' as $C_n^{\omega-1}$.

If the length of C is at most ω, then C itself induces a clique in G. If $x \in V(C)$, then each stable set in the unique coloring $X(G - x)$ of $G - x$ contains exactly one vertex of C. Let $x_0, \ldots, x_k$ be the vertices of C in order. Let $S_1, \ldots, S_k$ be the stable sets such that $S_i \cup \{x_i\} \in X(G - x_0)$. By Theorem 8.1.41 the members of $X(G - x_i)$ are the same as the members of $X(G - x_{i+1})$ except for the substitution of x_i and x_{i+1} for each other in one set. Hence $X(G - x_1)$ contains $S_i \cup \{x_i\}$ for $i \geq 2$, and it also contains $S_1 \cup \{x_0\}$. Continuing these substitutions while moving along the critical edges on C, we find that $X(G - x_k)$ contains $S_i \cup \{x_{i-1}\}$ for $1 \leq i \leq k$. Now we take one more step to return to x_0 and find that $X(G - x_0)$ contains $S_i \cup \{x_{i-1}\}$ for $2 \leq i \leq k$ and $S_i \cup \{x_k\}$. Since $k \geq 3$ and $\alpha \geq 2$, these sets are different from the sets in $X(G - x_0)$ we started with. Since the coloring $X(G - x_0)$ is unique, we have obtained a contradiction, and this case does not arise. □

8.1.48. Theorem. (Chvátal [1976]) If G is a p-critical graph such that the spanning subgraph G' obtained by deleting the edges of G belonging to no maximum clique is $C_{\alpha(G)\omega(G)+1}^{\omega(G)-1}$, then G is an odd hole or odd antihole (and equals G').

Proof. By Theorem 8.1.41, G' has the same maximum cliques and maximum stable sets as G. Letting $a = \alpha(G)$ and $w = \omega(G)$, we can index the vertices so that the maximum cliques of G consist of w cyclically consecutive vertices, and the maximum stable sets have the form $v_i, v_{i+w}, \ldots, v_{i+aw}$. In particular, vertices separated by a multiple of w on the cycle $v_0, \ldots, v_{aw}$ are nonadjacent in G' and in the full graph G. If $a = 2$, then $G - e$ has more maximum stable sets than G, and if $w = 2$, then $G - e$ has fewer maximum cliques than G. Hence if $G' \neq G$ we may assume $a, w \geq 3$.

If $a, w \geq 3$, we exhibit an imperfect proper induced subgraph H of G. Let $S = \{v_{aw}, v_1, v_w, v_{w+2}\} \cup \{v_{iw+1} : 2 \leq i \leq a - 1\}$, and let $T = \{v_{(a-1)w+1}, v_{aw}, v_1, v_w\} \cup \{v_{w+i} : 2 \leq i \leq w - 1\}$. The sets S and T have sizes

$a + 2$ and $w + 2$, and when $a, w \geq 3$ they share the five distinct vertices $\{v_{(a-1)w+}, v_{aw}, v_1, v_w, v_{w+2}\}$. Furthermore, S intersects every maximum clique of G' (and hence of G), and T intersects every maximum stable set of G' (and hence of G) (Exercise 41). Letting $H = G - (P \cup Q)$, this yields $\alpha(H) = a - 1$, $\omega(H) = w - 1$, and $n(H) \geq n(G) - (a + w - 1) > (a - 1)(w - 1)$. Hence H is imperfect. $\square$

8.1.49. Corollary. (Giles-Trotter-Tucker [1984]) If G is a p-critical graph and for each $v \in V(G)$ the minimum coloring $X(G - v)$ has (at least) two sets that each contain exactly one neighbor of v, then G is an odd hole or an odd antihole.

Proof. If $X(G - v)$ has a set that contains exactly one neighbor u of v, then the edge uv is critical. Hence the hypothesis states that the subgraph of critical edges has minimum degree at least 2 and therefore contains a cycle. By Theorem 8.1.47, the subgraph G' obtained by deleting the edges belonging to no maximum clique is $C_n^{\omega-1}$. Hence we have a circulant graph as a spanning subgraph of a p-critical graph. By Theorem 8.1.48, G is an odd hole or an odd antihole. $\square$

8.1.50. Corollary. (Parthasarathy-Ravindra [1976]) The SPGC holds for $K_{1,3}$-free graphs.

Proof. (Giles-Trotter-Tucker [1984]). Let G be a p-critical $K_{1,3}$-free graph. For each $v \in V(G)$, $N(v)$ induces a perfect subgraph having no stable set of size 3. This means that $N(v)$ can be covered by two cliques, which implies $d(v) \leq 2\omega(G) - 2$. Since $X(G - v)$ is a partition of $G - v$ into ω stable sets of size $\alpha(G)$, there must be at least 2 of these sets that contain only one neighbor of v (none can miss $N(v)$, because then adding v would yield a larger stable set in G). Hence G satisfies the hypothesis of Corollary 8.1.49, and G is an odd hole or antihole. $\square$

The SPGC remains open, but a statement intermediate between it and the PGT has been proved (it is immediately implied by the SPGC and immediately implies the PGT). Chvátal conjectured that if G and H have the same vertex set and have the same 4-tuples of vertices that induce P_4, then G is perfect if and only if H is perfect. Reed [1987] proved this; it is the *Semi-Strong Perfect Graph Theorem*.

EXERCISES

8.1.1. (–) Compute $\chi(G)$ and $\omega(G)$ for the complement of the odd cycle C_{2k+1}.

8.1.2. *Clique identification preserves perfection.* Suppose that $G = G_1 \cup G_2$, that $G_1 \cap G_2$ is a clique, and that G_1 and G_2 are perfect. Prove that G is perfect.

8.1.3. (!) P_4-free graphs are also called *co-graphs*, which stands for "complement reducible". A graph is *complement reducible* if it can be reduced to an empty graph by successively taking complements within components.

a) Prove that a graph G is P_4-free if and only if it is complement reducible.

b) Use part (a) and the Perfect Graph Theorem to prove that every P_4-free graph is perfect. (Seinsche [1974])

8.1.4. Suppose G is a Cartesian product of cliques. Prove that $\alpha(G) = \theta(G)$. Prove that $K_2 \square K_2 \square K_3$ is not perfect.

8.1.5. (+) Prove that G is an odd cycle if and only if $\alpha(G) = (n(G) - 1)/2$ and $\alpha(G - u - v) = \alpha(G)$ for all $u, v \in V(G)$. (Melnikov-Vizing [1971], Greenwell [1978])

8.1.6. Suppose $v_1, \ldots, v_n$ is a simplicial elimination ordering of G, and let $Q(v_i) = \{v_j \in N(v_i): j > i\}$; note that $Q(v_i)$ is the clique of neighbors of v_i at the time when v_i is deleted in the elimination ordering. Let $S = \{y_1, \ldots, y_k\}$ be the stable set obtained "greedily" from the ordering $v_1, \ldots, v_n$; i.e., set $y_1 = v_1$, discard $N(y_1)$ from the remainder of the ordering, and proceed iteratively, at each step adding the least remaining element x to the stable set and discarding what remains of $Q(x)$.

a) Prove that applying the greedy coloring algorithm to the construction ordering $v_n, \ldots, v_1$ yields an optimal coloring and that $\omega(G) = 1 + \max \sum_{x \in V(G)} |Q(x)|$. (Fulkerson-Gross [1965])

b) Prove that S is a maximum stable set and that the sets $\{y_i\} \cup Q(y_i)$ form a minimum clique covering. (Gavril [1972])

8.1.7. Add a test to the MCS algorithm to check whether the resulting ordering is a simplicial elimination ordering. (Tarjan-Yannakakis [1984])

8.1.8. Prove directly (without using a simplicial elimination ordering) that the intersection graph of a family of subtrees of a tree has no chordless cycle.

8.1.9. (–) Prove that every graph is the intersection graph of a family of subtrees of some graph.

8.1.10. Prove that every chordal graph has an intersection representation by subtrees of a host tree with maximum degree 3.

8.1.11. *Intersection graphs of subtrees of a graph.* A *fraternal orientation* of a graph is an orientation such that any pair of vertices with a common successor are adjacent.

a) (–) Prove that a graph is chordal if and only if it has an acyclic fraternal orientation.

b) (–) Obtain a graph with no fraternal orientation.

c) A family of trees in a graph is *rootable* if the trees can be assigned roots so that a pair of them intersects if and only if at least one of the two roots belongs to both subtrees. Prove that G has a fraternal orientation if and only if G is the intersection graph of a rootable family of subtrees of some graph. (Gavril-Urrutia [1994])

8.1.12. (!) Prove that a simple graph G is a forest if and only if every pairwise intersecting family of paths in G has a common vertex. (Hint: for sufficiency, use induction on the number of paths in the family.)

8.1.13. (!) *Forbidden subgraph characterization of split graphs.* A graph is a *split graph* if its vertices can be partitioned into a clique and a stable set.
   a) Prove that if G is a split graph, then G and $\overline{G}$ are chordal graphs. Observe that if G and $\overline{G}$ are chordal graphs, then G has no induced subgraph in $\{C_4, 2K_2, C_5\}$.
   b) Prove that if G is a simple graph with no induced subgraph in $\{C_4, 2K_2, C_5\}$, then G is a split graph. (Hint: Among the maximum-sized cliques, let Q be one such that $G - Q$ has the minimum number of edges. Prove that $G - Q$ is a stable set, using the choice of Q and the forbidden subgraph conditions.) (Hammer-Simeone [1981])

8.1.14. (–) Determine the trees that are split graphs, and construct a pair of nonisomorphic split graphs with the same degree sequence.

8.1.15. The *k-trees* are the graphs arise from a k-clique by 0 or more iterations of adding a new vertex joined to a k-clique in the old graph. Prove that G is a k-tree if and only if G satisfies the following three properties:
   1) G is connected.
   2) G has a k-clique but no $k + 2$-clique.
   3) Every minimal vertex separator of G is a k-clique.

8.1.16. Let G be an n-vertex chordal graph having no clique of order $k + 2$. Prove that $e(G) \le kn - \binom{k+1}{2}$, with equality if and only if G is a k-tree.

8.1.17. (+) Generalize Theorem 2.2.3 (Cayley's Formula) by proving that the number of k-trees with vertex set $[n]$ is $\binom{n}{k}[k(n-k)+1]^{n-k-2}$. (Hint: generalize the Prüfer sequence for *rooted* trees; this generates a sequence with $n-1$ entries and never deletes the root. In a k-tree, the vertices belonging to exactly one $k + 1$-clique are the *leaves*. A k-tree can be grown using any of its k-cliques as a root. The sequences generated from k-trees with a fixed root have as symbols 0 and pairs ij, where i comes from some k-set and j from some $n - k$-set.) (Greene-Iba [1975])

8.1.18. Suppose G is a chordal graph with $\omega(G) = r$. Prove that G has at most $\binom{r}{j} + \binom{r-1}{j-1}(n-r)$ cliques of order j, with equality (for all j simultaneously) if and only if G is an $r - 1$-tree.

8.1.19. *The Helly property of the real line.* Suppose that $I_1, \ldots, I_k$ are pairwise intersecting intervals in $\mathbb{R}$. Prove that $I_1, \ldots, I_k$ have a common point.

8.1.20. Prove directly that a tree is an interval graph if and only if it is a caterpillar (a tree having a path that contains at least one vertex of each edge).

8.1.21. *Characterization of interval graphs.* (see Definition 8.1.16 for (b)).
   a) Prove that every interval graph is a chordal graph and is the complement of a comparability graph.
   b) Prove that a graph G has an interval representation if and only if the clique-vertex incidence matrix of G has the consecutive 1's property.

8.1.22. Prove that G is an interval graph if and only if the vertices of G can be ordered $v_1, \ldots, v_n$ such that $v_i \leftrightarrow v_k$ implies $v_j \leftrightarrow v_k$ whenever $i < j < k$. (see Jacobson-McMorris-Mulder [1991])

8.1.23. An *asteroidal triple* in a graph is a triple of vertices x, y, z such that between any two there exists a path avoiding the neighborhood of the third. Prove that no asteroidal triple occurs in an interval graph. (Comment: interval graphs are precisely the chordal graphs that have no asteroidal triples (Lekkerkerker-Boland [1962]))

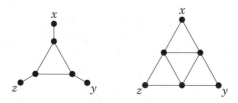

8.1.24. Six professors visited the library on the day the rare book was stolen. Each entered once, stayed for some time, and then left. For any two of them that were in the library at the same time, at least one of them saw the other. Detectives questioned the professors and gathered the following testimony:

PROFESSOR	CLAIMED TO HAVE SEEN
Abe	Burt, Eddie
Burt	Abe, Ida
Charlotte	Desmond, Ida
Desmond	Abe, Ida
Eddie	Burt, Charlotte
Ida	Charlotte, Eddie

In this situation, "lying" means providing false information, not omitting information. Assume that the culprit tried to frame another suspect by lying. If one professor lied, who was it? (Golumbic [1980, p20])

8.1.25. (+) Prove that G is a unit interval graph (representable by intervals of the same length) if and only if the matrix $A(G) + I$ has the consecutive ones property. (Roberts [1968])

8.1.26. (+) Prove that G is a proper interval graph (representable by intervals such that none properly contains another) if and only if the clique-vertex incidence matrix of G has the consecutive ones property for both rows and columns. (Fishburn [1985]

8.1.27. (–) Prove that every P_4-free graph is a Meyniel graph.

8.1.28. (!) Prove that every chordal graph is o-triangulated.

8.1.29. (+) Prove that the conditions below are equivalent.
   A) Every odd cycle of length at least 5 has a crossing pair of chords.
   B) For every pair $x, y \in V(G)$, chordless x, y-paths are all even or all odd.
(Hint: for A $\Rightarrow$ B, consider a pair P_1, P_2 of x, y-paths with opposite parity such that the sum of their lengths is minimal.) (Burlet-Uhry [1984])

8.1.30. (!) Prove that the graphs below are strongly perfect but are not perfectly orderable.

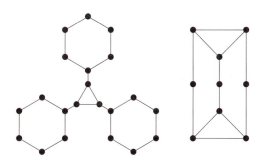

8.1.31. (–) Prove that the graph on the left above is a Meyniel graph but is not perfectly orderable. Prove that the graph $\overline{P}_5$ is perfectly orderable but is not a Meyniel graph.

8.1.32. Prove that every perfectly orderable graph is strongly perfect. (Hint: use Lemma 8.1.22.) (Chvátal [1984])

8.1.33. (!) *Weakly chordal graphs.*
    a) Prove that every chordal graph is weakly chordal.
    b) Prove that the graph below is weakly chordal but not strongly perfect.

8.1.34. (–) A *skew partition* of G is a partition of $V(G)$ into two nonempty parts X, Y such that $G[X]$ is disconnected and $\overline{G}[Y]$ is disconnected. Chvátal [1985] conjectured that no minimal imperfect graph has a skew partition. Prove that this implies the Star-Cutset Lemma and is implied by the SPGC.

8.1.35. Prove that the non-circulant graph in Example 8.1.34 is partitionable. (Chvátal-Graham-Perold-Whitesides [1979])

8.1.36. (+) Prove that no p-critical graph has *antitwins*, which are a pair of vertices such that every other vertex is adjacent to exactly one of them. (Hint: given a *p*-critical graph with antitwins $\{x, y\}$, let S be the stable set containing y in the unique optimal coloring of $G - x$. Find among the vertices of the $\omega - 1$-colorable subgraph $G - x - S$ an $\omega - 1$ clique in $N(x)$ that doesn't extend into $N(y)$. Similarly, find a stable set in $N(y)$ that doesn't extend into $N(x)$. Now build an induced 5-cycle.) (Note: the non-circulant partitionable graph of Example 8.1.34 has antitwins.) (Olariu [1988])

8.1.37. Vertices x, y form an *even pair* if every chordless x, y-path has even length (number of edges). *Twins* (nonadjacent vertices with the same neighborhood) are a special case.
    a) Suppose that S_1, S_2 are maximum stable sets in a partitionable graph

G. Prove that the subgraph of G induced by the symmetric difference of S_1 and S_2 is connected. (Bland-Huang-Trotter [1979])

b) Use (a) to prove that no p-critical graph has an even pair. (Comment: hence no p-critical graph has twins, which proves yet again that vertex duplication preserves perfection.) (Meyniel [1987], Bertschi-Reed [1988])

8.1.38. Suppose G is a partitionable graph, and S_1, S_2 are stable sets in the optimal coloring of $G - x$. Use part (a) of the preceding problem to prove that the subgraph of G induced by $S_1 \cup S_2 \cup \{x\}$ is 2-connected. (Buckingham-Golumbic [1983])

8.1.39. Prove that one graph below is a circle graph but not a circular-arc graph, and prove that the other is a circular-arc graph but not a circle graph.

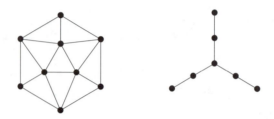

8.1.40. (!) The graph $K_{1,3} + e$ is the 4-vertex graph obtained by adding one edge to $K_{1,3}$. Using the perfection of Meyniel graphs, prove that $K_{1,3} + e$-free graphs satisfy the SPGC. (Meyniel [1976])

8.1.41. Suppose $G = C_{aw+1}^{w-1}$. Let $S = \{v_{aw}, v_1, v_w, v_{w+2}\} \cup \{v_{iw+1}: 2 \le i \le a - 1\}$, and let $T = \{v_{(a-1)w+1}, v_{aw}, v_1, v_w\} \cup \{v_{w+i}: 2 \le i \le w - 1\}$. Prove that S intersects every maximum clique of G, and that T intersects every maximum stable set of G. (Chvátal [1976])

8.1.42. (!) *SPGC for circle graphs.* (Buckingham-Golumbic [1983])

a) Use the Star-Cutset Lemma Lemma to prove that if x is a vertex in a partitionable graph G, then $G - N[x]$ is connected ($N[x]$ denotes the closed neighborhood $\{x\} \cup N(x)$).

b) Use part (a) to prove that a circle graph that is partitionable has no induced $K_{1,3}$.

c) Conclude from part (b) that the SPGC holds for circle graphs.

8.2 Matroids

Many results of graph theory extend or simplify in the theory of matroids. These include the use of the greedy algorithm to find minimum spanning trees, the strong duality between maximum matching and minimum vertex cover in bipartite graphs, and the geometric duality relating planar graphs and their duals.

Matroids arise in many contexts but are special enough to exhibit rich combinatorial structure. When a result from graph theory generalizes to matroids, it can then be interpreted in other special cases. Several difficult theorems about graphs have found easier proofs using matroids. Matroids were introduced by Whitney [1935] to generalize algebraic properties of graphs and to study planar graphs, by MacLane [1936] as an aspect of geometric lattices, and by Van der Waerden [1937] to study independence properties of vector spaces. In this presentation, we emphasize applications to graph theory.

HEREDITARY SYSTEMS AND EXAMPLES

In many situations we view certain subsets of a set as "independent". Examples include linearly independent sets of vectors, sets of pairwise nonadjacent vertices in a graph, acyclic sets of edges in a graph, etc.

8.2.1. Example. *Acyclic sets of edges.* Let E be the edge set of a graph G, and let $X \subseteq E$ be "independent" if it contains no cycle. Every subset of an independent set is independent, and the empty set is independent. The cycles are the minimal dependent sets.

Consider $K_4 - e$, which has five edges. Since spanning trees of this graph have three edges, every set having more than three edges is dependent. Also the two triangles are dependent; this yields eight dependent sets and 24 independent sets among the subsets of E. There are three minimal dependent sets (the cycles) and eight maximal independent sets (the spanning trees). □

8.2.2. Definition. A *hereditary family* is a non-empty collection of sets, $\mathbf{F}$, such that every subset of a set in $\mathbf{F}$ is also in $\mathbf{F}$. A *hereditary system* M on a set E consists of a hereditary family $\mathbf{I}_M$ of subsets of E and the various ways of specifying $\mathbf{I}_M$, called *aspects* of M. The family $\mathbf{I}_M$ is the collection of *independent sets* of M; the other subsets of E are *dependent*. The *bases* $\mathbf{B}_M$ are the maximal independent sets. The *circuits* $\mathbf{C}_M$ are the minimal dependent sets. The *rank* of a subset of E is the maximum size of an independent set contained in it, so the *rank function* r_M is defined by $r_M(X) = \max\{|Y|: Y \subseteq X, Y \in \mathbf{I}\}$.

8.2.3. Example. *Hereditary systems.* Label each vertex $a = (a_1, \ldots, a_n)$ of the hypercube Q_n by the corresponding set $X_a = \{i: a_i = 1\}$. Draw Q_n in the plane so that $X \cup \{e\}$ is above X for each set X and element $e \in [n] - X$. The diagram below illustrates the relationships among the independent sets, bases, circuits, and dependent sets of a hereditary system. The bases are the maximal elements of the family **I** and the circuits are the minimal elements not in **I**. In every hereditary system, $\varnothing$ belongs to **I**. If every set is independent, then there is no circuit, but there is always at least one base.

In the example on the right, the independent sets are the acyclic edge sets in a graph with three edges. The only dependent sets are $\{1, 2\}$ and $\{1, 2, 3\}$, the only circuit is $\{1, 2\}$, and the bases are $\{1, 3\}$ and $\{2, 3\}$. The rank of an independent set is its size. For the dependent sets, we have $r(\{1, 2\}) = 1$ and $r(\{1, 2, 3\}) = 2$. $\square$

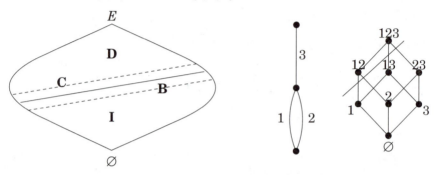

8.2.4. Remark. *Aspects of hereditary systems.* A hereditary system M can be specified by any of $\mathbf{I}_M$, $\mathbf{B}_M$, $\mathbf{C}_M$, r_M, etc., because each aspect determines the others. We have expressed $\mathbf{B}_M, \mathbf{C}_M, r_M$ in terms of $\mathbf{I}_M$. Conversely, if we know $\mathbf{B}_M$, then $\mathbf{I}_M$ must consist of the sets contained in members of $\mathbf{B}_M$. If we know $\mathbf{C}_M$, then $\mathbf{I}_M$ must consist of the sets containing no member of $\mathbf{C}_M$. If we know r_M, then we can determine $\mathbf{I}_M$ and the other aspects by setting $\mathbf{I}_M = \{X \subseteq E : r_M(X) = |X|\}$. $\square$

A matroid is a hereditary system satisfying an additional property. There are many equivalent choices for this property, partly because a hereditary system has many aspects. A constraint on one aspect translates into a constraint on any other aspect. Using various motivating examples, we will state several of these properties that characterize matroids. We will later summarize them and prove that they are equivalent. We begin with the fundamental example from graphs.

8.2.5. Definition. The *cycle matroid* $M(G)$ of a graph G is the hereditary system on $E(G)$ whose circuits are the cycles of G. A hereditary system that is $M(G)$ for some graph G is *graphic*.

8.2.6. Example. *Bases in a cycle matroid $M(G)$.* The bases are the edge sets of the maximal forests. Each maximal forest consists of a spanning tree from each component, so they have the same size.

Suppose $B_1, B_2 \in \mathbf{B}$ with $e \in B_1 - B_2$. Deleting e from B_1 disconnects some component of B_1; since B_2 contains a tree spanning the same component of G, some edge $f \in B_2 - B_1$ can be added to $B_1 - e$ to reconnect that component (Proposition 2.1.6). For a hereditary system M, the *base exchange property* is: if $B_1, B_2 \in \mathbf{B}_M$, then for all $e \in B_1 - B_2$ there exists $f \in B_2 - B_1$ such that $B_1 - e + f \in \mathbf{B}_M$. A hereditary system is a matroid if it satisfies the base exchange property.

The hereditary system in Example 8.2.1 is $M(K_4 - e)$.) □

8.2.7. Example. *Rank function of a cycle matroid $M(G)$.* Let G_X be the spanning subgraph of G having edge set $X \subseteq E(G)$. Suppose Y is a maximum independent subset of X. The graph G_Y consists of a spanning tree from each component of G_X; hence G_Y and G_X have the same number of components. Since G_Y is a forest, it has $n(G) - |Y|$ components. By the definition of the rank function, we have $r(X) = r(Y) = |Y|$. Hence $r(X) = n(G) - k$, where k is the number of components of G_X.

If $r(X \cup e) = r(X)$ for some $e \in E - X$, then e has both endpoints in a single component of X; adding it does not reduce the number of components. If we add two such edges, then again we do not reduce the number of components. Therefore, $r(X) = r(X \cup e) = r(X \cup f)$ implies $r(X) = r(X \cup e \cup f)$. For a hereditary system M on E, the *(weak) absorption property* is: if $X \subseteq E$ and $e, f \in E$, then $r(X) = r(X \cup e) = r(X \cup f)$ implies $r(X \cup e \cup f) = r(X)$. A hereditary system is a matroid if it satisfies the absorption property.

The term "absorption property" is due to A. Kezdy. In the illustration below, X and Y have rank 9, and the subgraphs G_X and G_Y have 13 vertices and four components.). □

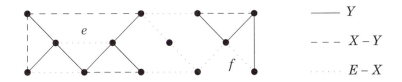

Graphs in general may have loops and multiple edges. We have corresponding terminology in hereditary systems, reflecting the behavior of loops and multiple edges when we consider cycle matroids.

8.2.8. Definition. In a hereditary system, a *loop* is an element forming a circuit of size 1. *Parallel elements* are distinct non-loop elements forming a set with rank 1. A hereditary system is *simple* if it has no loops or parallel elements.

8.2.9. Definition. The *vectorial matroid* on a set E of vectors in a vector space is the hereditary system whose independent sets are the linearly independent subsets of vectors in E. A matroid expressible in this way is *linear* (also *representable*). The *column matroid* $M(A)$ of a matrix A is the vectorial matroid defined on its columns. (The set E may have repeated vectors.)

8.2.10. Example. *Circuits in a vectorial matroid.* The circuits are the minimal sets $X = \{x_1, \ldots, x_k\} \subseteq E$ such that $\Sigma c_i x_i = 0$ using coefficients not all zero. Minimality of X requires that all $c_i \neq 0$. Let C_1, C_2 be distinct circuits containing x. Using the equations of dependence for C_1 and C_2, we can express x as linear combinations of the vectors in $C_1 - x$ and of those in $C_2 - x$. Equating these expressions yields an equation of dependence for $(C_1 \cup C_2) - x$. Hence $C_1 \neq C_2$ and $x \in C_1 \cap C_2$ implies that $(C_1 \cup C_2) - x$ contains a circuit. For a hereditary system M on E, the *(weak) elimination property* is: for distinct $C_1, C_2 \in \mathbf{C}_M$ and $x \in C_1 \cap C_2$, there is another member of $\mathbf{C}_M$ contained in $(C_1 \cup C_2) - x$. A hereditary system is a matroid if it satisfies the weak elimination property.

The column matroid of the matrix below is $M(K_4 - e)$.

$$\begin{pmatrix} 0 & 0 & 0 & 1 & 1 \\ 0 & 1 & 1 & 0 & 1 \\ 1 & 1 & 0 & 0 & 0 \end{pmatrix}$$

8.2.11. Definition. Given finite sets $A_1, \ldots, A_m$ with union E, let G be the bipartite graph with bipartition $E, [m]$ such that $e \leftrightarrow i$ if and only if $e \in A_i$. The *transversal matroid* associated with $A_1, \ldots, A_m$ is the hereditary system on E whose independent sets are the subsets of E that can be saturated by a matching in G.

8.2.12. Example. *Independent sets in transversal matroids.* When M, M' are matchings in G and $|M'| > |M|$, the symmetric difference $M \triangle M'$ contains an M-augmenting path P (as in the proof of Theorem 3.1.6). Replacing $M \cap P$ with $M' \cap P$ yields a matching of size $|M| + 1$ that saturates all the vertices of M plus the endpoints of P.

Suppose I_1, I_2 are independent sets in the transversal matroid generated by $A_1, \ldots, A_m$, with I_1 and I_2 saturated by matchings M_1 and M_2, respectively, in the associated bipartite graph. (In the illustration below, M_1 uses solid edges and M_2 uses dashed edges.) If $|I_2| > |I_1|$, then the matching obtained from M_1 by using an M_1-augmenting path in $M_2 \triangle M_1$ saturates I_1 plus an element $e \in I_2 - I_1$; the element e "augments" I_1. For a hereditary system M on E, the *augmentation property* is: for distinct $I_1, I_2 \in \mathbf{I}_M$ with $|I_2| > |I_1|$, there exists $e \in I_2 - I_1$ such that $I_1 \cup \{e\} \in \mathbf{I}$. A hereditary system is a matroid if it satisfies the augmentation property.

The transversal matroid of the family $\mathbf{A} = \{\{1,2\},\{2,3,4\},\{4,5\}\}$, illustrated by the bipartite graph on the right below, is $M(K_4 - e)$. □

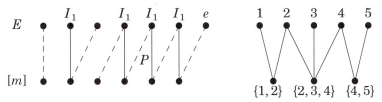

Suppose X is an independent set in the transversal matroid associated with $\mathbf{A} = \{A_1, \ldots, A_m\}$. If X can be matched to $J \subseteq [m]$ in the corresponding bipartite graph G, then X is a system of distinct representatives (SDR) for $\{A_j \colon j \in J\}$. Such a partial system of distinct representatives for $\mathbf{A}$ is a "partial transversal"; hence the name "transversal matroid". That these are matroids was discovered by Edmonds and Fulkerson [1965] and independently by Mirsky and Perfect [1967], who extended the result to infinite sets.

Every matroid must satisfy all properties of matroids. Once we show that the properties described above are equivalent for hereditary systems, we need only verify one to know that all hold. Before doing this, we verify that they all hold for cycle matroids.

8.2.13. Example. *Augmentation in cycle matroids.* When $I_1, I_2 \in \mathbf{I}_{M(G)}$, the spanning subgraph G_{I_1} has $k = n - |I_1|$ components (Example 8.2.7). The largest forest in G_{I_1} has $n - k = |I_1|$ edges. Therefore, the forest I_2 is not a subgraph of G_{I_1} and has some edge with endpoints in two components of G_{I_1}. This edge can be added to I_1 to obtain a larger independent set. Hence the augmentation property holds. □

8.2.14. Example. *Weak elimination in cycle matroids.* The circuits of $M(G)$ are the edge sets of cycles of G. Cycles have even degree at each vertex. If $C_1, C_2 \in \mathbf{C}$, then the symmetric difference $C_1 \triangle C_2$ also has even degree at each vertex. If $C_1 \neq C_2$, this implies that $C_1 \triangle C_2$ contains a cycle (Lemma 1.2.17). This is stronger than the weak elimination property, since $C_1 \triangle C_2 \subseteq (C_1 \cup C_2) - x$. In the picture below, C_1 and C_2 are face boundaries of length 9 sharing the dashed edges, and $C_1 \triangle C_2$ is the union of two disjoint cycles. □

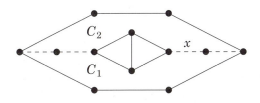

For transversal matroids, the base exchange property is similar to the augmentation property, and Exercise 6 considers the elimination property. For linear matroids, direct verification of the augmentation or base exchange property requires the algebraic result that k linearly independent vectors cannot all be expressed as linear combinations of a smaller set. Theorem 8.2.18 avoids this. Indeed, since we have verified the weak elimination property for linear matroids, these theorems of linear algebra follow from Theorem 8.2.18!

8.2.15. Remark. *Notational conventions*: Boldface $\mathbf{I}, \mathbf{B}, \mathbf{C}$ for families of subsets of E allows $I \in \mathbf{I}$, $B \in \mathbf{B}$, $C \in \mathbf{C}$ to denote generic members of the families. Roman letters I,B,C,R denote properties that yield matroids. We use e, f, x, y as generic elements of E and X, Y, F as generic subsets of E. We use addition [subtraction] to denote addition [deletion] of elements to [from] sets, and we usually drop the set brackets from singleton sets. $\square$

Every hereditary family is the collection of independent sets of a hereditary system. A collection $\mathbf{B}$ is realizable as the set of bases of a hereditary system if and only if $\mathbf{B}$ is non-empty and no element of $\mathbf{B}$ contains another. A collection $\mathbf{C}$ is realizable as the set of circuits of a hereditary system if and only if the elements of $\mathbf{C}$ are non-empty and no element of $\mathbf{C}$ contains another. The characterization of rank functions is more subtle. It includes two properties (r1,r2) below that we will need, plus an additional technical condition that forces r to be the rank function of the hereditary system M defined by $\mathbf{I}_M = \{X \subseteq E: r(X) = |X|\}$.

8.2.16. Lemma. If r is the rank function of a hereditary system on E, then
(r1) $r(\varnothing) = 0$.
(r2) $r(X) \leq r(X \cup e) \leq r(X) + 1$ whenever $X \subseteq E$ and $e \in E$.

Proof. From the definition $r(X) = \max\{|Y|: Y \subseteq X, Y \in \mathbf{I}\}$, we immediately have $r(\varnothing) = 0$. Because $X \cup e$ contains every independent subset of X, we have $r(X \cup e) \geq r(X)$. Because the independent subsets of $X \cup e$ not contained in X consist of e together with an independent subset of X, we have $r(X \cup e) \leq r(X) + 1$. $\square$

PROPERTIES OF MATROIDS

Many equivalent conditions on hereditary systems yield matroids. We can show that a hereditary system is a matroid by verifying any of them, after which we can employ them all without additional proof. We obtained the same benefit from equivalent characterizations of trees.

Adding one edge to a forest creates at most one cycle. More generally, adding one element to an independent set in a matroid creates at most one circuit. Our proof of the greedy algorithm for spanning trees (Section 2.3) used *only* this property of graphs. This "induced circuit" property is equivalent to the other conditions that characterize matroids, as is the effectiveness of the greedy algorithm itself! Both properties appear in our list.

Given weights on the elements of a matroid, we define the *greedy algorithm* to be the process of iteratively selecting the element of largest positive weight addition to the independent set of elements already selected yields a larger independent set. Rado [1957] proved that matroids are precisely the hereditary systems for which the greedy algorithm selects a maximum weighted independent set regardless of the choice of weights.

8.2.17. Definition. A hereditary system M on E is a *matroid* if it satisfies any of the following additional properties, where $\mathbf{I}, \mathbf{B}, \mathbf{C}, r$ are the independent sets, bases, circuits, and rank function of M.

I: *augmentation* - if $I_1, I_2 \in \mathbf{I}$ with $|I_2| > |I_1|$, then $I_1 \cup e \in \mathbf{I}$ for some $e \in I_2 - I_1$.

U: *uniformity* - for every $X \subseteq E$, the maximal subsets of X belonging to $\mathbf{I}$ have the same size.

B: *base exchange* - if $B_1, B_2 \in \mathbf{B}$, then for all $e \in B_1 - B_2$ there exists $f \in B_2 - B_1$ such that $B_1 - e + f \in \mathbf{B}$.

R: *submodularity* - for all $X, Y \subseteq E$, $r(X \cap Y) + r(X \cup Y) \le r(X) + r(Y)$.

A: *weak absorption* - for $X \subseteq E$ and $e, f \in E$, $r(X) = r(X \cup e) = r(X \cup f)$ implies $r(X \cup e \cup f) = r(X)$.

A': *strong absorption* - for $X, Y \subseteq E$, $r(X \cup e) = r(X)$ for all $e \in Y$ implies $r(X \cup Y) = r(X)$.

C: *weak elimination* - for any distinct $C_1, C_2 \in \mathbf{C}$ and $x \in C_1 \cap C_2$, there is another member of $\mathbf{C}$ contained in $(C_1 \cup C_2) - x$.

J: *induced circuits* - if $I \in \mathbf{I}$, then $I \cup e$ contains at most one circuit.

G: *greedy algorithm* - for each weighting of the elements of E the greedy algorithm selects an independent set of maximum total weight.

The base exchange property implies that all bases have the same size: if $|B_1| < |B_2|$ for some $B_1, B_2 \in \mathbf{B}$, then we can iteratively replace elements of $B_1 - B_2$ by elements of $B_2 - B_1$ to obtain a base of size $|B_1|$ contained in B_2, but no base is contained in another.

The rank of a set $X \subseteq E$ in a vectorial matroid is the dimension of the space spanned by X. Hence for vectorial matroids the submodularity inequality says that $\dim U \cap V + \dim U \oplus V \le \dim U + \dim V$, where $U, V, U \oplus V$ are the spaces spanned by $X, Y, X \cup Y \subseteq E$. The usual proof of this is the specialization to vector spaces of the proof of U $\Rightarrow$ R in Theorem 8.2.18. Exercises 7 and 8 discuss submodularity in cycle matroids and in transversal matroids.

Various of these properties (together with those implying that the relevant aspect yields a hereditary system) have been used as the defining condition for a matroid. Examples include I (Welsh [1976], Schrijver [to appear]), U (Edmonds [1965b,c], Bixby [1981], Nemhauser-Wolsey [1988]), A (Whitney [1935]), C (Tutte [1970]), G (Papadimitriou-Steiglitz [1982]), Exercise 30 (van der Waerden [1937]), and S of Theorem 8.2.26 (Rota [1964], Crapo-Rota [1970], Aigner [1979]).

Starting with hereditary systems yields a concise proof of equivalence (or simply a concise proof that matroids have all these properties). All properties of hereditary systems are available at all times, and we need not verify the properties of a hereditary system each time we switch to another aspect.

8.2.18. Theorem. For a hereditary system M, the conditions defining matroids in Definition 8.2.17 are equivalent.

Proof.

$U \Rightarrow B$. By uniformity for $X = E$, all bases have the same size. We then apply uniformity to the set $(B_1 - e) \cup B_2$; the independent set $B_1 - e$ can only be augmented from B_2 to reach size $|B_2|$.

$B \Rightarrow I$. Given independent sets $I_1, I_2 \in \mathbf{I}$ with $|I_2| > |I_1|$, choose $B_1, B_2 \in \mathbf{B}$ such that $I_1 \subseteq B_1$, $I_2 \subseteq B_2$. We can use base exchange to replace any elements of $B_1 - I_1$ outside B_2 with elements of B_2. Hence we may assume that $B_1 - I_1 \subseteq B_2$. If $B_1 - I_1 \subseteq B_2 - I_2$, then $|B_1| < |B_2|$, which is forbidden by the base exchange property (we would replace elements of $B_1 - B_2$ with elements of $B_2 - B_1$ to obtain a base inside B_2). Hence I_2 intersects $B_1 - I_1$, which yields an augmentation.

$I \Rightarrow A$. Suppose $r(X) = r(X \cup e) = r(X \cup f)$. If $r(X \cup e \cup f) > r(X)$, let I_1, I_2 be maximum independent subsets of X and of $X \cup e \cup f$. Then $|I_2| > |I_1|$, and we can augment I_1 from I_2. Since I_1 is a maximum independent subset of X, the augmentation can only add e or f, which contradicts the hypothesis $r(X) = r(X \cup e) = r(X \cup f)$.

$A \Rightarrow A'$. We use induction on $|Y - X|$. The statement is trivial when $|Y - X| = 1$. When $|Y - X| > 1$, choose $e, f \in Y - X$, and let $Y' = Y - e - f$. An application of weak absorption to $X \cup Y', e, f$ yields $r(X \cup Y) = r(X \cup Y')$. By the induction hypothesis, $r(X \cup Y') = r(X)$.

$A' \Rightarrow U$. If Y is a maximal independent subset of X, then $r(Y \cup e) = r(Y)$ for all $e \in X - Y$. By strong absorption, $r(X) = r(Y) = |Y|$. Hence all such Y have the same size.

$U \Rightarrow R$. Given $X, Y \subseteq E$, choose a maximum independent set I_1 from $X \cap Y$. By uniformity, I_1 can be enlarged to a maximum independent subset of $X \cup Y$; call this I_2. Consider $I_2 \cap X$ and $I_2 \cap Y$; these are independent subsets of X and Y, and each includes I_1. Hence $r(X \cap Y) + r(X \cup Y) = |I_1| + |I_2| = |I_2 \cap X| + |I_2 \cap Y| \le r(X) + r(Y)$.

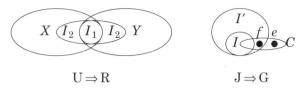

$$U \Rightarrow R \qquad\qquad\qquad J \Rightarrow G$$

$R \Rightarrow C$. Consider distinct circuits $C_1, C_2 \in \mathbf{C}$ with $x \in C_1 \cap C_2$. We have $r(C_1) = |C_1| - 1$ and $r(C_2) = |C_2| - 1$. Also $r(C_1 \cap C_2) = |C_1 \cap C_2|$, since every proper subset of a circuit is independent. If $(C_1 \cup C_2) - x$ does not contain a circuit, then $r((C_1 \cup C_2) - x) = |C_1 \cup C_2| - 1$, and hence $r(C_1 \cup C_2) \geq |C_1 \cup C_2| - 1$. Applying submodularity to C_1 and C_2 yields $|C_1 \cap C_2| + |C_1 \cup C_2| - 1 \leq |C_1| + |C_2| - 2$, which is impossible.

$C \Rightarrow J$. If $I \cup e$ contains $C_1, C_2 \in \mathbf{C}$ for some $I \in \mathbf{I}$, then C_1, C_2 both contain e. Now weak elimination guarantees a circuit in $(C_1 \cup C_2) - e$. On the other hand, $(C_1 \cup C_2) - e$ is independent, being contained in I.

$J \Rightarrow G$. For weight function w, let I be the output of the greedy algorithm. Among the maximum weight independent sets, let I^* be one having largest intersection with I. The algorithm cannot end with $I \subset I^*$. If $I \neq I^*$, we can therefore let e be the first element of $I - I^*$ chosen by the algorithm. If $I^* + e \in \mathbf{I}$, then $w(I^* + e) > w(I^*)$, so $I^* + e$ has a unique circuit C. If $f \in C$, then $I' = I^* + e - f \in \mathbf{I}$. The optimality of I^* implies $w(f) \geq w(e)$. By the induced circuit property, f was available when the algorithm selected e, so $w(f) \leq w(e)$. Hence $w(I') = w(I^*)$. Since $I' \cap I \supset I^* \cap I$, this contradicts the choice of I, and we have $I^* = I$.

$G \Rightarrow I$. Given $I_1, I_2 \in \mathbf{I}$ with $k = |I_1| < |I_2|$, we design a weight function for which the success of the greedy algorithm yields the desired augmentation. Let $w(e) = k + 2$ for $e \in I_1$, and let $w(e) = k + 1$ for $e \in I_2 - I_1$. Let $w(e) = 0$ for $e \notin I_1 \cup I_2$. Then $w(I_2) \geq (k+1)^2 > k(k+2) = w(I_1)$, so I_1 is not a maximum weighted independent set. However, the greedy algorithm chooses every element of I_1 before any element of $I_2 - I_1$. Because it finds a maximum weighted independent set, it must continue after absorbing I_1, which requires an element of $I_2 - I_1$ such that $I_1 \cup e \in \mathbf{I}$. $\square$

The property most often used to show that a hereditary system is a matroid is the augmentation property.

8.2.19. Example. *Uniform matroids.* The *uniform matroid* of rank k, denoted $U_{k,n}$ when $|E| = n$, is defined by $\mathbf{I} = \{X \subseteq E : |X| \leq k\}$. This immediately satisfies the base exchange and augmentation properties. The *free matroid* (or *trivial matroid*), is the uniform matroid of rank $|E|$. Uniform matroids are used in building more interesting matroids and in characterizing classes of matroids (Example: $M(K_4 - e)$ is *not* a uniform matroid; neither is $M(K_4)$.)

A linear matroid representable over the field $\mathbf{Z}_2$ or $\mathbf{Z}_3$ is *binary* or *ternary*, respectively. Every graphic matroid is binary, but $U_{2,4}$ is not binary (and hence is not graphic); $U_{2,4}$ is ternary (Exercise 34). $\square$

8.2.20. Example. *Partition matroids.* The *partition matroid* on E induced by a partition of E into blocks $E_1, \ldots, E_k$ is defined by $\mathbf{I} = \{X \subseteq E : |X \cap E_i| \le 1 \text{ for all } i\}$. Since $\varnothing \in \mathbf{I}$, and since $X \in \mathbf{I}$ if and only if its elements lie in distinct blocks, $\mathbf{I}$ is a hereditary family. Given $I_1, I_2 \in \mathbf{I}$ with $|I_2| > |I_1|$, the set I_2 must intersect more blocks than I_1; hence it intersects a block that I_1 misses, and the augmentation property holds. Alternatively, $r(X)$ is the number of blocks having elements in X; this satisfies the absorption property. (Example: $M(K_4 - e)$ is *not* a partition matroid.)

Given a bipartite graph G with bipartition U, V, the incidences with $U = u_1, \ldots, u_k$ define a partition matroid on $E(G)$ (this differs from the transversal matroid on U induced by G.) The blocks are the sets $E_i = \{e \in E(G) : u_i \in e\}$. A set $X \subseteq E(G)$ is a matching in G if and only if X is independent in the partition matroid induced by U *and* in the partition matroid induced by V.

When G has an odd cycle, G has no set of vertices whose incident sets partition $E(G)$. In a digraph, however, each edge has a head and a tail, and we can define the *head partition matroid* and the *tail partition matroid* using the edge partitions induced by incidences with heads and by incidences with tails. (Example: the matroid of Example 8.2.3 arises as the partition matroid on E induced by U in the bipartite graph below, as the head partition matroid in the first digraph, and as the tail partition matroid in the second digraph.) $\square$

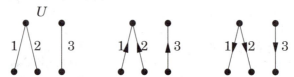

THE SPAN FUNCTION AND DUALITY

We begin by introducing several additional aspects of hereditary systems and properties of matroids involving these aspects. We do this partly to simplify our discussion of matroid duality, which in turn will lead to a characterization of planar graphs using matroids.

In linear algebra, we often consider the subspace "spanned" by a set of vectors. This concept extends to hereditary systems. The definition is suggested by cycle matroids; a set spans itself and the elements that complete circuits with its subsets.

8.2.21. Definition. The *span function* of a hereditary system M is the function σ_M on the subsets of E defined by $\sigma_M(X) = X \cup \{e \in E : Y \cup e \in \mathbf{C}_M$ for some $Y \subseteq X\}$. If $e \in \sigma(X)$, then X *spans* e.

A set in a hereditary system is dependent if and only if it contains a circuit, which by Definition 8.2.21 holds if and only if $e \in \sigma(X - e)$ for some $e \in X$. When M is a hereditary system, we can therefore obtain the independent sets (and all other aspects of M) from σ_M via $\mathbf{I}_M = \{X \subseteq E : (e \in X) \Rightarrow (e \notin \sigma_M(X - e))\}$. The properties of σ that we use in studying matroids are (s1,s2,s3) below (an additional technical condition is needed to characterize the span functions of hereditary systems). First we illustrate property (s3) using graphs.

8.2.22. Example. *Steinitz exchange in cycle matroids.* In the cycle matroid $M(G)$, the meaning of $e \notin \sigma(X)$ is that X contains no path between the endpoints of e. If also $e \in \sigma(X \cup f)$, then adding f completes such a path. The path completes a cycle with e, and hence also $f \in \sigma(X \cup e)$. In the illustration, the solid edges are the edges of X. □

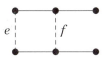

8.2.23. Proposition. If σ is the span function of a hereditary system on E, and $X, Y \subseteq E$, then
s1) $X \subseteq \sigma(X)$ (σ is *expansive*).
s2) $Y \subseteq X$ implies $\sigma(Y) \subseteq \sigma(X)$ (σ is *order-preserving*).
s3) $e \notin \sigma(X)$ and $e \in \sigma(X \cup f)$ imply $f \in \sigma(X \cup e)$ (*Steinitz exchange*).

Proof. Definition 8.2.21 implies immediately that σ is expansive and order-preserving. If $e \in \sigma(X \cup f)$, then e belongs to a circuit C in $X \cup f \cup e$. If also $e \notin \sigma(X)$, then $f \in C$. This circuit yields $f \in \sigma(X \cup e)$, and hence σ satisfies the Steinitz exchange property. □

Properties of the span function lead us to a short proof of a stronger form of the weak elimination property.

8.2.24. Example. *Strong elimination for cycle matroids.* Given two cycles C_1, C_2, we have remarked that the symmetric difference $C_1 \triangle C_2$ is an even graph. Every edge of an even graph belongs to a cycle, and hence inductively $C_1 \triangle C_2$ can be partitioned into pairwise edge-disjoint cycles. This statement is considerably stronger than the statement we will prove for all matroids: if e belongs to distinct circuits C_1, C_2 and $f \in C_1 \triangle C_2$, then f belongs to a circuit contained in $(C_1 \cup C_2) - e$. □

We will need a property that holds for all hereditary systems. The converse holds if and only if the hereditary system is a matroid (the converse is essentially the first property in Theorem 8.2.26). This lemma and theorem can be omitted without loss of continuity.

8.2.25. Lemma. In a hereditary system, $r(X \cup e) = r(X) \implies e \in \sigma(X)$.

Proof. Let Y be a maximum independent subset of X. We have $|Y| = r(X) = r(X \cup e)$, so Y is also a maximum independent subset of $X \cup e$. Hence e completes a circuit with some subset of X contained in Y, and $e \in \sigma(X)$. $\square$

8.2.26. Theorem. If M is a hereditary system, then each condition below is necessary and sufficient for M to be a matroid.
P: *incorporation* - $r(\sigma(X)) = r(X)$ for all $X \subseteq E$.
S: *idempotence* - $\sigma^2(X) = \sigma(X)$ for all $X \subseteq E$.
T: *transitivity of dependence* - if $e \in \sigma(X)$ and $X \subseteq \sigma(Y)$, then $e \in \sigma(Y)$.
C': *strong elimination* - whenever $C_1, C_2 \in \mathbf{C}$, $e \in C_1 \cap C_2$, and $f \in C_1 \triangle C_2$, there exists $C \in \mathbf{C}$ such that $f \in C \subseteq (C_1 \cup C_2) - e$.

Proof. A'$\Rightarrow$P. If we can show that $e \in \sigma(X)$ implies $r(X \cup e) = r(X)$, then strong absorption will yield $r(\sigma(X)) = r(X)$. Choose $e \in \sigma(X) - X$. Let Y be maximal among the independent subsets of X such that $Y \cup e$ is dependent. For each $f \in X \cup e$, the dependence of $Y \cup f \cup e$ and choice of Y implies that $Y \cup f$ is dependent. Since Y is independent, this implies that hence $r(Y \cup f) = r(Y)$. Applying strong absorption to Y and $X \cup e$ now yields $r(X \cup e) = r(Y)$. Using the monotonicity of r, we have $r(X \cup e) = r(Y) \le r(X) \le r(X \cup e)$, and hence $r(X \cup e) = r(X)$ as desired.

P$\Rightarrow$S. Since σ is expansive, we have $\sigma^2(X) \supseteq \sigma(X)$ and need only show that $e \in \sigma^2(X)$ implies $e \in \sigma(X)$. By the incorporation property, $r(\sigma(X) \cup e) = r(\sigma(X))$ and $r(\sigma(X)) = r(X)$. Since $X \subseteq \sigma(X)$, monotonicity of r yields $r(X) \le r(X \cup e) \le r(\sigma(X) \cup e) = r(X)$. Since equality holds throughout, Lemma 8.2.25 yields $e \in \sigma(X)$.

S$\Rightarrow$T. If $X \subseteq \sigma(Y)$, then the order-preserving and idempotence properties of σ imply $\sigma(X) \subseteq \sigma^2(Y) = \sigma(Y)$.

T$\Rightarrow$C'. Given distinct $C_1, C_2 \in \mathbf{C}$ with $e \in C_1 \cap C_2$ and $f \in C_1 - C_2$, we want $f \in \sigma(Y)$, where $Y = (C_1 \cup C_2) - e - f$. We have $f \in \sigma(X)$, where $X = C_1 - f$. By T, it suffices to show $X \subseteq \sigma(Y)$. Since $X - e \subseteq Y \subseteq \sigma(Y)$, we need only show $e \in \sigma(Y)$. Since σ is order-preserving, we have $e \in \sigma(C_2 - e) \subseteq \sigma(Y)$.

C'$\Rightarrow$C. C is a less restrictive statement than C'. $\square$

Idempotence occurs naturally for graphic and linear matroids. The span of a set of vectors contains nothing additional in its span; similarly, every edge that can be added to the span of a set of edges joins two components. This suggests related aspects of hereditary functions.

8.2.27. Definition. The *spanning sets* of a hereditary system on E are the sets $X \subseteq E$ such that $\sigma(X) = E$. The *closed sets* are the sets $X \subseteq E$ such that $\sigma(X) = X$ (also called *flats* or *subspaces*). The *hyperplanes* are the maximal proper closed subsets of E. $\square$

The span function of a matroid is often called its *closure function*. A *closure operator* is an expansive, order-preserving, idempotent function on the set of subsets of a set. A closure operator is the span function of a matroid if and only if it has the Steinitz exchange property. The span function of a hereditary system M is a closure operator if and only if M is a matroid. Matroids arose in lattice theory because the closed sets for a closure operator are the flats of a matroid if and only if they form a "geometric lattice" (MacLane [1936], see also Rota [1964]).

Spanning sets and hyperplanes simplify the treatment of some aspects of matroid duality.

8.2.28. Example. *Duality in cycle matroids.* Every connected plane graph G has a natural dual graph G^* such that $(G^*)^* = G$ (Exercise 7.1.3). A set of edges forms a spanning tree in G if and only if the duals to the remaining edges form a spanning tree in G^* (Exercise 7.1.7). Hence the bases in the cycle matroid $M(G^*)$ are the complements of the bases in $M(G)$. $\square$

We generalize the dual behavior of bases to define duality for matroids and for hereditary systems.

8.2.29. Definition. The *dual* of a hereditary system M on E is the hereditary system M^* whose bases are the complements of the bases of M. The aspects of the dual M^* are $\mathbf{B}^*, \mathbf{C}^*, \mathbf{I}^*, r^*, \sigma^*$, etc., called the cobases, cocircuits, etc., of M. The *subbases* $\mathbf{S}$ of M are the sets $X \subseteq E$ that contain a base. The *hypobases* $\mathbf{H}$ are the maximal subsets containing no base. We write $\overline{X}$ for $E - X$.

8.2.30. Lemma. If M is a hereditary system, then
a) $\mathbf{B}^* = \{\overline{B}: B \in \mathbf{B}\}$ and $M^{**} = M$.
b) $\mathbf{I}^* = \{\overline{S}: S \in \mathbf{S}\}$ and $\mathbf{S}^* = \{\overline{I}: I \in \mathbf{I}\}$.
c) $\mathbf{C}^* = \{\overline{H}: H \in \mathbf{H}\}$ and $\mathbf{H}^* = \{\overline{C}: C \in \mathbf{C}\}$.

Proof. The statement about $\mathbf{B}^*$ is the definition of M^*. It implies that $M^{**} = M$, that the complements of the subbases of M are the independent sets of M^*, and that the complements of the independent sets of M are the subbases of M^*. Also, X is a maximal (proper) subset of E containing no base (a hypobase of M) if and only if $\overline{X}$ is a minimal nonempty set contained in no cobase - a circuit of M^*. Similarly, the hypobases of M^* are the complements of the circuits of M. $\square$

We have chosen "supbase" and "hypobase" to share initials with "spanning" and "hyperplane", because for matroids the spanning sets and supbases are the same, and the hyperplanes and hypobases are the same; neither statement holds for all hereditary systems (Exercise 36).

8.2.31. Lemma. If M is a matroid, then the supbases are the spanning sets, and the hypobases are the hyperplanes.

Proof. A set X is spanning if and only if $\sigma(X) = E$. By the incorporation property, this is equivalent to $r(X) = r(E)$. By the uniformity property, this is equivalent to X containing a base. We leave the proof for hyperplanes to Exercise 36. $\square$

When **B** is a non-empty collection of subsets of E with none containing another, the same property holds for the set of complements of members of **B**. Therefore, the dual of a hereditary system is a hereditary system. The notion of duality becomes useful when we prove that the dual of a matroid is a matroid. This follows easily from a dual version of the base exchange property.

8.2.32. Lemma. If M is a matroid and $B_1, B_2 \in \mathbf{B}$, then for each $e \in B_1 - B_2$ there exists $f \in B_2$ such that $B_2 + e - f$ is a base.

Proof. Since B_2 is a base, $B_2 \cup e$ contains exactly one circuit C. Since B_1 is independent, C also contains an element $f \in B_2 - B_1$. Now $B_2 + e - f$ contains no circuit and has size $r(E)$. $\square$

8.2.33. Theorem. (Whitney [1935]) The dual of a matroid M o E is a matroid with rank function $r^*(X) = |X| - (r(E) - r(\overline{X}))$.

Proof. We have observed that M^* is a hereditary system; now we prove the base exchange property for M^*. If $\overline{B}_1, \overline{B}_2 \in \mathbf{B}^*$ and $e \in \overline{B}_1 - \overline{B}_2$, then $B_1, B_2 \in \mathbf{B}$, with $e \in B_2 - B_1$. By Lemma 8.2.32, there exists $f \in B_1 - B_2$ such that $B_1 + e - f \in \mathbf{B}$. Now $\overline{B}_1 - e + f \in \mathbf{B}^*$ is the desired exchange.

To compute $r^*(X)$, let Y be a maximal coindependent subset of X, so $r^*(X) = r^*(Y) = |Y|$. By Lemma 8.2.30, $\overline{Y}$ is a minimal superset of $\overline{X}$ that contains a base of M. Since $\overline{Y}$ arises from $\overline{X}$ by augmenting a maximal independent subset of $\overline{X}$ to become a base, we have $|\overline{Y}| - |\overline{X}| = r(E) - r(\overline{X})$. With $|\overline{Y}| - |\overline{X}| = |X| - |Y|$, this yields the desired formula $r^*(X) = |X| - (r(E) - r(\overline{X}))$. $\square$

We can restate any property of matroids using dual aspects; Exercise 25 requests a characterization of the set of hyperplanes of a matroid by this method. More subtle results involve relationships between a matroid and its dual.

8.2.34. Proposition. (Dual augmentation property) Suppose M is a matroid. If $X \in \mathbf{I}$ and $X' \in \mathbf{I}^*$ are disjoint, then there are disjoint $B \in \mathbf{B}$ and $B' \in \mathbf{B}^*$ such that $X \subseteq B$ and $X' \subseteq B^*$.

Proof. Since X' is coindependent in M, $\overline{X'}$ is spanning in M. Hence every maximal independent subset of $\overline{X'}$ is a base; we augment $X \subseteq \overline{X'}$ to a base B contained in $\overline{X'}$. The cobase $B' = \overline{B}$ contains X'. $\square$

We will use cycle matroids to characterize planar graphs. The next result enables us to describe the cocircuits of a cycle matroid.

8.2.35. Proposition. Given a matroid M, $\mathbf{C}^*$ consists of the minimal sets intersecting every base, and $\mathbf{B}$ consists of the minimal sets intersecting every cocircuit.

Proof. The cocircuits are the minimal sets contained in no cobase. Because $B \in \mathbf{B}$ if and only if $\overline{B} \in \mathbf{B}^*$, a set is contained in no cobase if and only if it intersects every base. Similarly, the cobases are the maximal sets containing no cocircuit, so the complements of the cobases are the minimal sets intersecting every cocircuit. $\square$

8.2.36. Corollary. The cocircuits of a cycle matroid $M(G)$ are the bonds of G.

Proof. By Proposition 8.2.35, the cocircuits are the minimal sets intersecting every maximal forest. Hence the cocircuits are the minimal edge sets whose deletion increases the number of components; these are the bonds. $\square$

8.2.37. Definition. The *bond matroid* or *cocycle matroid* of a graph G is the matroid $M^*(G)$ (the dual of the cycle matroid).

The weak elimination property now applies to bonds. Exercise 4.1.20 makes the stronger statement that the symmetric difference of two bonds is an edge-disjoint union of bonds (this also holds for cycles).

Since a cycle must return to its starting point, it cannot intersect a bond in exactly one edge. This generalizes to matroids as another characterization of cocircuits.

8.2.38. Theorem. The cocircuits of a matroid M on E are the minimal non-empty sets $C^* \subseteq E$ such that $|C^* \cap C| \neq 1$ for every $C \in \mathbf{C}$.

Proof. To show that every cocircuit has this property, suppose that $C \in \mathbf{C}$, $C^* \in \mathbf{C}^*$, $C^* \cap C = e$. Then $C - e \in \mathbf{I}$ and $C^* - e \in \mathbf{I}^*$, and the dual augmentation property yields $B \in \mathbf{B}$ and $\overline{B} \in \mathbf{B}^*$ such that $C - e \subseteq B$ and $C^* - e \subseteq \overline{B}$. Since e must appear in B or B^*, we obtain $C \in \mathbf{I}$ or $C^* \in \mathbf{I}^*$.

For the converse, it suffices to show that every non-empty coindependent set meets some $C \in \mathbf{C}$ in one element; since cocircuits do not, the cocircuits will then be the *minimal* non-empty sets that do not. Suppose $X^* \in \mathbf{I}^*$; let B^* be a cobase containing X^* and let $B = \overline{B^*}$. For each $e \in X^*$, $B \cup e$ contains a circuit C, and $X^* \cap C = \{e\}$. □

MINORS AND PLANAR GRAPHS

Given a graph G, we can obtain smaller graphs by repeatedly deleting and/or contracting edges. The resulting graphs, called *minors* of G, apply in various contexts. Wagner [1937] proved that G is planar if and only if it has no minor isomorphic to K_5 or $K_{3,3}$. Hadwiger [1943] conjectured that G is k-colorable if G has no minor isomorphic to K_{k+1}. A simple graph is a forest if and only if it does not have C_3 as a minor.

To generalize these operations to matroids, we need to know how deletion and contraction affect cycle matroids. The acyclic subsets of $E(G - e)$ are precisely the acyclic subsets of $E(G)$ that omit e. The acylic subsets of $E(G \cdot e)$ are the subsets of $E(G) - e$ whose union with e is acyclic in G. A dual description of contraction is more convenient: X contains a spanning tree of each component of $G \cdot e$ if and only if $X \cup e$ contains a spanning tree of each component of G.

In extending the definitions to matroids, we want the notation to extend in a natural way. This causes difficulty, because discussion of graph minors tends to focus on the edges removed, while discussion of matroid minors tends to focus on the set of elements that remains. We compromise by using common matroid notation to discuss the matroid on the set that remains, while extending the graph notation to describe matroids obtained by deleting or contracting a single element.

8.2.39. Definition. Given a hereditary system M on E, the *restriction of M to $F \subseteq E$*, denoted $M|F$ and obtained by *deleting $\overline{F}$*, is the hereditary system defined by $\mathbf{I}_{M|F} = \{X \subseteq F : X \in \mathbf{I}_M\}$. The *contraction of M to $F \subseteq E$*, denoted $M \cdot F$ and obtained by *contracting $\overline{F}$*, is the hereditary system defined by $\mathbf{S}_{M.F} = \{X \subseteq F : X \cup \overline{F} \in \mathbf{S}_M\}$. When $F = E - e$, we write $M - e = M|F$ and $M \cdot e = M.F$. The *minors* of a hereditary system M are the hereditary systems produceable from M by succession of deletions and contractions.

These definitions guarantee that $M|F$ and $M.F$ are hereditary systems. Our definition of contraction using subpases (Definition 8.2.39) yields the natural duality between contraction and deletion:

8.2.40. Proposition. For hereditary systems, restriction and contraction are dual operations: $(M.F)^* = (M^*|F)$ and $(M|F)^* = (M^*.F)$.

Proof. $\mathbf{I}_{(M.F)^*} = \{X \subseteq F: F - X \in \mathbf{S}_{M.F}\} = \{X \subseteq F: (F-X) \cup \overline{F} \in \mathbf{S}_M\} = \{X \subseteq F: \overline{X} \in \mathbf{S}_M\} = \{X \subseteq F: X \in \mathbf{I}_{M^*}\} = \mathbf{I}_{M^*|F}$. For the second statement, apply the first to M^* and take duals. $\square$

8.2.41. Corollary. The behavior of cycle matroids and bond matroids under deletion or contraction of an edge $e \in E(G)$ is

$$M(G - e) = M(G) - e \qquad M^*(G - e) = M^*(G) \cdot e$$
$$M(G \cdot e) = M(G) \cdot e \qquad M^*(G \cdot e) = M^*(G) - e$$

Proof. The definitions of deletion and contraction were made so that the statements in the first column describe the behavior of cycle matroids. Using these and Proposition 8.2.40, we compute

$$M^*(G - e) = [M(G - e)]^* = [M(G) - e]^* = M^*(G) \cdot e$$
$$M^*(G \cdot e) = [M(G \cdot e)]^* = [M(G) \cdot e]^* = M^*(G) - e. \; \square$$

Indeed, every restriction or contraction of a matroid is a matroid.

8.2.42. Theorem. Given $F \subseteq E$ and a matroid M on E, both $M|F$ and $M.F$ are matroids on F. In terms of r_M, their rank functions are $r_{M|F}(X) = r_M(X)$ and $r_{M.F}(X) = r_M(X \cup \overline{F}) - r_M(\overline{F})$.

Proof. Given two sets in $\mathbf{I}_{M|F}$, we can apply the augmentation property from M; hence $M|F$ satisfies the augmentation property and restrictions of matroids are matroids. Using duality, $M.F = (M^*|F)^*$ is also a matroid. The rank function follows immediately for $M|F$. Together with repeated application of Theorem 8.2.33 to $(M^*|F)^*$, this yields the rank function for $M.F$ (Exercise 39). $\square$

The formula for $r_{M.F}$ yields a description of the independent sets: $X \in \mathbf{I}_{M.F}$ if and only if adding X to $\overline{F}$ increases the rank by $|X|$.

When forming minors, restriction and contraction commute (Exercise 43). Minors are used in characterizing classes of matroids by forbidden substructures; for example, a matroid is binary if and only if it does not have $\mathbf{U}_{2,4}$ as a minor. Minors also are used to produce a winning strategy for a matroid generalization of Bridg-it (Theorem 2.1.15).

8.2.43. Definition. Given $e \in E$ and a matroid M on E, the *Shannon Switching Game* (M, e) is played by the Spanner and the Cutter. The Cutter deletes elements of $E - e$ and the Spanner seizes them, one per move. The Spanner aims to seize a set that spans e, and the Cutter aims to prevent this. The Cutter moves first.

Having the Spanner move first can be simulated by adding an element e' such that $\{e, e'\}$ is a circuit; the Cutter must begin by deleting e' to avoid losing immediately. Bridg-it occurs by letting M be the cycle matroid of the graph in Theorem 2.1.15 with e the "auxiliary edge" and e' an extra auxiliary edge. The spanning tree strategy for the Spanner results from the following sufficient condition for a winning strategy. The condition is also necessary, but proving that takes more effort.

8.2.44. Theorem. (Lehman [1964]) In the Shannon Switching Game (M, e), the Spanner has a winning strategy if there are disjoint subsets X_1, X_2 of $E - e$ such that $e \in \sigma(X_1) = \sigma(X_2)$.

Proof. We use X_1, X_2 to produce a winning strategy. Let $X = \sigma(X_1) = \sigma(X_2)$. Since the Spanner can ignore deletions outside X and play in $M|(X \cup e)$, we may assume that X_1, X_2 are disjoint bases. If the Cutter plays g and the Spanner plays f, then g is no longer available and f cannot be deleted; the effect is deletion and contraction. Letting $M' = (M - g) \cdot f$, we have $e \in \sigma_{M'}(X)$ if and only if $g \notin X$ and $e \in \sigma_M(X + f)$. The Spanner wins if e is a loop in M', which is equivalent to $e \in \sigma_M(F)$, where F is the set seized by the Spanner.

If $|E| = 1$, then e is a loop and the Spanner wins; we proceed by induction on $|E|$. It suffices to provide an immediate answer f to g so that $M' = (M - g) \cdot f$ has two disjoint bases. If the Cutter deletes g not in X_1 or X_2, then the spanner seizes an arbitrary f, and the two sets $X_1 - g - f$ and $X_2 - g - f$ are disjoint and spanning in M'. Hence we may assume $g \in X_1$. The base exchange property yields $f \in X_2$ such that $X' = X_1 - g + f \in \mathbf{B}$. Now $X' - f$ and $X_2 - f$ are disjoint bases avoiding e in the game (M', e). $\square$

The dual nature of deletion and contraction is most intuitive when we consider planar graphs. Deleting an edge in a plane graph G corresponds to contracting the corresponding dual edge in G^*, and contracting it corresponds to deleting the edge in the dual, as illustrated below.

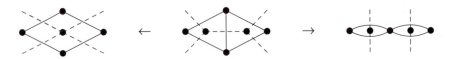

In Theorem 7.1.10, we proved that a set of edges in a plane graph G forms a cycle if and only if the corresponding dual edges form a bond in G^*. Using the natural bijection between edges and dual edges, this tells us that the cycle matroid of a plane graph G is (isomorphic to) the bond matroid of G^*. By Corollary 8.2.36, for each graph H the bond matroid of H is $[M(H)]^*$. Applying this to G and to G^* tells us that the bond matroid of G is (isomorphic to) the cycle matroid of G^*. In particular, the bond matroid of G is graphic. Using Kuratowski's Theorem, we will prove that this condition is also sufficient for planarity.

Whitney [1933a] approached this question by defining an "abstract dual" of a graph. Changing his definition slightly, we say that H is an *abstract dual* of G if there is a bijection $\phi \colon E(G) \to E(H)$ such that $X \subseteq E(G)$ is a bond in G if and only if $\phi(X)$ is the edge set of a cycle in H. With this definition, the statement that G has an abstract dual is the same as the statement that the bond matroid of G is graphic; the bijection ϕ establishes an isomorphism between $M^*(G)$ and $M(H)$.

8.2.45. Theorem. (Whitney [1933a]) A graph G is planar if and only if its bond matroid $M^*(G)$ is graphic.

Proof. We have demonstrated that the condition is necessary. For sufficiency, we first prove that existence of an abstract dual is preserved under deletion and contraction of edges. Suppose G has an abstract dual H, so that $M(H) \cong M^*(G)$. Let e' be the edge of H corresponding to e under the bijection. To prove that $H \cdot e'$ is an abstract dual of $G - e$ and that $H - e'$ is an abstract dual of $G \cdot e$, we use Corollary 8.2.36 to compute

$$M^*(G - e) = M^*(G) \cdot e \cong M(H) \cdot e' = M(H \cdot e').$$
$$M^*(G \cdot e) = M^*(G) - e \cong M(H) - e' = M(H - e').$$

To complete the proof, we suppose to the contrary that G is nonplanar and has an abstract dual. By Kuratowski's Theorem, G contains a subdivision G' of K_5 or $K_{3,3}$. We obtain G' from G by deleting some edges of G, and then we obtain K_5 or $K_{3,3}$ from G' by contracting some edges incident to vertices of degree 2. Hence K_5 or $K_{3,3}$ is a minor of G. It suffices to prove that neither of these has an abstract dual.

Suppose H is an abstract dual of G. Also G is an abstract dual of H, since $M^*(G) \cong M(H)$ if and only if $M(G) \cong M^*(H)$. If G has girth g, then bonds of H have size at least g, so $\delta(H) \geq g$. Also $e(H) = e(G)$, and the degree-sum formula implies $n(H) \leq \lfloor 2e(H)/\delta(H) \rfloor = \lfloor 2e(G)/g \rfloor$.

Suppose H is an abstract dual of K_5. Since K_5 has girth 3, $n(H) \leq \lfloor 20/3 \rfloor = 6$. Since all bonds of K_5 have size four or six, all cycles of H have length four or six, and H is a simple bipartite graph. However, no simple bipartite graph with at most six vertices has 10 edges.

Suppose H is an abstract dual of $K_{3,3}$. Since $K_{3,3}$ has girth 4, $n(H) \leq \lfloor 18/4 \rfloor = 4$. Since all bonds of $K_{3,3}$ have size at least three, all cycles of H have length at least three. However, no simple graph with at most four vertices has 9 edges. $\square$

Kuratowski's Theorem also immediately yields the result of Wagner [1937] that G is planar if and only if G has no minor isomorphic to K_5 or $K_{3,3}$. We used Kuratowski's Theorem above to observe that every nonplanar graph has K_5 or $K_{3,3}$ as a minor. Conversely, no planar graph has K_5 or $K_{3,3}$ as a minor, since planarity is preserved under deletion and contraction of edges.

The argument that bond matroids of plane graphs are graphic shows that every "geometric" dual of a planar graph is an abstract dual; we have seen that the dual need not be unique. Nevertheless, the cycle matroid of each graph dual to G must be $M^*(G)$, and hence all duals of G have the same cycle matroid. Whitney [1933b] determined when graphs have the same cycle matroid (see Exercise 46).

MATROID INTERSECTION

Matroid theory took a great step forward with the Matroid Intersection and Matroid Union Theorems. These provided a unified context for many known min-max relations; we will obtain various graph-theoretic results as applications. The Matroid Intersection Theorem is a min-max relation for the maximum size of a common independent set in two matroids M_1, M_2 on the same ground set. We can view the intersection of two matroids as a hereditary system, but *not* as a matroid. We write $\mathbf{I}_i$ and r_i instead of $\mathbf{I}_{M_i}$ and r_{M_i} for the independent sets and rank function of M_i. Again $\overline{X}$ denotes $E - X$.

8.2.46. Definition. Given hereditary systems M_1, M_2 on E, the *intersection* of M_1 and M_2 is the hereditary system whose independent sets are $\{X \subseteq E: X \in \mathbf{I}_1 \cap \mathbf{I}_2\}$.

For example, the intersection of the two natural partition matroids on the edges of a bipartite graph G has as its independent sets the matchings of G. These are generally not the independent sets of a matroid (Exercise 2), which explains why the greedy algorithm does not solve maximum weighted bipartite matching.

8.2.47. Theorem. (Matroid Intersection Theorem - Edmonds [1970]). Given two matroids M_1, M_2 on E, the size of a largest common independent set is
$$\max\{|I|: I \in \mathbf{I}_1 \cap \mathbf{I}_2\} = \min_{X \subseteq E}\{r_1(X) + r_2(\overline{X})\}.$$

Proof. (Seymour [1976]) For weak duality, consider arbitrary $I \in \mathbf{I}_1 \cap \mathbf{I}_2$ and $X \subseteq E$. The sets $I \cap X$ and $I \cap \overline{X}$ are also common independent sets, and $|I| = |I \cap X| + |I \cap \overline{X}| = r_1(I \cap X) + r_2(I \cap \overline{X}) \leq r_1(X) + r_2(\overline{X})$.

To show that equality is achievable, we use induction on $|E|$; when $|E| = 0$ both sides equal 0. Also, if every element of E is a loop in M_1 or M_2, then $\max|I| = 0 = r_1(X) + r_2(\overline{X})$, where X consists of all loops in M_1. Hence we may assume that $|E| > 0$ and that some $e \in E$ is not a loop in either matroid. Let $F = E - e$, and consider the matroids $M_1|F$, $M_2|F$, $M_1 . F$, and $M_2 . F$.

Let $k = \min_{X \subseteq E} r_1(X) + r_2(\overline{X})$; we want to find a common independent k-set in M_1 and M_2. If there is none, then $M_1|F$ and $M_2|F$ have no common independent k-set, and $M_1.F$ and $M_2.F$ have no common independent $k-1$-set. By the induction hypothesis and the expressions for $r_{M|X}$ and $r_{M.X}$ in terms of r_M for arbitrary matroids, we have

$$r_1(X) + r_2(F - X) \le k - 1 \qquad \text{for some } X \subseteq F.$$
$$r_1(Y + e) - 1 + r_2(F - Y + e) - 1 \le k - 2 \qquad \text{for some } Y \subseteq F.$$

We use $(F - Y) + e = \overline{Y}$ and $F - X = \overline{X \cup e}$ and sum the two inequalities:

$$r_1(X) + r_2(\overline{X \cup e}) + r_1(Y \cup e) + r_2(\overline{Y}) \le 2k - 1.$$

Now we apply submodularity of r_1 to X and $Y \cup e$ and submodularity of r_2 to $\overline{Y}$ and $\overline{X \cup e}$. For clarity, write $U = X \cup e$ and $V = Y \cup e$. Applying the result to the preceding inequality yields

$$r_1(X \cup V) + r_1(X \cap V) + r_2(\overline{Y} \cup \overline{U}) + r_2(\overline{Y} \cap \overline{U}) \le 2k - 1.$$

Since $\overline{Y} \cap \overline{U} = \overline{X \cup V}$ and $\overline{Y} \cup \overline{U} = \overline{X \cap V}$, the left side sums two instances of $r_1(Z) + r_2(\overline{Z})$, and the hypothesis $k \le r_1(Z) + r_2(\overline{Z})$ for all $Z \subseteq E$ yields $2k \le 2k - 1$. $\square$

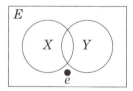

It can be helpful to restrict the range of the minimization.

8.2.48. Corollary. The maximum size of a common independent set in matroids M_1, M_2 on E is the minimum of $r_1(X_1) + r_2(X_2)$ over sets X_1, X_2 such that $X_1 \cup X_2 = E$ and each X_i is closed in M_i.

Proof. The incorporation property implies that $r_i(\sigma_i(X)) = r_i(X)$. $\square$

We have seen the result of the Matroid Intersection Theorem for various special classes of matroids. We proved the König-Egerváry Theorem directly in Theorem 3.1.11, and we proved the Ford-Fulkerson characterization of CSDR's from Menger's Theorem in Theorem 4.2.21. Whenever we have two matroids on the same set, the Matroid Intersection Theorem tells us that there must be a min-max relation for the maximum size of a common independent set, tells what the result should be, and provides a proof.

8.2.49. Corollary. (König [1931], Egerváry [1931]) In a bipartite graph, the largest matching and smallest vertex cover have equal size.

Proof. When M_1 and M_2 are the partition matroids on the edges induced by the partite sets U_1, U_2 of G, the matchings are the common independent sets. Given $X_1, X_2 \subseteq E$, the rank $r_i(X_i)$ is the number of vertices of U_i incident to edges in X_i. Hence if $X_1 \cup X_2 = E$, then G has a vertex cover of size $r_1(X_1) + r_2(X_2)$, using vertices of U_i to cover X_i. Conversely, if $T_1 \cup T_2$ is a vertex cover with $T_i \subseteq U_i$, let X_i be the set of edges incident to T_i; we have $X_1 \cup X_2 = E$ with X_i closed in M_i and $r_1(X_1) + r_2(X_2) = |T_1| + |T_2|$. We conclude that

$$\alpha'(G) = \max\{|I|: I \in \mathbf{I}_1 \cap \mathbf{I}_2\} = \min\{r_1(X_1) + r_2(X_2)\} = \beta(G). \quad \square$$

For the next corollary, we will need an expression for the rank function of a transversal matroid.

8.2.50. Example. *Transversal matroids.* Suppose $A_1 \cup \cdots \cup A_m = E$, and let G be the corresponding incidence graph with partite sets E and $[m]$. Consider $X \subseteq E$. If $N(Y) < |Y|$ for some $Y \subseteq X$, then Y forces at least $|Y| - |N(Y)|$ unsaturated elements in X. Hall's Condition yields $r(X) = \min_{Y \subseteq X} \{|X| - (|Y| - |N(Y)|)\}$ (Exercise 3.1.23).

Ideas of Ore [1955] yield an alternative expression. Let the *defect* of $X \in \mathbf{I}$ be $d(X) = m - |X|$, the number of sets not represented, and let $A(J) = \cup_{i \in J} A_i$, where $J \subseteq [m]$. Then $\mathbf{A}$ has a partial transversal with defect at most d if and only if $|A(J)| \geq |J| - d$ for all $J \subseteq [m]$ (see Exercise 51). Letting $k = m - d$, we obtain $r(M) = \max\{k: |A(J)| - |J| + m \geq k$ for all $J \subseteq [m]\}$. To determine $r(X)$, we restrict this to $A(J) \cap X$, obtaining $r(X) = \min_{J \subseteq [m]} \{|A(J) \cap X| - |J| + m\}$.

Below we illustrate the distinction between these formulas for $r(X)$. The first uses neighborhoods of subsets of X; the second uses neighborhoods of subsets of $[m]$. Further material on transversals appears in Mirsky [1971] and in Lovász-Plummer [1986]. $\square$

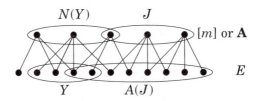

8.2.51. Corollary. (Ford-Fulkerson [1958]) There is a common system of distinct representatives (CSDR) for $\mathbf{A} = \{A_1, \ldots, A_m\}$ and $\mathbf{B} = \{B_1, \ldots, B_m\}$ if and only if, for each $I, J \subseteq [m]$,

$$|(\cup_{i \in I} A_i) \cap (\cup_{j \in J} B_j)| \geq |I| + |J| - m.$$

Proof. A common partial SDR is a common independent set in the two transversal matroids M_1, M_2 induced on E by $\mathbf{A}$ and $\mathbf{B}$. To determine when there is a complete CSDR, we need only restate the condition $r_1(X) + r_2(\overline{X}) \geq m$ to find the appropriate condition on the set systems.

The defect formula for r_1, r_2 yields

$$r_1(X) + r_2(\overline{X}) = \min_{I \subseteq [m]} \{|A(I) \cap X| - |I| + m\} + \min_{J \subseteq [m]} \{|B(J) \cap \overline{X}| - |J| + m\}.$$

Hence $r_1(X) + r_2(\overline{X}) \geq m$ for all X if and only if

$$|A(I) \cap X| + |B(J) \cap \overline{X}| \geq |I| + |J| - m \text{ for all } X \subseteq E \text{ and } I, J \subseteq [m].$$

Given I, J, consider the contributions to the left side from each element of E. Every element of $A(I) \cap B(J)$ is counted exactly once whether it belongs to X or $\overline{X}$. Elements of $A(I) - B(J)$ are counted if and only if they belong to X, and elements of $B(J) - A(I)$ are counted if and only if they belong to $\overline{X}$. Hence the left side is minimized when for a pair I, J when $A(I) - B(J)$ is placed in $\overline{X}$ and $B(J) - A(I)$ is placed in X. In this case the left side equals $|A(I) \cap B(J)|$, and we obtain the Ford-Fulkerson condition. $\square$

The augmenting path approach to maximum bipartite matching generalizes to the matroid intersection problem. The algorithm yields a common independent set of maximum size and a set X such that $r_1(X) + r_2(\overline{X})$ has that size as its value. Finding a maximum common independent set in three matroids is NP-complete (Exercises 55,56).

MATROID UNION

Although the intersection of two matroids is generally not a matroid, a natural concept of union does yield a matroid. The Matroid Union Theorem states this, but its more important part is a min-max relation for the rank function. The Matroid Intersection and Union Theorems are equivalent in that each can be derived from the other. Welsh [1976] begins with the Matroid Union Theorem; here we obtain it from the Matroid Intersection Theorem.

8.2.52. Definition. Given hereditary systems $M_1, \ldots, M_k$ on E, the *union* $M = M_1 \cup \cdots \cup M_k$ is the hereditary system on E with $\mathbf{I}_M = \{I_1 \cup \cdots \cup I_k : I_i \in \mathbf{I}_i\}$. Given hereditary systems $M_1, \ldots, M_k$ on disjoint sets $E_1, \ldots, E_k$, the *matroid sum* $M = M_1 \oplus \cdots \oplus M_k$ is the hereditary system defined on $E_1 \cup \cdots \cup E_k$ with $\mathbf{I}_M = \{I_1 \cup \cdots \cup I_k : I_i \in \mathbf{I}_i\}$.

The matroid sum $M_1 \oplus \cdots \oplus M_k$ on $E_1, \ldots, E_k$ can be expressed as the union of $M_1', \ldots, M_k'$ on $E' = E_1 \cup \cdots \cup E_k$ by letting M_i' be a copy of M_i with the additional elements of $E' - E_i$ added as loops. When each M_i is a uniform matroid, the matroid sum is called a *generalized partition matroid*. Here $E_1, \ldots, E_k$ partition E, there are positive integers

$r_1, \ldots, r_k$, and $X \in \mathbf{I}$ if $|X \cap E_i| \le r_i$. The partition matroids defined earlier arise when all $r_i = 1$.

8.2.53. Proposition. Given matroids $M_1, \ldots, M_k$ on disjoint sets $E_1, \ldots, E_k$, the direct sum $M = M_1 \oplus \cdots \oplus M_k$ is a matroid.

Proof. Since the E_i's are pairwise disjoint, the intersection of any $I \in \mathbf{I}$ with each E_i is independent in M_i (this holds whenever $\{M_i\}$ are hereditary systems). If $I_1, I_2 \in \mathbf{I}$ with $|I_2| > |I_1|$, then for some i, $|I_2 \cap E_i| > |I_1 \cap E_i|$. Since both sets are independent in M_i, we can augment $I_1 \cap E_i$ from $I_2 \cap E_i$ and therefore I_1 from I_2. Hence $M_1 \oplus \cdots \oplus M_k$ satisfies the augmentation property. $\square$

Using a matroid sum, we prove that the union of matroids is always a matroid, and we compute the rank function.

8.2.54. Theorem. (Matroid Union Theorem - Edmonds-Fulkerson [1965], Nash-Williams [1966]). If $M_1, \ldots, M_k$ are matroids on E with rank functions $r_1, \ldots, r_k$, then the union $M = M_1 \cup \cdots \cup M_k$ is a matroid with rank function $r(X) = \min_{Y \subseteq X}(|X - Y| + \Sigma r_i(Y))$.

Proof. (following Schrijver [to appear]). First we observe that we only need to compute $r(E)$. If we consider the restriction of the hereditary system M to the set X, we have $\mathbf{I}_{M|X} = \{Y \subseteq X : Y \in \mathbf{I}_M\}$ and $r_{M|X}(Y) = r_M(Y)$ for $Y \subseteq X$. We then have $M|X = \cup_i(M_i|X)$, which implies that it suffices to prove the rank formula for the full set E. After computing this, we verify the submodularity property to prove that M is a matroid.

Imagine a k by $|E|$ grid of elements E' in which the jth column E_j consists of k copies of the element $e_j \in E$. To compute $r(E)$, we will apply the Matroid Intersection Theorem to two matroids N_1, N_2 defined on E', because a set will be independent in both N_1 and N_2 if and only if there is an independent set of the same size in M. Let M_i' be a copy of M_i defined on the elements E^i of row i in E'. Let N_1 be the direct sum matroid $M_1' \oplus \cdots \oplus M_k'$, and let N_2 be the partition matroid induced on E' by the column partition $\{E_j\}$.

Each set $X \in \mathbf{I}_M$ has a decomposition as a *disjoint* union of subsets $X_i \in \mathbf{I}_i$, because $\mathbf{I}_i$ is an hereditary family. Given a decomposition $\{X_i\}$ of $X \in \mathbf{I}_M$, let X_i' be the copy of X_i in E^i. Since $\{X_i\}$ are disjoint, $\cup X_i'$ is independent in N_2, and $X_i \in \mathbf{I}_i$ implies that $\cup X_i'$ is also independent in N_1. From $X \in \mathbf{I}_M$, we have constructed $\cup X_i'$ of size $|X|$ in $\mathbf{I}_{N_1} \cap \mathbf{I}_{N_2}$. Conversely, any $X' \in \mathbf{I}_{N_1} \cap \mathbf{I}_{N_2}$ corresponds to a decomposition of a set in $\mathbf{I}_M$ of size $|X'|$ when the sets $X' \cap E^i$ are transferred back to E, because N_2 forbids multiple copies of elements.

Hence $r(E) = \max\{|I| : I \in \mathbf{I}_{N_1} \cap \mathbf{I}_{N_2}\}$. To compute this, let the rank functions of N_1, N_2 be q_1, q_2, and let r_i' be the rank function of the copy

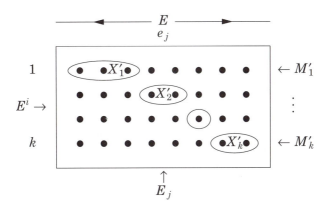

M_i' of M_i on E^i. We have $q_1(X') = \Sigma r_i'(X' \cap E^i)$, and $q_2(X')$ is the number of elements of E that have copies in X'. The Matroid Intersection Theorem yields $r(E) = \min_{X' \subseteq E'} \{q_1(X') + q_2(E' - X')\}$.

The minimum is achieved by a set X' such that $E' - X'$ is closed in N_2. The closed sets in the partition matroid N_2 are the sets that contain all or none of the copies of each element--the unions of full columns of E'. Given X' with $E' - X'$ closed in N_2, let $Y \subseteq E$ be the set of elements whose copies comprise X'. Then $q_2(E' - X') = |E - Y|$, and X' contains all copies of the elements of Y, so $q_1(X') = \Sigma r_i'(X' \cap E^i) = \Sigma r_i(Y)$. We conclude that $r(E) = \min_{Y \subseteq E} \{|E - Y| + \Sigma r_i(Y)\}$.

To show that M is a matroid, we verify submodularity for r. Given $X, Y \subseteq E$, the formula for r yields $U \subseteq X$, $V \subseteq Y$ such that $r(X) = |X - U| + \Sigma r_i(U)$ and $r(Y) = |Y - V| + \Sigma r_i(V)$. Since $U \cap V \subseteq X \cap Y$ and $U \cup V \subseteq X \cup Y$, the formula for r guarantees $r(X \cap Y) + r(X \cup Y) \leq |(X \cap Y) - (U \cap V)| + \Sigma r_i(U \cap V) + |(X \cup Y) - (U \cup V)| + \Sigma r_i(U \cup V)$. Perusal of a Venn diagram confirms that $|(X \cap Y) - (U \cap V)| + |(X \cup Y) - (U \cup V)| = |X - U| + |Y - V|$, and submodularity of the r_i yields $r_i(U \cap V) + r_i(U \cup V) \leq r_i(U) + r_i(V)$; we conclude that $r(X \cap Y) + r(X \cup Y) \leq r(X) + r(Y)$. $\square$

Because we applied the Matroid Intersection Theorem, we needed to know that N_1 was a matroid, which depended on $\{M_i\}$ being matroids. Hence this rank formula does not apply for unions of arbitrary hereditary systems.

The Matroid Union Theorem can be used to prove that the condition in Theorem 8.2.44 for the Spanner to win the Shannon Switching Game is both sufficient and necessary (Lehman [1964]). The theorem also yields formulas for the optimal number of bases in packing or covering a matroid. Each bound is an optimization over subsets of E in which the extreme is always attained by a closed set, since switching from X to $\sigma(X)$ improves the numerator without changing the denominator. The applications to graphs were noted by Edmonds [1965b,c].

8.2.55. Corollary. (Matroid Covering Theorem - Edmonds [1965b]). Given a loopless matroid M on E, The minimum number of independent sets whose union is E is $\max_{X \subseteq E} \lceil \frac{|X|}{r(X)} \rceil$.

Proof. E is the union of k independent sets in M if and only if E is an independent set in the union M' of k matroids $M_1, \ldots, M_k$ that are copies of M on E. This requires $r'(E) = |E|$, which by the Matroid Union Theorem requires $|\overline{Y}| + kr(Y) = |\overline{Y}| + \Sigma r_i(Y) \geq r'(E) = |E|$ for all $Y \subseteq E$. Collecting terms yields E as the union of k independent sets if and only if $kr(Y) \geq |Y|$ for all $Y \subseteq E$. $\square$

8.2.56. Corollary. (Nash-Williams [1964]) The minimum number of forests needed to cover the edges of a graph G, called its *arboricity*, is $\max_{H \subseteq G} \lceil \frac{e(H)}{(n(H) - 1)} \rceil$.

Proof. We apply Corollary 8.2.55 to $M(G)$. If G is connected, then the best lower bound comes from an induced subgraph plus isolated vertices (corresponding to a closed set in $M(G)$). This yields the formula for the arboricity, first proved by Nash-Williams [1964] using a difficult ad hoc argument. $\square$

8.2.57. Corollary. (Matroid Packing Theorem - Edmonds [1965c]).). Given a matroid M on E, the maximum number of pairwise disjoint bases equals $\min_{X: r(X) < r(E)} \lfloor \frac{|E| - |X|}{r(E) - r(X)} \rfloor$.

Proof. E contains k disjoint bases if and only if $r'(E) = kr(E)$ in the union M' of k matroids $M_1, \ldots, M_k$ that are copies of M on E. By the Matroid Union Theorem, this requires $|\overline{Y}| + kr(Y) = |\overline{Y}| + \Sigma r_i(Y) \geq r'(E) = kr(E)$ for all $Y \subseteq E$. Collecting terms yields k disjoint bases in E if and only if $(|E| - |Y|)/(r(E) - r(Y)) \geq k$ for all $Y \subseteq E$. $\square$

8.2.58. Corollary. (Nash-Williams [1961], Tutte [1961]) A graph G has k pairwise edge-disjoint spanning trees if and only if, for every vertex partition P, there are at least $k(|P| - 1)$ edges with endpoints in different parts of P.

Proof. The closed sets in $M(G)$ correspond to partitions of $V(G)$. For each such partition $V_1, \ldots, V_p$, the corresponding closed set X is $\cup E(G[V_i])$, which has rank $n - p$ if the subgraphs $G[V_i]$ are connected. Hence G has k disjoint spanning trees if and only if the condition holds. The original arguments were intricate. $\square$

The theory of matroids and their generalizations is an enormous subject; in these few pages we could only mention a few topics closely related to graph theory. Books that discuss these topics and others in matroid theory include: Korte-Lovász-Schrader [1991], Björner-Las Vergnas-Sturmfels-White-Ziegler [1992], Kung [1986a], Oxley [1992], Recski [1989], Truemper [1992], Tutte [1970], Welsh [1976], and White [1986, 1992].

EXERCISES

8.2.1. (–) Characterize the graphs G such that the family of stable sets is the family of independent sets of a matroid.

8.2.2. Characterize the graphs G whose matchings form the collection of independent sets of a matroid on $E(G)$.

8.2.3. Determine which uniform matroids are graphic. Characterize the graphs whose cycle matroids are uniform matroids.

8.2.4. Determine which partition matroids are graphic. Characterize the graphs whose cycle matroids are partition matroids.

8.2.5. Using only linear dependence, prove that adding an element to a linearly independent set of vectors creates at most one minimal dependent set.

8.2.6. Describe the circuits of a transversal matroid M in terms of the corresponding bipartite graph G. Using only properties of bipartite graphs, prove that M satisfies the weak elimination property.

8.2.7. Consider the cycle matroid $M(G)$. Let $k(X)$ be the number of components of the spanning subgraph G_X with edge set X; as in Example 8.2.7, $r(X) = n - k(X)$. Let H be the intersection graph of the family of sets consisting of the components of G_X and the components of G_Y.
   a) Count the vertices and components of H in terms of $k(X), k(Y), k(X \cap Y)$, and prove that $k(X \cup Y) \geq e(H)$.
   b) Use part (a) to prove the submodularity property for M directly (without using properties of matroids). (Aigner [1979])

8.2.8. Use the König-Egervary Theorem to prove directly that the rank function of a transversal matroid is submodular.

8.2.9. Let D be a digraph with distinguished source s and sink t. Let $E = V(D) - \{s, t\}$. For $X \subseteq E$, let $r(X)$ be the number of edges from $s \cup X$ to $\overline{X} \cup t$. Prove that r is submodular.

8.2.10. (–) Prove equivalence of the following characterizations of loops.
   a) $r(x) = 0$.
   b) $x \in \sigma(\varnothing)$.
   c) Every set containing x is dependent.
   d) x belongs to the closure of every $X \subseteq E$.
   e) x is a circuit.
   f) x belongs to no base.

8.2.11. (–) Prove equivalence of the following characterizations of parallel elements, assuming $x \neq y$ and neither is a loop.

    (a) $r(x, y) = 1$.

    (b) $\{x, y\} \in \mathbf{C}$.

    (c) $x \in \sigma(y)$, $y \in \sigma(x)$, $r(x) = r(y) = 1$.

Furthermore, show that if x, y are parallel and $x \in \sigma(X)$, then $y \in \sigma(X)$.

8.2.12. *Alternative matroid axiomatics.* Suppose M is a hereditary system. Prove the following implications directly.

    a) Submodularity (R) implies weak absorption (A).

    b) Base exchange (B) implies the induced circuit property (J).

    c) The induced circuit property (J) implies weak elimination (C).

    d) The induced circuit property (J) implies the augmentation property (I). (Hint: use induction on $|I_1 - I_2|$.)

8.2.13. (–) If r and σ are the rank function and span function of a matroid, prove that $r(X) = \min \{|Y|: Y \subseteq X, \sigma(Y) = \sigma(X)\}$.

8.2.14. (–) If $r(X) = r(X \cap Y)$ for some $X, Y \subseteq E$ in a matroid on E, prove that $r(X \cup Y) = r(Y)$.

8.2.15. Let M be a hereditary system with nonnegative weights on E. Prove directly that if M satisfies the base exchange property (B), then the greedy algorithm generates a maximum weighted base.

8.2.16. Prove that a hereditary system is a matroid if and only if it satisfies this "ultra-weak" augmentation property: If $I_1, I_2 \in \mathbf{I}$ with $|I_2| > |I_1|$ and $|I_1 - I_2| = 1$, then $I_1 \cup e \in \mathbf{I}$ for some $e \in I_2 - I_1$. (Chappell [1994a])

8.2.17. When e is an element outside a base B in a matroid, let $C(e, B)$ denote the unique circuit in $B \cup e$. Prove

    a) Prove that $B - f + e$ is a base if and only if $f \in C(e, B)$.

    b) Prove that if C is a circuit and $e \in C$, then there exists a base B such that $C = C(e, B)$.

8.2.18. (–) Suppose B_1, B_2 are bases of a matroid such that $|B_1 \triangle B_2| = 2$. Prove that there is a unique circuit C such that $B_1 \triangle B_2 \subseteq C \subseteq B_1 \cup B_2$.

8.2.19. (–) If $B_1, B_2 \in \mathbf{B}$ and $X_1 \subseteq B_1$, prove there exists $X_2 \subseteq B_2$ such that $(B_1 - X_1) \cup X_2$ and $(B_2 - X_2) \cup X_1$ are both bases of M. (Greene [1973])

8.2.20. Prove directly that the weak elimination property implies the strong elimination property, using induction on $|C_1 \cup C_2|$. (Lehman [1964])

8.2.21. *Stronger base properties.* Suppose B_1, B_2 are distinct bases of a matroid M. Prove that

    (a) There exists a bijection $\pi: B_1 \to B_2$ such that for all $e \in B_1$, $B_2 - \pi(e) + e$ is a base of M.

    (b) For any $e \in B_1$, there exists $f \in B_2$ such that $B_1 - e + f$ and $B_2 - f + e$ are bases of M.

    (c) Use $M(K_4)$ to show that there may be no bijection π satisfying (a) such that e and $f = \pi(e)$ always satisfy (b).

8.2.22. Let M be a matroid on a set of elements with nonnegative integer weights. Use the greedy algorithm to obtain the following min-max formula for the maximum weighted independent set: $\max_{I \in \mathbf{I}} \Sigma_{e \in I} w(e) = \min \Sigma_i r(X_i)$, where the minimum is taken over all chains (by inclusion) of sets in E such that each element $e \in E$ appears in at least $w(e)$ sets in the chain (sets may repeat in such a chain.)

8.2.23. (–) A set of $|E| - r(E)$ circuits of a matroid on E form a *fundamental set of circuits* if there is a permutation $e_1, \ldots, e_n$ of E such that C_i contains $e_{r(E)+i}$ but no higher-indexed element. Prove that every matroid has a fundamental set of circuits. (Whitney [1935])

8.2.24. (–) Given k distinct circuits $\{C_i\}$ such that none is contained in the union of the others, and given a set X with $|X| < k$, prove that $\cup_{i=1}^{k} C_i - X$ contains a circuit. (Welsh [1976, p27])

8.2.25. Use the weak elimination property to give a characterization of the family of hyperplanes of a matroid.

8.2.26. (–) Prove that a matroid is simple (no loops or parallel elements) if and only if 1) no element appears in every hyperplane, and 2) every distinct pair of elements intersects some hyperplane exactly once.

8.2.27. *Closed sets and hyperplanes.*
a) Suppose X, Y are closed sets in a matroid M, with $Y \subseteq X$ and $r(Y) = r(X) - 1$. Prove that there is a hyperplane H in M such that $Y = X \cap H$. (Hint: Given a maximal independent subset Z of Y, augment it by $e \in X$ and then to a base B, and let $H = \sigma(B - e)$.)
b) Suppose X is a closed set with rank k. Prove that X is the intersection of $r_M - k$ distinct hyperplanes.

8.2.28. Prove that a matroid of rank r has at least 2^r closed sets (Lazarson [1957])

8.2.29. *Properties of flats (closed sets) of a matroid.* Prove that
(a) The intersection of two flats is a flat.
(b) $\sigma(X)$ is the intersection of all flats containing X and hence is the unique minimal flat containing X.
(c) The union of two flats need not be a flat.

8.2.30. *Dependence properties.* Let $\sim$ be a relation between elements and subsets of E. Prove that the collection $\{X \subseteq E: e \sim X - e \text{ for some } e \in X\}$ is the collection of dependent sets of some matroid if and only if the following three axioms hold: (van der Waerden [1937])
1) $e \sim X$ for all $e \in X$,
2) $e \sim X \cup f$ and $e \not\sim X$ imply $f \sim X \cup e$, and
3) If $e \sim X$ and $x \sim Y$ for all $x \in X$, then $e \sim Y$.

8.2.31. A *refinement* of a matroid M on E is a matroid N on E such that every circuit of M is a circuit of N. Prove that M has a refinement $N \neq M$ if and only if no circuit of M has size $r(M) + 1$.

8.2.32. Let E be a finite set of vectors in $\mathbb{R}^m$. A set of vectors $X = \{x_i\}$ is *affinely dependent* if there exist $\{\lambda_i\}$ not all zero such that $\Sigma \lambda_i = 0$ and $\Sigma \lambda_i x_i = \bar{0}$; otherwise X is *affinely independent*. Prove that the hereditary system whose independent sets are the affinely independent subsets of E is a matroid.

8.2.33. Prove that the cycle matroid $M(G)$ is the column matroid over $\mathbf{Z}_2$ of the vertex-edge incidence matrix of G. (Hence every graphic matroid is binary.)

8.2.34. *The uniform matroid* $U_{2,4}$.
    a) Prove that the matrix with columns $\{(1,0)^T, (0,1)^T, (1,1)^T, (1,2)^T\}$ provides a representation of $U_{2,4}$ over $\mathbf{Z}_3$.
    b) Prove that $U_{2,4}$ has no representation over $\mathbf{Z}_2$. (Comment: contraction and deletion do not change the fields over which a matroid is representable. Hence a binary matroid cannot have $U_{2,4}$ as a minor. Tutte [1958] proved that a matroid M is binary if and only if it does not have $U_{2,4}$ as a minor.)

8.2.35. *The Vámos matroid.* Let M be the hereditary system on eight elements whose bases are all of the 4-element sets except $\{1234, 1256, 1278, 3456, 3478\}$. Prove that M is a nonlinear matroid.

8.2.36. *Supbases and hypobases.* Suppose M is a hereditary system.
    a) Prove that every supbase is a spanning set, but that the converse need not hold. (Comment: By Lemma 8.2.31, the converse holds when M is a matroid.)
    b) Prove that if M is a matroid, then the hypobases of M are the same as the hyperplanes of M.
    c) Provide examples to show that neither the equality of supbases and spanning sets nor the equality of hypobases and hyperplanes implies that M is a matroid.

8.2.37. Prove that the closed sets of a matroid are the complements of the unions of cocircuits.

8.2.38. (–) Show that the dual of a simple matroid (no loops or parallel elements) need not be simple, and determine whether a set can be both a circuit and a cocircuit in a matroid.

8.2.39. (–) Use Theorem 8.2.33 to derive the formula for the rank function of $M.F$.

8.2.40. (–) If M is a matroid on E and $A \subseteq E$, let $\mathbf{I}'$ be the collection of $X \subseteq E$ such that $X \in \mathbf{I}$ and $X \cap A = \emptyset$. Prove that $\mathbf{I}'$ is the collection of independent sets of a matroid on E.

8.2.41. Prove that $M.X$ has no loops if and only if $\overline{X}$ is a closed set in M.

8.2.42. Prove that when e belongs to a base B in a matroid M, there is exactly one cocircuit of M disjoint from $B - e$. Use this to prove that if C is a circuit of a matroid M and x, y are distinct elements of C, then there is a cocircuit $C^* \in \mathbf{C}^*$ with $C^* \cap C = \{x, y\}$. (Minty [1966])

8.2.43. Prove that any minor of a matroid obtained by restricting and then contracting can also be obtained by contracting and then restricting. In particular, if M is a matroid on E and $Y \subseteq X \subseteq E$, prove that $(M \mid X). Y = (M. \overline{X - Y}) \mid Y$ and $(M. X) \mid Y = (M \mid \overline{X - Y}). Y$.

8.2.44. A *pairwise balanced design* on a set E is a family **F** of subsets of E ("blocks"), such that every pair from E appears together in λ common blocks. Prove that a family **F** having no block of size 1 is a pairwise balanced design with $\lambda = 1$ if and only if **F** is the collection of hyperplanes of a simple matroid of rank 3. (Hint: describe the circuits of the matroid.) (Chappell [1995])

8.2.45. (+) The *matroid basis graph* is the graph whose vertices correspond to the bases of a matroid, with edges between the pairs of bases with symmetric difference of size 2. Prove that every matroid basis graph has a spanning cycle, and interpret the result for graphic matroids and for uniform matroids. (Hint: use contraction and restriction inductively to establish a spanning cycle through any edge.) (Holzmann-Harary [1972], see also Kung [1986b, p72])

8.2.46. Prove that the three operations below preserve the cycle matroid of G. (Comment: Whitney's 2-Isomorphism Theorem states that G and H have the same cycle matroid if and only if some sequence of these operations turns G into H. This implies Steinitz's Theorem: every 3-connected planar graph has essentially one one planar embedding; i.e. only one dual graph.)

a) Decompose G into its blocks $B_1, \ldots, B_k$, and reassemble them to form another graph G' with blocks $B_1, \ldots, B_k$.

b) In a block B of G that has a two-vertex cut $\{x, y\}$, interchange the neighbors of x and y in one of the components of $B - \{x, y\}$.

c) Add or delete isolated vertices.

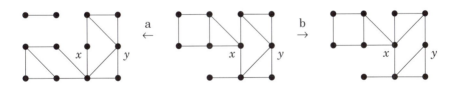

8.2.47. Prove that restrictions and unions of transversal matroids are transversal matroids, but that contractions and duals of transversal matroids need not be transversal matroids.

8.2.48. *Gammoids.* Let D be a digraph, and let F, E be subsets of $V(D)$. The *gammoid* on E induced by D, F is the hereditary system given by $\mathbf{I} = \{X \subseteq E:$ there exist $|X|$ pairwise disjoint paths from F to $X\}$; equivalently, $r(X)$ is the maximum number of pairwise disjoint F, X-paths.

a) Verify that every transversal matroid is a gammoid.

b) (+) Prove that every gammoid is a matroid. (Hint: Use Menger's Theorem to verify the submodularity property. Verifying the augmentation property is also possible but somewhat longer.) (Mason [1972])

8.2.49. *Strict gammoids.* Let D be a directed graph, let F, E be subsets of the vertices of D, and let M be the *gammoid* on E induced by D, F. If E consists of all vertices of D, the gammoid is a *strict gammoid*. Prove that a matroid is a strict gammoid if and only if it is the dual of a transversal matroid. (Hint: use a natural correspondence between directed graphs on n vertices and bipartite graphs on $2n$ vertices.) (Ingleton-Piff [1973])

8.2.50. Use weak duality of linear programming to prove the weak duality property for matroid intersection: $|I| \le r_1(X) + r_2(\overline{X})$ for any $I \in \mathbf{I}_1 \cap \mathbf{I}_2$ and $X \subseteq E$. (Hint: define appropriate linear programs.)

8.2.51. Hall's Theorem states that a set system $\mathbf{A} = A_1, \ldots, A_m$ has a transversal if and only if $|A(J)| \ge |J|$ for all $J \subseteq [m]$, where $A(J) = \cup_{i \in J} A_i$. Use this to prove that $\mathbf{A}$ has a partial transversal with defect at most d (i.e., having size $m - d$) if and only if $|A(J)| \ge |J| - d$ for all $J \subseteq [m]$.

8.2.52. Let M_1, M_2 be two matroids on E.

a) Prove that the minimum size of a set in E that is spanning in both M_1 and M_2 is $\max_{X \subseteq E}(r_1(E) - r_1(X) + r_2(E) - r_2(\overline{X}))$.

b) Apply (a) to prove that in a bipartite graph with no isolated vertices the minimum number of edges needed to cover all the vertices equals the maximum number of vertices with no edges among them. (König's "Other" Theorem)

c) From (a), prove that the maximum size of a common independent set plus the minimum size of a common spanning set equals $r_1(E) + r_2(E)$. In particular, conclude Gallai's Theorem for bipartite graphs: in a bipartite graph with no isolated vertices, the maximum size of a matching plus the minimum number of edges needed to cover the vertices equals the number of vertices.

8.2.53. Use the Matroid Intersection Theorem to prove that in every acyclic orientation of G the vertices can be covered with at most $\alpha(G)$ pairwise-disjoint paths. (Chappell [1994b]) (Comment: this is the special case of the Gallai-Milgram Theorem [1960] for acyclic digraphs.)

8.2.54. Suppose G is an n-vertex weighted graph, and $E_1, \ldots, E_{n-1}$ is a partition of $E(G)$ into $n - 1$ sets. Is there a polynomial-time algorithm to compute a spanning tree of minimum weight among those that have exactly one edge in each subset E_i?

8.2.55. (−) Use 3-D MATCHING to prove that 3-MATROID INTERSECTION is NP-complete. Given triples of the form (x_1, x_2, x_3) with $x_i \in V_i$ and disjoint sets V_1, V_2, V_3, 3-D MATCHING is the problem of finding the maximum number of triples such that each element appears in at most one selected triples (in contrast, bipartite matching is the 2-D matching problem with the sets being pairs).

8.2.56. (−) Use HAMILTONIAN PATH to prove that 3-MATROID INTERSECTION is NP-complete.

8.2.57. (−) Since the union of two matroids is a matroid, there is a dual operation yielding its dual. Given matroids M_1, M_2 with spanning sets $\mathbf{S}_1, \mathbf{S}_1$, let $M_1 \wedge M_2$ be the hereditary system whose spanning sets are $\{X_1 \cap X_2: X_1 \in \mathbf{S}_1, X_2 \in \mathbf{S}_2\}$. Use the Matroid Union Theorem to prove that $M_1 \wedge M_2$ is the matroid $(M_1^* \cup M_2^*)^*$.

8.2.58. *Matroid Intersection from Matroid Union* (the Matroid Intersection Theorem is not available for use in this problem).

a) Prove that the maximum size of a common independent set in M_1, M_2 on E equals $r_{M_1 \cup M_2^*}(E) - r_{M_2^*}(E)$.

b) Obtain the Matroid Intersection Theorem by applying the Matroid Union Theorem to $M_1 \cup M_2^*$.

8.2.59. (!) Use the characterization of graphs having k pairwise edge-disjoint spanning trees (Example 8.2.58) to prove that every $2k$-edge-connected graph has k pairwise edge-disjoint spanning trees. Exhibit for each k a $2k$-edge-connected graph that does not have $k + 1$ pairwise edge-disjoint spanning trees. (Nash-Williams [1961])

8.3 Ramsey Theory

"Ramsey theory" refers to a large body of deep results in mathematics concerning partitions of large collections. An informal way of describing this is Motzkin's statement that "Complete disorder is impossible". Although the results are deep, the objects we consider are merely sets and numbers, and the techniques are little more than induction. Ramsey's Theorem generalizes the pigeonhole principle, which itself concerns partitions of sets. We will consider applications of the pigeonhole principle, prove Ramsey's Theorem, and then focus on Ramsey-type questions for graphs. At the end of the section we will discuss Sperner's Lemma, a result about labeling vertices of a triangulation that has the flavor of a Ramsey-type question.

THE PIGEONHOLE PRINCIPLE REVISITED

The pigeonhole principle (Lemma 1.3.10) states that if m objects are partitioned into n classes, then some class has at least $\lceil m/n \rceil$ objects (and some class has at most $\lfloor m/n \rfloor$ objects). This is a discrete version of the statement that every set of numbers contains a number at least as large as the average (and one at least as small). The concept is very simple, but the applications can be surprisingly subtle. The difficulty is how to define a partitioning problem relevant to the desired application. We illustrate this with four examples.

8.3.1. Proposition. Among six persons it is possible to find three mutual acquaintances or three mutual non-acquaintances.

Proof. (Exercise 1.1.6). In terms of graph theory, we are asked to show that for every 6-vertex graph G, there is a triangle in G or in $\overline{G}$. By the pigeonhole principle, vertex x has degree at least three in G or in $\overline{G}$, since these degrees sum to 5. By symmetry, we may assume $d_G(x) \geq 3$. If some pair of x's neighbors are adjacent, then form a triangle in G with x; otherwise, three of them form a triangle in $\overline{G}$. □

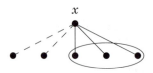

8.3.2. Theorem. (Graham-Entringer-Székely [1996]) If T is a spanning tree of the k-dimensional cube Q_k, then there is an edge of Q_k outside T whose addition to T creates a cycle of length at least $2k$. (In particular, every spanning tree of Q_k has diameter at least $2k - 1$ (Graham-Harary [1992]).

Proof. For each vertex v of Q_k, expressed as a binary k-tuple, there is an "antipodal" vertex v' obtained by complementing each coordinate. There is a unique v, v'-path in T; orient its first edge toward v'. Since $n(Q_k) = e(T) + 1$, doing this for each vertex yields an edge uv that receives two orientations, by the pigeonhole principle.

Since this edge receives an orientation from u and from v, we have v on the u, u'-path and u on the v, v'-path in T. Hence the u, v'-path and the v, u'-path in T are disjoint. Each has length at least $k - 1$, since the distance in Q_k between a vertex and its antipodal vertex is k. Finally, $u \leftrightarrow v$ in Q_k implies also $u' \leftrightarrow v'$, which completes a cycle of length at least $2k$. □

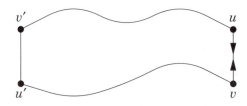

8.3.3. Theorem. (Erdős-Szekeres [1935]) Every sequence of more than n^2 distinct numbers has a monotone subsequence of length more than n.

Proof. Suppose that $a = a_1, \ldots, a_{n^2+1}$ is the sequence. Label position k with the ordered pair (x_k, y_k), where x_k is the length of the longest increasing subsequence ending at a_k, and y_k is the length of a longest decreasing subsequence ending at a_k. If the sequence has no monotone subsequence of length $n + 1$, then x_k and y_k never exceed n, and there are only n^2 possible labels. By the pigeonhole principle, two labels must be the same. This is impossible when the elements of a are distinct.

When $i < j$ and $a_i < a_j$, we can append a_j to the longest increasing sequence ending at a_i. When $i < j$ and $a_i > a_j$, we can append a_j to the longest decreasing sequence ending at a_i. A generalization of this argument applies to digraph coloring in Exercise 5.1.24. □

a:	7	4	1	8	5	2	9	6	3	0
x, y:	1,1	1,2	1,3	2,1	2,2	2,3	3,1	3,2	3,3	4,1

8.3.4. Theorem. (Graham-Kleitman [1973]) If the $\binom{n}{2}$ edges of a complete graph on n vertices are labeled arbitrarily with the integers $1, \ldots, \binom{n}{2}$, then there is a trail of length at least $n - 1$ with an increasing sequence of edge-labels.

Proof. We assign each vertex a weight equal to the length of the longest increasing trail ending there. If we can show the weights sum to at least $n(n - 1)$, then the pigeonhole principle guarantees a vertex with a large enough weight. The problem is how to compute the weights and accumulate their sum.

We accumulate the weights and their sum iteratively, growing the graph from the trivial graph by adding the edges in order. The vertex weights begin at 0. If the next edge joins two vertices whose weights were both i, then their weights both become $i + 1$; if it joins two vertices of weights i and j, with $i < j$, then their weights become $j + 1$ and j. In either case, when an edge is added, the sum of the weights of the vertices increases by at least 2. Therefore, when the construction is finished, the sum of the vertex weights is at least $n(n - 1)$. □

Finally, we note that the thresholds in the classes may differ.

8.3.5. Theorem. If $\Sigma p_i - k + 1$ objects are partitioned into k classes with quotas $\{p_i\}$, then some class must meet its quota.

Proof. If not, then at most $\Sigma(p_i - 1)$ objects can be accomodated. □

RAMSEY'S THEOREM

The pigeonhole principle guarantees a class with many objects when we partition the objects into classes. The famous theorem of F.P. Ramsey [1930] makes a similar statement about partitioning the r-subsets of the objects into classes. For clarity, we use the terminology of coloring when discussing partitioning problems. Roughly put, Ramsey's Theorem says the following: for every partition of the r-sets of a sufficiently large set S into k classes, there is a p-subset of S whose r-sets all go into the same class.

8.3.6. Definition. A *k-partition* or *k-coloring* of a set is a partition of it into k classes. A class or its label is a "color". We can refer to the ith class in each of several k-partitions as "color i". The collection of r-subsets of S is $\binom{S}{r}$; a k-coloring of these is a map $f\colon \binom{S}{r} \to [k]$. A *homogeneous* set is a set whose r-sets have received the same color; if color i, then it is *i-homogeneous*. The notation $N \to (p_1, \ldots, p_k)^r$ means that every k-coloring of the r-subsets of N elements has an i-homogeneous p_i-set for some i. If such an integer N exists, then the smallest such integer is the *Ramsey number* $R(p_1, \ldots, p_k; r)$.

The statements $N \to (p_1, \ldots, p_k)^r$ and $R(p_1, \ldots, p_k; r) \le N$ are equivalent. It is not immediately obvious that $R(p_1, \ldots, p_k; r)$ is finite for every choice of the parameters (also called "quotas" or "thresholds"); this is the assertion of the theorem. A thorough study of Ramsey's Theorem and other partitioning theorems appears in Graham-Rothschild-Spencer [1980, 1990].

Before proving the theorem, we consider the case $r = k = 2$, because the structure of the proof here is the same as in the general case. When $r = 2$, we can view the partition as an edge-coloring of a complete graph. Example 8.3.1 shows that $R(3, 3; 2) \le 6$. By examining the coloring of edges involving a single element x, we can generalize the argument of Example 8.3.1 to prove that $R(p_1, p_2; 2) \le R(p_1 - 1, p_2; 2) + R(p_1, p_2 - 1; 2)$. The threshold version of the pigeonhole principle implies that if there are $N + M - 1$ elements other than x, then x must have N incident edges of color 1 or M incident edges of color 2. Let $N = R(p_1 - 1, p_2; 2)$ and $M = R(p_1, p_2 - 1; 2)$, and consider a 2-coloring of $E(K_{N+M})$. By symmetry, we may assume that x has at least N incident edges of color 1. By the definition of N, the clique induced by the neighbors of x along these edges has a p_2-clique in color 2 or a $p_1 - 1$-clique in color 1 (which would combine with x to form a p_1-clique in color 1). In either case, we obtain an i-homogeneous set of size p_i for some i. We postpone discussion of the resulting bound on $R(p_1, p_2; 2)$.

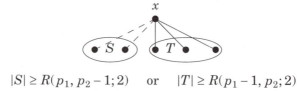

$$|S| \ge R(p_1, p_2 - 1; 2) \quad \text{or} \quad |T| \ge R(p_1 - 1, p_2; 2)$$

8.3.7. Theorem. (Ramsey [1930]). Given positive integers k, r, and $p_1, \ldots, p_k$, there exists a finite positive integer N such that $N \to (p_1, \ldots, p_k)^r$.

Proof. The proof uses induction on r, and the induction step for fixed r is a proof by induction on Σp_i. As a basis, we need to verify the hypothesis for $r = 1$ and for some initial threshold vectors when $r > 1$. The case $r = 1$ is the threshold version of the pigeonhole principle (Proposition 8.3.5). If $r > 1$ and some quota p_i is less than r, then a set of p_i objects has no r-subsets, so vacuously all its r-subsets belong to class i. Hence $R(p_1, \ldots, p_k; r) = \min\{p_1, \ldots, p_k\}$ if $\min\{p_1, \ldots, p_k\} < r$. Alternatively, $R(p_1, \ldots, p_k; r) = s$ if all $p_i = r$ except for one that equals s.

For clarity in the induction step, we describe only the case $k = 2$ in detail; the analogous argument works for the general case (Exercise 16). Write (p, q) for (p_1, p_2), and set $p' = R(p - 1, q; r)$, $q' = R(p, q - 1; r)$, and $N = 1 + R(p', q'; r - 1)$. By the induction hypothesis, the integers p', q', N all exist. We use a double induction because p' and q' may be enormously large. Let S be a set of N elements, and choose $x \in S$. Consider a 2-coloring f of $\binom{S}{r}$. We need to show that f has a 1-homogeneous p-set or a 2-homogeneous q-set. By time-honored tradition in Ramsey theory, color 1 is "red" and color 2 is "blue".

We use f to induce a 2-coloring f' of the $r - 1$-sets of $S' = S - x$. This motivates choosing $|S'|$ as a Ramsey number for $r - 1$-sets. Define f' by assigning color i to an $(r - 1)$-set of S' if its union with x has color i under f. Since $|S'| = R(p', q'; r - 1)$, the induction hypothesis implies that at least one of the colors meets its quota (p' or q') under f' (when $r = 2$, this step is the invocation of the pigeonhole principle). By symmetry, we may assume that the red quota is met, and let S'' be the p'-subset of S' whose $r - 1$-sets are red under f'.

We return to the original coloring f, restricted to $\binom{S''}{r}$. Since $|S''| = p' = R(p - 1, q; r)$, f must have a red-homogeneous $p - 1$-set or a blue-homogeneous q-set contained in S''. If it has a blue-homogeneous q-set, we are done. If it has a red-homogeneous $p - 1$-set P, consider $P \cup \{x\}$. From the definition of S'', the $(r - 1)$-sets of P are all red under f', which means their unions with x are red under f. Hence $P \cup \{x\}$ is a red-homogeneous p-set under f. □

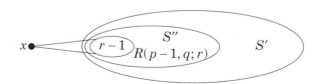

Like the pigeonhole principle, Ramsey's Theorem has subtle and fascinating applications. Ramsey's Theorem typically gives an elegant existence proof but a horrible bound. Careful combinatorial arguments are needed for exact solutions of extremal problems.

8.3.8. Theorem. (Erdős-Szekeres [1935]). Given an integer m, there exists a (least) integer $N(m)$ such that every set of at least $N(m)$ points in the plane with no three collinear contains an m-subset forming a convex m-gon.

Proof. We need two facts. *(1) Among five points in the plane, four determine a convex quadrilateral* (if no three are collinear). Construct the convex hull of the five points. If it is a pentagon or a quadrilateral, the result follows immediately. If it is a triangle, the other two points lie inside. By the pigeonhole principle(!), two of the vertices of the triangle are on one side of the line through the two inside points. These two vertices together with the two points inside form a convex quadrilateral, as illustrated below.

In a convex m-gon, any four points determine a convex quadrilateral; we need the converse: *(2) If every 4-subset of m points in the plane form a convex quadrilateral, then the m points form a convex m-gon.* If the claim fails, then the convex hull of the m points consists of t points, for some $t < m$. The remaining points lie inside the t-gon. When we triangulate the t-gon, as illustrated on the right above, a point inside lies in one of the triangles. With the vertices of that triangle, it forms a 4-set that does not determine a convex quadrilateral.

To prove the theorem, let $N = R(m, 5; 4)$. Given N points in a plane, color each 4-set by convexity: red if it determines a convex quadrilateral, blue if it does not. By fact (1), there cannot be 5 points whose 4-subsets are all blue. By Ramsey's Theorem, this means there are m points whose 4-subsets are all red. By fact (2), they form a convex m-gon. Hence $N(m)$ exists and is at most $R(m, 5; 4)$. □

The bound $R(m, 5; 4)$ is very loose. It is exact for $m = 4$, where fact (1) implies that $N(4) = 5 = R(4, 5; 4)$. In contrast, $N(5) = 9$ but $R(5, 5; 4)$ is enormous. Erdős and Szekeres conjectured that $N(m) = 2^{m-2} + 1$ and proved that $2^{m-2} \leq N(m) \leq \binom{2m-4}{m-2} + 1$ (Exercises 11-13).

Another elegant application concerns search strategies for storing numbers in tables. Suppose that n numbers are selected from a large universe of numbers and stored in a table of size n according to some strategy that specifies how to store each such set. Yao proved that when the universe is quite enormous, the strategy minimizing the worst-case number of probes required to check the presence of a number from the universe is to store them in sorted order and to test membership by binary search. Like most applications of Ramsey's Theorem, the bound is most likely much larger than needed.

RAMSEY NUMBERS

Ramsey's Theorem defines the Ramsey numbers $R(p_1, \ldots, p_k; r)$. There is no exact formula, and very few Ramsey numbers have been computed. To prove that $R(p_1, \ldots, p_k; r) = N$, one must exhibit a k-coloring of the r-sets on $N - 1$ points that meets no quota (or show that one exists without constructing it), *and* show that *every* coloring on N points meets some quota. In principle, one could run a computer through all the k-colorings of $\binom{[n]}{r}$ for various n till finding N such that every coloring of K_N meets a quota and some coloring of K_{N-1} does not. Even for 2-color Ramsey numbers, $2^{\binom{n}{2}}$ rapidly becomes too large to think about. Erdős has joked that if an alien being threatens to destroy the earth unless we provide it the exact value of $R(5,5)$, then we should set all the computers in the world to work on an exhaustive solution. If it asks for $R(6,6)$, then he advises us to leave the earth.

No Ramsey numbers are known exactly for $r > 2$, and only a few are known for $r = 2$. When $p = p_1 = \cdots = p_k$, we abbreviate the notation $R(p_1, \ldots, p_k; r)$ to $R_k(p; r)$, and when $r = 2$ we write $R(p_1, \ldots, p_k)$ or $R_k(p)$. Even for $r = 2$, only one Ramsey number is known exactly when $k > 2$, which is $R(3,3,3) = 17$. The table below contains all the known values of $R(p, q)$ and some bounds better than the general ones we will develop shortly.

	3	4	5	6	7	8	9
3	6	9	14	18	23	28	36
4		18	25	34/43			
5			43/52	58/94			
6				102/169			

The computations of $R(3,9)$ (Grinstead-Roberts [1982]), $R(3,8)$ (McKay-Zhang [1992]) and $R(4,5)$ (McKay-Radziszowski [1995]) are recent; the others are much older (due primarily to Greenwood-Gleason [1955], Kalbfleisch [1967], and Graver-Yackel [1968]). For $r > 2$, little is known other than $R(4,4;3) = 13$ (McKay-Radziszowski [1991]).

We prove only the first two of these results (see Exercise 15 for $R(3,5)$). For $r = 2$, the edge-coloring interpretation is natural. If $k = 2$, we further simplify terminology by defining the two colors to be "in" and "out". Then Ramsey's Theorem becomes: "There exists a minimum integer $R(p, q)$ such that every graph on $R(p, q)$ vertices has a clique of size p or an independent set of size q."

8.3.9. Example. $R(3,3) = 6$. We showed earlier that $R(3,3) \le 6$. Since the 5-cycle has no K_3 or $\overline{K}_3$, $R(3,3) \ge 6$. $\square$

8.3.10. Example. $R(3,4) = 9$. The graph below has no K_3 and no $\overline{K}_4$. (four independent vertices on an 8-cycle must have the same parity, but this graph has edges among those vertices). Hence $R(3,4) \ge 9$.

For the upper bound, we use $R(2,4) = 4$ and $R(3,3) = 6$. If some vertex in an arbitrary graph has four neighbors or six non-neighbors, then the graph must have a K_3 or a $\overline{K}_4$, by adding that vertex (if necessary) to what is forced among its neighbors or non-neighbors. Avoiding both possibilities requires at most three neighbors and at most five non-neighbors, so there are at most 9 vertices altogether. If this happens for a 9-vertex graph, then *every* vertex has exactly 3 neighbors. No 9-vertex graph has odd degree at every vertex, so $R(3,4) = 9$. $\square$

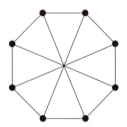

The lack of explicit formulae has stimulated research on bounds. The proof of Ramsey's Theorem yields a recursive upper bound on $R(p,q;r)$, but it is extremely large, especially for $r > 2$.

8.3.11. Theorem. $R(p,q) \le R(p-1,q) + R(p,q-1)$. If both summands on the right are even, then the inequality is strict.

Proof. If a vertex in an arbitrary graph has $R(p-1,q)$ neighbors or $R(p,q-1)$ non-neighbors, then the graph contains K_p or $\overline{K}_q$. With $R(p-1,q) + R(p,q-1)$ points altogether in the graph, the pigeonhole principle guarantees that one of these possibilities occurs. Equality in the bound requires a regular graph with $R(p-1,q) + R(p,q-1) - 1$ vertices. If both summands are even, this requires a regular graph of odd degree on an odd number of vertices, which is impossible. $\square$

Theorem 8.3.11 yields $R(p,q) \le \binom{p+q-2}{p-1}$. There have been some asymptotic improvements. For fixed q and large p, $R(p,q) \le cp^{q-1} \log\log p / \log p$ (Graver-Yackel [1968], Chung-Grinstead [1983]). For $q = 3$ and $p \ge 3$, there is a further improvement: $R(p,3) \le cp^2 / \log p$ (Ajtai-Komlós-Szemerédi [1980]).

The Ramsey numbers for equal quotas are called *diagonal Ramsey numbers*. Asymptotically, the upper bound of $\binom{2p-2}{p-1}$ for $R(p,p)$ is $c4^p/\sqrt{p}$. Exercise 14 presents a constructive lower bound that is polynomial in p. The best known constructive lower bound grows faster than every polynomial in p (Frankl-Wilson [1981]) but slower than every exponential in p (Exercise 28). An exponential lower bound can be proved by counting methods. It implies that $\sqrt{2} \leq \liminf R(p,p)^{1/p} \leq \limsup R(p,p)^{1/p} \leq 4$. Determination of this limit (and whether the limit exists) is the foremost open problem about Ramsey numbers.

8.3.12. Theorem. (Erdős [1947]). $R(p,p) > (e\sqrt{2})^{-1}p2^{p/2}(1+o(1))$.

Proof. Consider the graphs with vertex set $[n]$. Each particular p-clique occurs in $2^{\binom{n}{2}-\binom{p}{2}}$ of these $2^{\binom{n}{2}}$ graphs. Similarly, each particular set of p vertices occurs as an independent set in $2^{\binom{n}{2}-\binom{p}{2}}$ of these graphs. Discarding these leaves only graphs having no p-clique or independent p-set. Since there are $\binom{n}{p}$ ways to choose p vertices, the inequality $2\binom{n}{p}2^{-\binom{p}{2}} < 1$ implies $R(p,p) > n$. Rough approximations yield $\binom{n}{p}2^{1-\binom{p}{2}} < 1$ whenever $n < 2^{p/2}$. More careful approximations (using Stirling's formula to approximate the factorials) lead to the result claimed. □

GRAPH RAMSEY THEORY

Ramsey's Theorem for $r = 2$ says that k-coloring the edges of a large enough complete graph forces a monochromatic complete subgraph. A monochromatic p-clique contains a monochromatic copy of *every* n-vertex graph. Perhaps monochromatic copies of graphs with fewer edges can be forced by coloring a smaller graph than needed to force K_p. For example, 2-coloring the edges of K_3 always yields a monochromatic P_3, although 6 points are needed to force a monochromatic triangle. This suggests many Ramsey number questions, some easier to answer than the questions for cliques.

8.3.13. Definition. For graphs $G_1, \ldots, G_k$, we write $n \rightarrow (G_1, \ldots, G_k)$ to mean that every k-coloring of $E(K_n)$ contains a copy of G_i in color i for some i. The *(graph) Ramsey number* $R(G_1, \ldots, G_k)$ is the smallest integer n such that $n \rightarrow (G_1, \ldots, G_k)$. When $G_i = G$ for all i, we write $R_k(G) = R(G_1, \ldots, G_k)$, and we write $R(G) = R_2(G)$.

Burr [1983] determined $R(G)$, called the "Ramsey number of G", for all 113 graphs with six or fewer edges and no isolated vertices. Surprisingly, $R(G_1, G_2)$ has a simple formula for many examples. Again we use red and blue for the two colors.

8.3.14. Theorem. (Chvátal [1977]). If T is an m-vertex tree, then
$R(T, K_n) = (m - 1)(n - 1) + 1$.

Proof. For the lower bound, color $K_{(m-1)(n-1)}$ by letting the red graph be $(n - 1)K_{m-1}$. With red components of order $m - 1$, there is no red m-vertex tree. The blue edges form an $n - 1$-partite graph and hencehence cannot contain K_n.

The proof of the upper bound uses induction on each parameter, focusing on the neighbors of one vertex. Our presentation uses induction on n, invoking a property of trees proved in Chapter 2 by induction on m. The basis step is $n = 1$; no edges are needed to obtain K_1.

Given a 2-coloring of $E(K_{(m-1)(n-1)+1})$, consider a vertex x. If x has more than $(m - 1)(n - 2)$ neighbors along blue edges, then among them is a red T or a blue K_{n-1}, by the induction hypothesis, which yields a red T or a blue K_n (with x) in the original coloring. Otherwise, every point has at most $(m - 1)(n - 2)$ incident blue edges and at least $(m - 1)(n - 1) + 1 - 1 - (m - 1)(n - 2) = m - 1$ incident red edges. Now there must be a red T, because by Proposition 2.1.5 every graph having minimum degree at least $m - 1$ contains T. □

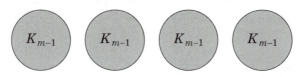

This lower bound holds more generally, with the same coloring. If the largest component of G has m vertices, and H has chromatic number n, then $R(G, H) \geq (m - 1)(n - 1) + 1$ (Chvátal-Harary [1972]). It would be interesting to know when this lower bound is exact. The general conjecture is that if H is "complete enough" or G is "sparse enough", then $R(G, H) = (n(G) - 1)(\omega(H) - 1) + 1$. Gould and Jacobson [1983] proved such results when G is a tree and $\omega(H) \geq n(H) - 2$. For the sparseness condition, Burr and Erdős [1983] conjectured that $R(G, K_n) = (m - 1)(n - 1) + 1$ for every m-vertex graph G that is sufficiently large in terms of its sparseness. More precisely, let x be the maximum edge density $(e(F)/n(F))$ over all subgraphs F of G; "sufficiently large" means that m is at least some number depending only on x and $n(H)$. The conjecture holds (Burr [1981]) when G has many vertices of degree 2.

In the lower bound, it is crucial that the color classes in H are single vertices. When this fails, the lower bound can be far from correct. If $G = H = mK_3$, then the largest component and chromatic number both equal 3, yielding a lower bound of 5, but the correct value is $5m$. Here the coloring for the lower bound is surprisingly non-symmetric, considering the symmetry of the inputs.

8.3.15. Theorem. (Burr-Erdős-Spencer [1975]). $R(mK_3, mK_3) = 5m$ for $m \geq 3$.

Proof. Let the red graph be $K_{3m-1} + K_{1,2m-1}$, as shown below. Every triangle in this graph uses three vertices from the $3m - 1$-clique, but the clique does not have enough vertices to make m disjoint triangles. The complementary blue graph is $(K_{2m-1} + K_1) \vee \overline{K}_{3m-1}$. Every blue triangle has at least 2 vertices in the copy of K_{2m-1}, so there cannot be m disjoint blue triangles.

The proof of the upper bound uses induction on m. The basis step $m = 2$ requires a case analysis that is fairly short if phrased carefully (Exercise 26). For the induction step, suppose $m \geq 3$. Since $5m > R(3, 3) = 6$, we know that every 2-coloring contains a monochromatic triangle. Discarding those three vertices, we can continue to find disjoint monochromatic triangles until we fall below 6 vertices. Since $5m - 3m \geq 6$ for $m \geq 2$, we find m disjoint monochromatic triangles. If these all have the same color, then we are done; otherwise, we have a triangle in each color.

Suppose that abc is a red triangle and def is a blue triangle disjoint from it. Of the 9 edges between them, we may assume by symmetry that at least 5 are red. Some pair of these must have a common endpoint in def. Now we have a red triangle and a blue triangle that intersect and hence have five vertices together. If $m > 2$, we can apply the induction hypothesis to the coloring on the remaining $5m - 5$ vertices to find an $(m - 1)K_3$ in one color, to which we add the appropriately colored triangle from the five special vertices. $\square$

For those worried about omission of the basis step above, consider coloring K_{11}. Avoiding $2K_3$ forces a bow tie (monochromatic triangles with a common vertex) as above, but then we find another monochromatic triangle among the remaining 6 points. Hence we have at least proved that $R(mK_3, mK_3) \leq 5m + 1$. A similar result with a simpler argument and no messy basis case is $R(mK_2, mK_2) = 3m - 1$ (Exercise 25). Without the bow tie argument, it is easy to abstract the rest of the argument to obtain an upper bound on $R(m_1 G_1, \ldots, m_k G_k)$ in terms of $R(G_1, \ldots, G_k)$ (Exercise 27).

We close by mentioning a remarkable result. The Ramsey number of an arbitrary graph may be exponential in the number of vertices, such as for K_n. Chvátal, Rödl, Szemerédi, and Trotter [1983] proved that for the class of graphs with maximum degree d, the Ramsey number grows at most linearly in the number of vertices! In other words, $R(G, G) \le cn(G)$, where c is a constant depending only on d. Of course, the constant is a fast-growing function of d, but it does not depend on $n(G)$. The proof uses the Szemerédi Regularity Lemma [1978], itself a difficult result with many applications.

SPERNER'S LEMMA AND BANDWIDTH

Although Sperner's Lemma is not generally considered part of Ramsey theory, we include this material in this section because Sperner's Lemma has the flavor of Ramsey theory: every labeling of a triangulation that satisfies certain boundary conditions contains a piece with a special labeling. Another aspect that Sperner's Lemma shares with Ramsey's Theorem is that it relies on very simple ideas but has subtle applications; Ramsey's Theorem relies on the pigeonhole principle and induction, while Sperner's Lemma uses only a parity argument (and induction for a generalization to higher dimensions).

8.3.16. Definition. A *simplicial subdivision* of a large triangle T is a partition of T into triangular *cells* such that every intersection of two cells is a common edge or corner. We call the corners of cells *nodes*. A *proper labeling* of a simplicial subdivison of T assigns labels from $\{0, 1, 2\}$ to the nodes, avoiding label i on the ith edge of T ($i \in \{0, 1, 2\}$). A *completely labeled cell* has all three labels.

In a proper labeling, each label appears at one corner of T, and label i avoids the edge of T joining the corners not labeled i. The figure below illustrates a simplicial subdivision and the graph we will obtain from it to prove that it has a completely labeled cell.

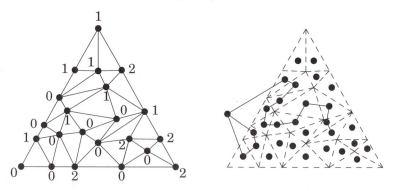

8.3.17. Theorem. (Sperner's Lemma [1928]) Every properly labeled simplicial subdivision contains a completely labeled cell.

Proof. We prove the stronger result that there are an odd number of completely labeled cells. Suppose we begin outside T and enter a cell by crossing a 0,1-edge. If we reach a cell whose third label is 2, we are finished. If not, then the third label is 0 or 1, and the cell has another 0,1-edge. By crossing it, we enter a new cell and can continue looking for a cell with the third label.

This suggests a graph G encoding our possible travels. We introduce a vertex for each cell, plus one for the outside region. Two vertices of G form an edge if those regions share a boundary whose endpoints have the labels 0 and 1. The graph on the right above results from the proper labeling on the left. There are several types of vertices in G. A completely labeled cell becomes a vertex of degree 1. A cell with no 0 or no 1 becomes a vertex of degree 0. The remaining cells have corners labeled $0,0,1$ or $0,1,1$ and become vertices of degree 2. Hence the desired cells become vertices of degree 1 in G, and these are the only cells that become vertices of odd degree. We have transformed the original problem into the problem of showing that G has such a vertex of degree 1.

What about the vertex v for the outside region? As we travel from the 0-corner to the 1-corner along the edge of T avoids label 2, we cross an edge of G involving v every time we switch from a 0 to a 1 or back again. Since we start with 0 and end with 1, we switch an odd number of times. Hence v also has odd degree. Since the number of vertices of odd degree in every graph is even, the number of vertices other than v having odd degree is odd, so there are an odd number of completely labeled cells. $\square$

8.3.18. Application. *(Brouwer Fixed-Point Theorem).* Brouwer's Theorem (for two dimensions) can be interpreted as saying that a continuous mapping from a triangular region T to itself must have a fixed point. Suppose the corners of T are the points (vectors) x_0, x_1, x_2. Just as we can express a point on a segment uniquely as a weighted average of its endpoints, so we can express each $x \in T$ uniquely as a weighted average of the corners: $x = a_0 x_0 + a_1 x_1 + a_2 x_2$, where $\Sigma a_i = 1$ and each $a_i \geq 0$ (Exercise 37). We can specify x by its vector of coefficients $a = (a_0, a_1, a_2)$.

Define sets S_0, S_1, S_2 for the mapping f by saying that $a \in S_i$ if $a_i' \leq a_i$, where $f(a) = a'$. Because the coefficients of each point sum to one, every point in T belongs to at least one of the sets, and a point belongs to all three sets if and only if it is a fixed point for f. We want to show that the three sets have a common point.

Given an arbitrary simplicial subdivision of T, for each node a choose a label i such that $a \in S_i$. Note that the points on the edge of T opposite x_i have ith coordinate 0. Since their ith coordinate cannot decrease under f, we can choose some label different from i for each point on that edge. The resulting labeling is proper, and Sperner's Lemma guarantees a completely labeled cell. If we repeat the process using triangulations with smaller and smaller cells, we obtain a sequence of smaller and smaller completely labeled triangles; call them $\{x_j, y_j, z_j\}$ receiving labels 0,1,2, respectively. In each S_i, we obtain an infinite sequence of points.

The remaining details are topological; we merely suggest the steps. Since f is continuous, each S_i is closed and bounded. Every infinite sequence of points in a closed and bounded set has a convergent subsequence. Hence $\{x_i\}$ has a convergent subsequence $\{x_{i_k}\}$. Because the distance from x_{i_k} to y_{i_k} and z_{i_k} approaches 0, these subsequences also converge to the same point. Since S_0, S_1, S_2 are closed and bounded, this limit point belongs to all three of them and is a fixed point of f. $\square$

We also apply Sperner's Lemma to compute a parameter of the "triangular grid".

8.3.19. Definition. When the vertices of G are numbered with distinct integers, the *dilation* is the maximum difference between integers assigned to adjacent vertices. The *bandwidth* $B(G)$ of a graph G is the minimum dilation of a numbering of G.

Dilation is always minimized when there are no gaps in the numbering, but it can be convenient to allow gaps (Exercise 41). The motivation for bandwidth is to minimize the delay between adjacent vertices when the vertices must be processed in a linear order. Computation of bandwidth is NP-hard even for trees with maximum degree 3 (Garey-Graham-Johnson-Knuth [1978]). The name "bandwidth" comes from matrix theory; the optimal numbering describes a permutation of the rows and columns of the adjacency matrix so that the 1's appear only in diagonal bands close to the main diagonal. There are two fundamental lower bounds on bandwidth:

8.3.20. Lemma. $B(G) \geq \max_{H \subseteq G} \dfrac{n(H) - 1}{\operatorname{diam} H}$.

Proof. Every numbering of G contains a numbering of each subgraph of G. For every graph H, the two numbers farthest apart are at least $n(H) - 1$ apart, and by the pigeonhole principle some step on a path between them has dilation at least $n(H) - 1$ divided by the distance between them. $\square$

8.3.21. Lemma. (Harper [1966]) If ∂S denotes the subset of vertices in a set $S \subseteq V(G)$ that have at least one neighbor outside S, then
$B(G) \geq \max_k \min \{|\partial S| : |S| = k\}$.

Proof. For every value of k, some set S of k vertices must be the first k vertices in the optimal numbering of G. The bandwidth of G must be at least $|\partial S|$, because the vertex among ∂S that has the least label has an edge of dilation at least $|\partial S|$ to its neighborhood above S. □

Chung [1988] named the first bound the *local density* bound. The computation of Harper's bound is usually difficult. For the cube Q_k, the value is $\Sigma_{i=0}^{n-1}\binom{i}{\lfloor i/2\rfloor}$. For the grid $P_m \square P_n$, the value of Harper's lower bound is min $\{m, n\}$, which can be achieved (Exercise 42).

8.3.22. Example. *The triangular grid.* The triangular grid T_l consists of vertices (i, j, k) such that i, j, k are non-negative integers summing to l, with two vertices adjacent if the total of the absolute differences in corresponding coordinates is 2. Below we show T_4. Numbering the vertices by rows produces an upper bound of $l + 1$ for $B(T_l)$. This is optimal, but the local density bound is only about $l/2$; Harper's bound completes the computation, using Sperner's Lemma. □

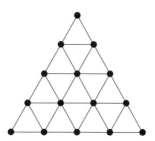

Suppose G is the graph formed by a simplicial subdivision of a triangle. The outer boundary of G is a cycle, the bounded regions are triangles, and the cycle is partitioned into three paths by the corners of the large triangle. We say that a *connector* is a vertex set inducing a connected subgraph that contains a vertex of each boundary path.

8.3.23. Lemma. (Hochberg-McDiarmid-Saks [1996]) Let T be a simplicial subdivision in which each vertex is assigned a color, red or blue. Let R, B be the subgraphs induced by the red and by the blue vertices, respectively. For each such coloring, exactly one of R, B contains a connector.

Proof. For each vertex v, consider the vertices reachable from v using vertices with the same color as v. If the three sides are not all reachable, label v with the smallest index of a side not reachable from v. For the vertices on the ith side, the label i does not appear. If there is no connector, then each node has a label, and this is a proper labeling of T. By Sperner's Lemma, there is a completely labeled cell. Since the cell has three corners and we only used two colors R, B, two of the corners of this cell have the same color. Since they are adjacent, they can reach the same same of vertices in their color. Hence the least side unreachable from them cannot be different. This contradiction means that we could not have constructed the specified labeling. Hence there is a vertex from which every side is reachable.

If one color has a connector, it partitions the remaining vertices into sets such that from each set at least one side is unreachable. Hence there cannot be connectors in both colors. □

8.3.24. Theorem. (Hochberg-McDiarmid-Saks [1996]) Suppose G is a graph that is a triangulation of a region bounded by a cycle C partitioned into three paths, and let k be the minimum over $v \in V(G)$ of the sum of the distances from v to each of the three paths. Then $B(G) \geq k + 1$. In particular, for the triangular grid T_l with side-length l, $B(T_l) \geq l + 1$.

Proof. Let f be a numbering of G. Let t be the maximum index such that the subgraph induced by the vertices numbered $1, \ldots, t$ does not contain a component meeting all three paths. Let R be this vertex set, let S be the set of vertices outside R having neighbors in R, and let T be the remaining vertices. By construction, the vertex v with $f(v) = t + 1$ belongs to S. Since $R \cup \{v\}$ contains a connector, $R \cup S$ contains a connector, and T does not. Since there is no edge between R and T and R contains no connector, $R \cup T$ contains no connector. But now, by the lemma, S contains a connector. The set S equals $\partial(S \cup T)$ for the terminal segment $S \cup T$ in the numbering. Therefore the numbering has difference at least $|S|$ on some edge.

A connector contains walks from each of its vertices to each of the three boundary paths. By hypothesis, the sum of the lengths of these walks from any fixed vertex is at least k. There exists a vertex in S for which these walks in S are disjoint paths. Hence $|S| \geq k + 1$.

For each vertex (i, j, k) in T_l, the distances to the three sides are i, j, k, respectively, so the sum of the distances is l. Hence the bandwidth is at least $l + 1$, which we have observed is achievable. □

The vertex numberings on which dilation is measured for bandwidth can be considered as embeddings of G into a path. This viewpoint leads to many generalizations of bandwidth, and there are other related parameters that optimize other aspects of vertex numberings.

EXERCISES

8.3.1. (–) Each of two concentric discs has 20 radial sections of equal size. For each disc, 10 sections are painted red and 10 blue, in some arrangement. Prove that the two discs can be aligned so that at least 10 sections on the inner disc that match colors with the corresponding sections on the outer disc.

8.3.2. Prove that every set of $n + 1$ distinct integers chosen from $\{1, \ldots, 2n\}$ contains a pair with greatest common factor 1 and a pair such that one divides the other. For each n, exhibit two sets of size n to show that these results are best possible.

8.3.3. Use partial sums and the pigeonhole principle to prove the following statements.

a) Every set of n integers contains a non-empty subset whose sum is divisible by n. (Also exhibit a collection of $n - 1$ integers with no such subset.)

b) Given $x \in \mathbb{R}$, prove that at least one of $\{x, 2x, \ldots, (n-1)x\}$ differs by at most $1/n$ from an integer.

8.3.4. A private club has 90 rooms and 100 members. Keys are given to the members so that any 90 members have access to the rooms in the sense that each of these 90 members will have a key to a different room. (They do not share their keys.) Prove that at least 990 keys are needed and 990 suffice.

8.3.5. Use the technique of Proposition 8.3.2 to prove that every tree has exactly one center or two adjacent centers. (Jordan [1869], Graham-Entringer-Székely [1996])

8.3.6. Prove that every set of $2^m + 1$ integer lattice points in $\mathbb{R}^m$ contains a pair of points whose centroid (mean vector) is also an integer lattice point.

8.3.7. Prove that in every 2-coloring of the integer lattice points in $\mathbb{R}^m$, there is a collection of n points with the same color whose centroid (mean vector) is an integer lattice point also having that color. (Hint: no need for Ramsey's Theorem; there is a short proof using no more than the pigeonhole principle.) (Bòna [1990])

8.3.8. Consider a collection S of $n + 1$ positive integers summing to k. If $k \le 2n + 1$, prove that S has a subset with sum i if $0 \le i \le k$. For each n, exhibit a collection that fails this if $k = 2n + 2$.

8.3.9. Prove that any nine points in the plane (no three colinear) contain the vertices of a convex 5-gon. Exhibit a set of 8 points without this property.

8.3.10. Let S be a set of $R(m, m; 3)$ points in the plane no three of which are colinear. Prove that S contains m points that form a convex m-gon. (Tarsi)

8.3.11. Consider a tournament with edge weights. A *nondecreasing [decreasing]* path is a (directed) path on which the edge weights are successively nondecreasing [decreasing]. Given p, q, prove that the maximum number of vertices in a weighted tournament having no nondecreasing $p + 1$-edge path or increasing $q + 1$-edge path is $\binom{p+q}{p}$. (Erdős-Szekeres [1935])

8.3.12. A sequence of points $\{v_i\}$, where $v_i = (x_i, y_i)$ with $x_1 < \cdots < x_n$, is *convex* if the segment $v_i v_j$ passes above $\{v_{i+1}, \ldots, v_{j-1}\}$ for all $i < j$. Such a sequence is *concave* if $v_i v_j$ passes below $\{v_{i+1}, \ldots, v_{j-1}\}$ for all $i < j$. Prove that the maximum size of a nondegenerate set of points in the plane that does not contain a convex sequence of r points or a concave sequence of s points is $\binom{r+s-4}{r-2}$. Conclude that $\binom{2m-4}{m-2}+1$ points force a convex m-gon. (Erdős-Szekeres [1935])

8.3.13. (+) Construct a nondegenerate set of 2^{m-2} points in the plane that span no convex m-gon. (Hint: Make use of the sets T_i for $0 \le i \le m-2$ (guaranteed by the preceding problem) that consist of $\binom{m-2}{i}$ points but contain no concave $i+2$-sequence or convex $m-i$-sequence.) (Erdős-Szekeres [1935])

8.3.14. (!) The *composition* or *lexicographic product* of two simple graphs G and H is the simple graph $G[H]$ whose vertex set is $V(G) \times V(H)$, and whose edges are given by $(u,v) \leftrightarrow (u',v')$ if and only if (1) uu' is an edge of G, or (2) $u = u'$ and vv' is an edge of H.
 a) Show that $\alpha(G[H]) \le \alpha(G)\alpha(H)$, where α is the independence number.
 b) Prove that the complement of $G[H]$ is $\overline{G}[\overline{H}]$.
 c) Use (a) and (b) to prove by construction that
$$r(pq+1, pq+1) - 1 \ge [r(p+1, p+1) - 1] \times [r(q+1, q+1) - 1]$$
 d) Deduce that $r(2^n + 1, 2^n + 1) \ge 5^n + 1$ for $n \ge 0$ and compare this lower bound to the non-constructive lower bound for $r(k, k)$. (Abbott [1972])

8.3.15. (−) Use the graph below to prove that $R(3, 5) = 14$.

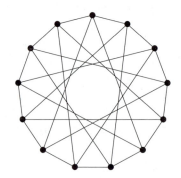

8.3.16. *Ramsey numbers with multiple colors.*
 a) Let $p = (p_1, \ldots, p_k)$, and let q_i be obtained from p by subtracting one from p_i but leaving the other coordinates unchanged. Prove that $r(p) \le \Sigma_{i=1}^{k} r(q_i) - k + 2$.
 b) Prove that $r(p_1 + 1, \ldots, p_k + 1) \le \dfrac{(p_1 + \cdots + p_k)!}{p_1! \cdots p_k!}$.

8.3.17. Let $r_k = R_k(3; 2)$ (this is the value of n such that k-coloring $E(K_n)$ forces a monochromatic triangle).
 a) Show that $r_k \le k(r_{k-1} - 1) + 2$.
 b) Use (a) to show that $r_k \le \lfloor k!e \rfloor + 1$, so that $r_3 \le 17$. (Comment: $r_3 = 17$, but the lower bound requires a clever 3-coloring of K_{16} that arises from the finite field $GF(2^4)$).

8.3.18. Prove that $R_k(p; r+1) \le r + k^M$, where $M = \binom{R_k(p;r)}{r}$.

8.3.19. (+) *Off-diagonal Ramsey numbers.*
a) Prove that $R(k,l) > n$ if $\binom{n}{k}p^{\binom{k}{2}} + \binom{n}{l}(1-p)^{\binom{l}{2}} < 1$ for some $p \in (0,1)$. Prove that $R(k,l) > n - \binom{n}{k}p^{\binom{k}{2}} - \binom{n}{l}(1-p)^{\binom{l}{2}}$ for all $n \in \mathbb{N}$ and $p \in (0,1)$.
b) Use part (a) to prove $R(3,k) > k^{3/2+o(1)}$. What lower bound on $R(3,k)$ can be obtained from the first part of (a)? (Spencer [1977])
c) Use part (a) to obtain a lower bound for $R_k(q)$.

8.3.20. (!) Determine the Ramsey number $R(K_{1,m}, K_{1,n})$. (Hint: the answer depends on whether m and n are even or odd.)

8.3.21. (!) Suppose $m-1$ divides $n-1$, and T is a tree with m vertices. Determine the Ramsey number $R(T, K_{1,n})$. (Burr [1974])

8.3.22. If $p > (m-1)(n-1)$, prove that every 2-coloring of $E(K_p)$ in which the red graph is transitively orientable contains a red m-clique or a blue n-clique, and prove that this is best possible. (Brozinsky-Nishiura) (Hint: use perfect graphs.)

8.3.23. If T is a tree on m vertices, show that $R(T, K_{n_1}, \ldots, K_{n_k}) = (m-1)(R(n_1, \ldots, n_k) - 1) + 1$ (Burr).

8.3.24. Computation of $R(C_4)$.
a) If a 2-colored complete graph with at least 6 vertices has a monochromatic C_5 or C_6, show that it also has a monochromatic C_4.
b) Given $R(P_4, P_4) = 5$, $R(P_4, C_4) = 5$, and part (a), prove that $R(C_4, C_4) = 6$.

8.3.25. (!) Prove that $R(mK_2, mK_2) = 3m - 1$.

8.3.26. Complete the proof that $R(2K_3, 2K_3) = 10$. (Hint: obtain a bow-tie and complementary monochromatic 5-cycles, then use symmetry.)

8.3.27. (!) For $1 \le i \le k$, suppose G_i is a graph on p_i vertices, and fix a multiplicity m_i. Prove that $R(m_1 G_1, \ldots, m_k G_k) \le \Sigma(m_i - 1)p_i + R(G_1, \ldots, G_k)$.

8.3.28. Frankl and Wilson [1981] gave an explicit construction of graphs with n vertices that have no clique or independent set with more than $2^{c\sqrt{\log n \log \log n}}$, where c is a particular constant. Prove that this gives a lower bound for $R(p,p)$ that grows faster than every polynomial in p but slower than every exponential in p.

8.3.29. (!) Determine $R(P_3, G)$ for every G, as a function only of the number of vertices of G and the maximum size of a matching in $\overline{G}$.

8.3.30. (!) Prove that every 2-coloring of $E(K_{2t-1,2t-1})$ has a monochromatic connected graph with at least $2t$ vertices. Exhibit a coloring having no monochromatic connected graph with more than $2t$ vertices.

8.3.31. Use the preceding problem to prove that every 3-coloring of $E(K_n)$ contains a monochromatic connected subgraph with more than $n/2$ vertices if $n \ne 0 \bmod 4$, but that this fails if 4 divides n.

8.3.32. (!) Bondy [1971a] proved that $x \not\leftrightarrow y$ implies $d(x) + d(y) \geq n(G)$, then $G = K_{t,t}$ or G has a cycle of each length from 3 to n. Use this to prove that $R(C_m, K_{1,n}) = \max\{m, 2n+1\}$, except possibly if m is even and at most $2n$. (Lawrence [1973]).

8.3.33. (!) Prove that every 2-coloring of $E(K_n)$ has a Hamiltonian cycle that is monochromatic or consists of two monochromatic paths. (Hint: Use induction on n.) (Lovász [1979, p85, p482 - attributed to H. Raynaud])

8.3.34. (+) *Ramsey numbers for cycles.* Let f be a 2-coloring of $E(K_n)$. Prove the following:

   a) If f contains a monochromatic C_{2k+1} for some $k \geq 3$, then f also contains a monochromatic C_{2k}.

   b) If f contains a monochromatic C_{2k} for some $k \geq 3$, then f also contains a monochromatic C_{2k-1} or $2K_k$.

   c) If $m \geq 5$, then $R_2(C_m) \leq 2m - 1$ (the case $m = 4$ is considered in Exercise 24). (Hint: Use (a),(b), and the result of Erdős-Gallai [1959] (Theorem 8.4.31) that $e(G) > (m-1)(n(G)-1)/2$ forces a cycle of length at least m in G. There is still one difficult case).

8.3.35. The *Ramsey multiplicity* of G is the minimum number of monochromatic copies of G in a 2-coloring of the edges of a clique on $R_2(G)$ vertices. Show that the Ramsey multiplicity of K_3 is 2.

8.3.36. Prove that each point in a triangular region has a unique expression as a convex combination of the vertices of the triangle (convex combinations are linear combinations where the coefficients are nonnegative and sum to 1).

8.3.37. *Sperner's Lemma in higher dimensions.* A *k-dimensional simplex* consists of all points in $\mathbb{R}^k$ that are weighted averages of $k+1$ points not lying in a hyperplane. A *simplicial subdivision* expresses a k-dimensional simplex as the union of cells that are k-dimensional simplices, such that any two intersecting cells intersect in a simplex of lower dimension. A *completely labeled* cell has $\{0, \ldots, k\}$ at its corners. State a general definition of "proper labeling" such that every proper labeling of a simplicial subdivision of a k-simplex contains a completely labeled cell. Prove this theorem. (Hint: the proof of Sperner's Lemma in 2 dimensions (Theorem 7.1.21) is an instance of the induction step for a proof by induction on k.)

8.3.38. (−) Compute the bandwidths of P_n, K_n, and C_n.

8.3.39. Compute the bandwidth of the complete multipartite graph $K_{n_1,\ldots,n_k}$ with $n_1 \geq \cdots \geq n_k$ and $\Sigma n_i = n$. (Eitner [1979])

8.3.40. (+) *Bandwidth of caterpillars.* Suppose G is a caterpillar (Theorem 2.2.19) and m is an integer such that $\lceil \frac{n(H)-1}{\mathrm{diam}H} \rceil \leq m$ for all $H \subseteq G$. Prove that $B(G) \leq m$. (Hint: Prove that G has a numbering f in which $f(v)$ is a multiple of m whenever v is on the spine and $|f(u) - f(v)| \leq m$ for all $u \leftrightarrow v$. (Syslo-Zak [1982], Miller [1981])

8.3.41. *Bandwidth of grids.*

a) Compute the local density bound on bandwidth of $P_m \square P_n$.

b) Let S be a k-set of vertices in $P_n \square P_n$ with a_i vertices in the ith row and b_j vertices in the jth column. Prove that $|\partial T| \le |\partial S|$ if T is the set consisting of the first a_i vertices in the ith row for each i.

c) Prove that $|\partial S|$ is minimized over k-sets in $V(P_n \square P_n)$ by some S such that $a_1 \ge \cdots \ge a_n$ and $b_1 \ge \cdots \ge a_n$. Conclude that Harper's lower bound for $B(P_n \square P_n)$ is n.

d) Conclude that $B(P_m \square P_n) = \min\{m, n\}$. (Chvátalová [1975])

8.3.42. *Bandwidth under edge addition.* (Wang-Yao-West [1995])

a) For $e \in \overline{G}$, prove that $B(G + e) \le 2B(G)$.

b) Prove that if $n \ge 6b$, then there is an n-vertex graph G with bandwidth b and an edge $e \in E(\overline{G})$ such that $B(G + e) = 2B(G)$. Comment: More generally, if $g(b, n)$ denotes the maximum value of $B(G + e)$ when G has order n and bandwidth q, then

$$g(b, n) = \begin{cases} b + 1 & \text{if } n \le 3b + 4 \\ \lceil \frac{n-1}{3} \rceil & \text{if } 3b + 5 \le n \le 6b - 2 \\ 2b & \text{if } n \ge 6b - 1 \end{cases}$$

8.4 More Extremal Problems

Extremal graph theory is a huge area. We mention two common types of problems. Given a graph as input, we may seek an extremal substructure; examples include diameter, maximum clique or independent set, maximum matching, connectivity, minimum coloring, longest cycle, etc. Extremal problems to be answered for each graph are often called "optimization problems". In contrast, what we usually call an "extremal graph problem" is determining the extreme value of some graph parameter over some class of graphs. The archetypal example is the Turán problem: find the maximum number of edges in a graph not containing H as a subgraph.

We list other examples: The maximum number of vertices in a clique coverable by k bipartite graphs is 2^k (Example 1.3.14 - generalized in Exercise 9). The maximum diameter among n-vertex graphs with minimum degree k is $\lfloor 3(n - 2)/(k + 1) \rfloor - 1$ (Exercise 2.1.38). The maximum vertex cover number among graphs with no matching of size $k + 1$ is $2k$ (Exercise 3.3.5). The minimum connectivity among n-vertex graphs with minimum degree δ is $2\delta + 2 - n$ (if $\delta \ge n/2 - 1$) (Exercise 4.1.11). The maximum chromatic number among graphs with clique number k and no induced $2K_2$ is $\binom{k+1}{2}$ (Exercise 5.2.8). The maximum number of edges in a non-Hamiltonian graph with n vertices is $\binom{n-1}{2} + 1$

(Exercise 6.2.21). The maximum number of guards needed to watch an art gallery with n corners is $\lfloor n/3 \rfloor$ (Exercise 7.3.5). The maximum number of vertices among graphs that have no p-clique and no independent set of size q is $R(p, q) - 1$ (Section 8.3).

With such enormous variety of extremal problems, we can only hope in this section to exhibit a small sample of interesting results.

ENCODINGS OF GRAPHS

We first consider three parameters that measure the difficulty of encoding a graph. Each model involves assigning vectors to vertices, and the parameter is the minimum length of vectors that suffice. We consider the maximum value over n-vertex graphs. The parameters are intersection number, product dimension, and squashed-cube dimension.

8.4.1. Definition. An *intersection representation* of *length* t assigns each vertex a 0,1-vector of length t such that $u \leftrightarrow v$ if and only if their vectors have a 1 in a common position. Equivalently, it assigns each $x \in V(G)$ a set $S_x \subseteq [t]$ such that $u \leftrightarrow v$ if and only if $S_u \cap S_v \neq \varnothing$. The *intersection number* $\theta'(G)$ is the minimum length of an intersection representation of G.

The elements of $[t]$ in a representation correspond to cliques that cover $E(G)$. This motivates our use of θ' for intersection number: $\theta(G)$ is the minimum number of cliques needed to cover $V(G)$.

8.4.2. Proposition. Erdős-Goodman-Pósa [1966] The intersection number equals the minimum number of cliques needed to cover $E(G)$.

Proof. We define a natural bijection between representations of length t and coverings of $E(G)$ by t cliques. Each $i \in [t]$ generates a clique $\{v \in V(G): i \in S_v\}$. Together, these cliques cover $E(G)$, since $u \leftrightarrow v$ if and only if $S_u \cap S_v \neq \varnothing$. Conversely, if cliques $Q_1, \ldots, Q_t$ cover $E(G)$, then we obtain an intersection representation by assigning to each vertex v the set $\{i: v \in Q_i\}$. $\square$

Hence $\theta'(G) = e(G)$ if G is triangle-free, and $\theta'(K_{\lfloor n/2 \rfloor, \lceil n/2 \rceil}) = \lfloor n^2/4 \rfloor$. In fact, this is the unique n-vertex graph maximizing $\theta'(G)$. Exercise 7 suggests a direct proof of the bound; here we present a stronger result.

The **F**-*decomposition* problem for G is that of partitioning $E(G)$ into the minimum number of graphs in **F**. If **F** is not closed under taking subgraphs, then **F**-decomposition may require more subgraphs of G than **F**-*covering*. For example, we can cover $E(K_4 - e)$ with two cliques, but three cliques are needed to partition $E(K_4 - e)$.

Hence decomposing every n-vertex graph into $\lfloor n^2/4 \rfloor$ cliques is a stronger result. Chung [1981] and Györi-Kostochka [1979] independently proved the still stronger result that every n-vertex graph has a clique decomposition $\{Q_i\}$ such that $\Sigma n(Q_i) \le \lfloor n(G)^2/2 \rfloor$ (the extremal example is not unique). McGuinness proved that successively taking the edges of a maximum remaining clique satisfies this bound. This is not algorithmically effective, because finding a maximum clique is NP-hard. Winkler [1992] conjectured that every decomposition obtained by successively taking maximal cliques also works. McGuinness proved that at least the *number* of cliques in such decompositions satisfies its desired bound.

8.4.3. Theorem. (McGuinness [1994]) If cliques $Q_1, \ldots, Q_m$ partition $E(G)$ so that each Q_i is a maximal clique in $G - \cup_{j<i} E(Q_j)$, then $m \le n(G)^2/4$.

Proof. We use induction on $n(G)$. The claim is trivial for $n(G) \le 2$. Suppose $n = n(G) > 2$, and let $\mathbf{Q} = Q_1, \ldots, Q_m$ be a greedy clique decomposition of G. If each Q_i has at least three vertices, then $m < n^2/6$, so we may assume some Q_j is an edge xy. Let R be the set of cliques in $\mathbf{Q} - \{Q_j\}$ containing x, and let S be those containing y. The set $\mathbf{Q}' = \mathbf{Q} - (R \cup S \cup \{Q_j\})$ is a greedy decomposition of $G - x - y$. By the induction hypothesis, $|\mathbf{Q}'| \le (n-2)^2/4$, and hence it suffices to prove that $|R| + |S| \le n - 2$.

We prove this by picking distinct representatives for the cliques in $R \cup S$ from $V(G) - \{x, y\}$. Since each edge is deleted exactly once, each $v \notin \{x, y\}$ appears once in R if $v \in N(x)$ and once in S if $v \in N(y)$. Suppose $Q \in R$. If Q contains a vertex $v \notin N(y)$, we choose v for Q. If $Q \subseteq N(y)$, then Q appears earlier than xy in $\mathbf{Q}$, since xy is a maximal clique when chosen. Since Q is maximal when chosen, some clique Q' preceding Q in $\mathbf{Q}$ contains an edge vy with $v \in Q$. For Q choose the vertex v such that vy is the first deleted edge from Q to y. For elements of S, choose vertices by reversing the roles of x and y. If v is chosen for $Q \in R$ and for $Q' \in S$, then the edge $vy \in Q'$ is deleted by a clique (Q') earlier than Q, and the edge $vx \in Q$ is deleted by a clique (Q) earlier than Q'. We cannot have Q' earlier than Q and Q earlier than Q', so no vertex is chosen twice, and $m \le n^2/4$. $\square$

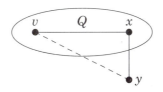

Now we consider the second scenario.

8.4.4. Definition. A *product representation* of *length* t assigns the vertices distinct vectors of length t so that $u \leftrightarrow v$ if and only if their vectors differ in every position. The *product dimension* pdim$G$ is the minimum length of such a representation of G. An *equivalence* on G is a spanning subgraph of G whose components are cliques.

A product representation expresses G as an induced subgraph of a "tensor product" of cliques. By devoting one coordinate to each $e \in E(\overline{G})$, in which the vertices of e have value 0 and other vertices have distinct positive values, we obtain pdim$(G) \leq e(\overline{G})$ (if G is not a clique).

8.4.5. Example. *Product dimension of cliques, stable sets, and* $K_1 + K_{n-1}$. A clique has product dimension 1. For $\overline{K}_n$, each pair of vertices must agree in some coordinate, but they can't all agree in the only coordinate if the vectors are distinct. Hence two coordinates are needed, and assigning $(0, j)$ to v_j suffices.

For $K_1 + K_{n-1}$, the vectors for the clique must be distinct in each coordinate. The vector for the isolated vertex must agree with each of the others somewhere, but it cannot agree with more than one in each coordinate. Hence at least $n - 1$ coordinates are needed. This suffices, by using $(1, 2, \ldots, n - 1)$ for the isolated vertex and $(i, i, \ldots, i)$ for the ith vertex of the clique. □

Again we can describe the parameter using cliques:

8.4.6. Proposition. The product dimension of G is the minimum number of equivalences $E_1, \ldots, E_t$ such that $\cup E_i = E(\overline{G})$ and $\cap E_i = \varnothing$.

Proof. Again there is a natural bijection. Given a product representation, the ith coordinate generates E_i, with a clique for each value in the ith coordinate. Every nonadjacent pair agrees in some coordinate, so every edge of $\overline{G}$ is covered. Conversely, given $\{E_i\}$, each clique in E_i becomes a fixed value in the ith coordinate of a representation. The requirement $\cap E_i = \varnothing$ is the requirement of using distinct vectors in the product representation. □

8.4.7. Lemma. If $\chi'(\overline{G}) > 1$, then pdim$(G) \leq \chi'(\overline{G})$, with equality if $\overline{G}$ is triangle-free.

Proof. Every matching is a disjoint union of cliques and becomes an equivalence by the addition of isolated vertices; hence $\chi'(\overline{G})$ equivalences cover $E(\overline{G})$. If $\chi'(\overline{G}) > 1$, these have empty intersection. If $\overline{G}$ is triangle-free, then every set of equivalences whose union is $\overline{G}$ is a covering of $E(\overline{G})$ by matchings. □

8.4.8. Corollary. For $n \geq 3$, the maximum product dimension of an n-vertex graph is $n - 1$.

Proof. Let G be an n-vertex graph. By the lemma and Vizing's Theorem (Theorem 6.1.7), $\mathrm{pdim}\, G \leq \chi'(\overline{G}) \leq n$. Furthermore, the bound improves to $n - 1$ unless $\Delta(\overline{G}) = n - 1$. Let S be the set of vertices of degree $n - 1$ in $\overline{G}$, and suppose $|S| = k \geq 1$. Since $\overline{G - S}$ has $n - k$ vertices, $\mathrm{pdim}(G - S) \leq n - k$. By duplicating coordinates if needed, we may consider a product representation of $G - S$ of length $n - k$, with vector x^i assigned to v_i.

Each vertex of S is isolated in G; to extend the representation, let the jth coordinate of the vector for each $v \in S$ be x_i^j. If $k = 1$, this completes a representation of length $n - 1$. If $k > 1$, we have assigned the same vector to all of S; we add one coordinate using distinct values to complete a representation of length $n - k + 1 \leq n - 1$. The graph $K_1 + K_{n-1}$ of Example 8.4.5 shows that the bound is best possible. $\square$

Lovász-Nešetřil-Pultr [1980] characterized the graphs achieving the maximum (Exercise 10). They also provided a general lower bound using a dimensionality argument in linear algebra.

8.4.9. Theorem. (Lovász-Nešetřil-Pultr [1980]) If G has two lists of r vertices, $A = \{u_1, \ldots, u_r\}$ and $B = \{v_1, \ldots, v_r\}$, such that $u_i \leftrightarrow v_j$ if and only if $i = j$, then $\mathrm{pdim}\, G \geq \lceil \lg r \rceil$.

Proof. Suppose G has a representation of length d. Let $x^1, \ldots, x^r$ and $y^1, \ldots, y^r$ be the vectors for the special vertices; x^i and y^i differ in every coordinate, but x^i and y^j agree in some coordinate if $i \neq j$. Hence $\Pi_{k=1}^d (x_k^i - y_k^j)$ is nonzero if and only if $i = j$. Using this, we construct r linearly independent vectors in $\mathbb{R}^{2^d}$; this proves $r \leq 2^d$ and hence $\mathrm{pdim}\, G \geq \lceil \lg r \rceil$.

Expansion of $\Pi_{k=1}^d (w_k - z_k)$ for vectors $w, z \in \mathbb{R}^d$ yields the sum $\Sigma_{S \subseteq [d]} \Pi_{i \in S} w_i \Pi_{j \in \bar{S}} (-z_j)$. In order to relate r to 2^d, we view this as a dot product in $\mathbb{R}^{2^d}$, with coordinates indexed by the subsets of $[d]$. For each $w \in \mathbb{R}^d$, define two vectors $\bar{w}$ and $\hat{w}$ in $\mathbb{R}^{2^d}$ by setting $\bar{w}_S = \Pi_{i \in S} w_i$ and $\hat{w}_S = \Pi_{i \notin S} (-w_i)$ for every $S \subseteq [d]$. Then $\Pi_{k=1}^d (w_k - z_k) = \bar{w} \cdot \hat{z}$, where $\cdot$ denotes inner product. The conditions on the x's and y's thus imply that $\bar{x}^i \cdot \hat{y}^j$ is nonzero if and only if $i = j$.

We claim that $\{\bar{x}^i\}$ are independent. Consider a linear dependence $\Sigma c_i \bar{x}^i = \bar{0}$. Taking the dot product of $\hat{y}^j$ with both sides kills all but one term of the sum, yielding $c_j \bar{x}^j \cdot \hat{y}^j = 0$. Since $\bar{x}^j \cdot \hat{y}^j \neq 0$, we have $c_j = 0$. Hence $\bar{x}^1, \ldots, \bar{x}^r$ are independent, which requires $2^d \geq r$. $\square$

8.4.10. Example. *The complete matching;* $\text{pdim}(n/2)K_2 = \lceil \lg n \rceil$. Given
k coordinates, the graph that results from using all 2^k binary sequences
as vertex encodings is $2^{k-1}K_2$, since each vector disagrees in every coordinate only with its complement. If n is not a power of 2, we can discard
some complementary pairs. The lower bound follows from the theorem,
using each vertex in each list ($u_i = v_{n+1-i}$, for example). $\square$

In our third model, we want to use the encoding to retrieve not
only adjacency but also the distance between two vertices. This arises
from an addressing problem in communication networks. We want each
message to travel a shortest path to its destination. Avoiding central
control, we want each vertex to compute the next vertex on the path
locally from the "address" of the destination. If the vectors for two vertices yield the distance between them in G, then a vertex can compare
the destination vector with the vectors for its neighbors and send the
message to a neighbor whose vector is closest to the destination vector.

We would like to assign each vertex an integer vector such that the
distance between vertices is the number of positions where the vectors
differ (assuming G is connected). This is an *isometric* or *"distance-
preserving"* embedding of G into $H = K_{n_1} \square \cdots \square K_{n_t}$, meaning a mapping
$f: V(G) \to V(H)$ such that $d_G(u,v) = d_H(f(u), f(v))$. Unfortunately,
many connected graphs have no isometric embedding in a Cartesian
product of cliques; C_{2k+1} for $k \geq 2$ is an example (Exercise 17).

Hence we introduce a "don't care" symbol $*$. Let $S = \{0, 1, *\}$, and
define a symmetric function d by $d(0,1) = 1$ and $d(0,*) = 0 = d(1,*)$. Let
S^N denote the N-tuples (vectors) with entries in S, and for $a, b \in S^N$ let
$d_S(a,b) = \Sigma d(a_i, b_i)$. Although this function does not satisfy the triangle
inequality and hence is not a metric ($d(0,*) + d(*,1) = 0 < 1 = d(0,1)$),
we will see that for some N an encoding $f: V(G) \to S^N$ exists so that
$d_G(u,v) = d_S(f(u), f(v))$ for all $u, v \in V(G)$.

Each $a \in S^N$ corresponds to a subcube of Q_N, the N-dimensional
cube; the dimension of the subcube is the number of $*$'s in a. For
$a, b \in S^N$, $d_S(a,b)$ is the minimum distance between vertices of the corresponding subcubes. The vectors assigned to distinct vertices correspond to disjoint subcubes, else their distance would be 0. If we contract the edges of each assigned subcube, we obtain a "squashed cube"
H. The distance-preserving map $f: V(G) \to S^N$ is an isometric embedding of G in H.

8.4.11. Definition. A *squashed-cube embedding* of *length* N is a map
$f: V(G) \to S^N$ such that $d_G(u,v) = d_S(f(u), f(v))$. The *squashed-
cube dimension* $\text{qdim} G$ is the minimum length of a squashed-cube
embedding of G.

8.4.12. Example. *Squashed-cube embeddings of cliques and paths.*
The vectors $000, 001, 01*, 1**$ form a squashed-cube embedding of K_4.
Two adjacent vertices of the 3-cube remain unchanged, an edge adjacent
to them collapses, and the entire opposite face collapses. The resulting
graph is K_4. The image subcubes appear as solid edges below. This
generalizes to embed K_n in a squashed $n-1$-dimensional cube.

The path P_n embeds isometrically in Q_{n-1} without squashings,
using $00 \cdots 00$, $10 \cdots 00$, $11 \cdots 00$,..., $11 \cdots 10$, $11 \cdots 11$. No shorter
embedding exists, because the distance between the endpoints of P_n is
$n-1$, and each coordinate of the embedding contributes at most one
toward the distance between the vectors. $\square$

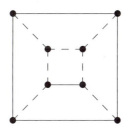

8.4.13. Example. $\operatorname{qdim}(G) \le \Sigma_{i<j} d_G(v_i, v_j)$. For each pair i, j with $i < j$,
we dedicate a block of $d_G(v_i, v_j)$ coordinates. Set these coordinates to 0
for v_i, to 1 for v_j, and to $*$ for other vertices. Given two vertices, the
only coordinates where neither contains $*$ are the coordinates dedicated
to the pair, so $d_G(v_i, v_j) = d_S(f(v_i), f(v_j))$. $\square$

Using an eigenvalue technique (Exercise 8.6.11), Graham and Pol-
lak [1971, 1973] proved a general lower bound on $\operatorname{qdim}(G)$ that yields
$\operatorname{qdim} K_n = n - 1$. Hence K_n and P_n both have squashed-cube dimension
$n - 1$; Graham and Pollack conjectured that $\operatorname{qdim} G \le n - 1$ for every n-
vertex connected graph. Graham offered 100 dollars for a proof, and
Winkler found an encoding scheme to prove this "Squashed Cube Conjecture".

Winkler's procedure generates an explicit $n - 1$-dimensional
squashed-cube encoding for each connected n-vertex graph G. First we
number the vertices. Choose v_0 arbitrarily, and find a spanning tree T
such that $d_T(v, v_0) = d_G(v, v_0)$ for all $v \in V(G)$, such as by breadth-first
search from v_0. Now, number the vertices by a *depth-first* numbering of
T. In other words, having chosen the vertices through v_i, let v_{i+1} be an
unvisited neighbor in T of v_i, if one exists; otherwise backtrack toward
the root v_0 until a vertex with an unvisited neighbor in T is found. The
resulting indices increase along each path from v_0 in T.

8.4.14. Example. *Depth-first numbering of a breadth-first spanning tree.* Below, the solid edges belong to T and the dashed edges to $G - T$. We will use this example to illustrate several steps in the proof. □

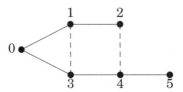

We henceforth refer to vertices by their index in this ordering. Let P_i be the vertices on the path between i and 0 in T (inclusive), let $i \wedge j = \max(P_i \cap P_j)$ (the vertex at which these paths meet), and let $i' = \max(P_i - \{i\})$ (i.e. the next vertex after i on the path from i to 0 in T).

8.4.15. Lemma. (Winkler [1983]). Given a depth-first numbering of a breadth-first tree T in G, let $c(i, j) = d_T(i, j) - d_G(i, j)$. Then
a) $c(i, j) = c(j, i) \geq 0$.
b) If $i \in P_j$, then $c(i, j) = 0$.
c) If neither $i \in P_j$ nor $j \in P_i$, then $c(i, j') \leq c(i, j) \leq c(i, j') + 2$.

Proof. (a): distance is symmetric, and it cannot decrease when edges are deleted to form a subgraph. (b): every subpath of a shortest $i, 0$-path in G (such as the $i, 0$-path in T) is a shortest path between its endpoints. (c): The i, j-path and i, j'-path in T pass through $i \wedge j$; hence $d_T(i, j) - d_T(i, j') = 1$. Since $j \leftrightarrow j'$ in G, we have $|d_G(i, j) - d_G(i, j')| \leq 1$. Hence $c(i, j) - c(i, j')$ is 0, 1, or 2. □

8.4.16. Theorem. (Winkler [1983]). Every connected n-vertex graph G has squashed-cube dimension at most $n - 1$.

Proof. Choose a tree T and numbering $0, \ldots, n - 1$ as described above. For the encoding $f(i) = (f_1(i), \ldots, f_{n-1}(i))$ defined below, we prove that $d_G(i, j) = d_S(f(i), f(j))$:

$$f_k(i) = \begin{cases} 1 & \text{if } k \in P_i \\ * & \text{if } c(i, k) - c(i, k') = 2 \\ * & \text{if } c(i, k) - c(i, k') = 1, \, i < k, \, c(i, k) \text{ even} \\ * & \text{if } c(i, k) - c(i, k') = 1, \, i > k, \, c(i, k) \text{ odd} \\ 0 & \text{otherwise} \end{cases}$$

In Example 8.4.14, the encoding is $f(0) = 00000$, $f(1) = 10000$, $f(2) = 110{*}0$, $f(3) = {*}0100$, $f(4) = {*}{*}110$, and $f(5) = {*}{*}111$.

To compute $d_S(f(i), f(j)) = d_G(i, j)$, we count the coordinates k such that $\{f_k(i), f_k(j)\} = \{0, 1\}$. Such coordinates must satisfy $k \in P_i \cup P_j$, where all the 1's are located in $f(i)$ and $f(j)$. By symmetry, we may assume $i < j$. Hence $j \notin P_i$. If $i \in P_j$, then $d_G(i, j) = d_T(i, j) = |P_j - P_i|$. Since $c(i, k) = 0$ for $i, k \in P_j$, we have $f_k(j) = 1$ and $f_k(i) = 0$ if and only if $k \in P_j - P_i$, as desired (note that $f_k(i) = f_k(j) = 1$ if $k \in P_i$).

Hence we may assume that $j \notin P_i$ and $i \notin P_j$. The set $\{f_k(i), f_k(j)\}$ has exactly one 1 if and only if $k \in (P_j - P_i) \cup (P_i - P_j)$. We prove that, among these coordinates k, the other vector has the value $*$ exactly $c(i, j)$ times. This yields $d_S(f(i), f(j)) = |P_j - P_i| + |P_i - P_j| - c(i, j) = d_T(i, j) - c(i, j) = d_G(i, j)$. In Example 8.4.14, $(P_5 - P_2) \cup (P_2 - P_5)$ is $\{1, 2, 3, 4, 5\}$; there are three coordinates where $*$ matches 1, and we have $d_S(f(2), f(5)) = d_G(2, 5) = 2$.

To locate the $*$'s among these coordinates, consider the change in c as we bring either of i, j to the point $i \wedge j$. We have $c(i, j) \geq c(i, j') \geq \cdots \geq c(i, i \wedge j) = 0$ and $c(i, j) \geq c(i', j) \geq \cdots \geq c(i \wedge j, j) = 0$. We obtain one $*$ in $f(i)$ for each even m with $0 < m \leq c(i, j)$ and one $*$ in $f(j)$ for each odd m with $0 < m \leq c(i, j)$.

For even m with $0 < m \leq c(i, j)$, let j_m be the unique vertex in the first sequence such that $c(i, j_m) \geq m$ but $c(i, j'_m) < m$. Even when the difference between consecutive values is 2 and the value m does not appear, j_m is well-defined. Because c changes by at most 2 with each step, the values of j_m are distinct. Furthermore, the depth-first ordering guarantees $i < k$ for all $k \in P_j - P_i$. Thus $f_k(i) = *$ for $k \in P_j - P_i$ if and only if $k = j_m$ for some even m. In Example 8.4.14, for $(i, j) = (2, 5)$ we have $j_2 = 4$ and $f_4(i) = *$.

Similarly, for odd m with $0 < m \leq c(i, j)$, let i_m be the unique vertex in the second sequence such that $c(i_m, j) \geq m$ but $c(i'_m, j) < m$. As before, the values of i_m are distinct and well-defined. The depth-first ordering guarantees $j > k$ for all $k \in P_i - P_j$, so $f_k(j) = *$ for $k \in P_i - P_j$ if and only if $k = i_m$ for some odd m. In Example 8.4.14, for $(i, j) = (2, 5)$ we have $i_1 = 1$, $i_3 = 2$, and $f_1(j) = f_3(j) = *$.

Thus we have counted the coordinates in $(P_i - P_j) \cup (P_j - P_i)$ having $*$'s in $f(i)$ or $f(j)$. The total is the number of even integers between 1 and $c(i, j)$ plus the number of odd integers between 1 and $c(i, j)$, which together equal $c(i, j)$. $\square$

BRANCHINGS AND GOSSIP

We have studied the problem of finding the maximum number of pairwise edge-disjoint spanning trees in a graph; this equals the maximum k such that for every vertex partition P, there are at least $k(|P| - 1)$ edges crossing between parts of P (Corollary 8.2.58). Here we consider an analogous problem for digraphs that is related to Menger's Theorem (Exercise 20). Menger's Theorem is a min-max theorem that focuses on vertex pairs. We examine "connectedness" from a single vertex to the rest of the digraph.

8.4.17. Definition. An *r-branching* in a digraph is a rooted tree "branching out" from r. Vertex r has indegree 0, all other vertices have indegree 1, and all other vertices are reachable from r. Let $\kappa(r; G)$ denote the minimum number of edges whose deletion makes some vertex unreachable from r.

Observe that $\kappa(r; G)$ also equals the minimum, over nonempty $X \subseteq V(G) - \{r\}$, of the number of edges entering X. For one inequality, deleting the edges entering X makes each vertex of X unreachable. For the other, Menger's Theorem yields a cut of size k when deletion of $k = \kappa(r; G)$ edges makes v unreachable from r.

We cannot have more than $\kappa(r; G)$ pairwise edge-disjoint r-branchings in G; Edmonds proved that this bound is achievable.

8.4.18. Theorem. (Edmonds' Branching Theorem [1973]). In a directed multigraph containing vertex r, the maximum number of pairwise edge-disjoint r-branchings is $\kappa(r; G)$.

Proof. (Lovász [1976]) The upper bound holds since each subset of $V(G) - \{r\}$ must be entered by at least one edge in every r-branching. We prove the existence of $\kappa(r; G)$ edge-disjoint r-branchings by induction on $k = \kappa(r; G)$. For $k = 1$, a breadth-first search suffices to grow an r-branching, since every vertex is reachable. For $k > 1$, we seek an r-branching T such that $\kappa(r; G - E(T)) = k - 1$; the induction hypothesis then supplies the other $k - 1$ r-branchings.

To build the desired r-branching, we consider r-branchings T of induced subgraphs of G such that $\kappa(r; G - E(T)) \geq k - 1$. The vertex r itself is such a branching, with $E(T) = \varnothing$. Hence we can let T be such a branching with the maximum number of vertices; if $V(T) = V$ we are done. Suppose $V(T) \neq V$, and let e_X denote the number of edges entering $X \subseteq V - r$ in $G - E(T)$. If $e_X \geq k$ for every $X \subseteq V - r$ that contains a vertex of $V - V(T)$, then we can extend T by adding any edge from $V(T)$ to $V - V(T)$. Hence we can choose a smallest set $U \subseteq V - r$ that contains a vertex of $V - V(T)$ and has exactly $k - 1$ edges entering it. (In the illustration, $E(T)$ are the solid edges.)

Because $\kappa(r; G) = k$ and we have deleted no edge entering $U - V(T)$, we still have $e_{U-V(T)} \geq k$. However, $e_U = k - 1$, so there must be an edge xy from $V(T) \cap U$ to $U - V(T)$. We claim that xy can be added to enlarge T, contradicting the maximality of T. We need only verify that at least $k - 1$ edges still enter each $W \subseteq V - r$ when we delete xy from $G - E(T)$. This holds trivially unless $x \in V - W$ and $y \in W$. Hence it suffices to show $e_W \geq k$ for such a W. Rewrite e_W by $e_W = e_{W \cup U} - e_U + e_{W \cap U}$. We have $e_{W \cup U} \geq k - 1$ by the defining property of T, $e_U = k - 1$ by construction, and $e_{W \cap U} \geq k$ by $x \in U - W$ and the minimality of U. Hence $e_W \geq k - 1 - (k - 1) + k = k$, as desired. $\square$

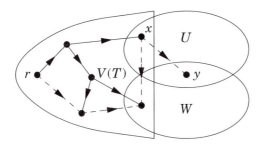

Lovász's proof can be converted to an algorithm for finding the maximum number of pairwise disjoint r-branchings; Tarjan [1975] gave another algorithm. We might call $\kappa(r; G)$ the *local-global connectivity*. Theorem 8.4.18 has several equivalent forms:

8.4.19. Corollary. If G is a directed graph, r is a vertex of G, and $k \geq 0$, then the following statements are equivalent.
A) G has k pairwise edge-disjoint r-branchings.
B) $\kappa(r; G) \geq k$; equivalently, $|[\overline{X}, X]| \geq k$ for all $X \subseteq V(G) - \{r\}$.
C) For each $s \neq r$ there exist k pairwise edge-disjoint r, s-paths.
D) There exist k pairwise edge-disjoint spanning trees of the underlying graph that for each $s \neq r$ contain among them exactly k edges entering s.

Proof. A$\Leftrightarrow$B is Edmonds' theorem, B$\Leftrightarrow$C is Menger's Theorem, and A$\Rightarrow$D is immediate. For D$\Rightarrow$B, assume that the trees exist and consider $U \subseteq V - r$. Each spanning tree has at most $|U| - 1$ edges within U, so the trees together have at most $k(|U| - 1)$ edges within U. By hypothesis they have exactly $k|U|$ edges with heads in U, so at least k edges enter U. $\square$

Schrijver has observed that Edmonds' Branching Theorem can also be proved using matroid union and matroid intersection. Discard the edges entering the root r. Let M_1 be the union of k copies of the cycle matroid on the underlying undirected graph. Let M_2 be the matroid in which a set of edges is independent if and only if no $k + 1$ of them have the same head (this is the direct sum of uniform matroids of rank k). There exist k disjoint r-branchings if and only if these two matroids have a common independent set of size $k(n(G) - 1)$.

Pairwise edge-disjoint r-branchings can be viewed as a fault-tolerant static protocol for message transmissions from r; there are alternative trees to use. Next we consider a static protocol for transmissions from each vertex to every other. We consider two-way transmissions, but the transmissions are performed in a specified order.

The resulting question is the *gossip problem*. Consider n gossips, each having a tidbit of information. Being gossips, each wants to know all the information, and when two communicate they tell each other everything they know. How many telephone calls are needed to transmit all the information? Several solutions were published in the early 1970's; we present one using trees and matchings.

Succeeding with $2n - 3$ calls is easy: everyone calls x, and then x calls everyone back, saving one by combining the last call in and first call out. When $n \geq 4$, $2n - 4$ calls suffice: first the others call in to a set S of four people, then S shares the information in two successive pairings, and finally the others receive calls back from S, using a total of $(n - 4) + 4 + (n - 4) = 2n - 4$ calls. Using a graph model, we show that this is optimal.

8.4.20. Definition. An *ordered multigraph* is a multigraph with a linear ordering on the edges. An x, y-path on which the edge indices successively increase is a path transmitting information from x to y; this is an *increasing path*. A *gossip scheme* is an ordered multigraph that has an increasing path from each vertex to every other vertex. A gossip scheme satisfies *NOHO* ("No One Hears hir Own information") if it has no increasing path from a vertex x followed by a call between x and the other endpoint of that path.

8.4.21. Theorem. (Gossip Theorem) For $n \geq 4$, the minimum number of edges in a gossip scheme is $2n - 4$.

Proof. (Baker-Shostak [1972]). We freely use "calls" in place of "edges" to emphasize the ordering. The scheme described above uses $2n - 4$ calls, and case analysis shows that it is optimal for $n = 4$. This provides the basis for a proof by induction on n. For $n > 4$, we may assume that every gossip scheme with $n - 1$ vertices uses at least $2n - 6$ calls. If $2n - 4$ is not optimal for n vertices, then we can add calls to the optimal scheme (if necessary) to obtain an n-vertex gossip scheme G with exactly $2n - 5$ calls.

Claim 1. G satisfies NOHO. Otherwise, G has an increasing path $x, v_1, \ldots, v_k$ along edges $e_1 < \cdots < e_k$, followed by a call $e_{k+1} = v_k x$. Delete e_1 and e_{k+1}. Partition the other calls involving x into $k + 2$ sets: E_0 consists of those before e_1, E_i for $1 \leq i \leq k$ consists of those between e_i and e_{i+1}, and E_{k+1} consists of those after e_{k+1}. In each edge $e \in E_i$, replace x by v_1, v_i, or v_k if $i = 0$, $1 \leq i \leq k$, or $i = k + 1$, respectively (see illustration). Now $E(G) - \{e_1, e_{k+1}\}$ is a gossip scheme on $V(G) - \{x\}$, because every increasing path through x is replaced by an increasing path that consists of the same edges and perhaps additional edges from $\{e_i\}$. The scheme has $2(n - 1) - 5$ edges, which contradicts the induction hypothesis.

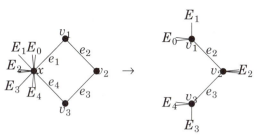

Claim 2. $d(x) - 3$ calls are useless to x, and hence $\delta(G) \geq 3$. Let $O(x)$ be the set of calls on which some vertex is reached for the first time by an increasing path "Out" from x; viewing these calls in order shows that they form a spanning tree. The tree $I(x)$ of edges useful In to x is $O(x)$ for the reverse order on $E(G)$. If some increasing x, y-path reaches $y \in N(x)$ before the edge xy, then x violates NOHO; hence $xy \in O(x)$. Similarly, xy cannot be omitted from $I(x)$. Conversely, if $O(x) \cap I(x)$ contains an edge e not incident to x, then an increasing path from x containing e and an increasing path to x containing e combine to violate NOHO for x. Hence $|O(x) \cap I(x)| = d(x)$. The edges "useless to x" are those not in $O(x) \cup I(x)$, and $|\overline{O(x) \cup I(x)}| = 2n - 5 - (n-1) - (n-1) + d(x) = d(x) - 3$. Since this counts a set of edges, $\delta(G) \geq 3$.

Claim 3. The subgraph obtained by deleting the first call and the last call made by each vertex has at least 5 components and no isolated vertex. Suppose xy is the first call involving x. If the first call by y goes to z, then it happens before xy, and these two calls do not communicate from x to z. After yz and xy, an increasing x, z-path violates NOHO at z. Hence the set F of first calls is a matching, and there are $n/2$ of them. Similarly, the set L of last calls is a matching of size $n/2$. The graph $G - F - L$ has $n - 5$ edges and hence at least 5 components (Exercise 1.2.12). It has no isolated vertex since $\delta(G) \geq 3$.

The contradiction. Since $e(G) = 2n - 5 < 2n$, some vertex x has degree at most 3. Let C_1, C_2, C_3 be the components of $G - F - L$ containing x, its first neighbor, and its last neighbor, respectively (its middle neighbor is also in C_1). Edges of $G - F - L$ can belong to $O(x)$ only via paths that start with the first or middle edge involving x and thereafter use no first or last edge, so they must belong to C_1 or C_2. Similarly, edges of $G - F - L$ belonging to $I(x)$ can appear only in C_1 or C_3. Hence the edges of the remaining components, of which there are at least two, are useless to x. This contradicts the count that there are $d(x) - 3 = 0$ edges useless to x. □

Many variations of gossiping have been considered. In practical applications, one might be concerned about minimizing the total length of the messages or about minimizing the total time under the assumption that each vertex participates in at most one call per time unit. We can also restrict the pairs that are allowed to call each other. If the graph of available calls is connected and has a 4-cycle, then $2n - 4$ calls still suffice. Retaining a 4-cycle is also necessary for gossiping in $2n - 4$ calls, as shown by Bumby [1981] and by Kleitman-Shearer [1980].

LIST COLORINGS AND CHOOSABILITY

List coloring is a more general version of the usual coloring problem. We still want to pick a single color for each vertex, but the set of colors available at each vertex may be restricted.

8.4.22. Definition. For each vertex v in a graph G, let $L(v)$ denote a list of colors available for v. A *list coloring* or *choice function* from a given collection of lists is a proper coloring f such that $f(v)$ is chosen from $L(v)$. A graph G is *k-choosable* or *k-list-colorable* if it has a proper list coloring from every assignment of k-element lists to the vertices. The *list chromatic number* or *choosability* $\hat{\chi}(G)$ is the minimum k such that G is k-choosable.

Since the lists can be chosen to be identical, $\hat{\chi}(G) \geq \chi(G)$. If the lists have size at least $1 + \Delta(G)$, then coloring the vertices in succession leaves an available color at each vertex. This argument is analogous to the greedy coloring algorithm and proves that $\hat{\chi}(G) \leq 1 + \Delta(G)$ (see Exercise 28 for other analogues with $\chi(G)$). It is not possible, however, to place an upper bound on $\hat{\chi}(G)$ in terms of $\chi(G)$; there are bipartite graphs with arbitrarily large list-chromatic number.

8.4.23. Theorem. (Erdős-Rubin-Taylor [1979]) If $m = \binom{2k-1}{k}$, then $K_{m,m}$ is not k-choosable.

Proof. Let X, Y be the bipartition of $G = Kmm$. For each of X, Y, let the lists be the k-subsets of $[2k - 1]$. Suppose G has a choice function f. If f uses fewer than k distinct choices in X, then there is a k-set $K \subseteq [2k - 1]$ not used, which means that no color was chosen for the vertex of X having K as its list. If f uses at least k colors on vertices of X, then there is a k-set $K \subseteq [2k - 1]$ of colors used in X, and no color can be properly chosen for the vertex Y with list K. □

List chromatic number is more difficult to compute than chromatic number; the statements of the upper bound and lower bound both involve universal quantifiers. Determining the 3-choosable complete bipartite graphs was difficult; for $m \leq n$, $K_{m,n}$ is 3-choosable if and only if $m \leq 2$, $m = 3$ and $n \leq 26$ (Erdős-Rubin-Taylor [1979]), $m = 4$ and $n \leq 20$ (Mahadev-Roberts-Santhanakrishnan [1991]), $m = 5$ and $n \leq 12$ (Shen-de-Tesman [1994]), or $m = 6$ and $n \leq 10$ (O'Donnell [1996]).

Alon and Tarsi [1992] used a polynomial associated with a graph to obtain upper bounds on $\hat{\chi}(G)$. Fleischner and Stiebitz [1992] used this to prove the conjecture of Du, Hsu, Hwang, and Erdős that adding n disjoint triangles to a $3n$-cycle yields a 3-chromatic graph, by showing the stronger result that the resulting graph is 3-choosable.

There is also an edge-coloring variant, where we assign k-lists to the edges and seek to choose a proper edge-coloring.

8.4.24. Definition. Let $L(e)$ denote the list of colors available for e. A *list edge-coloring* is a proper edge-coloring f with $f(e)$ chosen from $L(e)$ for each e. The *list chromatic index* or *edge-choosability* $\hat{\chi}'(G)$ is the minimum k such that G has a proper list edge-coloring for each assignment of lists of size k to the edges. Equivalently, $\hat{\chi}'(G) = \hat{\chi}(L(G))$, where $L(G)$ is the line graph of G.

The argument that implies $\chi'(G) \le 2\Delta(G) - 1$ also yields $\hat{\chi}'(G) \le 2\Delta(G) - 1$ (Exercise 28) and thus $\hat{\chi}'(G) < 2\chi'(G)$. The List Coloring Conjecture, which Bollobás [1986] reports is attributed variously to Gupta, Dinitz, Albertson, and Tucker, asserts that $\hat{\chi}'(G) = \chi'(G)$.

Since Vizing's Theorem [1964] implies $\chi'(G) \le \Delta(G) + 1$ when G is simple, it suffices to compare $\hat{\chi}'(G)$ with $\Delta(G)$, at least asymptotically. Bollobás and Harris [1985] proved that $\hat{\chi}'(G) < c\Delta(G)$ for each constant c exceeding 11/6, if $\Delta(G)$ is sufficiently large. This and subsequent improvements use probabilistic methods. Kahn [1996] proved that the conjecture holds asymptotically: $\hat{\chi}'(G) \le (1 + o(1))\Delta(G)$. Häggkvist and Janssen [1996] sharpened the error term, proving that $\hat{\chi}'(G) \le d + O(d^{2/3}\sqrt{\log d})$ for $d = \Delta(G)$. They also proved $\hat{\chi}'(K_n) = n$ for n odd, and earlier $\hat{\chi}'(K_n) \le n + 1$.

The special case $\hat{\chi}'(K_{n,n}) = \chi'(K_{n,n})$ was specifically known as the Dinitz Conjecture (1977 - see Erdős [1979]). Janssen [1993] proved the slightly weaker statement that equality holds for $K_{n,n-1}$. The Dinitz Conjecture can be phrased as follows: If each position of an n by n grid contains a set of size n, then it is possible to choose one element from each set so that the elements chosen in each row are distinct and the elements chosen in each column are distinct.

Galvin [1995] proved the List Coloring Conjecture for bipartite graphs, which includes the Dinitz Conjecture. Here we prove only the Dinitz Conjecture, using the Stable Matching Problem (Section 3.2).

8.4.25. Definition. A *kernel* of a digraph is an independent set S having an edge to every vertex outside S - an independent "out-dominating" set. A digraph is *kernel-perfect* if every induced subdigraph has a kernel. Given a function $f: V(G) \to \mathbb{N}$, the graph G is *f-choosable* if a proper coloring can be chosen from the lists at the vertices whenever $|L(x)| = f(x)$ for each x.

A directed 3-cycle or 3-vertex path, for example, has no kernel. An f-choosable graph is k-choosable for $k = \max f(x)$, since adding colors to a list cannot make the choice more difficult.

8.4.26. Lemma. (Bondy-Boppana-Siegel) If D is a kernel-perfect orientation of G and $f(x) = d_D^-(x)$ for all $x \in V(G)$, then G is $1 + f$-choosable.

Proof. By induction on $n(G)$; trivial for $n(G) = 1$. Suppose $n(G) > 1$, and consider an assignment of lists, with the list $L(x)$ for each x having size $1 + f(x)$. Choose a color c appearing in some list. Let $U = \{v: c \in L(v)\}$. Let S be the kernel of the induced subdigraph $D[U]$. Assign color c to all of S, which is permissible since S is independent. Delete c from $L(v)$ for each $v \in U - S$. Delete additional colors arbitrarily from other lists to reduce $L(x)$ for each $x \in V(D) - S$ to size $1 + f'(x)$, where $f'(x) = d_{D-S}^-(x)$. By the induction hypothesis, D' is $1 + f'(x)$-choosable, so we can complete a list coloring for the original graph from the remaining colors at $V(G) - S$. $\square$

8.4.27. Theorem. (Galvin [1995]) $\hat{\chi}'(K_{n,n}) = n$.

Proof. Since $\hat{\chi}'(G) = \hat{\chi}(L(G))$, it suffices by the lemma to prove that $L(K_{n,n})$ has a kernel-perfect orientation with each vertex having indegree and outdegree $n - 1$. The graph $L(K_{n,n})$ is the cartesian product $K_n \square K_n$; placed in an n by n grid, vertices are adjacent if and only if they share a row or share a column.

Label the vertices with labels $1, 2, \ldots, n$ so that vertex (r, s) has label $r + s - 1 \bmod n$. Define an orientation D of $K_n \square K_n$ by directing edges from vertex (r, s) with label i to the vertices in row r with higher labels and the vertices in column s with lower labels. Since label j is higher than $j - 1$ other labels, a vertex with label j has $j - 1$ predecessors in its row and $n - j$ predecessors in its column. Hence each vertex has indegree $n - 1$ and outdegree $n - 1$.

We prove that D is kernel-perfect. Given $U \subseteq V(D)$, we obtain a kernel for the subdigraph $D[U]$ by solving a stable matching problem. For each row r, the preferences among the columns begin with $\{s: (r, s) \in U\}$ in increasing order of the labels on these vertices, and the order among $\{s: (r, s) \notin U\}$ is irrelevant. For each column s, the preferences among the rows begin with $\{r: (r, s) \in U\}$ in decreasing order of the labels on these vertices, and the order among $\{r: (r, s) \notin U\}$ is irrelevant.

Theorem 3.2.13 yields a stable matching M for these preferences. Viewing the matched pairs in M as positions in the grid, let $S = M \cap U$. Because M is a matching, S has no two positions in the same row or column; hence S is independent. We show S has an edge to each $x \in U - S$. Let i be the label of position $x = (r, s) \in U - S$. The matching M has a position $y = (r, b)$ with some label j and a position $z = (a, s)$ with some label k. Because M is stable, we cannot have both of (1) $j > i$ or $y \notin U$, and (2) $k < i$ or $z \notin U$, because then r prefers s to b and s prefers r to a, which violates the definition of stable matching. We conclude that (1) $y \in S$ and $j < i$, or (2) $z \in S$ and $k > i$. In either case, we have an edge from S to x. □

The List Coloring Conjecture has been studied for planar graphs. Using results described by Alon [1993], it was observed independently by Jaeger and Tarsi (unpublished) that $\hat{\chi}'(G) = 3$ when G is a 2-connected cubic planar graph (this uses the Four Color Theorem). Ellingham and Goddyn [1996] extended this, proving that every k-regular k-edge-colorable planar multigraph is k-edge-choosable.

Concerning choosability, Vizing [1976] and Erdős-Rubin-Taylor [1979] conjectured that the maximum choice number among planar graphs is 5. Voigt [1993] constructed a planar graph with 238 vertices that is not 4-choosable and showed how to extend this to an infinite family of examples. Thomassen [1994b] completed the proof of the conjecture. As common in constructive proofs for plane graphs, the vertices of the unbounded face ("external vertices") play a special role in a loaded induction hypothesis. Thomassen [1995] also proved that planar graphs of girth 5 are 3-choosable.

8.4.28. Theorem. (Thomassen [1994b]) Planar graphs are 5-choosable.

Proof. Adding an edge never reduces the list-chromatic number, so we may restrict our attention to plane graphs in which the outer face is a cycle and every bounded face is a triangle. By induction on $n(G)$, we prove the stronger result that a list-coloring can be chosen even when two adjacent external vertices have distinct lists of size 1 and the other external vertices have lists of size 3. When $n = 3$, this leaves a color available for the third vertex.

For the induction step, suppose $n > 3$, and let v_p, v_1 be the vertices with fixed colors on the external cycle C with vertices $v_1, \ldots, v_p$ in clockwise order. Suppose C has a chord $v_i v_j$ for some $1 \le i \le j - 2 \le p - 2$. We apply the induction hypothesis to the graph consisting of the cycle $v_1, \ldots, v_i, v_j, \ldots, v_p$ and its interior. This selects a proper coloring in which v_i, v_j receive some fixed colors. Next we apply the induction hypothesis to the graph consisting of the cycle $v_i, v_{i+1}, \ldots, v_j$ and its interior to complete the list coloring of G.

Hence we may assume that C has no chord. Let $v_1, u_1, \ldots, u_m, v_3$ be the neighbors of v_2 in counterclockwise order ($3 = p$ is possible). Because bounded faces are triangles, G contains the path P with vertices $v_1, u_1, \ldots, u_m, v_3$. Since C is chordless, $u_1, \ldots, u_m$ are internal vertices, and the outer face of $G' = G - v_2$ is bounded by a cycle C' in which P replaces v_1, v_2, v_3.

Let c be the color assigned to v_1. Since $|L(v_2)| \geq 3$, we may choose distinct colors $x, y \in L(v_2) - \{c\}$. We reserve x, y for possible use on v_2 by forbidding x, y from $u_1, \ldots, u_m$. Since $|L(u_i)| \geq 5$, this leaves $|L(u_i) - \{x|$ ≥ 3. Hence we can apply the induction hypothesis to G', with $u_1, \ldots, u_m$ having lists of size at least 3 and other vertices having the same lists as in G. In the resulting coloring, v_1 and $u_1, \ldots, u_m$ have colors outside $\{x, y\}$. We extend this coloring to G by choosing for v_2 a color in $\{x, y\}$ that does not appear on v_3 in the coloring of G'. $\square$

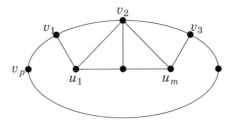

PARTITIONS USING PATHS AND CYCLES

We have considered the **F**-*decomposition* problem; partitioning $E(G)$ into the minimum number of subgraphs in a family **F**. This has been studied for many families **F**, such as cliques (Theorem 8.4.3), bipartite graphs (Exercise 9), complete bipartite graphs (Theorem 8.6.18), stars (vertex cover number - Section 3.1), and forests (arboricity - Corollary 8.2.56). Before considering the extremal problems for edge-decomposition into paths and cycles, we discuss the easier problem of covering the *vertices* of a digraph using the minimum number of paths.

Comparability graphs are those having transitive orientations; a digraph is *transitive* if $x \to y$ and $y \to z$ imply $x \to z$. The vertices of a path in a transitive digraph induce a tournament. In Section 5.3, we proved that comparability graphs are perfect, meaning that a transitive digraph D in which the maximum tournament has ω vertices can be properly ω-colored. By the Perfect Graph Theorem (Section 8.1), we also know that $V(D)$ can be covered using $\alpha(D)$ tournaments in D, where $\alpha(D)$ is the maximum size of an independent set.

Letting paths be "chains" and independent sets be "antichains", we have Dilworth's Theorem for transitive loopless digraphs: The maximum size of an antichain equals the minimum number of chains needed to partition $V(D)$. In addition to following from the Perfect Graph Theorem, Dilworth's Theorem is equivalent to the König-Egerváry Theorem (Exercise 31), and a generalization of it follows from the Matroid Intersection Theorem (Exercise 8.2.59). Here we present a further generalization that has a short and self-contained proof.

8.4.29. Theorem. (Gallai-Milgram [1970]) The vertices of a digraph D can be covered using at most $\alpha(D)$ disjoint paths.

Proof. Since $V(D)$ can be covered using n disjoint paths of length 0, the theorem follows immediately from a stronger claim: If **C** is a collection of disjoint paths covering $V(D)$, and S is the collection of sources (initial vertices) of these paths, then $V(D)$ can be covered using at most $\alpha(D)$ disjoint paths with sources in S. The proof is by induction on $n(D)$, trivial for $n(D) = 1$, and the added statement about the sources helps the induction step go through.

Suppose $n > 1$ and **C** is as specified. If $|\mathbf{C}| = k > \alpha(D)$, we will construct a cover using fewer disjoint paths, all with sources in S. Since $k > \alpha$, there exists an edge xy with $x, y \in S$. Let A, B be the paths in **C** starting with x, y. We may assume that A has an edge xz, else we could add x to the beginning of B and save one path. By deleting x from the start of A, we obtain a cover **C'** of $V(D - x)$ by k paths having sources in $S' = S - x + z$. Since $\alpha(D - x) \le \alpha(D)$, the induction hypothesis yields a cover **C''** of $V(D - x)$ by fewer than k paths with sources in S'. All elements of S' belong to S except z.

If z is the source of a path in **C''**, then we add x at the beginning of that path. If z is not a source but y is, then we add x at the beginning of the path starting with y. If neither y nor z is a source, then at most $|S'| - 2 = k - 2$ paths have been used, and we can add x as a path by itself to obtain a cover of $V(D)$ using $k - 1$ paths. In all cases, the resulting paths are disjoint and have sources in S. By repeating this argument while $k > \alpha$, we can reduce the number of paths to α. $\square$

We return to the decomposition problem. Gallai conjectured that every n-vertex graph can be decomposed using $\lceil n/2 \rceil$ paths. Equality holds for cliques (Exercise 32). Other graphs have fewer edges, but the lack of connections could require more paths. Hajós conjectured analogously that an n-vertex even graph can be decomposed into $\lfloor n/2 \rfloor$ cycles. Both conjectures remain open, but Lovász proved the optimal bound when paths *and* cycles are allowed. The *size* of a decomposition is the number of subgraphs used.

8.4.30. Theorem. (Lovász [1968]) Every n-vertex graph can be decomposed into $\lfloor n/2 \rfloor$ paths and cycles.

Proof. Let **F** be the family of all paths and cycles, and let $n'(G)$ denote the number of non-isolated vertices in G. We use induction on $\lambda(G) = 2e(G) - n'(G)$ to prove that every graph G has an **F**-decomposition of size at most $\lfloor n'(G)/2 \rfloor$. Each component of G with more than one edge contributes positively to $\lambda(G)$. Hence $\lambda(G) \geq 0$, with equality only when each nontrivial component is an edge. The claim holds with equality when $\lambda(G) = 0$.

For the induction step, suppose $\lambda(G) > 0$. We consider two cases. *Case 1*: If G has a vertex y of positive even degree, choose $x \in N(y)$, and let $W = \{z \in N(x): d(x)$ is even$\}$. In this case, let $G' = G - \{xz: z \in W\}$. In obtaining G', we lose at least one edge (xy) and we isolate at most one vertex (x), so $\lambda(G') < \lambda(G)$. *Case 2*: If G has no vertex of positive even degree, then $\lambda(G) > 0$ forces $\Delta(G) > 1$. Let x be a vertex of degree at least 3, and form G^+ by introducing a new vertex y to subdivide an edge xx'. Let $W = \{y\}$, and let $G' = G^+ - xy$. Now $e(G') = e(G)$, but $n'(G') > n'(G)$, so $\lambda(G') < \lambda(G)$.

In each case, the induction hypothesis yields an **F**-decomposition **D** of G' with $|\mathbf{D}| \leq \lfloor n'(G')/2 \rfloor$. We will convert **D** into an **F**-decomposition of size $|\mathbf{D}|$ for the graph H obtained from G' by adding edges from x to W. In Case 1, $H = G$ and $n'(G') \leq n'(G)$, so this is the desired decomposition. In Case 2, $H = G^+$ and $n'(G') = n'(G^+)$. Since $n'(G)$ is even, $\lfloor n'(G)/2 \rfloor = \lfloor n'(G^+)/2 \rfloor$. In an **F**-decomposition of G^+, the $n'(G)$ vertices of odd degree must all be endpoints of paths; in particular, the added vertex y of degree 2 cannot be the end of a path. This means that xy and yx' belong to the same subgraph and can be replaced by xx' to obtain the desired decomposition of G.

The two cases now combine; we need only obtain the decomposition of H from **D**. Let $U = N_H(x)$. Every vertex of U has odd degree in G', so for each $u \in U$ there is a path $P(u)$ in **D** with endpoint u. For $u \in W$, we would like to extend $P(u)$ to absorb ux. This cannot be done if $P(u)$ passes through but does not end at x, since the resulting subgraph would not be in **F**. The idea is to cut the edge $u'x$ on which $P(u)$ reaches x, re-establishing a path, and use $u'x$ to extend $P(u')$ instead. This generates a sequence of changes from each $u \in W$. We must show that the sequences terminate and do not conflict with each other.

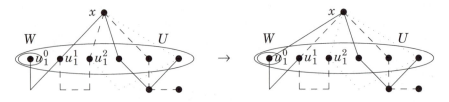

Let $W = w_1, \ldots, w_t$. For $w_i \in W$, we form a sequence $u_i^0, u_i^2, \cdots$ with $u_i^0 = w_i$ and each $u_i^j \in U$. If in the ith sequence we have chosen a vertex u_i^j, we check whether $x \in P(u_i^j)$. If $x \notin P(u_i^j)$, then we stop and do not define u_i^{j+1}. If $x \in P(u_i^j)$, then we set u_i^{j+1} to be the vertex on $P(u_i^j)$ just before x; this is the "u'" suggested above. (The picture on the left shows a sequence with three paths: $P(u_1^0)$ solid, $P(u_1^1)$ dashed, $P(u_1^2)$ dotted. The path $P(u_i^j)$ for $j \geq 1$ cannot start along the edge $u_i^j x$, because that edge belongs to $P(u_i^{j-1})$.

Since $xu_i^j \in E(G')$ if $j \geq 1$, the vertices of W appear only as initial vertices in the sequences. This is the basis for a proof by induction on $\min\{j, l\}$ that $u_i^j = u_k^l$ only if $i = k$ and $j = l$. For $j, l \geq 1$, the induction hypothesis yields $u_i^{j-1} \neq u_k^{l-1}$, and hence the paths $P(u_i^{j-1})$ and $P(u_k^{l-1})$ start at distinct vertices. If $u_i^j = u_k^l$, then the two paths share the edge $u_i^j x$ and must be the same path. This happens only if u_i^{j-1} and u_k^{l-1} are opposite ends of the path, but then they cannot both visit u_i^j before x. We conclude that no vertex of U appears twice in the sequences. Let $W' = \{u_i^j\}$. If $u = u_i^j$ and u is not the end of its sequence, let $u' = u_i^{j+1}$.

We define an **F**-decomposition of G consisting of one path or cycle Q' corresponding to each $Q \in \mathbf{D}$. If $Q \neq P(u)$ for some $u \in W'$, let $Q' = Q$. If $Q = P(u)$, let $Q' = Q + ux$ or $Q' = Q + ux - u'x$ depending on whether u is or is not the last vertex in its sequence. Always Q' is a path, except that Q' is a cycle when Q ends at x (and then u' is not defined). The union of the new paths corresponding to $\{P(u_i^j)\}$ is the same as $\cup P(u_i^j)$, except that the edges $\{xw_i\}$ are absorbed. Since $u \in W'$ appears only once in the sequences, the edge ux winds up in only one of the new paths, and $\{Q': Q \in \mathbf{D}\}$ is a decomposition of H. $\square$

Note that in this proof Q may be the selected path from each of its endvertices $u, v \in W'$. This is not a problem, because the adjustments to Q made from the two ends do not conflict. The path may visit x (thus defining u' and v') or not, as illustrated below.

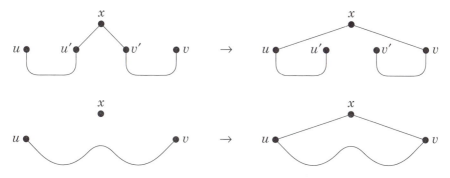

CIRCUMFERENCE

When a sufficient condition for Hamiltonian cycles fails slightly, it is reasonable to expect that the graph still must have a fairly long cycle. The length of the longest cycle in G is the *circumference* $c(G)$. We first consider the number of edges needed to force a cycle of length at least c in an n-vertex graph. In this section, $P(i, j)$ denotes the v_i, v_j-portion of a path $P = v_1, \ldots, v_l$, and P, Q denotes the concatenation of paths P and Q when the last vertex of P is the first vertex of Q.

8.4.31. Theorem. (Erdős-Gallai [1959]) If G has more than $(c - 1)(n - 1)/2$ edges, for $3 \le c \le n$, then $c(G) \ge c$.

Proof. (Woodall [1972]) We use induction on n for fixed c. When $n = c$, the number of edges missing is less than $(n - 1)/2$, which implies $\delta(G) \ge n/2$, and G is Hamiltonian. Suppose $n > c$ and $c(G) < c$. If $d(x) \le (c - 1)/2$, then $e(G - x) \ge (c - 1)(n - 2)/2$. Applying the induction hypothesis to $G - x$ yields $c(G - x) \ge c$. Hence we may assume that $\delta(G) > (c - 1)/2$.

Among all longest paths in G, choose $P = v_1, \ldots, v_l$ to maximize the degree d of v_1. Let $W = \{v_i: v_1 \leftrightarrow v_{i+1}\}$; note that $|W| = d$. For $v_k \in W$, the path $P(k, 1), v_1 v_{k+1}, P(k + 1, l)$ also has length l; hence $N(v_k) \subseteq V(P)$ and the choice of P implies $d(v_k) \le d$. Furthermore, no $v_k \in W$ has a neighbor v_j such that $j \ge c$, since adding $v_j v_k$ to $P(k, 1), v_1 v_{k+1}, P(k + 1, j)$ would complete the long cycle.

By limiting the edges incident to W, we force many edges into $G - W$. Let $Z = \{v_1, \ldots, v_r\}$, where $r = \min\{l, c - 1\}$. For each $v_k \in W$, we have shown that $N(v_k) \subseteq Z$. Hence there are $|[W, Z - W]| + e(G[W])$ edges incident to W. For fixed degree-sum in W, this is maximized when $[W, Z - W]$ is a complete bipartite graph. We further maximize by letting each vertex of W have degree d. The resulting count is $\frac{1}{2}|W|[d + |Z - W|] = dr/2 \le d(c - 1)/2$. Therefore, $G - W$ has $n - d$ vertices and more than $(c - 1)(n - d - 1)/2$ edges. By the induction hypothesis $c(G - W) \ge c$. (If the number of edges forced into $G - W$ is too large to exist, then this case cannot occur, and an earlier case applies.) $\square$

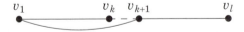

Most of the sufficient conditions for Hamiltonian cycles have "long cycle" versions. The long cycle version of Dirac's Theorem says that a 2-connected graph G has a cycle of length at least $\min\{n, 2\delta(G)\}$ (proved in Dirac [1952b]). The requirement of 2-connectedness eliminates the example formed by two copies of $K_{\delta+1}$ sharing a vertex, in which the circumference is $\delta + 1$.

The long cycle version of Ore's Theorem [1960] came much later. It is implicit in Bondy [1971b] and was made explicit in Bermond [1976] and in Linial [1976]. The fundamental argument used in many long cycle results appears in Bondy [1971b]. It strengthens the Ore/Dirac switching argument by considering "gaps".

8.4.32. Lemma. (Bondy [1971b]) If $P = v_1,\ldots,v_l$ is a longest path in a 2-connected graph G, then $c(G) \geq \min\{n(G), d(v_1) + d(v_l)\}$.

Proof. (See also Linial [1976]). Let $m = d(v_1) + d(v_l)$, and suppose that $c(G) < \min\{n(G), m\}$. Since G is connected, we may assume that $v_1 \not\leftrightarrow v_l$, else we obtain a longer path. If $v_1 \leftrightarrow v_j$ and $v_i \leftrightarrow v_l$ for some $i < j$, then i, j is a *crossover* with *gap* $j - i$. If we add $v_1 v_j$ and $v_l v_i$ to $P(j,l)$ and $P(i,1)$, we obtain a cycle with length $1 + (l - j) + 1 + (i - 1) = l + 1 - (j - i)$. Hence $l + 1 - (j - i) < m$ when i, j is a crossover.

$$x = v_1 \quad v_i \quad v_j \quad y = v_l$$

Let $x = v_1$ and $y = v_l$. If P has a crossover, let i, j be a crossover with smallest gap. The set $N(y)$ omits all of $v_{i+1},\ldots,v_{j-2}$ and $\{v_{r-1}: v_r \leftrightarrow x\}$. Since $N(y) \subseteq V(P) - \{y\}$, this yields $d(y) \leq (l - 1) - (j - 2 - i) - d(x)$. Since $l + 1 - (j - i) < m$, we have $d(x) + d(y) < m$, which contradicts the hypothesis. Hence there is no crossover.

With $t_0 = \max\{i: x \leftrightarrow v_i\}$ and $u = \min\{i: y \leftrightarrow v_i\}$, we have proved that $t_0 \leq u$. We use this to construct a cycle containing all of $N(x) \cup N(y) \cup \{x, y\}$. Since $t_0 \leq u$, $|N(x) \cap N(y)| \leq 1$, and hence such a cycle has length at least $d(x) + d(y) + 1 > m$.

We iteratively define a sequence of paths $P_1, P_2, \cdots$. Given the vertex t_{i-1}, we choose integers $s_i < t_{i-1} < t_i$ such that G has a v_{s_i}, v_{t_i}-path internally disjoint from P and t_i is as large as possible. This path becomes P_i. Such a path exists because G is 2-connected. The sequence $\{t_i\}$ is strictly increasing. If some P_i shares a vertex with a later path P_j, then we can choose an s_i, t_j-path for P_i, contradicting the maximality of t_i. Hence the paths $P_1, P_2, \cdots$ are pairwise disjoint.

Let r be the smallest index such that $t_r > u$. Also set $a = \min\{j: x \leftrightarrow v_j$ and $j > s_1\}$ and $b = \max\{j: y \leftrightarrow v_j$ and $j < t_r\}$. Since $s_1 < t_0$ and $t_r > u$, the indices a, b are well-defined. Now we use the even-indexed paths P_i to build one x, y-path and the odd-indexed paths to build another. If r is odd, the two paths are formed by the following concatenations, as illustrated.

$$xv_a, P(a, s_2), P_2, P(t_2, s_4), P_4, \ldots, P(t_{r-1}, b), v_b y$$
$$P(1, s_1), P_1, P(t_1, s_3), P_3, P(t_3, s_5), \ldots, P_r, P(t_r, l)$$

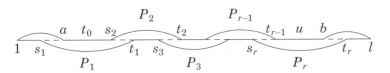

If r is even, then the path starting with xv_a reaches t_r and ends with $P(t_r, l)$, while the other path reaches v_b and ends with $v_b y$.

By the maximality of t_i, we have $s_{i+1} \geq t_{i-1}$; otherwise we would choose P_{i+1} instead of P_i. Hence

$$s_1 < a \leq t_0 \leq s_2 < t_1 \leq s_3 < t_2 \cdots < t_{r-1} \leq u \leq b < t_r.$$

This implies that the two paths described are indeed paths and that their union is a cycle. By the definition of a, $N(x) \subseteq P(1, s_1) \cup P(a, t_0)$, and similarly $N(y) \subseteq P(u, b) \cup P(t_r, l)$. Hence x, y, and $N(x) \cup N(y)$ belong to the cycle and it has length at least $2 + d(x) + d(y) - 1 > m$. □

Bondy's Lemma implies the long cycle version of Ore's theorem, which is stronger than the long cycle version of Dirac's theorem.

8.4.33. Theorem. If G is 2-connected and $d(u) + d(v) \geq s$ for every non-adjacent pair $u, v \in V(G)$, then $c(G) \geq \min\{n(G), s\}$.

Proof. Ore's Theorem guarantees a Hamiltonian cycle if $s \geq n$, so we may assume $s < n$. Suppose that $P = v_1, \ldots, v_l$ is a longest path in G, with endpoints $x = v_1$ and $y = v_l$, and suppose $c(G) < s$. If $l < c(G)$, then a longest cycle contains a longer path. If $l = c(G) < n$, then the connectedness of G allows us to depart from a longest cycle to obtain a longer path. Hence $l > c(G)$, which implies $x \not\leftrightarrow y$. Now the condition $d(x) + d(y) \geq s$ allows us to invoke Bondy's lemma. □

Bermond proved a degree sequence version of this result, analogous to a "long cycle" combination of Chvátal's condition and Las Vergnas' condition. The technique of edge-switches involving an endpoint of a longest path was used in Theorem 8.4.31. The technique of extracting a bad pair of indices was used in Theorem 6.2.10. Our statement is slightly weaker than that of Bermond but has a simpler proof.

8.4.34. Theorem. (Bermond [1976]) Suppose G is a 2-connected graph with degree sequence $d_1 \leq \cdots \leq d_n$. If G has no nonadjacent pair x, y with degrees i, j such that $d_i \leq i < c/2$, $d_{j+1} \leq j$, and $i + j < c$, then $c(G) \geq c$.

Proof. Among the longest paths in G, let $P = v_1, \ldots, v_l$ with endpoints $x = v_1$ and $y = v_l$ be chosen to maximize $d(v_1) + d(v_l)$. If $d(x) + d(y) \geq c$, then we apply Bondy's Lemma. If $d(x) + d(y) < c$, then we claim that the pair x, y contradicts the hypotheses. Because G is connected,

$x \not\leftrightarrow y$. We may assume $d(x) \le d(y)$ and set $i = d(x)$ and $j = d(y)$.

All neighbors of x and y lie in P. If $x \leftrightarrow v_k$, then $P(v_{k-1}, x)$, xv_k, $P(v_k, y)$ is also a longest path with y at the end, so $d(v_{j-1}) \le d(x) = i$. Since this holds for each of the i neighbors of x, we have $d_i \le i$. Similarly, each of the j neighbors of y yields a vertex with degree at most j. Also $d(x) \le j$, so $d_{j+1} \le j$. By hypothesis $i + j = d(x) + d(y) < c$, which completes the contradiction. $\square$

G.-H. Fan [1984] strengthened the long cycle version of Ore's result by weakening the condition on a pair of nonadjacent vertices and by requiring it only for pairs with common neighbors. T. Feng [1988] used Bondy's Lemma to shorten the proof. The result includes a sufficient condition for Hamiltonian cycles that does not require the closure to be complete.

8.4.35. Example. *A Hamiltonian graph.* Given n even, let $G_1 = K_{n/2}$ and $G_2 = (n/4)K_2$, and form G by adding a matching between disjoint copies of G_1 and G_2. The Hamiltonian closure of G is G itself, so our previous sufficient conditions do not apply. Even though G has $n/2$ 2-valent vertices, Fan's Theorem implies that G is Hamiltonian. $\square$

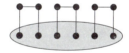

8.4.36. Theorem. (Fan [1984]) If G is a 2-connected graph, and $\max\{d(u), d(v)\} \ge c/2$ for every pair u, v of vertices at distance 2 in G, then $c(G) \ge c$.

Proof. (Feng [1988]) Given such a graph G, let $U = \{v \in V(G): d(v) \ge c/2\}$. By Bondy's Lemma, it suffices to find a longest path that has both endpoints in U. We examine a longest path $P = v_1, \ldots, v_m$ that has the maximum number of endpoints in U and show that, if P fails to have both endpoints in U, then we can find a longer path or a path of the same length with more endpoints in U.

Since $d(v) < c/2$ for all $v \notin U$, the hypothesis implies that $G - U$ is a disjoint union of cliques. Let Y be one such clique, and let $X = \{v \in U: N(v) \cap Y \ne \varnothing\}$. By the hypotheses, there is no edge from X to any other component of $G - U$. Also $|X| \ge 2$, because G is 2-connected.

We may assume that $v_1 \in Y$. Let $r = |Y|$. We first show that P begins with all of Y, i.e. $Y = \{v_1, \ldots, v_r\}$. If P omits any vertex of Y, we can absorb it before the first exit from Y. If P leaves and returns to Y, then it returns via an edge xy. Because $G[Y]$ is complete, we have $y \leftrightarrow v_1$, and we can replace xy in P with $v_1 y$, obtaining an x, v_m-path having the same length as P but more endpoints in U.

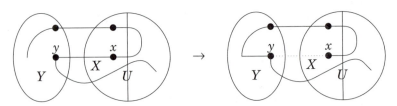

Suppose $x \in X - v_{r+1}$. If $x \notin V(P)$, it suffices to add x on to the beginning of a path with the same vertices as P. If $x \in V(P)$, let x' be the vertex before x on P. If x has a neighbor in Y, then x' has distance 2 from that vertex and must lie in U. In this case we find a path of the same length as P with endpoints x', v_m instead of v_1, v_m.

Suppose $x \in X - v_{r+1}$ exists with $y \in N(x) \cap (Y - v_r)$. Since P visits all of Y before v_r and $G[Y]$ induces a clique, we can rearrange the initial portion of P to start at y. If $x \notin V(P)$, we can add xy at the beginning to lengthen P. If $x \in V(P)$, we replace $x'x$ with yx to obtain a path of the same length with endpoints x', v_m.

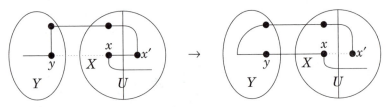

We know that $|X| \geq 2$, so there exists $x \in X - v_{r+1}$, but we have now shown that its only neighbor in Y is v_r. If also G has an edge yv_{r+1} with $y \in Y - v_r$, then we rearrange P to start with $v_r, \ldots, y, v_{r+1}$ instead of $v_1, \ldots, v_r, v_{r+1}$. If $x \notin V(P)$, we add xv_r at the beginning to lengthen P. If $x \in V(P)$, then we replace $x'x$ with $v_r x$.

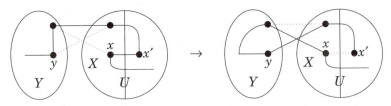

Now every edge between Y and X is incident to v_r. Since G is 2-connected, this implies that $|Y| = r = 1$. With $x \in X - v_{r+1}$ as before, we append x to the beginning of P or replace $x'x$ with xv_1. $\square$

Finally, we present one result about digraphs, strengthening Ghouila-Houri's sufficient condition (Theorem 6.2.15) for Hamiltonian cycles. We consider only loopless digraphs having at most one copy of each ordered pair as an edge; call these *strict* digraphs. Woodall [1972] proved that the following is a sufficient condition for a digraph G to be Hamiltonian: $uv, vu \notin E(G)$ implies $d^+(u) + d^-(v) \geq n(G)$. This immediately yields Ore's Theorem for undirected graphs (Exercise 37). Let $d(v) = d^+(v) + d^-(v)$. Ghouila-Houri [1960] proved that every digraph G having $d(v) \geq n(G)$ for each v is Hamiltonian; this already is stronger than Theorem 6.2.15. Meyniel [1973] proved that a strong digraph G in which $uv, vu \notin E(G)$ implies $d(u) + d(v) \geq 2n(G) - 1$ must be Hamiltonian, and this is best possible. For digraphs, we use "u, v nonadjacent" to mean $uv, vu \notin E(G)$. Meyniel's Theorem implies Ghouila-Houri's Theorem and Woodall's Theorem (Exercise 37). Bondy and Thomassen [1977] gave a short proof of Meyniel's Theorem.

8.4.37. Example. *Meyniel's Theorem is best possible.* Let G consist of two doubly-directed cliques sharing a vertex. The digraph is strongly connected, and the only pairs of nonadjacent vertices consist of one vertex from each clique. If the cliques have order k and order $n + 1 - k$, then the total degrees for any nonadjacent pair are $2k - 2$ and $2n - 2k$, which sum to $2n - 2$. □

8.4.38. Theorem. (Meyniel [1973]) If G is a strict strongly connected digraph such that $uv, vu \notin E(G)$ implies $d(u) + d(v) \geq 2n - 1$ (when $u \neq v$), then G is Hamiltonian.

Proof. (Bondy-Thomassen [1977]). We first observe that if $T = v_1, \ldots, v_k$ is a path that cannot absorb the vertex v internally (between two of its vertices), then the number of edges from v to T plus the number of edges from T to v is at most $k + 1$. This follows from a simple count. For $1 \leq i \leq k - 1$, only one of the edges $v_i v$ and $v v_{i+1}$ is permitted. Also $v v_1$ and $v_k v$ are permitted; there is no restriction on absorption at the end.

We use this to prove the following result: If G is a strict strong non-Hamiltonian digraph, and S is a maximal vertex subset having a spanning cycle $(x_1, \ldots, x_m)$ in G, then there exist $v \in \bar{S}$ and integers a, b with $1 \leq a \leq m$ and $1 \leq b < m$ such that $x_a v \in E(G)$, v is not adjacent to any x_{a+i} with $1 \leq i \leq b$, and $d(v) + d(x_{a+b}) \leq 2n - 1 - b$. Since $b \geq 1$, the conclusion of this statement is impossible under the hypothesis of the theorem, so we will conclude that the only maximal vertex set having a spanning cycle is $V(G)$.

Suppose first that no path leaves S and returns to it. Since G is strong and $S \neq V(G)$, some cycle S' of length at least 2 shares exactly one vertex with S. Let this vertex be x_a, and let v be the successor of x_a on S'. By the path condition, there is no path between v and $S - \{x_a\}$ in either direction. In particular, each vertex outside $S \cup \{v\}$ is incident to at most two edges also incident to v or v_{a+1}. Also, v is incident to at most two edges also incident to S (the other endpoint must be v_a.

Finally, each vertex of $S - v_{a+1}$ is incident to at most two edges also incident to v_{a+1}. Summing the allowed contributions yields $d(v) + d(x_{a+1}) \leq 2n - 2$. Hence the desired condition holds with $b = 1$.

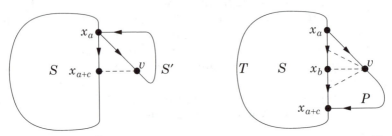

Now suppose there is a path leaving S and returning to it. Choose such a path P so that the distance c along S from the start of P to the end of P is minimal. Let x_a be the start of P, and let v be its successor on P. The maximality of S implies $c > 1$. Let T be the portion of S from x_{a+c} to x_a; this has $m - c + 1$ vertices. The maximality of S implies that v cannot be absorbed internally by T, and hence our initial observation implies that v belongs to at most $m - c + 2$ edges incident to T. The minimality of c makes v nonadjacent to $x_{a+1}, \ldots, x_{c-1}$.

Now let b be the largest integer in $[c]$ such that G contains a path from x_{a+c} to x_a with vertex set $S - \{x_{a+b}, \ldots, x_{a+c-1}\}$, and let R be such a path. (The path T with $b = 1$ guarantees the existence of R.) Since $P \cup R$ is a cycle, the maximality of S implies $b < c$. By the maximality of b, x_{a+b} cannot be absorbed internally by R; hence our initial observation implies that x_{a+b} belongs to at most $m - c + b + 1$ edges incident to R.

Now we count $d(v) + d(x_{a+b})$. Each vertex outside $S \cup \{v\}$ is incident to at most two edges also incident to $\{v, x_{a+b}\}$, because the minimality of c prevents a path of length 2 between v and x_{a+b} (in either direction) using a vertex not in S. We have observed that v belongs to at most $m - c + 2$ edges incident to S. We have observed that v_{a+b} belongs to at most $m - c + b + 2$ edges incident to R. Finally, x_{a+b} belongs to at most $2(c - b - 1)$ edges incident to $S - R$. Hence $d(v) + d(x_{a+b}) \leq 2(n - m - 1) + (m - c + 2) + (m - c + b + 1) + 2(c - b - 1) = 2n - 1 - b$. Again we have obtained the desired condition. $\square$

EXERCISES

8.4.1. (+) Determine the minimum number of edges in a connected n-vertex graph such that every edge belongs to a triangle. (Erdős [1988a])

8.4.2. Let G be a graph on n vertices such that $\overline{G}$ has no triangles. Prove that G has at least $\binom{\lfloor n/2 \rfloor}{3} + \binom{\lceil n/2 \rceil}{3}$ triangles, and that the result is best possible if $n > 9$.

8.4.3. (+) Determine the minimum number of edges in a triangle-free graph on $2n$ vertices whose complement contains no n-clique. (Erdős [1988b])

8.4.4. (+) Determine the maximum number of edges in a 10-vertex graph with no 4-cycle. (Bialostocki-Schönheim)

8.4.5. (+) *Extensions of Turán's Theorem.* Let $t_r(n) = e(T_{r,n})$.
 a) If $n \geq r + 1$, prove that every n-vertex graph with $t_{r-1}(n) + 1$ edges contains $K_{r+1} - e$. (Hint: remove a minimum degree vertex.)
 b) If $n \geq n' > r \geq 2$, prove that every n-vertex graph with $t_r(n) + \varepsilon$ edges has an n'-vertex subgraph with $t_r(n') + p$ edges, where $p \leq 1$.
 c) If $1 \leq p \leq r - 1$, prove that every n-vertex graph with $t_r(n)$ edges other than $T_{r,n}$ contains an $r + p$-vertex subgraph H with $e(H) > \binom{r+p}{2} - p$. (Dirac)

8.4.6. (+) Suppose a graph G with n vertices and $\lfloor n^2/4 \rfloor - k$ edges has a triangle, where $k \geq 0$. Prove that G in fact has at least $\lfloor n/2 \rfloor - k - 1$ triangles. Show this is best possible for all n.

8.4.7. Let $m = \lfloor n^2/4 \rfloor$. Prove that every n-vertex graph has an intersection representation using subsets of $[m]$ such that each element of $[m]$ appears in at most three sets. Equivalently, prove that every n-vertex graph decomposes into at most m edges and triangles.

8.4.8. Prove that the following conditions on a graph G with no isolated vertices are equivalent. (Choudom-Parthasarathy-Ravindra [1975])
 a) $\theta'(G) = \alpha(G)$.
 b) $\theta'(G \vee G) = (\theta'(G))^2$.
 c) $\theta'(G) = \theta(G)$.
 d) Every clique in a minimum clique cover of $E(G)$ contains
   a simplicial vertex of G.

8.4.9. (+) Let $b(G)$ be the minimum number of bipartite graphs needed to partition $E(G)$ (called *biparticity*). Let $a(G)$ denote the minimum number of classes needed to partition $E(G)$ such that every cycle of G contains a non-zero even number of edges from some class. Prove that these parameters both equal $\lceil \lg \chi(G) \rceil$. (Hint: Prove $\lg \chi(G) \leq b(G) \leq a(G) \leq \lceil \lg \chi(G) \rceil$.) (Harary-Hsu-Miller [1977], Alon-Egawa [1985])

8.4.10. Determine all the n-vertex graphs that are extremal for product dimension; i.e. have product dimension $n - 1$. (Lovász-Nešetřil-Pultr [1980])

8.4.11. Prove that $\mathrm{pdim}\, G \leq 2$ if and only if G is the line graph of a bipartite graph (Lovász-Nešetřil-Pultr [1980])

8.4.12. Given r, compute $\mathrm{pdim}(K_r + mK_1)$ for all $m \geq 1$. (Loász-Nešetřil-Pultr [1980])

8.4.13. (−) Compute $\mathrm{pdim}(K_2 \square K_2 \square K_2)$ and $\mathrm{pdim}(2K_m)$.

8.4.14. Obtain upper and lower bounds on the product dimension of the Petersen graph that differ by 1 (the upper bound will most likely be the correct value, but showing that it cannot be improved is tedious).

8.4.15. Let $f(n)$ be the maximum value of pdim$G \cdot$ pdim$\overline{G}$ over all graphs on n vertices. Prove that $\lfloor n^2/4 \rfloor \le f(n) \le (n-1)^2$.

8.4.16. For $n \ge 3$, prove that pdim$P_n = \lceil \lg n \rceil$, pdim$C_{2n} = 1 + \lceil \lg(n-1) \rceil$, and $1 + \lceil \lg n \rceil \le$ pdim$C_{2n+1} \le 2 + \lceil \lg n \rceil$. (Lovász-Nešetřil-Pultr [1980])

8.4.17. By obtaining a contradiction from a presumed embedding, prove directly that C_{2k+1} is not isometrically embeddable in any cartesian product of cliques if $k > 1$.

8.4.18. Determine the squashed-cube dimension of C_5.

8.4.19. Determine the squashed-cube dimension of $K_{3,3}$. (Hint: use symmetry!)

8.4.20. (!) Use Edmonds' Branching Theorem to prove the edge version of Menger's Theorem in digraphs: $\lambda'(x, y) = \kappa'(x, y)$. (Hint: Devise an appropriate graph transformation to obtain a short proof.)

8.4.21. (!) The gossip problem is also called the "telephone problem", and the corresponding problem for directed graphs is called the "telegraph problem". As a function of n, determine the minimum number of one-way transmissions among n people so that each person has a transmission path to every other. (Harary-Schwenk [1974])

8.4.22. Suppose D is a digraph solving the telegraph problem in which each vertex receives information from each other vertex exactly once. Prove that in D at least $n-1$ vertices hear their own information. For each n, construct such a D in which only $n-1$ vertices hear their own information but for each $x \ne y$ there is exactly one increasing x, y-path. (Seress [1987])

8.4.23. *The NOHO property.*

a) Suppose G is a connected linearly ordered multigraph with $2n-4$ edges that satisfies the NOHO property (no increasing cycle). Suppose also that G has more than eight vertices, of which at most two have degree 2. Prove that the graph obtained by deleting the first calls and last calls of vertices in G has 4 components, of which two are isolated vertices and two are caterpillars having the same size. (West [1982a])

b) For every even $n \ge 4$, construct a connected ordered multigraph with $2n-4$ edges that satisfies the NOHO property. (Hint: make use of the structural properties proved in part (a) to guide your search.)

8.4.24. A connected ordered multigraph has the NODUP property (no duplicate transmission) if there is exactly one increasing path from each vertex to every other.

a) - Prove that the NODUP property implies the NOHO property.

b) Prove that there is no NODUP scheme when $n \in \{6, 10, 14, 18\}$. (Comment: Seress [1986] proved that these are the only even values of n for which NODUP schemes do not exist, constructing them for all other values. For $n = 4k$, West [1982b] constructed NODUP schemes with $9n/4 - 6$ calls, and Seress [1986] proved that these are optimal.)

8.4.25. Suppose a vertex $v \in V(G)$ wishes to broadcast information to every vertex in a simple graph G. In each time unit, each vertex that already knows the information can make one call to a neighbor that does not know the information. The time required to broadcast from v is the minimum number of these rounds after which all vertices can know the information. Construct an n-vertex graph G with fewer than $2n$ edges such that every vertex of G can broadcast in time at most $1 + \lg n$. (Grigni-Peleg [1991])

8.4.26. (−) Prove that the graph below is not 2-choosable.

8.4.27. Prove that $K_{k,m}$ is k-list-colorable if and only if $m < k^k$ (Erdős-Rubin-Taylor [1979])

8.4.28. Prove $\hat{\chi}(G) \le 1 + \max_{H \subseteq G} \delta(G)$ and $\hat{\chi}(G) + \hat{\chi}(\overline{G}) \le n + 1$. Prove also that $\hat{\chi}'(G) \le 2\Delta(G) - 1$.

8.4.29. Prove that every chordal graph G is $\chi(G)$-choosable.

8.4.30. Suppose the list of colors assigned to each vertex of G has size at least $d(v)$, with strict inequality for at least one vertex. Prove that G can be colored properly from these lists.

8.4.31. (−) A *total coloring* of G assigns a color to each vertex and to each edge so that no colored objects have different colors if they are adjacent vertices, incident edges, or an incident vertex and edge. Prove that G has a total coloring with at most $\hat{\chi}'(G) + 2$ colors. (Comment: The Total Coloring Conjecture, due to Behzad [1965], states that every simple graph G has a total coloring with at most $\Delta(G) + 2$ colors. With this exercise, the List Coloring Conjecture and Vizing's Theorem state that every simple graph has a total coloring with at most $\Delta(G) + 3$ colors. Rosenfeld [1971] proved the Total Coloring Conjecture for bipartite graphs, complete tripartite graphs, complete r-partite graphs with equal part-sizes, and graphs with $\Delta(G) \le 3$. See also Behzad [1971].)

8.4.32. *Equivalence of Dilworth and König-Egerváry Theorems.*
      a) Given a bipartite graph G, apply Dilworth's Theorem to a transitive orientation of it to obtain the König-Egerváry Theorem.
      b) Given a transitive digraph D, form a bipartite graph G whose partite sets X, Y are copies of $V(D)$. The edges of G correspond to the edges of D; $x_i \leftrightarrow y_j$ in G if and only if $v_i \to v_j$ in D. Apply the König-Egerváry Theorem to G to obtain Dilworth's Theorem for D.

8.4.33. Prove that K_n can be decomposed into $\lceil n/2 \rceil$ paths. Prove that K_n can be decomposed into $\lfloor n/2 \rfloor$ cycles if n is odd.

8.4.34. Determine whether the Petersen graph can be partitioned into paths of length 3.

8.5.35. Prove that every 2-edge-connected 3-regular planar graph can be partitioned into paths of length 3. Prove the same statement for planar triangulations. (Jünger-Reinelt-Pulleyblank [1985])

8.5.36. Prove that T is a tree with m edges and G has no cycle of length less than m, then G has a decomposition into copies of T. (Häggkvist)

8.5.37. Theorem 8.4.31 states that G has a cycle of length at least c if $e(G) > (c-1)(n-1)/2$. Prove that this inequality is best possible when $n-1$ is a multiple of $c-2$.

8.5.38. Use Woodall's Theorem to prove Ore's Theorem, and use Meyniel's Theorem to prove Woodall's Theorem.

8.5.39. Use Meyniel's Theorem to prove that an n-vertex digraph has a Hamiltonian path if $d(u) + d(v) \geq 2n - 3$ for every pair u, v of distinct nonadjacent vertices.

8.5 Random Graphs

The study of random graphs is motivated by the modeling of physical properties and by the analysis of algorithms in computer science.

8.5.1. Example. *Melting points.* The behavior of random graphs suggests a mathematical explanation for the existence of melting points. Think of a solid as a three-dimensional grid of molecules, with neighboring molecules joined by bonds. We can model this using the grid graph $P_l \square P_m \square P_n$, with bonds corresponding to edges.

Adding energy excites molecules and breaks bonds. We assume that bonds break at random as we raise the temperature (energy level), with a given temperature corresponding to a given fraction of bonds broken. When the graph remains largely connected, the material acts like a solid. Breaking off small pieces doesn't change this, but if all the components are small the global nature of the material changes. Small components of molecules float freely, like a liquid or gas.

Mathematically, there is a threshold for the number of bonds to be broken (in terms of the size of the grid) such that almost every way of breaking somewhat fewer bonds leaves a giant component, and almost every way of breaking somewhat more bonds results in all components being tiny. This suggests that just below the threshold temperature the material will almost certainly be a solid, and just above the threshold the material will almost certainly not be a solid. □

8.5.2. Example. *Analysis of algorithms.* Worst-case complexity is the maximum time an algorithm takes over all inputs of size n; this allows comparison of algorithms for problems in P. For more difficult problems, an algorithm may take many steps on a few bizarre graphs while running quickly on the remaining graphs. We need a way to describe the usefulness of such algorithms.

The answer is *probabilistic analysis*. We assume a probability distribution on the inputs and study the expected running time with respect to this distribution. Choosing a realistic distribution can be difficult. In practice, we choose a probability distribution that makes the analysis feasible. We cannot define a probability distribution over infinitely many graphs, so we define a distribution on the graphs of each order. This is consistent with viewing the expected running time as a function of the input size. □

We also want to study the properties of random graphs. For each n, we have a probability distribution on the graphs with vertex set n, and there is a probability q_n that a graph chosen from this distribution has property Q. Since the number of graphs increases rapidly as n grows, we say that "almost every graph has property Q" if q_n approaches 1 as $n \to \infty$.

For random graphs, naive algorithms may become good. For example, finding a maximum clique is NP-hard. If we know that almost every graph has clique number about $2\lg n$, then we can test all vertex subsets up to size $3\lg n$ for being cliques. If $\omega(G) < 3\lg n$, then this computes $\omega(G)$, since every set of size $\omega(G)+1$ is not a clique. If $\omega(G) \geq 3\lg n$, then the algorithm fails to compute $\omega(G)$, but this rarely happens. There are too many subsets of size $2\lg n$ for this to be a polynomial-time algorithm, but it's close, and it illustrates one way in which the properties of random graphs can be used algorithmically.

Erdős and Rényi introduced random graphs in 1959. The subject developed rapidly in the 1980's, with books by Bollobás [1985], by Palmer [1985], and by Alon and Spencer [1992] (the last treats broader combinatorial applications of probabilistic methods). More sophisticated probabilistic techniques than we can develop here are now being applied to random graphs. We describe the basic techniques and suggest the flavor of the subject, with no attempt at exhaustive treatment.

EXISTENCE AND EXPECTATION

Before discussing properties of random graphs, we illustrate the use of probabilistic methods to prove existence statements. Suppose we want to prove existence of an object with some desired structure. We define a probability space in which occurrence of the desired structure is an event A. If A has positive probability, then an object with the desired structure exists. In this section, we use $P(A)$ to denote the probability of an event A.

Erdős popularized the probabilistic method in 1947 by using it to prove lower bounds on Ramsey numbers (Definition 8.3.6). We phrased this combinatorially in Theorem 8.3.12; here we present the same proof in probabilistic language.

8.5.3. Theorem. (Erdős [1947]) If $\binom{n}{p}2^{1-\binom{p}{2}} < 1$, then $R(p, p) > n$.

Proof. It suffices to show that for such n there exists an n-vertex graph G with $\omega(G) < p$ and $\alpha(G) < p$. We define a probability model on graphs with vertex set $[n]$ by letting each edge appear independently with probability .5. If the probability of the event Q = "no p-clique or independent p-set" is positive, then the desired graph exists.

Each possible p-clique occurs with probability $2^{-\binom{p}{2}}$, since edges outside that p-set are irrelevant. Hence the probability of having at least one p-clique is bounded by $\binom{n}{p}2^{-\binom{p}{2}}$. The same bound holds for independent p-sets. Hence the probability of "not Q" is bounded by $\binom{n}{p}2^{1-\binom{p}{2}}$, and the given inequality guarantees $P(Q) > 0$. □

8.5.4. Definition. A *random variable* is a function assigning a real number to each element of a probability space. The *expectation* $E(X)$ of a random variable X is the weighted average $\Sigma_k kP(X = k)^\dagger$. Given a random variable X, the *pigeonhole property* of the expectation is the statement that there exists an element of the probability space for which the value of X is as large as (or as small as) $E(X)$.

The pigeonhole property of the expectation holds for the same reason as the pigeonhole principle itself; if X is always less than α, then so is $E(X)$, so $E(X) = \alpha$ implies an occurrence where X is at least α.

Applying the pigeonhole property requires computing $E(X)$. Often the computation uses an expression for X in terms of simpler random variables. This rests on the *linearity* of expectation.

8.5.5. Lemma. If X and $X_1, \dots, X_n$ are random variables on the same space and $X = \Sigma X_i$, then $E(X) = \Sigma E(X_i)$. (Also $E(aX) = aE(X)$.)

Proof. In a discrete probability space, each sample point contributes the same amount to each side of the equation. □

Sometimes we can write $X = \Sigma X_i$, where each variable X_i takes only the value 0 or 1. Such random variables are called *indicator variables*, because they indicate whether a particular event has occurred.

---

†We are considering only probability models on finite sets. Analogous concepts hold in continuous probability spaces.

For example, the random variable X may count subgraphs of a particular type. We can express X as a sum of indicator variables, each indicating whether one of the possible subgraphs being counted actually occurs. The expectation of an indicator variable X_i equals $P(X_i = 1)$. The next result uses the linearity and pigeonhole properties.

8.5.6. Theorem. (Szele [1943]) There is an n-vertex tournament having at least $n!/2^{n-1}$ Hamiltonian paths.

Proof. For each pair $\{i, j\} \subseteq [n]$, we choose $i \to j$ or $j \to i$ with equal probability, generating a random tournament. If X denotes the number of Hamiltonian paths, then X is the sum of $n!$ indicator variables for the possible Hamiltonian paths. Each path occurs with probability $1/2^{n-1}$, so $E(X) = n!/2^{n-1}$, and there must be a tournament where X is as large as its expectation. $\square$

When a randomly generated object is close to having a desired property, we may be able to change it slightly to reach the desired goal. This has been called the *deletion method*, the *alteration principle* or the *two-step method*. To illustrate, we prove the existence of small dominating sets when $\delta(G) = k$. The proof includes our first use of the important inequality $1 - p < e^{-p}$ (Exercise 2).

8.5.7. Definition. A *dominating set* of vertices in a graph G is a set $S \subseteq V(G)$ such that every vertex of G belongs to S or has a neighbor in S.

8.5.8. Theorem. (Alon [1990]) If G is an n-vertex graph with minimum degree $k > 1$, then G has a dominating set with at most $n \dfrac{1 + \ln(k + 1)}{k + 1}$ vertices.

Proof. Form a random vertex subset $S \subseteq V(G)$ by including each vertex independently with probability $p = \ln(k + 1)/(k + 1)$. Given S, let T be the set of vertices outside S having no neighbor in S; adding T to S yields a dominating set. The experiment gives us both S and T, so we seek the expected size of the union. Since each vertex appears in S with probability p, $E(|S|) = np$. The random variable $|T|$ is the sum of n indicator variables for whether individual vertices belong to T. We have $v \in T$ if and only if v and its neighbors all fail to be in S, the probability of which is bounded by $(1 - p)^{k+1}$, since v has degree at least k. Since $(1 - p)^{k+1} < e^{-p(k+1)}$, we have $E(|S| + |T|) \leq np + ne^{-p(k+1)} = n \dfrac{1 + \ln(k + 1)}{k + 1}$. By the pigeonhole property of the expectation, there must be some S for which $S \cup T$ is a dominating set no larger than this. $\square$

A striking example of the deletion method is the proof (Erdős [1959]) that for fixed g, k there are graphs with girth g and chromatic number k (recall that *girth* is the length of the shortest cycle; Section 5.2 discusses this for $g = 4$). Explicit constructions of such graphs were not found until nine years later (Lovász [1968], Nešetřil-Rödl [1979]). The proof we present here is a simplification of the Erdős proof, due to Alon and Spencer [1992, p35].

8.5.9. Theorem. (Erdős [1959]). Given $k \geq 3$ and $g \geq 3$, there exists a graph with girth g and chromatic number k.

Proof. It suffices to obtain girth at least g and chromatic number at least k, since we can delete vertices from such a graph until the chromatic number decreases to k and then add a cycle of length g.

Given a large value of n, we generate graphs with vertex set $[n]$ by letting each pair be an edge with probability p, independently. A graph with no large independent set has large chromatic number, since $\chi(G) \geq n(G)/\alpha(G)$. We therefore choose p large enough to make the existence of large independent sets unlikely. We also want to choose p small enough that the expected number of short cycles (length less than g) is small. When we have such a graph satisfying both conditions, we can delete a vertex from each short cycle to obtain the desired graph.

To make it unlikely that the graph generated will have more than $n/2$ short cycles, we let $p = n^{t-1}$, where $t < 1/g$. Let X denote the number of cycles of length less than g that result. Since there are $n_{(j)}/2j = (2j)^{-1}\Pi_{i=0}^{j-1}(n - i)$ potential cycles of length j, linearity of expectation yields $E(X) = \Sigma_{i=3}^{g-1}n_{(i)}p^i/(2i) \leq \Sigma_{i=3}^{g-1}n^{ti}/(2i)$. Since $tg < 1$, this implies that $E(X)/n \to 0$ as $n \to \infty$. Since $E(X)/n \to 0$, we also have $P(X \geq n/2) \to 0$ as $n \to \infty$ (see Lemma 8.5.14 for further detail). In particular, we can make n large enough so that $P(X \geq n/2) < 1/2$.

Since we will retain at least $n/2$ vertices, it suffices to show that there will be such graphs with $\alpha(G) \leq n/(2k)$; $\alpha(G)$ cannot grow when we delete vertices, and hence at least k independent sets will be needed to cover the remaining vertices. With $r = \lceil 3\ln n/p \rceil$, we have $P(\alpha(G) \geq r) \leq \binom{n}{r}(1 - p)^{\binom{r}{2}} < [ne^{-p(r-1)/2}]^r$. This approaches 0 as n grows.

Since $r = \lceil 3n^{1-t}\ln n \rceil$ and k is fixed, we can choose n large enough that $r < n/(2k)$. If we also make n large enough that $P(X \geq n/2) < 1/2$ and $P(\alpha(G) \geq r) < 1/2$, there will exist an n-vertex graph G such that 1) $\alpha(G) \leq n/(2k)$ and 2) G has fewer than $n/2$ cycles of length less than g. We delete a vertex from each short cycle and retain a graph with girth at least g and chromatic number at least k. $\square$

PROPERTIES OF ALMOST ALL GRAPHS

The model where edges arise independently with the same probability is the most common model for random graphs, since it leads to the simplest computations. We allow this probability to be a function of n.

8.5.10. Definition. *Model A:* Given n and $p = p(n)$, generate graphs with vertex set $[n]$ by letting each pair be an edge with probability p, independently. Each graph with m edges has probability $p^m(1 - p)^{\binom{n}{2}-m}$. The random variable G^p denotes a graph drawn from this probability space. *The random graph* means Model A with $p = 1/2$, making the graphs with vertex set $[n]$ equally likely.

We study graphs with a fixed vertex set because computations are much more difficult for random isomorphism classes, and it is hard to generate these at random. Since algorithms take graphs with specified vertex sets as inputs, this choice does not hamper applications. Because we often express running time of algorithm in terms of the number of vertices and number of edges, we may also want to control the number of edges generated. This suggests a model in which the n-vertex labeled graphs with m edges are equally likely.

8.5.11. Definition. *Model B:* Given n and $m = m(n)$, generate each graph having vertex set $[n]$ and having m edges with probability $\binom{N}{m}^{-1}$, where $N = \binom{n}{2}$. The random variable G^m denotes a graph drawn from this sample space.

We treat graph properties as events; suppose property Q has probability q_n in the model for n-vertex graphs. The statement "almost every G^p (or G^m) has property Q" means $\lim_{n \to \infty} q_n = 1$. Applications seem more interesting in Model B than in Model A. We may ask "as a function of n, how many edges make a graph almost surely connected?" In Model A we would ask, "as a function of n, what edge probability makes a graph almost surely connected?"

Corresponding calculations are more tedious and messy in Model B than in Model A. Fortunately, Model B is accurately represented by Model A when n is large and p is chosen so that $m \sim p\binom{n}{2}$, since the actual number of edges generated under Model A is almost always close to the expectation. By the theorem stated next, the correspondence is valid for most properties of interest. A graph property Q is *convex* if F, H satisfying Q and $F \subseteq G \subseteq H$ imply that G satisfies Q.

8.5.12. Theorem. (Bollobás [1985, p34-35]) If Q is a convex property and $p(1 - p)\binom{n}{2} \to \infty$, then almost every G^p satisfies Q if and only if, for every fixed x, almost every G^m satisfies Q, where $m = \lfloor p\binom{n}{2} + x[p(1 - p)\binom{n}{2}]^{1/2} \rfloor$. $\square$

This explains why the simpler Model A is adequate for study and why we let p be a function of n in Model A. To study graphs with a linear number of edges, we must let p die off at a rate like $1/n$; if p remains constant, we almost always have dense graphs.

It is crucial that we need not compute $P(Q)$ to prove that almost every G^p has property Q; we need only show $P(Q) \to 1$. Probabilistic analysis uses asymptotic statements; exact computation of probabilities is difficult, unnecessary, and avoided wherever possible. The language of asymptotics rests on limits. We write $a_n \to L$ for $\lim_{n \to \infty} a_n = L$. To compare sequences, we use "big O" and "little o" notation. Recall that we write $a_n \in O(f(n))$ if there are constants k, N such that $|a_n| \le kf(n)$ for all $n > N$ (equivalently, $|a_n|/f(n)$ is bounded). We write $a_n \in o(f(n))$ if $a_n/f(n) \to 0$. We use " $\in$ " because $O(f(n))$ and $o(f(n))$ denote sets of functions. The notation can be read as "a_n is Big Oh of $f(n)$". We often use " $=$ " to discuss leading behavior; for example $a_n = b_n(1 + o(1))$ means that a_n and b_n differ by a sequence in $o(b_n)$, which is equivalent to $a_n/b_n \to 1$. Equivalently, we say that a_n is *asymptotic* to b_n, written $a_n \sim b_n$.

Proving that G^p is almost surely connected means showing that $P(G^p$ is disconnected$) \to 0$. We need not compute the probability; we need only show it is *bounded* by something tending to 0. Asymptotic arguments are "sloppy"; we don't care how loose the bound is as long as it approaches 0. The trick is to simplify the approximation without letting the bound become too big. Experience refines one's intuition about what can be discarded without getting into trouble.

8.5.13. Theorem. (Gilbert [1959]). For constant p, almost every G^p is connected.

Proof. We can make G disconnected by picking a vertex bipartition and forbidding edges between the parts. Occurrence of edges within the parts is irrelevant. We obtain an upper bound on the probability q_n that G^p is disconnected by summing over all bipartitions $S, \bar{S}$ the probability that $[S, \bar{S}] = \varnothing$. Graphs with many components are counted many times. When $|S| = k$, there are $k(n-k)$ possible edges in $[S, \bar{S}]$. Each has probability $1 - p$ of not appearing, independently, so $P([S, \bar{S}] = \varnothing) = (1-p)^{k(n-k)}$. By considering all S, we generate each partition from each side, so $q_n \le \frac{1}{2}\sum_{k=1}^{n-1}\binom{n}{k}(1-p)^{k(n-k)}$.

This formula is symmetric in k and $n - k$; hence q_n is bounded by $\sum_{k=1}^{\lfloor n/2 \rfloor}\binom{n}{k}(1-p)^{k(n-k)}$. Now we loosen the bound to simplify it. Immediately $\binom{n}{k} < n^k$; also $(1-p)^{n-k} \le (1-p)^{\lceil n/2 \rceil}$, since $(1-p) < 1$ and $n - k \ge \lceil n/2 \rceil$. Hence $q_n < \sum_{k=1}^{\lfloor n/2 \rfloor}(n(1-p)^{n/2})^k$. For large enough n, we have $n(1-p)^{n/2} < 1$. In this case our bound is the initial portion of a convergent geometric series, and we obtain $q_n < x/(1-x)$, where $x = n(1-p)^{n/2}$.

Since $n(1 - p)^{n/2} \to 0$ when p is constant, our bound on q_n approaches 0 as $n \to \infty$. $\square$

Integer-valued random variables lead to simpler arguments for stronger results. If we define an X so that $X = 0$ when G^p has property Q, then $E(X) \to 0$ implies that almost every G^p has Q. This is a special case of the following lemma. We prove it only for integer-valued variables, but it also holds for continuous variables.

8.5.14. Lemma. (Markov's Inequality). If X takes only nonnegative values, then $P(X \geq t) \leq E(X)/t$. In particular, if X is integer-valued, then $E(X) \to 0$ implies $P(X = 0) \to 1$.

Proof. $E(X) = \Sigma_{k \geq 0} k p_k \geq \Sigma_{k \geq t} k p_k \geq t \Sigma_{k \geq t} p_k = t P(X \geq t)$. $\square$

For connectedness, we can define $X(G^p)$ by $X = 1$ if G is disconnected and $X = 0$ otherwise. The expectation of an indicator variable is the probability it equals 1. We proved $P(X = 1) \to 0$ (when p is constant) to prove that almost every G^p is connected. With a different random variable we can simplify the proof and obtain a stronger result. We still want G to satisfy Q if $X = 0$ (in order to apply Markov's Inequality), but we don't need $(X = 0) \Leftrightarrow (G$ satisfies $Q)$. We define X to be the sum of many indicator variables, such that G satisfies Q if $X = 0$. The linearity of expectation and convenience of $E(X_i) = P(X_i = 1)$ for the indicator variables simplifies the task of proving $E(X) \to 0$.

8.5.15. Theorem. If p is constant, then almost every G^p has diameter 2 (and hence is connected).

Proof. Let $X(G^p)$ be the number of unordered vertex pairs with no common neighbor. If there are none, then G^p is connected and has diameter 2. By Markov's Inequality, we need only show $E(X) \to 0$. We express X as the sum of $\binom{n}{2}$ random variables X_{ij}, one for each vertex pair $\{v_i, v_j\}$, with $X_{ij} = 1$ if and only if v_i, v_j have no common neighbor. When $X_{ij} = 1$, the $n - 2$ other vertices fail to have edges to both of these, so $P(X_{ij} = 1) = (1 - p^2)^{n-2}$ and $E(X) = \binom{n}{2}(1 - p^2)^{n-2}$. When p is fixed, $E(X) \to 0$, and hence almost every G^p has diameter 2. $\square$

The intuition behind defining this random variable, made precise by Markov's Inequality, is that if we expect almost no bad pairs, then almost every graph has none. The summation disappears, and for the limit we need only know that $(1 - p^2)^{n-2}$ approaches 0 faster than any polynomial function of n.

THRESHOLD FUNCTIONS

Intuitively, random graphs with constant edge probability are connected because they have many more edges than needed to be connected. To improve Theorem 8.5.15, we want to make $p(n)$ as small as possible to have almost every G^p connected. We need a precise notion of a "threshold" probability function. By the relationship between Model A and Model B, a threshold edge probability also yields a threshold number of edges.

8.5.16. Definition. A *monotone property* is a graph property preserved by addition of edges. A *threshold probability function* for a monotone property Q is a function $t(n)$ such that $p(n)/t(n) \to 0$ implies that almost no G^p satisfies Q, and $p(n)/t(n) \to \infty$ implies that almost every G^p satisfies Q. A similar definition of *threshold edge function* holds for Model B.

This is the broadest notion of threshold function; it allows a monotone graph property to have many threshold functions. We obtain a sharper threshold function when we can show that almost no or almost every G^p satisfies Q when the ratio of $p(n)$ to $t(n)$ approaches a constant below or above 1 rather than approaching 0 or ∞. Sharper still is a threshold function $t(n)$ such that this behavior occurs when $p(n)$ differs from $t(n)$ by the subtraction or addition of a lower order term.

Markov's Inequality yields part of a derivation of a threshold function. If $X = 0$ implies property Q and we prove that $E(X) \to 0$, then $P(Q) \to 1$. We can obtain a candidate for a threshold function by determining which functions $p(n)$ yield $E(X) \to 0$. Often this yields a constraint on the value of a parameter such that $E(X) \to \infty$ when the constraint fails. The property $E(X) \to \infty$ suggests that $P(X = 0) \to 0$. This does not hold for all random variables; for example, $E(X) \to \infty$ when $P(X = 0) = 1/2$ and $P(X = n) = 1/2$. To obtain $P(X = 0) \to 0$, we need to show $E(X) \to \infty$ and prevent the probability from spreading out in this way. The name of the method comes from the moments of a random variable:

8.5.17. Definition. The kth-*moment* of X is the expectation of X^k. The quantity $E[(X - E(X))^2]$ is the *variance* of X, written $Var(X)$.

8.5.18. Lemma. (Second Moment Method). If X is a random variable, then $P(X = 0) \leq \dfrac{E(X^2) - E(X)^2}{E(X)^2}$. In particular, $E(X^2)/E(X)^2 \to 1$ implies $P(X = 0) \to 0$.

Proof. Applied to the random variable $(X - E(X))^2$ and the value t^2, Markov's Inequality yields $P[(X - E(X))^2 \geq t^2] \leq E[(X - E(X))^2]/t^2$. We rewrite this as $P[|X - E(X)| \geq t] \leq Var(X)/t^2$ (Chebyshev's Inequality). Since $Var(X) = E[(X - E(X))^2] = E[X^2 - 2XE(X) + (E(X))^2] = E(X^2) - (E(X))^2$, we can rewrite the inequality as $P[|X - E(X)| \geq t] \leq (E(X^2) - E(X)^2)/t^2$. Since $X = 0$ lies within the event $|X - E(X)| \geq E(X)$, setting $t = E(X)$ completes the proof. $\square$

Intuitively, if the mean grows and the standard deviation grows more slowly, then all the probability is pulled away from 0, and we obtain $P(X = 0) \to 0$. We illustrate the method by asking how large p must be to prevent isolated vertices. Since a connected graph has no isolated vertices, a threshold for connectedness must be as large as a threshold for disappearance of isolated vertices. The computations for disappearance of isolated vertices are simpler than for connectedness, because we can express this condition using a sum of identically distributed indicator variables with easily computed expectations. In fact, the two properties have the same threshold, because it is possible to show that at the threshold for isolated vertices, almost every graph consists of one huge component plus isolated vertices.

8.5.19. Theorem. In Model A, $\ln n/n$ (natural logarithm) is a threshold probability function for the disappearance of isolated vertices (i.e., for $\delta(G) \geq 1$). (The corresponding threshold in Model B is $\frac{1}{2} n \ln n$.)

Proof. Let X be the number of isolated vertices, with X_i indicating whether vertex i is isolated. Then $E(X) = \Sigma E(X_i) = n(1 - p)^{n-1}$, and to find a threshold we need a function $p(n)$ in the range between $E(X) \to 0$ and $E(X) \to \infty$. Since $(1 - p)^n = e^{n \ln(1-p)} = e^{-np} e^{-np^2[1/2 + p/3 + \cdots]}$, our expression for $E(X)$ simplifies asymptotically if $np^2 \to 0$. This is equivalent to $p \in o(1/\sqrt{n})$ and implies $(1 - p)^n \sim e^{-np}$ and $(1 - p)^{-1} \sim 1$, yielding $E(X) \sim ne^{-np}$. To simplify further, put $p = c \ln n/n$ to obtain $ne^{-np} = n^{1-c}$. This rewrites p by treating c as a function of n. When c is constant, $p \in o(1/\sqrt{n})$, as needed for the earlier approximation. If $c > 1$, we have $E(X) \sim n^{1-c} \to 0$, and the proof is half done.

When $c < 1$, we have $E(X) \to \infty$ and use the second moment method; we need only show $E(X^2) \sim E(X)^2$. In computing $E(X^2)$, we discover another helpful property of indicator variables: $X_i^2 = X_i$. By the linearity of the expectation, $E(X^2) = \Sigma_{i=1}^n E(X_i^2) + \Sigma_{i \neq j} E(X_i X_j) = E(X) + n(n - 1)E(X_i X_j)$. Since $X_i X_j$ is also an indicator variable, we have $X_i X_j = 1$ if and only if both vertices are isolated. That requires forbidding $2(n - 2) + 1$ edges, so $E(X_i X_j) = (1 - p)^{2n-3}$. Again $(1 - p)^n \sim e^{-np}$ when $p \in o(1/\sqrt{n})$, so $E(X_i X_j) \sim e^{-2np}$, and $E(X^2) \sim E(X) + n(n - 1)e^{-2np} \sim E(X) + E(X)^2$. Finally, $E(X) + E(X)^2 \sim E(X)^2$ whenever $E(X) \to \infty$. $\square$

We have proved a stronger result about isolated vertices than the definition of a threshold function requires. As discussed earlier, this is a sharper threshold, since already we guarantee or forbid isolated vertices when the ratio between $p(n)$ and $\ln n/n$ approaches a nonzero constant rather than 0 or ∞.

As a side remark, we comment briefly on a very sharp sort of threshold. Suppose our probability function of n involves a parameter x such that when x is fixed, $E(X)$ approaches a constant $c(x)$ (our property Q corresponds to $X = 0$). In this situation, it often happens that the distribution of X approaches a Poisson distribution. (The *Poisson distribution with mean* μ is defined by $P(X = k) = e^{-\mu}\mu^k/k!$.) For example, if $p = \ln n/n + x/n$, then the number of isolated vertices in G^p is asymptotically Poisson distributed with mean $E(X) \to \mu = e^{-x}$ (that is, $P(X = k) \sim e^{-\mu}\mu^k/k!$). The derivation of sharp thresholds requires additional techniques, which we omit.

Next we obtain a threshold function for the appearance of a graph H as a subgraph. A graph is *balanced* if the average vertex degree in every induced subgraph is no larger than the average degree of the entire graph. Trees, cycles, cliques, and all regular graphs are balanced.

8.5.20. Theorem. If H is a balanced graph with k vertices and l edges, then $p = n^{-k/l}$ is a threshold function in Model A for the appearance of H as a subgraph of almost every G^p.

Proof. Let X be the number of copies of H in G^p. We express X as a sum of indicator variables $\{X_i\}$ for the possible copies $\{H_i\}$ of H in K_n. There are $n(n-1)\cdots(n-k+1)$ ways to map $V(H)$ into $[n]$. The mappings produce each H_i a total of A times, where A is the number of automorphisms of H. We thus have $\Pi_{i=0}^{k-1}n - i/A$ possible copies of H. Since H_i occurs if its edges occur, $P(X_i = 1) = p^l$. By linearity, $E(X) \sim n^k p^l/A$, since k is fixed.

Setting $p = c_n n^{-k/l}$ yields $E(X) \sim c_n/A$. Hence $c_n \to 0$ implies $E(X) \to 0$ and $c \to \infty$ implies $E(X) \to \infty$, and it remains to show that $E(X^2) \sim E(X)^2$ when $c_n \to \infty$. Again $E(X^2) = E(X) + \Sigma_i\Sigma_{j\neq i}E(X_i X_j)$. These terms are not equal; the expectation $E(X_i X_j)$ depends on $H' = H_i \cap H_j$. We group the terms by the choice of $H' \subset H$. Suppose H' has r vertices and s edges. The number of edges needed to create H_i and H_j is $2l - s$, so $E(X_i X_j) = p^{2l-s}$.

We need the number of ordered pairs i, j such that $H' = H_i \cap H_j$. To specify such pairs, we choose r vertices for H', $k - r$ vertices for $H_i - H'$ and for $H_j - H'$, and an extension of H' to each of those sets. Choosing $r, k - r, k - r, n - 2k + r$ vertices for four specified sets can be done in $n!/[r!(k-r)!(k-r)!(n-2k+r)!] \sim n^{2k-r}/r!(k-r)!^2$ ways. The number M of ways to extend H' to obtain copies of H in both specified k-sets depends only on H and H'; it is independent of n and p. Let $\alpha_{H'}$

$= M/r!(k-r)!^2$. The resulting contribution $E_{H'}$ to $\Sigma E(X_i X_j)$ is asymptotic to $\alpha_{H'} n^{2k-r} p^{2l-s}$.

When $r = s = 0$, we obtain $M = (k!/A)^2$, and the contribution of the empty graph H' to $\Sigma E(X_i X_j)$ is asymptotic to $n^{2k} p^{2l}/A^2$, which is asymptotic to $E(X)^2$. The proof is completed by showing that the total contribution from all other choices of H' has lower order. We have $E_{H'} \sim \alpha_{H'} A^2 E(X)^2 n^{-r} p^{-s}$. Since $2s/r$ is the average degree of H', the requirement that H is balanced forces $2r/s \geq 2k/l$, or $pn^{r/s} \geq pn^{k/l} \to \infty$ when $c > 0$. But $pn^{r/s} \to \infty$ is equivalent to $n^{-r} p^{-s} \to 0$, which implies $E_{H'} \in o(E(X)^2)$ for $H' \neq \varnothing$, as desired. Since the number of possible subgraphs H' is bounded (by an exponential function of k and l, but these are constants), this implies $E(X^2) \sim E(X) + E_\varnothing \sim E(X)^2$. $\square$

This result generalizes for arbitrary fixed subgraphs. The ratio $d(G) = e(G)/n(G)$ is the *density* of G, and $\rho(G) = \max_{F \subseteq G} d(F)$ is the *maximum density* of G. Maximum density equals density precisely when G is balanced, and $p = n^\varepsilon n^{-1/\rho}$ is our threshold for the appearance of a balanced graph. If H is not balanced, then H has a balanced subgraph F such that $d(F) = \rho(H)$. We have seen that $pn^{\rho(H)} \to 0$ implies that almost every G^p has no copy of F and hence no copy of H. Rucínski-Vince [1985] showed that the second moment argument above can be reworded slightly to prove that $p = n^\varepsilon n^{-1/\rho(H)}$ is the threshold for the appearance of any graph H, simplifying the original argument of Bollobás [1981a].

EVOLUTION AND PROPERTIES OF RANDOM GRAPHS

In the subtitle to his book, Palmer [1985] tells us that random graphs involve study of

> "THRESHOLD FUNCTIONS, which facilitate the careful study of the structure of a graph as it grows, and specifically reveal the mysterious circumstances surrounding the abrupt appearance of the UNIQUE GIANT COMPONENT, which systematically absorbs its neighbors, devouring the larger first and ruthlessly continuing until the last ISOLATED VERTICES have been sucked up, whereupon the Giant is suddenly brought under control by a SPANNING CYCLE."

The evolutionary viewpoint generates random graphs with m edges by a process that yields the same probability space as Model B but makes intuitive reasoning easier. Almost everything that intuition or numerical experimentation suggests about random graphs is true. The evolutionary viewpoint develops this intuition.

Generating m edges simultaneously or one-by-one yields the same probability distribution, making the graphs with m edges equally likely. By studying the likely effect of a new edge on the present structure, we can make intuitive hypotheses about the properties of the graph at any stage. A *stage* of evolution is a range of values for $m(n)$ (or $p(n)$) in which the structural description of the typical graph doesn't change much. We have studied the basic techniques for verifying these descriptions, but the computations can be difficult. Hence we will only describe the stages using the evolutionary intuition.

Beginning with many vertices and no edges, each new edge is likely to be isolated. The random graph is a matching until a substantial fraction of the vertices are involved in edges. The thresholds $p \sim cn^{-k/(k-1)}$ for appearance of fixed subtrees generalize this. Let $t_k(n) = n^{-k/(k-1)}$. If $p/t_k \to \infty$ but $p/t_{k+1} \to 0$, then every fixed subtree on k vertices appears, but none on $k + 1$ vertices appears. (Although the threshold arguments guarantee individual k-vertex trees, we can make statements about all of them simultaneously because the number of these trees is bounded.) Furthermore, this p is also below the threshold for appearance of fixed cycles (density 1, length bounded by k), so G^p is a forest of trees of order at most k, and every tree on k vertices appears as a component.

Intuitively, the random graph has no cycles in this stage of evolution, because when there is no large component a random added edge is much more likely to join two components than to lie in one component. To make the intuition precise, we let X be the number of cycles in G^p and compute that $E(X) = \sum_{k=3}^{n} \binom{n}{k} \frac{1}{2} (k - 1)! p^k < \sum_{k=3}^{n} (np)^k/2k$. If $pn \to 0$, then $E(X) \to 0$.

The next major stage of evolution is $p = c/n$ with $0 < c < 1$. With X counting cycles, we can no longer say $E(X) \sim \sum_{k=3}^{n} (np)^k/2k$, because when k is a substantial fraction of n the ratio $n^k/(n)_k$ does not approach 1. We must break $E(X)$ into two sums, and the arguments become more difficult. When $pn \to c$, we find that $E(X)$ approaches a constant c', and the number of cycles in G^p is asymptotically Poisson distributed. With cycles in a few components and all components small, we still expect the next edge to join two components or create a cycle in a component that doesn't have one. In this range, the size of the largest component is about $\log n$, there are many components, and every component is a tree or is unicyclic. Most vertices still belong to acyclic components.

When c reaches and passes 1, the structure of G^p changes radically. This is called the *double jump* because the structure when $c < 1$, $c \sim 1$, or $c > 1$ is significantly different. At $pn = 1$ the expected number of cycles approaches ∞, and the second moment method guarantees that almost every G^p has a cycle. Also, the order of the largest component jumps from $\log n$ to $n^{2/3}$. For $pn = c > 1$ the number of vertices outside the "giant component" is $o(n)$. Also G^p is likely to have some cycle with three crossing chords and be non-planar.

Next, let p approach $c\ln n/n$. With $c < 1$, we have proved that almost every G^p has isolated vertices. With $c > 1$, these disappear. As we add edges to a disconnected graph, the edges may go within a component or connect two components. When the components are all small, added edges will almost surely join components. Eventually, this results in the creation of a giant component. At this point, added edges are likely to lie within the giant component or to join it to one of the small components. Of the small components, those most likely to receive such edges are the larger ones. In other words, as c passes through 1 the last remaining small components swallowed by the giant component are isolated vertices, so the threshold for connectivity is the same as the threshold for the disappearance of isolated vertices. With $c > 1$, suddenly almost every G^p is also Hamiltonian; hence Palmer's "spanning cycle". Minimum degree k (and the appearance of the Hamiltonian cycle when $k = 2$) has a threshold that involves a lower order term: $\ln n/n + (k-1)\ln\ln n/n$.

The last stages of evolution are those where $pn/\ln n \to \infty$ but $p = o(1)$, and then finally $p = c$; this brings us back to where we began our study.

As $c \to \infty$ when $p = c\ln n/n$, we leave the domain of sparse graphs. The evolutionary viewpoint becomes less valuable, and we study properties of the random graph. We pay less attention to probability threshold functions and concentrate on the likely value of graph parameters, especially when p is constant. Given a parameter μ, we want to show that $\mu(G^p) \sim f(n)$ for almost every G^p. We can view this as a threshold in the following sense; for almost every G^p, $\mu(G^p)$ is between $(1-\varepsilon)f(n)$ and $(1+\varepsilon)f(n)$. If $\mu(G^p)$ is almost always between $f(n) - \varepsilon g(n)$ and $f(n) + \varepsilon g(n)$, where $g(n) = o(f(n))$, then we write this stronger statement as $\mu(G^p) \in f(n)(1 + o(1))$.

We can also test conjectures on random graphs. Hadwiger's Conjecture (Section 5.2) is true for almost all graphs, but Hajos' Conjecture is true for almost none. The Strong Perfect Graph Conjecture is true for almost all graphs, but not in a very illuminating way, since for constant p almost every G^p has a chordless 5-cycle and fails to be perfect.

There are properties that are true for almost all graphs but occur in no known examples! For nine years, having large chromatic number and large girth was such a property. For the known lower bound on diagonal Ramsey numbers, there is still no construction of an infinite class of graphs such that $\alpha(G) < \log_{\sqrt{2}}(n(G))$ and $\omega(G) < \log_{\sqrt{2}}(n(G))$, even though we know that almost all graphs have this property.

Properties of the random graph can lead to fast algorithms for hard problems that will work almost always work. For example, after stating two results about vertex degrees in random graphs, we present an application to isomorphism testing. We will show how to use properties of the degree sequence to design a fast algorithm to test isomorphism "almost always".

In the literature of random graphs, ω_n denotes a function that is unbounded but grows arbitrarily slowly.

8.5.21. Theorem. (Erdős-Rényi [1966]). If $p = \omega_n \log n/n$ and $\varepsilon > 0$ is fixed, then almost every G^p satisfies

$$(1 - \varepsilon)pn < \delta(G^p) \leq \Delta(G^p) \leq (1 + \varepsilon)pn. \ \square$$

One expects that most vertices have degree near the average, but some have degree farther away. Bollobas [1982] showed that for $p \leq 1/2$, the vertex of maximum degree is unique in almost every G^p if and only if $pn/\log n \to \infty$. When we complete evolution by returning to the realm of constant edge probability, more detailed results are known about the degree distribution. There will almost always be some vertices with isolated high degrees before the degrees begin to bunch up. Bollobás determined how many distinct degrees can be guaranteed:

8.5.22. Theorem. (Bollobas [1981b]) In Model A with p fixed and $t \in o(n/\log n)^{1/4}$, almost every G^p has different degrees for its t vertices of highest degree. If $t \notin o(n/\log n)^{1/4}$, then almost every G^p has $d_i = d_{i+1}$ for some $i < t$. $\square$

We apply this result to isomorphism testing. No known polynomial-time algorithm tests isomorphism for all pairs of graphs, but Babai-Erdős-Selkow [1980] used the degree results for the random graph to develop an algorithm that works almost always and runs in $O(n^2)$ time. The idea is to specify a set **H** that contains almost all graphs and has the property that isomorphism with a graph in **H** can be tested quickly. The testing is done by a *canonical labeling algorithm*, which accepts a graph and labels it in a canonical way if it belongs to **H**. A canonical labeling will have the property that, when it labels vertices as $v_1, \ldots, v_n$ in one graph and $w_1, \ldots, w_n$ in another, only the bijection mapping v_i to w_i is a possible isomorphism. Isomorphism can be tested by comparing the adjacency matrices under this labeling.

8.5.23. Corollary. (Babai-Erdős-Selkow [1980]). There is a quadratic algorithm that tests isomorphism for almost all pairs of graphs.

Proof. Given a graph G on n vertices, presented by its adjacency matrix, compute and sort the vertex degrees, labeling the vertices in decreasing order of degree. Fix $r = \lfloor 3 \lg n \rfloor$. If $d(v_i) = d(v_{i+1})$ for any $i < r$, reject G. Using $p = 1/2$ in Theorem 8.5.22 implies that almost every graph successfully passes this test.

Let $U = \{v_1, \ldots, v_r\}$. With $r = \lfloor 3\lg n \rfloor$, there are about n^3 distinct subsets of the vertices of U. Since only $n - r$ vertices remain outside U, there is a chance that they can be distinguished by their neighborhoods in U. The set **H** will be all the graphs reaching this stage for which this holds: the vertices of $V - U$ have distinct neighborhoods in U. To test this in $O(n^2)$ time and complete the labeling, for each $x \in V - U$ encode $N(x) \cap U$ as a 0,1-vector of length r. Evaluate these as binary integers, and sort them! These steps take $O(n \log n)$ time. Relabel the vertices v_{r+1} to v_n as $w_{r+1}, \ldots, w_n$ in decreasing order of these values. If two consecutive values are the same, reject G.

If G has passed this far, then G has no nontrivial automorphisms. A graph isomorphic to such a G has only one isomorphism to G, given by applying the canonical labeling algorithm to it. The last stage in the algorithm, if both graphs pass canonical labeling, is to compare the adjacency matrices with rows and columns indexed according to the canonical labeling. The graphs are isomorphic if and only if the matrices are now identical. This comparison takes $O(n^2)$ time.

We must show that for almost every G^p, the adjacency vectors within a specified set of r vertices are distinct for the remaining vertices. If $p \leq 1/2$, then the probability for any pair x, y that x, y have the same adjacencies in U is bounded approximately by $(1 - p)^r$. We say approximately because U is not chosen at random; the fact that U are the vertices of highest degree impairs randomness, increasing the probability of a specified edge incident to these vertices. Nevertheless, it doesn't change by much, and the expected number of pairs of vertices outside U with identical adjacencies in U is bounded by $O(\binom{n-r}{2}(1 - p)^r)$. Given our choice of r, we can bound the base 2 logarithm of this by $2\lg n - 3\lg b \lg n$, where $b = 1/(1 - p) \geq 2$ (if $p \leq 1/2$). This approaches $-\infty$, so almost all graphs have distinct adjacency vectors in this set. $\square$

The probability of rejection in this labeling algorithm is bounded by $n^{-1/7}$ for sufficiently large n. A series of later improvements led to an algorithm running in time $O(n^2)$ with rejection probability c^{-n} (Babai-Kučera [1979]).

Some NP-hard decision problems become trivial for random graphs. Although $\Delta(G) \leq \chi'(G) \leq \Delta(G) + 1$ for every simple G (Vizing [1964]), deciding between these values is NP-hard (Holyer [1981]). Vizing proved that $\chi'(G) = \Delta(G) + 1$ requires at least 3 vertices of maximum degree. Thus Erdős and Wilson [1977], who noted the uniqueness of the vertex of maximum degree when $p = 1/2$, also observed that $\chi'(G) = \Delta(G)$ for the random graph.

CONNECTIVITY, CLIQUES, AND COLORING

We have claimed that for $k = 1$ the thresholds for $\delta(G) \geq k$ and $\kappa(G) \geq k$ are the same: $p(n) = \ln n/n$. This also holds for any fixed k, with sharp threshold $p(n) = \ln n/n + k \ln\ln n/n + x/n$ (the number of vertices with (minimum) degree k is asymptotically Poisson with mean $\mu = e^{-x}/k!$). When we return to constant edge probability, the vertex degrees are near pn, but what about connectivity? When p and k are fixed, vertex pairs in almost every G^p have k common neighbors, and hence G^p is k-connected (Exercise 12). Now we ask *how quickly* we can make k grow as a function of n and still be able to guarantee that almost every G^p is k-connected. A closer look at the argument for common neighbors shows that if $k \in o(n/\log n)$ and p is fixed, then almost every G^p is k-connected (Exercise 20). Requiring k common neighbors is too restrictive; Bollobás [1981b] showed that even for constant p, almost every G^p has connectivity equal to minimum degree.

What about clique number? For fixed k, we have derived edge-probability thresholds for the appearance of a k-clique, but for constant p the clique number will grow with n. Amazingly, for G^p we can guess the exact value with high probability without ever looking at the graph! The approach is to find a function $r(n)$ such that almost every G^p has an r-clique and almost none has an $r + 1$-clique.

8.5.24. Theorem. (Matula [1972]). For fixed $p = 1/b$ and fixed $\varepsilon > 0$, almost every G^p has clique number between $\lfloor d - \varepsilon \rfloor$ and $\lfloor d + \varepsilon \rfloor$, where $d = 2\log_b n - 2\log_b \log_b n + 1 + 2\log_b(e/2)$.

Proof. (sketch) If X_r is the number of r-cliques, then $E(X_r) = \binom{n}{r}p^{\binom{r}{2}}$, and an application of Stirling's approximation for $r!$ yields $E(X_r) \sim (2\pi r)^{-1/2}(enr^{-1}p^{(r-1)/2})^r$. We have $E(X_r) \to 0$ if $r \to \infty$ and $(enr^{-1}p^{(r-1)/2}) \leq 1$ for large n. To determine $r(n)$ such that this holds, take logarithms (base b) in the inequality and solve for r to find $r \geq 2\log_b n - 2\log_b r + 1 + 2\log_b e$. To satisfy this bound, r must be asymptotic to $2\log_b n$ (or larger), so we substitute this for r on the right to obtain $r \geq d(n)$ as defined above. This yields $d(n) + \varepsilon$ as an upper bound on the clique number of almost every G^p; if $r > d + \varepsilon$, then almost every G^p has no clique of size r.

The lower bound comes from careful application of the second moment method, as in Theorem 8.5.20, but the growth of r with n makes this analysis more difficult. The expectation of X_r^2 sums the probability of common occurrence for all ordered pairs of r-cliques. Since this probability depends only on the number of common vertices, we obtain $E(X_r^2) = \binom{n}{r}\Sigma_{k=0}^{r}\binom{r}{k}\binom{n-r}{r-k}p^{2\binom{r}{2}-\binom{k}{2}}$. Since r is small compared to n, we expect the term for $k = 0$ (disjoint cliques) to dominate. We split the contributions by writing $E(X_r^2)/E(X_r)^2 = \alpha_n + \beta_n$, where $\alpha_n =$

$\binom{n}{r}^{-1}\binom{n-r}{r}$ and $\beta_n = \binom{n}{r}^{-1}\Sigma_{k=1}^{r}\binom{r}{k}\binom{n-r}{r-k}b^{\binom{k}{2}}$. We want to show $\alpha_n \sim 1$ and $\beta_n \to 0$. Although all the functions in α_n are growing, when $r \sim 2\log_b n$ we can use an asymptotic formula for $\binom{a}{k}/\binom{b}{k}$ to show $\alpha_n \sim e^{-r^2/(n-r)} \to 1$. The discussion of β_n is more difficult; Palmer [1985, p75-80] presents further details. $\square$

The analysis of connectivity and clique number can be applied to measure the strength of conditions for Hamiltonian cycles, as discussed in Palmer [1985, p81-85]. A theorem proves nothing if its hypotheses are never satisfied, so it is reasonable to say that such a theorem has strength 0. A theorem is strong if the conclusion is satisfied only when the hypothesis is satisfied, because it is best possible in that the hypotheses cannot be weakened. Define the *strength* of a theorem to be the probability that its hypotheses are satisfied divided by the probability that its conclusion is satisfied.

Consider sufficient conditions for Hamiltonian cycles. We know that $p = \ln n/n$ is a threshold for a Hamiltonian cycle. Hence almost every G^p is Hamiltonian when p is fixed. Dirac [1952b] showed that G is Hamiltonian when every vertex degree is at least $n/2$ (Theorem 6.2.5). If $p > 1/2$ this is true for almost every G^p, but if $p \le 1/2$ it is almost never true. Hence the asymptotic strength of Dirac's Theorem is 0 when p is a constant at most $1/2$. The same fate befalls the other degree conditions of Chapter 6. Meanwhile, Chvátal and Erdős [1972] proved that G is Hamiltonian whenever its connectivity exceeds its independence number (Theorem 6.2.13). It follows from our thresholds for these parameters that this result is strong for every constant $p > 0$. We know that almost every G^p has $\alpha < 2(1+\varepsilon)\log_b n$, and we know that almost every G^p has connectivity at least k whenever $k = o(n/\log n)$ (since this many common neighbors can be guaranteed). Hence almost every G^p has $\kappa > \alpha$, and the asymptotic strength of the theorem is 1.

Finally, we consider chromatic number for constant p. Since $1 - p$ is also constant, we can apply the results on clique number: Almost every G^p has no stable set (independent set of vertices) with more than $(1+o(1))2\log_b n$ vertices, where $b = 1/(1-p)$. Hence $\chi(G^p) \ge (1/2 + o(1))n/\log_b n$ almost always. Achieving this bound requires finding many disjoint stable sets with near-maximum sizes. For a decade, the best result was an algorithmic guarantee of a coloring with at most twice the number of colors in the lower bound. Eventually, Bollobás [1988] proved that the lower bound is achievable, by using another probabilistic technique that guarantees finding enough large stable sets. He proved that, in almost every G^p, *every* set with at least $n/(\log_b n)^2$ vertices has a clique of order at least $2\log_b n - 5\log_b \log_b n$. This allows

stable sets of near-maximum size to be extracted until there are too few
vertices remaining to cause trouble; the remainder can be given distinct
colors.

Before developing Bollobás approach, we present the earlier result
because it has algorithmic interest; the greedy algorithm uses at most
$(1 + \varepsilon)n/\log_b n$ colors on almost every G^p. Thus it "almost always
works" as an approximation algorithm in the same sense that our earlier isomorphism algorithm almost always works. Garey and Johnson
[1976] showed there is no fast algorithm that uses at most 2χ colors on
every graph unless P = NP. Bollobás' proof does not give a fast algorithm for coloring almost every graph with an asymptotically optimal
number of colors; it is an existence proof only.

8.5.25. Theorem. (Grimmett-McDiarmid [1975]). Given fixed edge-
probability p and $\varepsilon > 0$, set $b = 1/(1 - p)$. Then almost every G^p
has chromatic number satisfying

$$(1/2 - \varepsilon)n/\log_b n \le \chi(G) \le (1 + \varepsilon)n/\log_b n.$$

Proof. The lower bound follows from the clique argument as suggested
above. For the upper bound, we show that the greedy coloring of
$v_1, \ldots, v_n$ in order almost never uses more than $(1 + \varepsilon)n/\log_b n$ colors. To
show that the greedy algorithm uses at most $f(n)$ colors on almost every
G^p, let $\mathbf{B}_{m+1}$ be the collection of labeled n-vertex graphs for which v_{m+1}
is the first vertex to use color $f(n) + 1$. We need only show $\Sigma_{m=1}^{n-1} P(\mathbf{B}_{m+1})$
$\to 0$. For each $G \in \mathbf{B}_{m+1}$, let G^m be the subgraph induced by $v_1, \ldots, v_m$.
Before using color $f(n) + 1$, color $f(n)$ must be used, so for any $G \in \mathbf{B}_{m+1}$
the greedy coloring of G^m uses $f(n)$ colors. Let k_i be the number of
times color i appears in this coloring. To require use of color $f(n) + 1$,
v_{m+1} must have at least one edge to a vertex of each color $1, \ldots, f(n)$, so
the probability of the graphs in $\mathbf{B}_{m+1}$ with this H on $v_1, \ldots, v_m$ is
$\Pi_{i=1}^{f}[1 - (1 - p)^{k_i}]P(G^m = H \mid G \in \mathbf{B}_{m+1})$. Hence the probability of requiring $f + 1$ colors is bounded by

$$\Sigma_{m=0}^{n-1}\Sigma_H \Pi_{i=1}^{f}[1 - (1 - p)^{k_i}]P(G^m = H \mid G \in \mathbf{B}_{m+1}).$$

To simplify this summation, Bollobás and Erdős [1976] used a con-
vexity inequality: $\Pi_{i=1}^{f}[1 - (1 - p)^{k_i}] \le [1 - (1 - p)^{n/f}]^f$. Now this quantity
is independent of H and m and can be moved outside. The various H's
for a given m have total probability 1, which bounds the probability of
needing more than $f(n)$ colors by $n[1 - (1 - p)^{n/f}]^f \le ne^{-f(1-p)^{n/f}}$. To
choose $f(n)$ to make this probability go to 0, set $f(n) = cn/\log_b n$. The
log of the bound becomes $\log n - cn^{1-1/c}/\log_b n$, which goes to $-\infty$ if and
only if $1 - 1/c > 0$. $\square$

MARTINGALES

Bollobás' result on chromatic number uses a special type of sequence of random variables We need the notion of conditional variables. If Y, X are random variables, then the expression $Y \mid X$, read as "Y given X", is itself a random variable. Actually, it is many random variables, one for each value of X; we treat X as a constant and normalize the distribution for Y by the probability that X takes on that value.

8.5.26. Definition. A *martingale* is a sequence of random variables $X_1, \ldots, X_n$ such that $E(X_i|X_0, \ldots, X_{i-1}) = X_{i-1}$.

A random walk on a line provides a simple example: at each time unit, there is probability p of taking one step to the left, probability p of taking one step to the right, and probability $1 - 2p$ of not moving. Intuitively, the random walk does not move very far in relation to the number of steps. We shall see that this depends its inability to move more than one unit in any one step.

Martingales are applied to combinatorial structures to show that a random variable is highly concentrated around its expectation. When the technique applies, we are relieved of the need for a Second Moment Method computation. The hard work is eliminated by invoking Azuma's Inequality, also called the Martingale Tail Inequality. It implies that if successive random variables in a martingale differ by at most one, then the probability that X_n exceeds X_0 by more than $\lambda\sqrt{n}$ is bounded by $e^{-\lambda^2/2}$. We first prove two lemmas. Like other statements we have made, these hold for continuous random variables, but our main application uses only discrete variables.

8.5.27. Lemma. Suppose Y is a random variable such that $E(Y) = 0$ and $|Y| \le 1$. If f is a convex function on $[-1, 1]$, then $E(f(Y)) \le \frac{1}{2}[f(-1) + f(1)]$. In particular, $E(e^{tY}) \le \frac{1}{2}(e^t + e^{-t})$ for any $t > 0$.

Proof. When Y takes only the values ± 1, each with probability .5, we have $E(f(Y) = \frac{1}{2}[f(-1) + f(1)]$. For any other distribution, pushing probability "out to the edges" increases $E(f(Y))$. For discrete variables, we can use induction on the number of values with nonzero probability. Convexity implies that $f(a) \le \frac{1-a}{2}f(1) + \frac{a+1}{2}f(-1)$. If $P(Y = a) = \alpha$, then we can decrease the probability at a to 0, increase $P(Y = 1)$ by $\alpha\frac{1-a}{2}$ and increase $P(Y = -1)$ by $\alpha\frac{a+1}{2}$ to obtain a new variable Y' with the same expectation. By the convexity inequality, $E(f(Y)) \le E(f(Y'))$. By the induction hypothesis, $E(f(Y')) \le \frac{1}{2}[f(-1) + f(1)]$. $\square$

The proof of Azuma's Inequality uses the expectation of a conditional variable. For any given value of X, we can compute $E(Y \mid X = i)$ over the portion of the sample space where $X = i$. The expectation $E(E(Y \mid X = i))$ is an expectation over the entire sample space. The value of $E(Y \mid X)$ when $X = i$ is weighted by $P(X = i)$; this removes the effect of conditioning, and we obtain $E(E(Y \mid X)) = E(Y)$.

8.5.28. Lemma. $E(E(Y \mid X)) = E(Y)$.

Proof. Letting $p_{ij} = P(X = i, Y = j)$, we have $P(Y = k \mid X = i) = p_{i,k}/P(X = i)$, and hence $E(Y \mid X = i) = (\Sigma_k k p_{i,k})/P(X = i)$. Overall, we compute $E(E(Y \mid X)) = \Sigma_i E(Y \mid X = i) P(X = i) = \Sigma_i \Sigma_k k p_{i,k} = E(Y)$. □

8.5.29. Theorem. (Azuma's Inequality) If $X_0, \ldots, X_n$ is a martingale with $|X_i - X_{i-1}| \le 1$, then $P(X_n - X_0 \ge \lambda \sqrt{n}) \le e^{-\lambda^2/2}$.

Proof. By translation, we may assume $X_0 = 0$. For $t > 0$, we have $X_n \ge \lambda \sqrt{n}$ if and only if $e^{tX_n} \ge e^{t\lambda\sqrt{n}}$, and hence $P(X_n \ge \lambda\sqrt{n}) = P(e^{tX_n} \ge e^{t\lambda\sqrt{n}})$. Applied to e^{tX_n}, Markov's Inequality yields $P(e^{tX_n} \ge e^{t\lambda\sqrt{n}}) \le E(e^{tX_n})/e^{\lambda t\sqrt{n}}$. This bound holds for each $t > 0$, and later we will choose t to minimize the bound.

First we prove by induction on n that $E(e^{tX_n}) \le \frac{1}{2}(e^t + e^{-t})$. We introduce X_{n-1} to condition on it. Lemma 8.5.28 yields $E(e^{tX_n}) = E(e^{tX_{n-1}} e^{t(X_n - X_{n-1})}) = E(E(e^{tX_{n-1}} e^{t(X_n - X_{n-1})} \mid X_{n-1}))$. When we condition on X_{n-1}, the value of X_{n-1} is constant, and we can rewrite the expression as $E(e^{tX_{n-1}} E(e^{tY} \mid X_{n-1}))$, where $Y = X_n - X_{n-1}$. Because $\{X_n\}$ is a martingale, $E(Y) = 0$, and by hypothesis $|Y| \le 1$. Hence Lemma 8.5.27 applies, yielding $E(e^{tY} \mid X_{n-1}) \le \frac{1}{2}(e^t + e^{-t})$. This itself is now a constant, and the induction hypothesis completes the proof.

We weaken the bound to a more usable form by observing that $\frac{1}{2}(e^t + e^{-t}) \le e^{t^2/2}$. This holds because the left side is $\Sigma t^{2k}/(2k)!$ and the right side is $\Sigma t^{2k}/(2^k k!)$. Hence our original probability is bounded by $e^{nt^2/2 - \lambda t\sqrt{n}}$ for each $t > 0$. We obtain the best bound by minimizing over t. The exponent is quadratic; we minimize it by choosing t to solve $tn - \lambda\sqrt{n} = 0$, or $t = \lambda/\sqrt{n}$. The resulting bound is $e^{-\lambda^2/2}$. □

Azuma's inequality is a one-sided tail inequality, meaning it is a bound on the probability that X_n is much larger than X_0. Since the conditions are symmetric in sign, we could apply the inequality to the sequence $-X_0, \ldots, -X_n$ to obtain the same inequality for the other tail, where X_n is much smaller than X_0.

8.5.30. Example. *The pragmatic gambler.* A gambler has the option n times of playing a game, where n is fixed. Each time he chooses to play, he wins or loses 1 with equal probability. His aim is to win $\lambda\sqrt{n}$, so he stops playing if he reaches that value. Letting X_i be his winnings after i games, we have $X_i = X_{i-1}$ if $X_{i-1} \geq \lambda\sqrt{n}$, and otherwise $X_i = X_{i-1} \pm 1$, each with probability .5. Hence the sequence X_i is a martingale that changes by at most one at each step, and Azuma's inequality applies. The probability that the gambler will be successful at earning $\lambda\sqrt{n}$ is bounded by $e^{-\lambda^2/2}$. If $\lambda = 1$, then there is a reasonable chance of success, but if $\lambda = 10$, then there is little chance. $\square$

For applications to random graphs and other random structures, we consider martingales that arise in a special way. We have an underlying probability space, and the value X_0 is the expectation of some random variable X on this space. The value X_n will be the value at an individual sample point. In such a setting, $P(|X_n - X_0| \geq t)$ measures how tightly the random variable X is concentrated around its expectation. Along the way from X_0 to X_n, we gradually learn more information about where in the probability space the sample point is located.

8.5.31. Lemma. Let X be a random variable defined on a probability space. Let $F_0 \supseteq F_1 \supseteq \cdots \supseteq F_n$ be a chain of subsets of the space, where F_0 is the full space, F_n is a single outcome, and F_i is a random variable that is a block in a partition of F_{i-1}. The probability that F_i is a particular block in the partition of F_{i-1} is proportional to the probability of that event in the full space. If $X_i = E(X|F_i)$, then the sequence $X_0, \ldots, X_n$ is a martingale.

Proof. We have $X_0 = E(X)$ and $X_n = X$, since F_n is a single outcome in the probability space. To compute $E(X|F_i)$, consider the the subspace defined by a particular instance of F_{i-1}. The variable F_i is a block within F_{i-1} and hence is conditioned on F_{i-1}. To compute the expectation of X_i, we weight each $E(X|F_i)$ by the probability of being in F_i. The distribution of F_i is determined by the value of F_{i-1}. Hence $E(X_i) = E(E(X|F_i)|F_{i-1}) = E(X \mid F_{i-1}) = X_{i-1}$. $\square$

Such martingales, which we call *restriction martingales*, arise whenever we reveal information gradually about a randomly generated object. Let F_i denote the subset of the probability space to which the object is confined after i steps (F for "inFormation"). In a coin-flipping problem, F_i could be given by the first i values in the sequence of flips. In a random graph problem, F_i is often the subgraph induced by the vertices $\{v_1, \ldots, v_i\}$, or perhaps the information about the existence of the first i edges.

To apply Azuma's Inequality, we need to bound $|X_i - X_{i-1}|$. There is a class of restriction martingales where this is easy. When we generate a graph at random, the knowledge of which edges arise incident to a fixed vertex v_i can change the chromatic number by at most 1; $\chi(G - v_i)$ equals $\chi(G)$ or $\chi(G) - 1$. As shown by the next lemma, such behavior allows us to conclude that $|X_i - X_{i-1}| \le 1$ in the restriction martingale defined by revealing vertices (and their edges to previously revealed vertices) one by one.

8.5.32. Lemma. Suppose a random structure is specified by the outcomes of independent steps $S_1, \ldots, S_n$. Let F_i be the subspace determined by knowing the results of steps $S_1, \ldots, S_i$, and let $X_0, \ldots, X_n$ be the corresponding restriction martingale for a random variable X. Let A be an event consisting of specified outcomes of S_j for all $j \ne i$, with S_i unknown. If the extreme values of X over the sample points in A differ by at most 1 for all such events H, then $|X_i - X_{i-1}| \le 1$ for all $i > 0$, and hence $P(X - E(X)) > \lambda\sqrt{n}) \le e^{-\lambda^2/2}$.

Proof. Suppose F_{i-1} and hence $X_i = E(X \mid F_{i-1})$ are given. We arrange the sample points in F_{i-1} in a grid. Each row is a choice for F_i, i.e. a block in the partition of F_{i-1} from which F_i is chosen. Each column is an event A in which $S_{i+1}, \ldots, S_n$ are fixed and only S_i varies. By hypothesis, in each column the maximum and minimum values of X differ by at most 1. Let m_s, M_s be the minimum and maximum values of X in column s.

Because S_i and $S_{i+1}, \ldots, S_n$ are specified independently, the probability of the outcome in row r and column s is $q_r p_s$, where q_r is the probability that the specification of S_i corresponds to this row and p_s is the probability that $S_{i+1}, \ldots, S_n$ yields this column. The computation of X_i is the expectation across a single row: $\Sigma m_s p_s \le E(X \mid F_i) \le M_s p_s \le 1 + \Sigma m_s p_s$. When we take the expectation over the entire grid to compute X_{i-1}, we obtain the same inequalities. Hence X_{i-1} and X_i are confined to a single interval of length 1 and differ by at most 1. Hence Azuma's Inequality applies. □

The event F_{i-1}
choices of A (or $S_{i+1}, \ldots, S_n$)

choices				
of				
F_i				
(or S_i)				

When the conditions of Lemma 8.5.32 hold, we can conclude that the value of X is highly concentrated around its mean. As mentioned earlier, we then do not need the second moment method to conclude that almost all the random structures (e.g., graphs) have value of X

asymptotic to $E(X)$.

8.5.33. Example. *Chromatic number of random graphs.* Fix n, and consider Model A with edge-probability p. Suppose we reveal the random n-vertex graph one vertex at a time. At stage i, we learn the edges from v_i to the previous vertices; this is S_i, and Model A specifies the outcomes of the stages S_i independently. The event A in which all but S_i are specified is the subgraph $G - v_i$ of the random graph G. Since $\chi(G - v_i) \leq \chi(G) \leq \chi(G - v_i) + 1$, the value of X differs by at most one across all possibilities in A. The hypotheses of Lemma 8.5.32 hold, and we can conclude immediately that the chromatic number is highly concentrated around its expectation. In particular,

$$P(|\chi(G) - E(\chi(G))|) \geq \lambda \sqrt{n}) \leq 2e^{-\lambda^2/2} \quad \square$$

The result of Example 8.5.33 says nothing about the value of $E(\chi(G))$. To approximate this we again use Azuma's Inequality. With constant edge probability p, we know that the clique number of G^p is almost always within one of $d = 2 \log_b n - 2 \log_b \log_b n + 1 + 2 \log_b(e/2)$, where $b = 1/p$. The same result holds for stable sets using the base $c = 1/(1 - p)$ for the logarithm. To show that the chromatic number of G^p is almost always close to $n/(2 \log_c n)$, we would like to extract a maximum stable set for each color until we have colored all the vertices. We cannot do this, since the graph induced by the remaining vertices is not a random graph in the same model, being conditioned on the knowledge that it had no larger stable set. Bollobás found a way around this difficulty. He proved that if we use stable sets a bit smaller than the maximum, then there will be so many of them that *every* sufficiently large vertex set will have a stable set that big. This enables us to pick stable sets for the desired coloring.

8.5.34. Theorem. (Bollobás [1988]) For almost every G^p with constant $p = 1 - 1/c$, the following property holds: every set of at least $m = \lceil n/\log_c^2 n \rceil$ vertices induces a subgraph having a stable set of size at least $r = 2 \log_c n - 5 \log_c \log_c n$.

Proof. (sketch) We use *r-staset*, by analogy with *r-clique*, to denote a stable set of size r. Let S be a set of m vertices. We will bound the probability that S has no r-staset by $e^{-dm^{1+\varepsilon}}$ for some d, ε. This in turn bounds the probability that there exists an m-set with no r-staset by $\binom{n}{m}e^{-dm^{1+\varepsilon}} < 2^n e^{-dm^{1+\varepsilon}}$. Since $n = m^{1+o(1)}$, this bound goes to 0, and the first moment method implies that almost every G^p has no bad m-set.

It suffices to study the subgraph G induced by $[m]$. Let X be the maximum number of pairwise pair-disjoint r-stasets in this subgraph, where *pair-disjoint* means they share at most one vertex. We will show that $X \geq 1$ almost always. To do this, it suffices to show 1) X is highly concentrated around its mean, and 2) $E(X)$ is bigger than something large (and growing).

We invoke Azuma's Inequality for (1). Consider the restriction martingale that results from revealing G *one edge-slot at a time* instead of one vertex at a time. At each step, we learn the presence or absence of the edge between one additional pair of vertices. We have $X_0 = E(X)$, $X_{\binom{m}{2}} = X$, and the sequence is a martingale. The status of one edge slot changes the value of X by at most one, so Lemma 8.5.32 applies, and $P(X - E(X) \leq -\lambda\binom{m}{2}^{1/2}) \leq e^{-\lambda^2/2}$. Setting $\lambda = E(X)/\binom{m}{2}^{1/2}$, we have $P(X = 0) = P(X - E(X) \leq -E(X)) \leq e^{-E(X)^2/(m^2 - m)}$. Hence it suffices to show that $E(X)/m \to \infty$.

To prove this, we consider another random variable $\hat{X}$, the number of r-stasets in G that have no pair in common with *any* other r-staset. Such a collection forms a pairwise pair-disjoint collection of r-stasets, so $X \geq \hat{X}$. We introduced X because the restriction martingale for $\hat{X}$ does not satisfy $|\hat{X}_i - \hat{X}_{i-1}| \leq 1$. In the drawing of $\overline{G}$ in the figure below, for example, we have $r = 4$ and seek 4-cliques; if the last (dotted) edge is present in $\overline{G}$ (absent in G), then $\hat{X} = 0$, but if it is absent from $\overline{G}$ (present in G), then $\hat{X} = 3$.

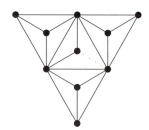

It is easier to compute $E(\hat{X})$ than $E(X)$. Expressing $\hat{X}$ as the sum of $\binom{m}{r}$ indicator variables, we obtain $E(\hat{X})$ as $\binom{m}{r}$ times the probability that $[r]$ induces an r-staset that is pair-disjoint from all others. This is $(1 - p)^{\binom{r}{2}}$ times the conditional probability that $[r]$ does not conflict with other r-stasets, given the event Z that $[r]$ is in fact independent. Let Y be the number of other r-stasets overlapping $[r]$ in at least two elements. By Markov's inequality, $E(Y|Z) \to 0$ implies $P(Y = 0|Z) \to 1$. Since the overlap is at least two vertices, we have $E(Y \mid Z) = \Sigma_{i \geq 2}\binom{r}{i}\binom{m-r}{r-i}(1 - p)^{\binom{r}{2}-\binom{i}{2}}$. From the expression for r in terms of m, this can be shown to approach 0 as $m \to \infty$.

Hence $E(\hat{X})$ is asymptotic to $\binom{m}{r}(1 - p)^{\binom{r}{2}}$. Using the expression for r in terms of m, we obtain $E(\hat{X}) \in \Omega(m^{5/3})$. This implies $E(X)/m \to \infty$, which completes the proof. $\square$

8.5.35. Corollary. (Bollobás [1988]) For constant edge probability $p = 1 - 1/c$, almost every G^p has chromatic number between $(1 + \varepsilon)n/(2 \log_c n)$ and $(1 + \varepsilon')n/(2 \log_c n)$, where $\varepsilon = \log_c \log_c n / \log_c n$ and $\varepsilon' = 5 \log_c \log_c n / \log_c n$.

Proof. The lower bound follows from the fact that almost every G^p has no stable set larger than $2 \log_c n - 2 \log_c \log_c n$. The upper bound follows from the theorem above, because we can almost always select stable sets of size $2 \log_c n - 5 \log_c \log_c n$ until we have only $n/\lg_c^2 n$ vertices left. Since $n/\lg_c^2 n \in o(n/\log_c n)$, we can complete the coloring by using distinct new colors on the remaining vertices. $\square$

EXERCISES

8.5.1. (–) *Expectation*.

a) Determine the expected number of fixed points in a random permutation of $[n]$.

b) Determine the expected number of vertices of degree k in a random n-vertex graph with edge-probability p.

8.5.2. Prove that $1 - p < e^{-p}$ for $p > 0$.

8.5.3. (!) *Bipartite subgraphs*.

a) Use a random partition of the vertices to prove that every graph has a bipartite subgraph with at least half its edges.

b) Use a balanced partition of the vertices to improve part (a): if G has m edges and n vertices, then G has a bipartite subgraph with at least $m\lceil n/2 \rceil/[2\lceil n/2 \rceil - 1]$ edges.

8.5.4. (!) A *hypergraph* consists of a collection of vertices and a collection of edges; if the vertex set is V, then the edges are subsets of V. The *chromatic number* $\chi(H)$ of a hypergraph H is the minimum number of colors needed to label the vertices so that no edge is monochromatic. A hypergraph is k-*uniform* if its edges all have size k. Prove that every k-uniform hypergraph with fewer than 2^{k-1} edges is 2-colorable. (Erdős [1963])

8.5.5. Suppose that G is a graph with p vertices, q edges, and automorphism group of size s. Let $n = (sk^{q-1})^{1/p}$. Prove that there exists a k-coloring of $E(K_n)$ that has no monochromatic copy of G. (Comment: this proves that $R_k(G) > n$ - see Section 8.3.) (Chvátal-Harary [1973]).

8.5.6. Use the deletion method to prove that $R(k, k) > n - \binom{n}{k}2^{1-\binom{k}{2}}$, where $n \in \mathbb{N}$. (Comment: with n chosen to asymptotic to $n \sim e^{-1}k2^{k/2}(1 - o(1))$, this implies that $R(k, k) > (1/e)(1 + o(1))k2^{k/2}$. This improves the lower bound from the simple existence argument, but only by a factor of $\sqrt{2}$.

8.5.7. (!) Use the deletion method to prove that a graph with n vertices and average degree $d \geq 1$ has an independent set with at least $n/(2d)$ vertices. (Hint: choose a random subset by including each vertex independently with a probability p chosen later. Compute the expected number of edges induced.)

8.5.8. The largest number of edges in an n-vertex graph not containing a fixed graph H as a subgraph is written as $ex(n; H)$. Use the deletion method to prove that $ex(n; C_k) \in \Omega(n^{1+1/(k-1)})$. (Bondy-Simonovits) (Note: it is not hard to show $ex(n; C_k) \in O(n^{1+2/k})$ by considering the average degree.)

8.5.9. Determine the smallest simple graph that is not balanced.

8.5.10. (!) Reword the second moment argument in the proof of Theorem 8.5.20 to complete the proof of the threshold for appearance of arbitrary graphs.

8.5.11. *Dominating set algorithm.* Suppose $\delta(G) = k$, and let $N[v] = N(v) \cup \{v\}$ be the set of vertices dominated by v.

a) Given $S \subseteq V(G)$, let $U = V(G) - \cup_{v \in S} N[v]$. Prove that some vertex of $G - S$ dominates at least $|U|(k + 1)/n$ vertices in U.

b) Suppose S is constructed by iteratively adding a vertex with the maximum number of neighbors undominated by vertices already chosen. Prove that at most $n/(k + 1)$ vertices remain undominated after $n\ln(k + 1)/(k + 1)$ steps, so adding them yields a dominating set of size at most $n \frac{1 + \ln(k + 1)}{k + 1}$.

8.5.12. *Common neighbors and non-neighbors.*

a) Fix k, s, t, p. Prove that almost every G^p has the property that for every choice of disjoint vertex sets S, T of sizes s, t, there are at least k vertices adjacent to every vertex of S and to no vertex of T. (Blass-Harary [1979])

b) Conclude that almost every G^p is k-connected.

c) Apply the same argument to random tournaments: almost all have the property that for every choice of disjoint vertex sets S, T of sizes s, t, there are at least k vertices with edges to every vertex of S and from every vertex of T.

8.5.13. (!) A random labeled tournament can be generated by orienting each edge $v_i v_j$ as $v_i \to v_j$ or $v_j \to v_i$ independently with probability 1/2.

a) Prove that almost every tournament is strongly connected.

b) By Proposition 1.4.12, every tournament contains a king (a vertex having a path of length at most 2 to every other vertex). Is it true that in almost every tournament every vertex is a king? (Palmer [1985, p17])

8.5.14. Find a candidate for a sharp threshold function for the property of having diameter at most 2. I.e., find a probability function with a parameter x such that the expected number of pairs of vertices not having a common neighbor approaches a constant (in terms of x).

8.5.15. Find a threshold probability function for the property that at least half the possible edges of a graph are present. How sharp is the threshold?

8.5.16. *Threshold for complete matching in a random bipartite graph.* Let G be a random labeled subgraph of $K_{n,n}$, with partite sets A, B and independent edge probability $p = (1 + \varepsilon)\ln n/n$, where ε is a nonzero constant. Call S a *violated set* if $|N(S)| < |S|$.

a) Prove that almost every G has no complete matching if $\varepsilon < 0$.

b) Suppose that S is a minimal violated set. Prove that $|N(S)| = |S| - 1$ and $G[S \cup N(S)]$ is connected.

c) Suppose that G has no complete matching. Prove that A or B contains a violated set with at most $\lceil n/2 \rceil$ elements.

d) Prove that almost every G has a complete matching if $\varepsilon > 0$.

8.5.17. (!) With $p = (1 - \varepsilon) \log n/n$, how large can m be such that almost every graph has at least m isolated vertices? (Hint: use Chebyshev's Inequality.)

8.5.18. Given a graph G, say that a k-set S is *bad* if G has no vertex v such that $S \subseteq N(v)$. For fixed p, how large can k be so that almost every G^p has no bad k-set. How slowly can k grow so that almost every G^p has a bad k-set?

8.5.19. (!) For $p = 1/n$ and fixed $\varepsilon > 0$, show that almost every G^p has no component with more than $(1 + \varepsilon)n/2$ vertices. (Hint: Instead of trying to bound the probability of this property directly, bound it by the probability of another property, which tends to 0.)

8.5.20. (!) By examining common neighbors, prove that if p is fixed and $k = k(n) \in o(n/\log n)$, then almost every G^p is k-connected.

8.5.21. Prove that if $k = \lg n - (2 + \varepsilon)\lg\lg n$, then almost every n-vertex tournament has the property that every set of k vertices has a common successor.

8.5.22. *The Coupon Collector.* Suppose f is a random function from $[m]$ to $[n]$. If all such functions are equally likely, find a threshold function $m = m(n)$ for the property that f is surjective. Also find a threshold function for the property that at least two members of $[m]$ map to each member of $[n]$. More generally, what is a threshold function on the length of a random n-ary sequence for the property of having $k + 1$ copies of each value. (Karp)

8.5.23. Prove that the length of the longest constant sequence in a random 0,1-sequence of length n is $(1 + o(1))\lg n$.

8.5.24. Compute explicitly the probability that the Hamiltonian closure of a random graph on 5 vertices is complete.

8.5.25. The *boxicity* of G is the minimum t such that G is the intersection of t interval graphs; equivalently, it is the minimum t such that G is the intersection graph of rectilinear boxes in t dimensions. The *interval number* of G is the minimum t such that G is the intersection graph of sets on the real line composed of at most t intervals each. Prove that the interval number and boxicity of the random graph (i.e., $p = 1/2$) are each at least $(1 + o(1))n/(4\lg n)$. (Hint: otherwise, the number of possible representations is a vanishing fraction of the total number of graphs.)

8.5.26. *Tail inequality for binomial distribution.* Let $X = \Sigma X_i'$, where each X_i' is an indicator variable with success probability $P(X_i' = 1) = .5$, so $E(X) = n/2$. Applying Markov's Inequality (Lemma 8.5.14) to the random variable $Z = (X - E(X))^2$ yields $P(|Z| \geq t) \leq Var(X)/t^2$. Setting $t = \alpha\sqrt{n}$ yields a bound on the tail probability: $P(|X - np| \geq \alpha\sqrt{n}) \leq 1/(2\alpha^2)$.

Let $Y_i' = X_i' - .5$. Let F_i be the knowledge of $Y_1', \ldots, Y_i'$, and let $Y_i = E(Y \mid F_i)$. Use Azuma's Inequality to prove the stronger bound that $P(X > np + \alpha\sqrt{n}) < e^{-2\alpha^2}$.

8.5.27. *Bin-packing.* Suppose the numbers $S = \{a_1, \ldots, a_n\}$ are drawn uniformly and independently from the interval $[0, 1]$. The numbers must be placed in bins, each having capacity 1. Let X be the number of bins needed. Use Lemma 8.5.32 to prove that $P(|X - E(X)| \geq \lambda\sqrt{n}) \leq 2e^{-\lambda^2/2}$.

8.5.28. (+) *Azuma's Inequality and the Traveling Salesman Problem.*

a) Prove the generalization of Azuma's Inequality to general martingales: If $E(X_i) = X_{i-1}$ and $|X_i - X_{i-1}| \leq c_i$, then $\mathrm{Prob}(X_n - X_0 \geq \lambda\sqrt{\Sigma c_i^2}) \leq e^{-\lambda^2/2}$.

b) Let Y be the distance from a given point z in the unit square to the nearest of n points chosen uniformly and independently in the unit square. Prove that $E(Y) < c/\sqrt{n}$, for some constant c. (Hint: for a nonnegative continuous random variable Y, $E(Y) = \int_0^\infty P(Y \geq y)dy$, which can be verified using integration by parts. In order to bound this integral, use (somewhere) the inequality $1 - a < e^{-a}$ and the definite integral $\int_0^\infty e^{-t^2}dt = \sqrt{\pi}/2$.)

c) Apply parts (a) and (b) to prove that the smallest length of a cycle through a random set of n points in the unit square is highly concentrated around its expectation. In particular, the probability that this deviates from the expected length by more than $\lambda c\sqrt{\ln n}$ is bounded by $2e^{-\lambda^2/2}$, for some appropriate c. (Hint: for the martingale in which X_i is the expected length of the tour when the first i points are known, prove that $|X_i - X_{i-1}| < c(n - i)^{-1/2}$.)

8.6 Eigenvalues of Graphs

Techniques from group theory and linear algebra help in solving many structural and enumerative problems in graph theory. Relevant aspects of linear algebra include both vector spaces and determinants. Associated with a graph is its *cycle space*, which is the nullspace of its incidence matrix (over the field $\{0, 1\}$); the *bond space* is the orthogonal complement of the cycle space. This interaction is closely related to the duality between cycles and bonds in Section 8.2 and to the use of determinants in the Matrix Tree Theorem of Section 2.2. We also used invertibility of matrices in the counting arguments used to characterize partitionable graphs in Section 8.1.

Although they do not appear in this book, there are also many interactions between groups and graphs. The automorphisms of a graph form a group of permutations of its vertices. Group-theoretic ideas lead to algorithms for testing isomorphism and to constructions for embedding graphs on surfaces. Conversely, every group can be modeled using graphs. An introduction to the interplay between groups and graphs appears in White [1973]. Groups arise also in counting isomorphism classes of graphs; Harary-Palmer [1973] discusses many topics in graph enumeration.

We confine our algebraic efforts in this section to the study of eigenvalues of adjacency matrices. We interpret the coefficients of the characteristic polynomial in terms of the graph, derive relationships between the eigenvalues and other graph parameters, and characterize the eigenvalues of bipartite graphs and of regular graphs. We close with applications to two problems, one of great practical importance and the other perhaps a frivolous curiosity. The first relates eigenvalues to "expansion" properties; in many applications, we seek regular graphs of small degree such that every set of vertices has many neighbors relative to its size. The second application is the "Friendship Theorem": if S is a set of people such that every two people in S have exactly one common friend in S, then there is some person in S who is everyone's friend. Further material on eigenvalues of graphs and other aspects of algebraic graph theory appears in Biggs [1993]. Cvetkovic-Doob-Sachs [1979] focuses entirely on graph eigenvalues.

THE CHARACTERISTIC POLYNOMIAL

8.6.1. Definition. The *eigenvalues* of a matrix A are the numbers λ such that the equation $Ax = \lambda x$ has a non-zero solution vector, in which case the solution vector is the corresponding *eigenvector*. The *eigenvalues* of a graph are the eigenvalues of its adjacency matrix A. These are the roots $\lambda_1, \ldots, \lambda_n$ of the *characteristic polynomial* $\phi(G; \lambda) = \det(\lambda I - A) = \Pi(\lambda - \lambda_i)$. The *spectrum* is the list of distinct eigenvalues with their respective multiplicities $m_1, \ldots, m_t$; we write $\text{Spec}(G) = \binom{\lambda_1 \cdots \lambda_t}{m_1 \cdots m_t}$.

8.6.2. Remark. *Elementary properties of eigenvalues.*

0) The eigenvalues are the values λ for which the square matrix $\lambda I - A$ is singular, which is equivalent to $\det(\lambda I - A) = 0$.

1) $\Sigma \lambda_i = \text{Trace} A$. This is the sum of the diagonal elements of A, and it is the coefficient of λ^{n-1} in $\Pi(\lambda - \lambda_i)$. For simple graphs, it is 0.

2) $\Pi \lambda_i = \det A = (-1)^n \phi(G; 0)$.

3) The total muliplicity of non-zero eigenvalues is the rank of A.

4) Adding a constant c to the diagonal shifts the eigenvalues by c, since $\alpha + c$ is a root of $\det(\lambda I - (cI + A))$ if and only if α is a root of $\det(\lambda I - A)$.

8.6.3. Example. *Spectra of complete bipartite graphs and cliques.* The adjacency matrix of $K_{m,n}$ has rank 2, so it has two non-zero eigenvalues λ_1, λ_2. The trace is 0, so $\lambda_1 = -\lambda_2 = c$. Hence $\phi(K_{m,n}, \lambda) = \lambda^2 - c^2 \lambda^{n-2}$. We compute c using $\phi(G; \lambda) = \det(\lambda I - A)$. Since λ appears only on the diagonal, the contributions to the coefficient of λ^{n-2} arise from permuations that select $n - 2$ positions from the diagonal. The remaining two positions must be $-a_{ij}$ and $-a_{ji}$ for some i, j. There are mn nonzero contributions of this form, and all are negative since one row interchange puts these positions on the diagonal. Hence $c^2 = mn$, and $\text{Spec}(K_{m,n}) = \begin{pmatrix} \sqrt{mn} & 0 & -\sqrt{mn} \\ 1 & m+n-2 & 1 \end{pmatrix}$.

For K_n, the adjacency matrix is $J - I$, where J is the matrix of all 1's. Hence the eigenvalues of K_n are shifted down one from those of J. Since $\text{Spec}J = \begin{pmatrix} n & 0 \\ 1 & n-1 \end{pmatrix}$, we have $\text{Spec}K_n = \begin{pmatrix} n-1 & -1 \\ 1 & n-1 \end{pmatrix}$. $\square$

We index the coefficients of the characteristic polynomial so that $\phi(G; \lambda) = \Sigma_{i=0}^{n-1} c_i \lambda^{n-i}$. Since $\phi(G; \lambda) = \det(\lambda I - A)$, we always have $c_0 = 1$ and $c_1 = -\text{Trace}A = 0$. Our computation of c_2 for $K_{m,n}$ extends to all graphs. A *principal submatrix* of a square matrix A is a submatrix selecting the same rows as columns. Since contributions to $c_2 \lambda^{n-2}$ involve $n-2$ factors of λ from the diagonal, the coefficient c_2 is the sum of all principal 2×2 subdeterminants of $-A$. For a simple graph, the off-diagonal elements are both -1 when $v_i \leftrightarrow v_j$ and both 0 otherwise. As before, the permutation selecting these positions is odd. Summing over all pairs of vertices, we have $c_2 = -e(G)$.

Similarly, c_3 is the sum of the principal 3×3 subdeterminants of $-A$. For each triple $v_i, v_j, v_k \in V(G)$, this determinant depends only on the number of edges among these three vertices. The subdeterminant is 0 unless the three vertices form a triangle, in which case it is -2. Hence c_3 is -2 times the number of 3-cycles in G. Since principal submatrices are the adjacency matrices of induced subgraphs, we can write the computation for c_i in general as $c_i = (-1)^i \Sigma_{|S|=i} \det A(G[S])$. Harary generalized these observations about c_2 and c_3 to describe the determinant of each adjacency matrix in terms of subgraphs of the graph.

8.6.4. Theorem. (Harary [1962b]). Let **H** be the collection of spanning subgraphs of a simple graph G such that every component is an edge or a cycle. If $k(H)$ and $c(H)$ denote the number of components of H and the number of components that are cycles, respectively, then $\det A(G) = \Sigma_{H \in \mathbf{H}} (-1)^{n(H)-k(H)} 2^{c(H)}$.

Proof. The general determinant formula is $\det A = \Sigma_{\sigma \in S_n} (-1)^{t(\sigma)} \Pi a_{i,\sigma(i)}$, where $t(\sigma)$ is the number of row interchanges (transpositions) need to put the positions $i, \sigma(i)$ on the diagonal. When A is a 0,1-matrix, nonzero contributions are products of n independent 1's. We can view such a σ as a vertex permutation mapping each i to $\sigma(i)$. The permutation σ partitions the vertices into orbits. Since $a_{i,\sigma(i)} = 1$ means $v_i \leftrightarrow v_{\sigma(i)}$, there are no orbits of size 1, orbits of size 2 correspond to edges, and longer orbits correspond to cycles. Thus the permutation makes a nonzero contribution when it describes a spanning subgraph H of G in which the components are edges and cycles.

The sign of the contribution is determined by the number of trans-positions needed to move the entries to the diagonal. Row exchanges move one element of an orbit at a time to the diagonal, but the last switch moves the last two elements to the diagonal. Hence $t(\sigma) = n(H) - k(H)$. Finally, each cycle of length at least 3 in H can appear in one of two ways in the permutation matrix, since we can follow the cycle in one of two directions. Hence the number of permutations that give rise to H is $2^{s(H)}$. $\square$

8.6.5. Corollary. (Sachs [1967]). Let $\mathbf{H}_i$ denote the collection of i-vertex subgraphs of G whose components are edges or cycles. If $\phi(G; \lambda) = \Sigma c_i \lambda^{n-i}$ is the characteristic polynomial of G, then $c_i = \Sigma_{H \in \mathbf{H}_i} (-1)^{k(H)} 2^{s(H)}$.

Proof. This follows from Theorem 8.6.4 and the earlier observation that $c_i = (-1)^i \Sigma_{|S| = i} \det A(G[S])$. $\square$

This formula leads to a recursive expression for the characteristic polynomial (Exercise 2). The formula can be used to construct non-isomorphic trees with the same characteristic polynomial (and only eight vertices). Schwenk [1973] proved that, as $n \to \infty$, the fraction of trees that are uniquely determined by their spectrum goes to 0. Nevertheless, eigenvalues can sometimes be used to determine when a graph belongs to a particular class. We will do this for bipartite graphs and for regular graphs.

8.6.6. Proposition. The (i, j)th entry of A^k counts the v_i, v_j-walks of length k. The eigenvalues of A^k are $(\lambda_i)^k$.

Proof. The statement about walks holds easily by induction on k (Exercise 1.2.19). For the second statement, $Ax = \lambda x$ implies $A^k x = \lambda^k x$, by repeated multiplication. Using an arbitrary eigenvector x ensures that the multiplicities of the eigenvalues don't change. $\square$

8.6.7. Lemma. If G is bipartite and λ is an eigenvalue of G with multiplicity m, then $-\lambda$ is also an eigenvalue with multiplicity m.

Proof. Adding isolated vertices to give the partite sets equal size merely adds rows and columns of 0's to the adjacency matrix, which does not change the rank and hence changes the spectrum only by including one extra 0 for each vertex added. Hence we may assume that the partite sets have equal sizes.

Since G is bipartite, we can permute the rows and columns of A to obtain the form $A = \begin{pmatrix} 0 & B \\ B^T & 0 \end{pmatrix}$, where B is square. Let λ be an eigenvalue for eigenvector $v = \begin{pmatrix} x \\ y \end{pmatrix}$, partitioned according to the bipartition of G. Then $\lambda v = Av = \begin{pmatrix} 0 & B \\ B^T & 0 \end{pmatrix}\begin{pmatrix} x \\ y \end{pmatrix} = \begin{pmatrix} By \\ B^T x \end{pmatrix}$. Hence $By = \lambda x$ and $B^T x = \lambda y$.

Let $v' = \begin{pmatrix} x \\ -y \end{pmatrix}$. We compute $Av' = \begin{pmatrix} B(-y) \\ B^T x \end{pmatrix} = \begin{pmatrix} -\lambda x \\ \lambda y \end{pmatrix} = -\lambda v'$. Hence v' is an eigenvector of A for the eigenvalue $-\lambda$. Furthermore, given a set of independent eigenvectors for λ, the set of eigenvectors for $-\lambda$ obtained in this way are also independent. Hence $-\lambda$ is an eigenvector of A with the same multiplicity as λ. $\square$

8.6.8. Theorem. The following are equivalent statements about a graph G.
A) G is bipartite.
B) The eigenvalues of G occur in pairs λ_i, λ_j such that $\lambda_i = -\lambda_j$.
C) $\phi(G; \lambda)$ is a polynomial in λ^2.
D) $\Sigma_{i=1}^{n} \lambda_i^{2t-1} = 0$ for any positive integer t.

Proof. We proved A $\Rightarrow$ B in the lemma.

B $\Leftrightarrow$ C: $(\lambda - \lambda_i)(\lambda - \lambda_j) = (\lambda^2 - a)$ if and only if $\lambda_j = -\lambda_i$. Hence the roots occur in such pairs if and only if $\phi(G; \lambda)$ can be factored into linear factors in λ^2.

B $\Rightarrow$ D: If $\lambda_j = -\lambda_i$, then $\lambda_j^{2t-1} = -\lambda_i^{2t-1}$.

D $\Rightarrow$ A: Because $\Sigma \lambda_i^k$ counts the closed k-walks in the graph (from each starting vertex), condition D forbids closed walks of odd length. This forbids odd cycles, since an odd cycle is an odd closed walk, and hence G is bipartite. $\square$

LINEAR ALGEBRA OF REAL SYMMETRIC MATRICES

Relating the eigenvalues to other graph parameters requires several results from linear algebra, including the Spectral Theorem and Cayley-Hamilton Theorem for real symmetric matrices. These are usually stated in more generality, but our focus is adjacency matrices. For real symmetric matrices, the theorems have shorter proofs. We begin with a lemma that follows *from* the Spectral Theorem when the latter is proved using complex matrices. The proofs of these results may be skipped, especially by readers well-versed in linear algebra.

8.6.9. Lemma. If $f(x) = x^T A x$, where A is a real symmetric matrix, then f attains its maximum and minimum over unit vectors x at eigenvectors of A, where it equals the corresponding eigenvalues.

Proof. The function f is continuous in $x_1, \ldots, x_n$. For constrained optimization, we use Lagrangian multipliers. Given the constraint $x^T x = 1$, we let $g(x) = x^T x - 1$. Forming $L(x, \lambda) = f(x) - \lambda g(x)$, the extreme values occur where all partial derivatives of L are 0. The derivative with respect to λ is 0 when $x^T x = 1$. Let ∇ denote the vector of partial derivatives with respect to the variables $x_1, \ldots, x_n$. We compute $\nabla L(x, \lambda) =$

$\nabla f(x) - \lambda \nabla g(x) = 2Ax - 2\lambda x$. The statement $\nabla f(x) = 2Ax$ uses the symmetry of A. We have $\nabla L = 0$ precisely when $Ax = \lambda x$, which requires x to be an eigenvector of A for eigenvalue λ. In this situation, $f(x) = x^T A x = \lambda x^T x = \lambda$. $\square$

Since our variables in the optimization are real, we have found at least one real eigenvector and eigenvalue. We can use this inductively to show that all eigenvectors have this property.

8.6.10. Theorem. (The Spectral Theorem). A real symmetric $n \times n$ matrix has real eigenvalues and n orthonormal eigenvectors.

Proof. We use induction on n; the claim holds easily for $n = 1$. Suppose $n > 1$, and let v_n be the eigenvector maximizing $x^T A x$. Let W be the orthogonal complement of the space spanned by v_n; it has dimension $n - 1$. If $w \in W$, we have $v_n^T A w = w^T A v_n = \lambda_n w^T v_n = 0$. Hence $Aw \in W$. Viewing multiplication by A as a mapping f_A, we have $f_A \colon W \to W$. Let S be a matrix whose columns are the vectors of an orthonormal basis of $\mathbb{R}^n$ with v_n being the last vector. Since the basis is orthonormal, $S^{-1} = S^T$. The matrix for f_A with respect to this basis is $S^T A S$. Since the basis is orthonormal and v_n is an eigenvector, the last column of this matrix is 0, except for λ_n in the last position. Furthermore, the matrix is symmetric. Hence we can take the upper $n - 1$ by $n - 1$ symmetric matrix as the matrix for f_A on W with respect to this basis. By induction, it has an orthonormal basis of eigenvectors $v_1, \ldots, v_{n-1}$ in W, with real eigenvalues. Using S, we retrieve the eigenvectors of the original matrix A; all real, orthonormal, and with the same real eigenvalues. $\square$

Next we consider polynomial fuctions of the adjacency matrix. Viewed as members of $\mathbb{R}^{n^2}$, the matrices $I, A, A^2, \ldots, A^{n^2}$ cannot be independent, since there are $n^2 + 1$ of them. Using the coefficients from an equation of linear dependence, we obtain a polynomial p such that $p(A)$ is the zero matrix. We do not need many of these n^2 powers; the characteristic polynomial itself, with degree n, suffices. This holds for all matrices, but again we consider only real symmetric matrices.

8.6.11. Theorem. (The Cayley-Hamilton Theorem). If $\phi(\lambda)$ is the characteristic polynomial of a real symmetric matrix A, then $\phi(A)$ is the zero matrix (A "satisfies" its own characteristic polynomial).

Proof. Suppose the eigenvalues of A are $\lambda_1, \ldots, \lambda_n$, so that $\phi(\lambda) = \Pi(\lambda - \lambda_i)$. Since the powers of A commute, the matrix polynomial obtaining by using A for λ also factors as $\phi(A) = \Pi(A - \lambda I)$. To prove $\phi(A) = 0$, we need only show that the matrix $\phi(A)$ maps every vector to 0. Write an arbitrary vector x as a linear combination of the basis of eigenvectors guaranteed by the Spectral Theorem. Applying $A - \lambda_i I$ kills the coefficient of v_i. Successively multiplying by all the factors $A - \lambda_i I$ produces the zero vector. $\square$

The polynomial of minimum degree satisfied by A, scaled to have leading coefficient 1, is the *minimum polynomial* ψ of the matrix. If A is the adjacency matrix of G, we also call this the minimum polynomial $\psi(G; \lambda)$ of G. The minimum polynomial is unique: if there are two monic polynomials of the same degree satisfied by A, then their difference is a polynomial of lower degree satisfied by A.

8.6.12. Theorem. The minimum polynomial of A is $\psi(A) = \Pi(\lambda - \lambda_i)$, where $\{\lambda_i\}$ are the distinct eigenvalues of A.

Proof. The minimum polynomial must divide every polynomial satified by A, since otherwise the remainder would be a polynomial of lower degree satisfied by A. With the Cayley-Hamilton Theorem, we conclude that ψ divides ϕ and must be the product of some of its factors. In order to kill the vectors in the subspace of eigenvectors for eigenvalue λ_i, we must have a factor of the form $A - \lambda_i I$. Since this factor kills all vectors in that subspace, we don't need more than one copy of this factor. $\square$

8.6.13. Lemma. (Sylvester's Law of Inertia) Let A be a real symmetric matrix. If $x^T A x$ can be written as a sum of N products involving two linear factors, that is $x^T A x = \Sigma_{m=1}^N (\Sigma_{i \in S_m} a_{i,m} x_i)(\Sigma_{j \in T_m} b_{j,m} x_j)$, then N is at least the maximum of the number of positive eigenvalues and number of negative eigenvalues of A.

Proof. (Tverberg [1982]) Write the linear expressions as $u_m(x)$ and $v_m(x)$. For each m, we have $u_m v_m = L_m^2(x) - M_m^2(x)$, where the expressions $L = \frac{1}{2}(u + v)$ and $M = \frac{1}{2}(u - v)$ are also linear combinations of the variables $x_1, \ldots, x_n$. This writes the quadratic form as $x^T A x = \Sigma_{m=1}^N L_m^2(x) - M_m^2(x)$.

On the other hand, A is a real symmetric matrix and thus has a full set of orthonormal eigenvectors $\{w^i\}$. Using this, we write $x^T A x = x^T S \Lambda S^T x$, where Λ is the diagonal matrix of eigenvalues $\lambda_1 \geq \cdots \geq \lambda_n$ and S has columns $w^1, \ldots, w^n$. If S has p positive and q negative eigenvalues, then this becomes $x^T A x = \Sigma_{i=1}^p (y^i \cdot x)^2 - \Sigma_{i=p+1}^{p+q} (z^i \cdot x)^2$, where each y^i or z^i is $|\lambda_i|^{1/2} w^i$.

Now we consider a homogeneous system of linear equations. We require $L_m(x) = 0$ for $1 \leq m \leq N$, also $z^i \cdot x = 0$ for $p < i \leq p + q$, and $w^i \cdot x = 0$ for $p + q < i \leq n$. This places $N + n - p$ homogeneous linear constraints on n variables. If $N < p$, then these equations have a nonzero simultaneous solution x'. Setting x to x' in the two expressions for $x^T A x$ yields $\Sigma_{i=1}^p (y^i \cdot x')^2 = -\Sigma_{m=1}^N M_m^2(x')$. Since x' is orthogonal to all eigenvectors with non-positive eigenvalues, the left side is positive, while the right is non-positive. The contradiction implies $N \geq p$; an analogous argument yields $N \geq q$. $\square$

EIGENVALUES AND GRAPH PARAMETERS

Eigenvalues provide bounds on various parameters, or alternatively graph parameters yield bounds on the eigenvalues. Our first result uses only the minimum polynomial.

8.6.14. Theorem. The diameter of a graph is less than its number of distinct eigenvalues.

Proof. The matrix A satisfies a polynomial of degree r if and only if there is a linear combination of $A^0, \ldots, A^r$ that is 0. Since the number of distinct eigenvalues equals the degree of the minimum polynomial, we need only show that $A^0, \ldots, A^k$ are linearly independent when $k \leq \operatorname{diam}(G)$. This we do by showing for each such k that A^k does not belong to the space spanned by $A^0, \ldots, A^{k-1}$. Choose $v_i, v_j \in V(G)$ such that $d(v_i, v_j) = k$. By counting walks, we have $A_{ij}^k \neq 0$ but $A_{ij}^t = 0$ for $t < k$. Therefore, A^k cannot be expressed as a linear combination of the smaller powers. $\square$

Since the Spectral Theorem guarantees real eigenvalues, we can index our eigenvalues as $\lambda_1 \geq \cdots \geq \lambda_n$. We also refer to λ_1 and λ_n as $\lambda_{\max}(G)$ and $\lambda_{\min}(G)$.

8.6.15. Lemma. If G' is an induced subgraph of G, then

$$\lambda_{\min}(G) \leq \lambda_{\min}(G') \leq \lambda_{\max}(G') \leq \lambda_{\max}(G) .$$

Proof. Since A is a real symmetric matrix, Lemma 8.6.9 yields $\lambda_{\min}(A) \leq x^T A x \leq \lambda_{\max}(A)$ for every unit vector x. Consider the adjacency matrix A' of G'. By permuting the vertices of G, we can view A' as an upper left principal submatrix of $A = A(G)$. Let z' be the unit eigenvector of A' such that $A'z' = \lambda_{\max}(G')z'$. Let z be the unit vector in R_n obtained by appending zeros to z'. Then $\lambda_{\max}(G') = z'^T A' z' = z^T A z \leq \lambda_{\max}(G)$. Similarly, $\lambda_{\min}(G') \geq \lambda_{\min}(G)$. $\square$

The behavior of the extreme eigenvalues under vertex deletion is a special case of the "Interlacing Theorem": If the eigenvalues of G are $\lambda_1 \geq \cdots \geq \lambda_n$ and the eigenvalues of $G - x$ are $\mu_1 \geq \cdots \geq \mu_{n-1}$, then these interlace as $\lambda_1 \geq \mu_1 \geq \lambda_2 \geq \cdots \geq \mu_{n-1} \geq \lambda_n$. We will not need this theorem and hence omit the proof, which involves only linear algebra.

8.6.16. Lemma. For every graph G, $\delta(G) \le \lambda_{\max}(G) \le \Delta(G)$.

Proof. Let x be an eigenvector for eigenvalue λ, and let $x_j = \max_i x_i$ be the largest coordinate value in x. Then $\lambda \le \Delta(G)$ follows from

$$\lambda x_j = (Ax)_j = \Sigma_{v_i \in N(v_j)} x_i \le d(v_j) x_j \le \Delta(G) x_j .$$

For the lower bound, we apply Lemma 8.6.9 to the unit vector with equal coordinates. Since the sum of the entries in the adjacency matrix is twice the number of edges of G, we have

$$\lambda_{\max} \ge \frac{\mathbf{1}_n^T}{\sqrt{n}} A \frac{\mathbf{1}_n}{\sqrt{n}} = \frac{1}{n} \Sigma\Sigma a_{ij} = \frac{2e(G)}{n} ,$$

which shows that in fact $\lambda_{\max}$ is at least the average degree. $\square$

Lemma 8.6.16 enables us to improve the trivial bound $1 + \Delta(G)$ on $\chi(G)$ that results from the greedy coloring algorithm. It is not possible to reduce $\Delta(G)$ to the average degree; $K_n + K_1$ has chromatic number n and average degree less than $n - 1$. Since $\lambda_{\max}$ is always at least the average degree, $1 + \lambda_{\max}(G)$ has a chance to work and can't be much improved.

8.6.17. Theorem. (Wilf [1967]). For every graph G, $\chi(G) \le 1 + \lambda_{\max}(G)$.

Proof. If $\chi(G) = k$, then we can successively delete vertices until we obtain a vertex k-critical subgraph H: $\chi(H - v) = k - 1$ for all $v \in V(H)$. As observed in Lemma 5.1.13, $\delta(H) \ge k - 1$. Since H is an induced subgraph of G, Lemma 8.6.16 and then Lemma 8.6.15 yield

$$k \le 1 + \delta(H) \le 1 + \lambda_{\max}(H) \le 1 + \lambda_{\max}(G) . \quad \square$$

From Sylvester's Law of Inertia, we can obtain a lower bound on the number of complete bipartite graphs (*bicliques*) needed to partition the edges of a graph. Because stars are bicliques and every subgraph of a star is a star, the number of bicliques needed is at most the vertex cover number $\beta(G) = n(G) - \alpha(G)$. Erdős conjectured that equality almost always holds, but this remains open. Graphs with special structure may have efficient partitions using non-stars. There is a general lower bound using the eigenvalues of the graph. This lower bound appears explicitly in (Reznick-Tiwari-West [1985]), but it is implicit in the work of earlier authors (Tverberg [1982], Peck [1984]), who used it for the complete graph.

8.6.18. Theorem. For any simple graph G, the number of bicliques needed to partition $E(G)$ is at least $\max\{p, q\}$, where p is the number of positive eigenvalues and q is the number of negative eigenvalues of the adjacency matrix $A(G)$.

Proof. When G decomposes into subgraphs $G_1, \ldots, G_t$, we may write $A(G) = \Sigma_{i=1}^{t} B_i$, where B_i is the adjacency matrix of the spanning subgraph of G with edge set $E(G_i)$. If G_i is the complete bipartite subgraph with bipartition S, T, we have $x^T B_i x = 2\Sigma_{j \in S} x_j \Sigma_{k \in T} x_k$. Writing these linear expressions as $u = \Sigma_{j \in S} x_j$ and $v = \Sigma_{k \in T} x_k$, we have $x^T B_i x = 2uv = L^2(x) - M^2(x)$, where $L = \frac{1}{\sqrt{2}}(u + v)$ and $M = \frac{1}{\sqrt{2}}(u - v)$ are linear combinations of the variables $x_1, \ldots, x_n$. If G decomposes into t bicliques, then applying this computation to each biclique yields $x^T A(G)x = \Sigma_{i=1}^{t} L_i^2(x) - M_i^2(x)$. The claim now follows from Sylvester's Law of Inertia, which says that $t \geq \max\{p, q\}$. $\square$

8.6.19. Example. *Biclique decomposition of $C_{(2t+1)n} \square C_n$.* There are simple formulas for the eigenvalues of a cycle (Exercise 3) and for computing the eigenvalues of a Cartesian product from the eigenvalues of the factors (Exercise 7). These yield simple formulas for the number of positive and negative eigenvalues of $C_m \square C_n$ when m is an odd multiple of n. In particular, $C_{(2t+1)n} \square C_n$ has $(2t + 1)(n^2 + 1)/2$ positive eigenvalues and $(2t + 1)(n^2 - 1)/2$ negative eigenvalues if n is odd (no 0 eigenvalues). Furthermore, such a product decomposes into $(2t + 1)(n^2 + 1)/2$ complete bipartite subgraphs, consisting of $(2t + 1)(n - 1)/2$ 4-cycles and $(2t + 1)(n + 1)/2$ stars. Indeed, 4-cycles and stars are the only bicliquesubgraphs of $C_m \square C_n$. The optimal decomposition of $C_{15} \square C_5$ appears below. Edges wrap around from top to bottom and right to left, and all grid points indicate vertices. The heavy dots indicate vertices that are centers of stars in the decomposition, and the circles indicate 4-cycles in the decomposition (those "wrapping around" appear twice).

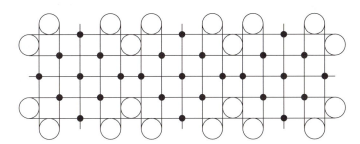

EIGENVALUES OF REGULAR GRAPHS

Like bipartite graphs, regular graphs can be characterized using spectra. The n-vector $\mathbf{1}_n$ with all coordinates 1 plays a special role in this and many other arguments involving eigenvalues, as does the matrix J of all 1's.

8.6.20. Theorem. The eigenvalue of G having largest absolute value equals $\Delta(G)$ if and only if some component of G containing a vertex of maximum degree is regular. In this case, its multiplicity equals the number of $\Delta(G)$-regular components.

Proof. If G is k-regular with adjacency matrix A, then the ith entry of $A\mathbf{1}_n$ is $d(v_i) = k$; hence $A\mathbf{1}_n = k\mathbf{1}_n$, and k is an eigenvalue with eigenvector $\mathbf{1}_n$. In general, suppose x is an eigenvector x for eigenvalue λ, and let x_j be a coordinate of largest absolute value among coordinates corresponding to the vertices of some component H of G. For the jth coordinate of Ax, we have

$$|\lambda|\,|x_j| = |(Ax)_j| = |\Sigma_{v_i \in N(v_j)} x_i| \le d(v_j)|x_j| \le \Delta(G)|x_j| \,.$$

Hence $|\lambda| \le \Delta(G)$. Equality requires $d(v_j) = \Delta(G)$ and $x_i = x_j$ for all $v_i \in N(v_j)$. We can iterate this argument to reach all coordinates for vertices in H. Hence the eigenvalue associated with x has absolute value as large as $\Delta(G)$ only if H is $\Delta(G)$-regular. Looking at all the components of G, we find that the eigenvalue associated with x has absolute value as large as $\Delta(G)$ if and only if each component of H where x has a non-zero coordinate is $\Delta(G)$-regular, and x is constant on each such coordinate. We can choose the constant independently for each $\Delta(G)$-regular component, so the dimension of the space of eigenvectors for eigenvalue $\Delta(G)$ is the number of $\Delta(G)$-regular components. □

When G is connected but possibly not regular, it remains true that an eigenvalue of largest absolute value has multiplicity 1 and that the coordinates of the associated eigenvector have the same sign. This is closely related to the Perron-Frobenius Theorem of linear algebra and uses arguments like those above; we omit the proof.

Under matrix multiplication, the matrices arising as polynomials in $A(G)$ form the *adjacency algebra* of G. Several interesting classes of graphs are characterized by the presence of particular matrices in the adjacency algebra. The connected regular graphs are characterized by the presence of the matrix J.

8.6.21. Theorem. (Hoffman [1963]). A graph G is regular and connected if and only if J is a linear combination of powers of $A(G)$.

Proof. *Sufficiency.* If J can be so expressed, then for each i, j we have $(A^k)_{ij} \ne 0$ for some $k \ge 0$, which requires a v_i, v_j-walk of length k. Hence G is connected. For regularity, consider the matrices JA and AJ. The i, jth position of AJ is $d(v_i)$ (constant on rows), and the i, jth position of JA is $d(v_j)$ (constant on columns). If $JA = AJ$, then the i, jth position is both $d(v_i)$ and $d(v_j)$ and the graph is regular. Indeed J commutes with A, because J is a linear combination of powers of A, each of which commutes with A.

Necessity. Since G is k-regular, k is an eigenvalue, and the minimum polynomial can be written as $\psi(G; \lambda) = (\lambda - k)g(\lambda)$ for some polynomial g. Since A satisfies its minimum polynomial, setting $\lambda = A$ yields $Ag(A) = kg(A)$. Hence each column of $g(A)$ is an eigenvector of A with eigenvalue k. Since G is regular and connected, each such eigenvector is a multiple of $\mathbf{1}_n$. Hence the columns of $g(A)$ are constant. However, $g(A)$ is a linear combination of powers of a symmetric matrix and therefore must itself be symmetric. Hence each column is the same multiple of $\mathbf{1}_n$, and $g(A)$ is a multiple of J. $\square$

When G is regular, $\overline{G}$ is also regular, and the eigenvalues of $\overline{G}$ can be obtained from the eigenvalues of G. This rests on the matrix expression for complementation: $A(\overline{G}) = J - I - A(G)$.

8.6.22. Lemma. Complementary graphs G and $\overline{G}$ are related by

$$\phi(\overline{G}; \lambda) = (-1)^n \det[(-\lambda - 1)I - A(G) + J] \ .$$

Proof. Direct computation yields $\det(\lambda I - A(\overline{G})) = \det(\lambda I - (J - I - A))$ $= \det[(\lambda + 1)I - J + A] = (-1)^n \det[(-\lambda - 1)I - A + J]$. $\square$

8.6.23. Theorem. If G is k-regular, then G and $\overline{G}$ have the same eigenvectors. The eigenvalue associated with $\mathbf{1}_n$ is k in G and $n - k - 1$ in $\overline{G}$. If $x \neq \mathbf{1}_n$ is an eigenvector of G for eigenvalue λ of G, then its associated eigenvalue in $\overline{G}$ is $-1 - \lambda$.

Proof. Since $\overline{G}$ is $n - k - 1$-regular, $\mathbf{1}_n$ is an eigenvector for both G and $\overline{G}$, with eigenvalue k for G and $n - k - 1$ for $\overline{G}$. Let x be another eigenvector of G in an orthonormal basis of eigenvectors, and let $\overline{A} = A(\overline{G})$. Since $\mathbf{1}_n \cdot x = 0$, $\Sigma x_i = 0$. We compute $\overline{A}x = Jx - x - Ax = 0 - x - Ax = (-1 - \lambda)x$. $\square$

This yields a lower bound on the smallest eigenvalue of a regular graph and another derivation of the spectrum of K_n.

8.6.24. Corollary. For a k-regular graph, $\lambda_n \geq k - n$.

Proof. If G is k-regular and $\lambda_1 \geq \cdots \geq \lambda_n$, then the eigenvalues of $\overline{G}$ are $(n - k - 1, -1 - \lambda_n, \ldots, -1 - \lambda_2)$, by Theorems 8.6.20 and 8.6.23. In particular, $n - k - 1 \geq -\lambda_n - 1$. $\square$

The eigenvalues of a connected regular graph G can be used to compute $\tau(G)$, the number of spanning trees. The eigenvalues need not be rational, yet the result is an integer. The Matrix Tree Theorem (Theorem 2.2.9) says that $\tau(G)$ equals each minor of the matrix $Q = D - A$, where A is the adjacency matrix and D is the diagonal matrix of degrees. When G is k-regular, $D = kI$. Letting $\mathrm{Adj}Q$ denote the adjugate matrix of Q (the matrix of signed cofactors), the Matrix Tree Theorem is the statement that $\mathrm{Adj}Q = \tau(G)J$. Using Cayley's Formula (Theorem 2.2.3) for spanning trees in K_n, we have $\mathrm{Adj}(nI - J) = n^{n-2}J$.

8.6.25. Lemma. Let D be the diagonal matrix of vertex degrees in a graph G, let $A = A(G)$, and let $Q = D - A$. The number of spanning trees of G is $\tau(G) = \det(J + Q)/n^2$.

Proof. Observe that $J^2 = nJ$, $JQ = 0$, and $\mathrm{Adj}(AB) = \mathrm{Adj}(A)\mathrm{Adj}(B)$. We apply the product formula the matrix $J + Q$ and the matrix $nI - J$ that arises from K_n. We have

$$\mathrm{Adj}(nI - J)\mathrm{Adj}(J + Q) = \mathrm{Adj}[(nI - J)(J + Q)] = \mathrm{Adj}(nQ),$$

since $J^2 = nJ$ and $JQ = 0$. We have computed that $\mathrm{Adj}(nI - J) = n^{n-2}J$. Also, $\mathrm{Adj}(nQ) = n^{n-1}\mathrm{Adj}Q$ for any matrix Q. Canceling common factors of n yields $J\mathrm{Adj}(J + Q) = n\tau(G)J$. Multiplying both sides of this on the right by $(J + Q)^T$ yields $J(\det(J + Q)I) = n\tau(G)nJ$. Both sides are multiples of J, so the desired equality holds. $\square$

We can now compute $\tau(G)$ from eigenvalues of G if G is regular.

8.6.26. Theorem. If G is a k-regular connected graph with spectrum $\left(\begin{smallmatrix} k & \lambda_2 \cdots \lambda_t \\ 1 & m_2 \cdots m_t \end{smallmatrix}\right)$, then $\tau(G) = n^{-1}\phi'(G; k) = n^{-1}\Pi_{j=2}^{t}(k - \lambda_j)^{m_j}$.

Proof. Since $J + Q = J + kI - A$, the determinant of $J + Q$ is the value at k of the characteristic polynomial of $A - J$. Since G is k-regular and connected, it has $\mathbf{1}_n$ as an eigenvector with eigenvalue k, and the other eigenvectors are orthogonal to $\mathbf{1}_n$. Every such eigenvector of A is also an eigenvector of $A - J$, with the same eigenvalue, since $(A - J)x = Ax - Jx = Ax = \lambda x$. By the analogous computation, $\mathbf{1}_n$ is an eigenvector of $A - J$ with eigenvalue $k - n$. This gives us a full set of eigenvalues for $A - J$. Evaluating the characteristic polynomial at k yields $\det(J + Q) = n\Pi_{j=2}^{t}(k - \lambda_j)$. The product term is precisely $\phi'(G; k)$, since $\phi(G; \lambda)$ has $\lambda - k$ as a non-repeated factor when G is k-regular and connected. By Lemma 8.6.25, we obtain $\tau(G)$ upon division by n^2. $\square$

EIGENVALUES AND EXPANDERS

Many applications in computer science require "expanding" graphs. The basic notion of expansion is that all small sets should have large neighborhoods. The aim is to establish good connectivity properties without using many edges.

8.6.27. Definition. An (n, k, c)-*expander* is a bipartite graph G with partite sets X, Y of size n, such that $\Delta(G) \le k$ and that $|N(S)| \ge (1 + c(1 - |S|/n) \cdot |S|$ for every $S \subseteq X$ with $|S| \le n/2$. An (n, k, c)-*magnifier* is an n-vertex graph G (not necessarily bipartite), such that $\Delta(G) \le k$ and that $|N(S) \cap \bar{S}| \ge c \cdot |S|$ for every $S \subseteq V(G)$ with $|S| \le n/2$. An n-*superconcentrator* is an acyclic digraph with n sources and n sinks such that for every set A of sources and every set B of $|A|$ sinks, there are $|A|$ disjoint A, B-paths.

Expanders appear in the parallel sorting network of Ajtai, Komlós, and Szemerédi [1983]. The condition for expansion strengthens Hall's Condition; we have not one matching but many. This facilitates using expanders to construct superconcentrators. Applications of superconcentrators are discussed in Alon [1986a] and its references. The importance of the bound on maximum degree is that the number of edges is linear in n, thereby reducing the cost of constructing the network.

Using probabilistic methods (Exercise 14), one can prove existence of expanders (and superconcentrators) with large n and bounded average degree (Pinsker [1973]), Pippenger [1977], Chung [1978]), but this yields no construction. Using algebraic ideas, Margulis [1973] constructed an explicit example, modified in Gabber-Galil [1981].

Using the probabilistic method, one can generate an object randomly that will almost always have good expansion properties, but it is hard to measure the expansion properties from the definition. Tanner [1984] and Alon and Milman [1984, 1985] independently used eigenvalues of graphs to remedy this. They proved that graphs have good expansion properties when their two largest eigenvalues are well separated. Since eigenvalues are easy to compute (or approximate), one can generate the graph randomly and then compute its eigenvalues to see whether it has a sufficiently good guarantee of expansion.

We will consider only the simpler case of regular graphs. Expanders are more useful than magnifiers in applications, but it is easy to obtain an $(n, (k + 1), c)$-expander from an (n, k, c)-magnifier (Exercise 13). Hence we consider the relationship between eigenvalues and magnification. Our presentation follows that of Alon-Spencer [1992, p119ff], which discusses additional properties of the eigenvalues of regular (and random) graphs.

8.6.28. Theorem. If G is a k-regular graph with vertex set $[n]$ and second-largest eigenvalue λ, and $\varnothing \subset S \subset V(G)$, then $|[S, \bar{S}]| \ge (k - \lambda)|S| |\bar{S}|/n$.

Proof. Let $s = |S|$. Since G is k-regular, its largest eigenvalue is k. The statement is trivial unless $k - \lambda$ is positive, so we may assume that G is connected. We compute

$$x^T(kI - A)x = k\Sigma x_i^2 - 2\Sigma_{ij \in E(G)} x_i x_j = \Sigma_{ij \in E(G)}(x_i - x_j)^2.$$

Now set $x_i = -(n - s)$ if $i \in S$ and $x_i = s$ if $i \notin S$. The sum on the right above becomes $n^2 |[S, \bar{S}]|$.

Because $\Sigma x_i = 0$, this vector x is orthogonal to the eigenvector of A corresponding to its largest eigenvector k, which is also the eigenvector for the smallest eigenvalue of $kI - A$. Hence x is a linear combination of the other eigenvectors of $kI - A$, the smallest of which is $k - \lambda$. By the argument in Lemma 8.6.9-10, the minimum value of $x^T(kI - A)x/x^T x$ over vectors orthogonal to 1_n is $k - \lambda$. Hence

$$x^T(kI - A)x \geq (k - \lambda)x^T x = (k - \lambda)(s(n - s)^2 + (n - s)s^2) = (k - \lambda)s(n - s)n.$$

Since $x^T(kI - A)x = n^2 |[S, \bar{S}]|$, we have $|[S, \bar{S}]| \geq (k - \lambda)s(n - s)/n$. □

8.6.29. Corollary. If G is a k-regular n-vertex graph with second-largest eigenvalue λ, then G is an (n, k, c)-magnifier, where $c = (k - \lambda)/2k$.

Proof. If S is a set of $s \leq n/2$ vertices in G, then Theorem 8.6.28 yields $|[S, \bar{S}]| \geq k - \lambda)s(n - s)/n$. Each vertex of $\bar{S}$ receives at most k of these edges, so S must have at least $(k - \lambda)s(n - s)/(nk)$ neighbors in $\bar{S}$. Since $(n - s)/n \geq 1/2$, the result follows. □

The greater the separation between the two largest eigenvalues, the greater the magnification. Alon and Milman [1984] improved the lower bound on the magnification to $c \geq (2k - 2\lambda)/(3k - 2\lambda)$. Alon [1986b] proved a partial converse: If a k-regular graph G is an (n, k, c)-magnifier, then the separation $k - \lambda$ is at least $c^2/(4 + 2c^2)$.

Explicit constructions of regular graphs are known with separation between the largest eigenvalues nearly as large as possible. The second largest eigenvalue of a k-regular graph with diameter d is at least $2\sqrt{k - 1}(1 - O(1/d))$ (see Nilli [1991]). Lubotzky-Phillips-Sarnak [1986] and Margulis [1988] constructed infinite families of regular graphs for which the second largest eigenvalue is at most $2\sqrt{k - 1}$ (where k is one more than a prime congruent to 1 mod 4).

STRONGLY REGULAR GRAPHS

Since eigenvalues say so much about regular graphs, we next consider their implications for special classes of regular graphs.

8.6.30. Definition. An n-vertex graph G is *strongly regular* if there are parameters k, λ, μ such that G is k-regular, every adjacent pair of vertices have λ common neighbors, and every non-adjacent pair of vertices have μ common neighbors.

Properties of eigenvalues of strongly regular graphs provide a short proof of a curious result called the "Friendship Theorem". This theorem says that at any party at which every pair of people have exactly 1 common acquaintance, there is one person who knows everyone (presumably the host). The resulting graph of the acquaintance relation consists of some number of triangles sharing a vertex. Another motivation for studying strongly regular graphs is their connection with the theory of designs. Strongly regular graphs with $\lambda = \mu$ correspond to symmetric balanced incomplete block designs. Other regular graphs with rich algebraic structure appear in Biggs [1993, part 3].

8.6.31. Theorem. If G is a strongly regular graph with n vertices and parameters k, λ, μ, then $\overline{G}$ is strongly regular with parameters $k' = n - k - 1$, $\lambda' = n - 2 - 2k + \mu$, and $\mu' = n - 2k + \lambda$.

Proof. For each adjacent pair $v \leftrightarrow w$ in G, there are $2(k-1) - \lambda$ other vertices in $N(v) \cup N(w)$, so v, w have $n - 2 - 2(k-1) + \lambda$ common non-neighbors. On the other hand, any non-adjacent pair $v \nleftrightarrow w$ has $2k - \mu$ vertices in $N(v) \cup N(w)$ and thus $n - 2k + \mu$ common non-neighbors. $\square$

8.6.32. Theorem. If G is a strongly regular graph with n vertices and parameters k, λ, μ, then $k(k - \lambda - 1) = \mu(n - k - 1)$.

Proof. We count induced copies of P_3 with a fixed vertex v as an endpoint. The middle vertex w can be picked in k ways. For each such w, the third vertex can be any neighbor of w not adjacent to v. With v unavailable, there are always $k - \lambda - 1$ ways to pick the third vertex. On the other hand, the third vertex can be picked in $n - k - 1$ ways as a non-neighbor of v, and for each such choice there are μ common neighbors with v that can serve as w. $\square$

8.6.33. Example. *Degenerate cases.* If G is strongly regular and $\mu = 0$ or $\lambda = k - 1$ or $k = n - 1$, then G is a disjoint union of copies of K_{k+1}. By the preceding theorem, $\lambda = k - 1$ if and only if $\mu = 0$ or $k = n - 1$. Hence we may assume $\lambda = k - 1$. This means every neighbor of v must be adjacent to every other, which forbids an induced P_3, which forces G to be a disjoint union of cliques. $\square$

Henceforth, we assume $\mu > 0$ and $\lambda < k - 1$. The preceding theorem gives a necessary condition on the choice of parameters for a strongly regular graph. Another condition arises from the eigenvalues. The argument for Theorem 8.6.34 includes an alternative derivation of Theorem 8.6.32.

8.6.34. Theorem. (Integrality Condition). If G is a strongly regular graph with n vertices and parameters k, λ, μ, then the numbers below are non-negative integers:

$$\frac{1}{2}\left(n - 1 \pm \frac{(n-1)(\mu - \lambda) - 2k}{\sqrt{(\mu - \lambda)^2 + 4(k - \mu)}}\right)$$

Proof. These are non-negative integers by virtue of being the multiplicities of eigenvalues. Consider A^2. The ijth entry of A^2 is k if $i = j$, λ if $v_i \leftrightarrow v_j$, and μ if $v_i \nleftrightarrow v_j$. Since $v_i \leftrightarrow v_j$ marks the 1's in the adjacency matrix and $v_i \nleftrightarrow v_j$ marks the 1's in the adjacency matrix of the complement, we have $A^2 = kI + \lambda A + \mu(J - I - A)$. Rearranging terms yields $A^2 = (k - \mu)I + (\lambda - \mu)A + \mu J$.

Multiplying $\mathbf{1}_n$ by both expressions for A^2 produces $k^2 \mathbf{1}_n = (k - \mu)\mathbf{1}_n + (\lambda - \mu)k\mathbf{1}_n + \mu n \mathbf{1}_n$, which reduces to $k(k - \lambda - 1) = \mu(n - k - 1)$. Let x be another eigenvector, for eigenvalue $\theta \neq k$. Since x is orthogonal to $\mathbf{1}_n$, we have $Jx = 0_n$. Multiplying x by both expressions for A^2 produces $\theta^2 - (\lambda - \mu)\theta - (k - \mu) = 0$. This quadratic equation for θ has two roots r, s, which must be the values of all the other eigenvalues. The values are $\frac{1}{2}(\lambda - \mu \pm \sqrt{(\lambda - \mu)^2 + 4(k - \mu)})$.

Now let a, b be the respective multiplicities of the eigenvalues r, s. Example 8.6.33 describes all cases when $\mu = 0$. Hence we may assume that $\mu > 0$, which implies that G is connected. Because G is connected, k is a simple eigenvalue, and we have $1 + a + b = n$. Since the eigenvalues sum to 0, we have $k + ra + sb = 0$. The solution to these two linear equations in two unknowns is $a = -(k + s(n-1))/(r - s)$ and $b = (k + r(n-1))/(r - s)$. These are the values claimed above to be non-negative integers. $\square$

The argument above can also be traced in the opposite direction.

8.6.35. Theorem. A k-regular connected graph G is strongly regular with parameters k, λ, μ if and only if it has exactly three eigenvalues $k > r > s$, which satisfy $r + s = \lambda - \mu$ and $rs = -(k - \mu)$. $\square$

8.6.36. Example. *Classes of strongly regular graphs.* For the multiplicities of r, s to be integers, two possibilities are $(n-1)(\mu - \lambda) = 2k$ or $(n-1)(\mu - \lambda) \neq 2k$. Excluding the uninteresting cases, the first case requires $\mu = \lambda + 1$, because $0 < 2k < 2n - 2$. By Theorem 8.6.30, this means that G and $\overline{G}$ are strongly regular graphs with the same parameters. In this case, we also know that $n = 4\mu + 1$ and that n is the sum of two perfect squares. Furthermore, the multiplicities of r, s are equal.

In the second case, rationality requires $(\mu - \lambda)^2 + 4(k - \mu) = d^2$ for some positive integer d, and d must divide $(n-1)(\mu - \lambda) - 2k$. Here the eigenvalues must be integers. Various examples of such strongly regular graphs are known. For $\lambda = 0$ and $\mu = 2$, three such graphs are known, but it is not known whether the list is finite! The examples are the square $((n, k, \lambda, \mu) = (4,2,0,2))$, the Clebsch graph $(16,5,0,2)$, and the Gewirtz graph $(56,10,0,2)$ (see Cameron-van Lint [1992], p43). Other constructions appear in the exercises. □

Finally, we prove the Friendship Theorem. It is startling that such a combinatorial-sounding result seems to have no short combinatorial proof. There do exist proofs avoiding eigenvalues (see Hammersley [1981]), but they require complicated numerical arguments to eliminate regular graphs.

8.6.37. Theorem. (Friendship Theorem). If G is a graph in which any two distinct vertices have exactly one common neighbor, then G has a vertex joined to all others.

Proof. The symmetry of the condition suggests that G might be regular. If G is a regular graph, then it is be strongly regular with $\lambda = \mu = 1$. By the integrality theorem, this requires that $\frac{1}{2}(n - 1 \pm k/\sqrt{k-1})$ is an integer. Hence $k/\sqrt{k-1}$ is an integer, which happens only when $k = 2$. However, K_3 is the only 2-regular graph satisfying the condition, which does have vertices of degree $n - 1$.

Now suppose G is not regular. We show that any non-adjacent pair of vertices have the same degree. Insistence on unique common neighbors forbids 4-cycles. Suppose $v \nleftrightarrow w$, and let u be their common neighbor. Let a be the common neighbor of u, v and b the common neighbor of u, w. We want to show w has as many neighbors as v. Any $x \in S = N(v) - u - a$ has a common neighbor $f(x)$ with w. If $f(x) = b$ for any $x \in S$, then x, b, u, v is a 4-cycle. If $f(x) = f(x')$ for distinct $x, x' \in S$, then $x, v, x', f(x)$ is a 4-cycle. Hence w has distinct neighbors for each neighbor of v, and $d(w) \geq d(v)$. By symmetry, $d(v) \geq d(w)$.

Since G is not regular, it has two vertices v, w with $d(w) \neq d(v)$. By the preceding argument, we know $v \leftrightarrow w$. Let u be their common neighbor. Since u cannot have the same degree as each of them, we may assume $d(u) \neq d(v)$. Now suppose G has a vertex $x \nleftrightarrow v$. Then $d(x) = d(v)$, but this requires $x \leftrightarrow w$ and $x \leftrightarrow u$. This creates the 4-cycle v, u, x, w. Hence $d(v) = n - 1$. □

EXERCISES

8.6.1. *Counting cycles using eigenvalues.* Let σ_k be the number of k-cycles in G. Let $L_k = \Sigma \lambda_i^k$ and $D_k = \Sigma d_i^k$ be the sum of the kth powers of the eigenvalues and the vertex degrees. Obtain formulas for σ_3 and σ_4 in terms of $\{L_k\}$ and $\{D_k\}$.

8.6.2. *Deletion formulas for the characteristic polynomial.* For clarity in this problem, we write $\phi(G;\lambda)$ as ϕ_G. Let v [xy] be an arbitrary vertex [edge] of G, and let $Z(v)$ [$Z(xy)$] be the collection of cycles containing v [xy]. Prove that the characteristic polynomial satisfies the following recurrences.

　　a) $\phi_G = \lambda\phi_{G-v} - \Sigma_{u \in N(v)}\phi_{G-v-u} - 2\Sigma_{C \in Z(v)}\phi_{G-V(C)}.$

　　b) $\phi_G = \phi_{G-xy} - \phi_{G-x-y} - 2\Sigma_{C \in Z(xy)}\phi_{G-V(C)}.$

(Hint: Induction or Sach's formula can be used. Also, the edge-deletion formula can be proved from the vertex-deletion formula. Comment: When G is a forest and v is a leaf with neighbor u, the formulas reduce to $\phi_G = \lambda\phi_{G-v} - \phi_{G-v-u}$ and $\phi_G = \phi_{G-xy} - \phi_{G-x-y}.$)

8.6.3. *Characteristic polynomial for paths and cycles.*

　　a) Use the deletion formulas to find recurrences relations for $\phi(Gn;\lambda)$ when $G_n = P_n$ and when $G_n = C_n$.

　　b) Without solving the recurrence, prove that the eigenvalues of C_n are $\{2\cos(2\pi j/n): 0 \le j \le n-1\}$.

　　c) Given $\text{Spec}(C_n)$, compute $\text{Spec}G$, where G is the graph obtained from C_n by adding edges joining vertices at distance 2 in C_n.

8.6.4. For a tree, prove that the coefficient of λ^{n-2k} in the characteristic polynomial is $(-1)^k\mu_k(G)$, where $\mu_k(G)$ is the number of matchings of size k. Use this to construct a pair of non-isomorphic "co-spectral" trees on 8 vertices; both have characteristic polynomial $\lambda^8 - 7\lambda^6 + 9\lambda^4$.

8.6.5. (+) Suppose T is a tree. Prove that $\alpha(T)$ is the number of nonnegative eigenvalues of T. (Hint: see Theorem 8.6.18.) Cvetkovic-Doob-Sachs [1979, p233].

8.6.6. Show that the eigenvalues of a graph are bounded in magnitude by $\sqrt{2e(n-1)/n}$.

8.6.7. Let $\{\lambda_i\}$ and $\{\mu_j\}$ be the collections of eigenvalues of G and H, respectively. Show that the mn eigenvalues of the product $G \square H$ are $\{\lambda_i + \mu_j\}$. Use this to derive the spectrum of the k-cube. (Hint: given an eigenvector of $A(G)$ associated with λ_i and an eigenvector of $A(H)$ associated with μ_j, construct an eigenvector for $A(G \square H)$ associated with $\lambda_i + \mu_j$.)

8.6.8. Compute the spectrum of the complete p-partite graph with m vertices in each part. (Hint: use the complement.)

8.6.9. If $\phi(G;x) = x^8 - 24x^6 - 64x^5 - 48s^4$, determine G.

8.6.10. Prove that G is bipartite if G is connected and $\lambda_{\max}(G) = -\lambda_{\min}(G)$.

8.6.11. (!) Given a graph G, let $R(G)$ be the matrix whose i,jth entry is $d_G(v_i, v_j)$. Prove that the squashed cube dimension of a graph (Definition 8.4.11) is at least the maximum of the number of positive eigenvalues and the number of negative eigenvalues of $R(G)$. Conclude that the squashed cube dimension of K_n is $n-1$. (Hint: rewrite the quadratic form x^TRx as a sum of squares of linear functions, and apply Sylvester's Law of Inertia.)

8.6.12. Given a real symmetric matrix partitioned as $M = \begin{pmatrix} P & Q \\ Q^T & R \end{pmatrix}$ with P, R square, a lemma in linear algebra yields $\lambda_{\max}(M) + \lambda_{\min}(M) \le \lambda_{\max}(P) + \lambda_{\max}(R)$.

a) Suppose A is a real symmetric matrix partitioned into t^2 submatrices A_{ij} such that the diagonal submatrices A_{ii} are square. Prove that $\lambda_{\max}(A) + (t-1)\lambda_{\min}(A) \le \Sigma_{i=1}^m \lambda_{\max}(A_{ii})$.

b) Prove that $\chi(G) \ge 1 + \lambda_{\max}(G)/(-\lambda_{\min}(G))$ when G is nontrivial. (Wilf)

c) Use the Four Color Theorem to prove that $\lambda_1(G) + 3\lambda_n(G) \le 0$ for planar graphs.

8.6.13. (–) Suppose G is an (n, k, c)-magnifier with vertices $v_1, \ldots, v_n$. Let H be the bipartite graph with partite sets $X = \{x_1, \ldots, x_n\}$ and $Y = \{y_1, \ldots, y_n\}$ such that $x_i y_j \in E(H)$ if and only if $i = j$ or $v_i v_j \in E(G)$. Prove that H is an $(n, k+1, c)$-expander.

8.6.14. *Existence of expanders of linear size.*

a) Suppose X is a random variable giving the size of the union of k s-subsets of $[n]$ chosen at random from $\binom{[n]}{s}$. Prove that $P(X \le l) \le \binom{n}{l}(l/n)^{ks}$.

b) (+) Suppose $\alpha\beta < 1$. Prove that there exists a constant k such that, for all n sufficiently large, there exists a subgraph of $K_{n,n}$ with maximum degree at most k such that $|N(S)| \ge \beta|S|$ whenever $|S| \le \alpha n$. (Hint: Generate bipartite subgraphs of $K_{n,n}$ by taking the union of k random complete matchings.)

c) Conclude the existence of k such that n, k, c-expanders exist for all sufficiently large n. An (n, α, β, d)-*expander* is a bipartite graph $G \subseteq K_{A,B}$ with $|A| = |B| = n$, $\Delta(G) \le d$, and $|N(S)| \ge \beta|S|$ whenever $|S| \le \alpha n$.

8.6.15. Suppose that G is a triangle-free graph on n vertices in which every pair of nonadjacent vertices has exactly two common neighbors. Prove that G is regular and that $n = 1 + \binom{k+1}{2}$, where k is the degree of the vertices in G. Prove that G is strongly regular. What constraints on k are implied by the integrality conditions? Construct examples for $k = 1, 2, 5$. A realization for $k = 10$ is known, using the existence of an appropriate Steiner system (see Chapter 19).

8.6.16. (+) Prove that the Petersen graph is strongly regular and determine its spectrum (this part is easy with properties of strongly regular graphs and not hard without them). Apply the spectrum to show that edges of the complete graph K_{10} cannot be partitioned into 3 disjoint copies of the Petersen graph. (Hint: Use the spectrum to prove that two copies of the Petersen matrix have a common eigenvector other than the constant vector.) (Schwenk [1983])

8.6.17. The *subconstituents* of a graph are the induced subgraphs of the form $G[U]$ where $v \in V(G)$ and $U = N(v)$ or $U = \overline{N[v]}$. A. Vince defined G to be *super-regular* if G has no vertices or if G is regular and every subconstituent of G is superregular. Let **S** be the class consisting of $\{aK_b : a, b \ge 0\}$ (disjoint unions of isomorphic cliques), $\{K_m \square K_m : m \ge 0\}$, C_5, and the complements of these graphs.

a) Prove that every graph in **S** is superregular and that every disconnected superregular graph is in **S**. (Comment: in fact, every superregular graph is in **S**, but the complete inductive proof of this requires several pages (Maddox [1996], West [1996])

b) Prove that every superregular graph is strongly regular.

8.6.18. (+) *Automorphisms and eigenvalues.*

a) Prove that σ is an automorphism of G if and only if the permutation matrix corresponding to σ commutes with the adjacency matrix of G; i.e., $PA = AP$.

b) If x is an eigenvector of G corresponding to an eigenvalue of multiplicity one and P is the permutation matrix corresponding to an automorphism of G, prove that $Px = \pm x$.

c) If every eigenvalue of G has multiplicity one, prove that every automorphism of G is an involution (i.e., its square is the identity). (Mowshowitz [1969], Petersdorf-Sachs [1969])

8.6.19. (+) Light bulbs $l_1, \ldots, l_n$ are controlled by switches $s_1, \ldots, s_n$. The ith switch changes the on/off status of the ith light and possibly others, but s_i changes the status of l_j if and only if s_j changes the status of l_i. Initially all the lights are off. Prove that it is possible to turn all the lights on. (Peled [1992]) (Hint: this uses vector spaces, not eigenvalues.)

Glossary of Terms

This glossary is intended as a resource, not merely as an index. It contains terms not used in this volume that the reader may encounter in further study of graph theory. Some terms apply to more than one model, such as to graphs and to digraphs. Some terms used by other authors are included.

Absorption property (matroids): $r(X) = r(X \cup e) = r(X \cup f)$ implies $r(X) = f(X \cup f \cup e)$
Acyclic: without cycles
Adjacency matrix A: entry a_{ij} is number of edges from vertex i to vertex j
Adjacency relation: set of ordered pairs forming edges in graph or digraph
Adjacency set $N(v)$: the set of vertices adjacent to v
Adjacent: vertices joined by an edge, sometimes used to describe intersecting edges
Adjoins: is adjacent to
Adjugate: matrix of cofactors
Alphabetic code: codeword order respects message order
M-alternating path: a path alternating between edges in M and not in M
Almost always: having asymptotic probability 1
Ancestor: in a rooted tree, a vertex along the path to the root
Antichain: family of pairwise incomparable items (under an order relation)
Anticlique: stable set
Antihole: induced subgraph isomorphic to the complement of a cycle
Approximation algorithm: polynomial-time algorithm with bounded performance ratio
Approximation scheme: family of approximation algorithms with arbitrarily good performance ratio
Arboricity Υ: minimum number of forests covering the edges of G
Arc: directed edge (ordered pair of vertices)
k-arc-connected: same as k-edge-connected for digraphs
Arborescence: a directed forest in which every vertex has out-degree at most one
Articulation point: a vertex whose deletion increases the number of components
Assignment problem: minimize (or maximize) the sum of the edge weights in a perfect matching of a complete bipartite graph with equal part-sizes
Asteroidal triple: three distinct vertices with each pair connected by a path avoiding the neighborhood of the third
Asymmetric: having no automorphisms other than the identity
Asymptotic: having ratio approaching 1
Augmentation property (matroids): $I_1, I_2 \in \mathbf{I}$ with $|I_2| > |I_1|$ implies the existence of $e \in I_2 - I_1$ such that $I_1 \cup e \in \mathbf{I}$
Augmenting path: for a matching, an alternating path that can be used to increase the size of the matching; for a flow, increases the flow value
Automorphism: a permutation of the vertices that preserves the adjacency relation
Automorphism group Γ: the group of automorphisms under composition

Average degree: $\Sigma d(v)/n(G) = 2e(G)/n(G)$.

Azuma's Inequality: a bound on the probability in the tail of a distribution

Backtracking: depth-first-search

Balanced k-partite: having part-sizes differing by at most one (see equipartite)

Balanced: average vertex degree in a subgraph of G is maximized by G itself

Bandwidth: the minimum, over vertex numberings by distinct integers, of the maximum difference between labels of adjacent vertices

Barycenter: vertex minimizing the sum of distances to other vertices

Base: maximal independent set of a matroid

Base exchange property (matroids): for all $B_1, B_2 \in \mathbf{B}$ and $e \in B_1 - B_2$, there exists an element $f \in B_2 - B_1$ such that $B_1 - e + f$ is a base.

Best possible: fails to be true when some condition (usually on a parameter) is loosened

Bicentral tree: a tree with two centers

Biclique: complete bipartite graph

Biconnected: 2-connected

Bigraphic: a pair of sequences realizable as the vertex degrees of a bipartite graph with that bipartition

Binary matrix (or vector): having all entries 0 or 1

Binary matroid: representable over the two-element field

Binary tree: rooted tree in which every non-leaf vertex has at most two children

Binomial coefficient $\binom{n}{k}$: the number of ways to choose a subset of size k from an n-element set, equal to $n!/[k!(n-k)!]$.

Biparticity: number of bipartite subgraphs needed to partition the edges

Bipartite graph: a graph whose vertices can be covered by two independent sets

Bipartite Ramsey number: for a bipartite G, the minimum n such that 2-coloring the edges of $K_{n,n}$ forces a monochromatic G

Bipartition: a partition of the vertex set into two independent sets

Birkhoff diamond: a particular reducible configuration for the Four Color Problem

Block: 1) a maximal subgraph with no cutvertex, 2) a graph with no cutvertex, 3) a class in a partition of a set

Block graph: intersection graph of blocks

Blossom: an odd cycle arising in Edmonds' algorithm for general matching

Bond: a minimal edge cut

Bond space: orthogonal complement to the cycle space; linear combinations of bonds (over $GF(2)$)

Bond matroid: dual of the cycle matroid of a graph

Book embedding: a decomposition of G into outerplanar graphs with a consistent ordering of the vertices (as on the spine of a book)

Box: a rectilinear parallelopiped with sides parallel to the coordinate axes

Boxicity: minimum dimension in which G is the intersection graph of boxes

Bouquet: a graph consisting of one vertex and some number of loops

Branch vertex: a vertex of degree at least 3

Branching: a directed graph in which each vertex has indegree one except for a single vertex (root) with indegree 0

r-branching: branching rooted at r

Breadth-first search: search procedure exploring vertices of a component in order by distance from root

Breadth-first tree: tree generated by a breadth-first search from a root

H-bridge of G: a component of $G - H$ together with the edges to its vertices of attachment (some authors use bridge to mean cut-edge)

Brooks' Theorem: $\chi(G) \le \Delta(G)$ for connected graphs, except for cliques and odd cycles

deBruijn graph: graph of possible transitions between words of fixed length in a stream of letters from an alphabet

Cactus: a graph in which every edge appears in at most one cycle

Cage: a regular graph of given degree and girth having minimal number of vertices

Capacity: a limit on flow 1) through an edge in a network; 2) across a cut

Cartesian product $G_1 \times G_2$: the graph with vertex set $V(G_1) \times V(G_2)$ and edges given by $(u_1, u_2) \leftrightarrow (v_1, v_2)$ if 1) $u_1 = v_1$ and $u_2 \leftrightarrow v_2$ in G_2 or 2) $u_2 = v_2$ and $u_1 \leftrightarrow v_1$ in G_1

Caterpillar: a tree with a single path (the spine) that includes or is incident to every edge

Cayley's Formula: there are n^{n-2} labeled trees on n vertices

2-cell: on a surface, a region homeomorphic to a disc, i.e. having no handle and having a simple closed curve as boundary

2-cell embedding: an embedding in which every region is a 2-cell

Center: a vertex that minimizes the maximum distance to all other vertices

Central tree: a tree with one center

α, β-chain: a path alternating between colors α and β

Characteristic polynomial $\phi(G; \lambda)$: characteristic polynomial of the adjacency matrix of the graph; whose roots are the eigenvalues

Children: neighbors of a rooted tree vertex not on the path to the root

Chinese postman problem: problem of finding the cheapest closed walk covering all the edges in an edge-weighted graph

Choice number: choosability

Choosability: minimum k such that G is k-choosable

k-choosable: for all lists of size k assigned to vertices of G, there is a proper coloring that selects a color for each vertex from its list

Chord: edge joining two nonconsecutive vertices of a path or cycle

Chordal: having no chordless cycle

Chordless cycle: an induced cycle of length at least 4

Chromatic index: edge-chromatic number

Chromatic number $\chi(G)$: minimum number of colors in a proper coloring.

Chromatic polynomial $\chi(G; k)$: a polynomial whose value at k is the number of proper colorings of G using colors from $[k]$. $1, \ldots, k$.

Chromatic recurrence: recurrence relation for chromatic polynomial

k-chromatic: having chromatic number k

Circle graph: an intersection graph of chords of a circle

Circuit: a closed trail (no repeated edge)

Circulant graph: having an adjacency matrix with constant diagonals, constructed as equally-spaced vertices on a circle with adjacency depending only on distance

Circular arc graph: an intersection graph of arcs of a circle

Circulation: a flow in a network such that the net flow at each vertex is 0

Circumference: the length of the longest cycle

Clause: a collection of literals in a logical (Boolean) formula

Claw: the graph $K_{1,3}$

Claw-free: having no induced $K_{1,3}$

Clique: a complete (sub)graph

Clique cover: a set of cliques covering the vertices (minimum size = $\theta(G)$)

Clique decomposition: a partition of the edges into cliques

Clique edge cover: a set of cliques covering the edges (see intersection number)

Clique identification: a perfection-preserving operation

Clique number ω: maximum order of a clique in G

Clique partition number: minimum number of cliques to partition $E(G)$

Clique-vertex incidence matrix: 0,1-matrix in which entry (i, j) is 1 if and only if vertex j belongs to maximal clique i

Clique tree: an intersection representation of a chordal graph, consisting of a host tree with a bijection between its vertices and the maximal cliques of G such that the cliques containing each vertex form a subtree of the host

Closed ear: a path between two (possibly equal) old vertices through new vertices

Closed ear decomposition: construction of a graph from a cycle by successive addition of closed ears

Closed neigborhood: a vertex and all its neighbors

Closed set: in matroids, a set whose span is itself

Closed walk: first vertex is the same as the last

Closure: 1) the graph $C(G)$ obtained from G by iteratively adding edges joining nonadjacent vertices with degree-sum at least $n(G)$; 2) image under a closure operator

Closure operator: an operator that is expansive, order-preserving, and idempotent

Cobase: a base of the dual matroid

Cocircuit: a circuit of the dual matroid

Cocritical pair: a non-edge whose addition increases the clique number

Cocycle matroid: the dual of a cycle matroid

Cograph: P_4-free graph (equivalent to complement reducible graph)

Color-critical: a graph such that every proper subgraph has smaller chromatic number

Color class: in a coloring, a set of objects receiving the same color

k-colorable: chromatic number $\leq k$

k-coloring: a vertex partition into k independent sets

P coloring: a vertex partition into subsets inducing graphs with property **P**

Column matroid $M(A)$: matroid whose independent sets are the linearly independent subsets of the matrix A

Comma-free code: no code word is a prefix of another

Common system of distinct representatives (CSDR): for set systems **A** and **B**, a set of elements that is an SDR of **A** and is an SDR of **B**

Comparability graph: graph having a transitive orientation

Competition graph: graph obtained from a directed graph D by $u \leftrightarrow v$ if u, v have a common out-neighbor in D

Complement $\overline{G}$ (or G^c): simple graph or digraph with the same vertex set as G, and $uv \in E(\overline{G})$ if and only if $uv \notin E(G)$

Complement reducible: reducible to the trivial graph by iteratively complementing components

Complete graph K_n: every pair of vertices forms an edge

Complete k-partite graph $K_{n_1,\ldots,n_k}$: k-partite graph in which every pair of vertices not belonging to the same partite set is adjacent (part-sizes are $n_1, \ldots, n_k$)

Complete matching: a 1-factor

Completely labeled cell: simplicial region with distinct labels on corners

Complexity: number of operations that may be needed by an algorithm

Component: maximal connected subgraph

S-component of G: subgraph of G induced by $S \cup V_i$, where V_i is the vertex set of a component of $G - S$

Composition $G_1[G_2]$: a graph product whose vertex set is the cartesian product of the vertices of the factors, in which $(u_1, u_2) \leftrightarrow (v_1, v_2)$ if and only if $u_1 \leftrightarrow v_1$ in G_1, or $u_1 = v_1$ and $u_2 \leftrightarrow v_2$ in G_2

Condensation of D: the digraph obtained by contracting each strong component of D to a vertex

Conflict graph: graph whose vertices are the bridges of a cycle (bridges conflict if they have three common endpoints or four alternating endpoints on the cycle)

Conflicting chords: two chords whose endpoint alternate on a specified cycle

Conjugate partition: two partitions of n such that one gives the row sizes and the other the column sizes of a Ferrers diagram

Connected: having a (u, v)-path for every pair of vertices u, v

k-connected: having connectivity $\geq k$

Connection relation: relation on the vertex set of a graph satisfied by any two vertices that are connected by a path

Connectivity $\kappa(G)$: the minimum number of vertices whose deletion disconnects the graph (sometimes called "vertex connectivity" for clarity)

Consecutive-ones property: having a permutation of columns so ones appear consecutively in rows

Conservation constraint: for a flow, the condition of net flow 0 at a vertex

Consisting rounding: conversion of the data and row/column sums in a matrix to nearest integers up or down such that row and column sums remain correct

Construction procedure: a procedure for iteratively building members of a class of graphs from a small base graph or graphs

Contracted vertex: the new vertex resulting from the contraction of an edge

Contraction: replacement of an edge uv by a vertex w incident to the union of the edge sets formerly incident to u or to v

Converse D^{-1}: $u \to v$ in D^{-1} iff $v \to u$ in D

Convex embedding: a plane graph in which every bounded face is a convex set and the outer boundary is a convex polygon

Convex function: satisfies the inequality $f(\lambda a + (1 - \lambda)b) \leq \lambda f(a) + (1 - \lambda)f(b)$ for all a, b and $0 \leq \lambda \leq 1$

Convex quadrilateral: does not have one point in the triangle formed by the other three

Convex sequence: a sequence of points in the plane such that the segments joining consecutive points yield the graph of a convex function

Cocycle space: name: bond space

Cost: name of the objective function for many weighted minimization problems

Cotree: with respect to a graph, the edges not belonging to a given spanning tree

F-covering: covering by subgraphs in the family **F**

Critical edge: edge whose deletion increases the independence number

Critical graph: used with respect to many graph properties, indicating that the deletion of any vertex (or edge, depending on context) destroys the property

k-critical graph: most commonly color-critical with chromatic number k

Critically 2-connected: deletion of an edge destroys 2-connectedness

Crossing: in a drawing of a graph, an internal intersection of two edges

Crossing number $v(G)$: minimum number of edge crossings when drawing G in the plane

k-cube Q_k: graph whose vertices are indexed by binary sequences of length k, in which vertices are adjacent if and only if theifr sequences differ in exactly one place

Cubic graph: a regular graph of degree 3

Cut $[S, \bar{S}]$: the edges from a vertex subset to its complement (used especially in network flow problems)

Cut-edge: an edge whose deletion increase the number of components

Cutset: a separating set of vertices

Cutter: the player in the Shannon Switching Game who deletes edges

Cut-vertex: vertex whose deletion increases the number of components

Cycle: a closed walk in which no vertex appears twice (except end = beginning)

k-cycle: a cycle of length k, consisting of k vertices and k edges

Cycle matroid $M(G)$: the matroid whose circuits are the cycles of G

Cycle rank: dimension of cycle space; number of edges – number of vertices + number of components

Cycle space: the nullspace of the incidence matrix, the set of even subgraphs

Cyclic edge-connectivity: number of edges that must be deleted to disconnect a component so that every remaining component contains a cycle

Cyclically k-edge-connected: cyclic edge-connectivity at least k

DeBruijn graph: digraph encoding possible transitions between k-ary n-tuples as additional characters are received

Decomposition: an expression of G as a union of edge-disjoint subgraphs

F-decomposition: decomposition using subgraphs in the family **F**

F-decomposition number of G: minimum number of graphs in an **F**-decomposition of G

Decision problem: a computational problem with a YES/NO answer

Degree $d(v)$: 1) for a vertex, the number of times it appears in edges (may be modified by "in-" or "out-" in a digraph); 2) for a regular graph, the degree of each vertex

Degree matrix $D(G)$: a diagonal matrix whose diagonal entries are the vertex degrees

Degree sequence $d_1 \geq \cdots \geq d_n$: the sequence of vertex degrees, usually indexed in non-increasing order regardless of vertex order

Degree set: the set of vertex degrees (appearing once each)

Degree-sum Formula: $\Sigma d(v) = 2e(G)$

Deletion method: a strengthening of the existence argument in the probabilistic method

Demand: sink constraint in transportation network

Density: ratio of number of edges to number of vertices

Dependent set: in matroids, a set containing a circuits

Depth-first search: backtracking search from a vertex

Descendants of x: members of a subtree rooted at x

Detachment of G: graph obtained from G by a sequence of splits: splitting a vertex v into an independent set, with the k new vertices inheriting the k sets of a partition of the edges incident to v

Diagonal Ramsey number: Ramsey number for an instance where the thresholds (numbers or graphs) are equal

Diameter: the maximum of $d(u, v)$ over vertex pairs u, v

Dicomponent: strongly connected component

Diconnected: a digraph having a (u, v)-path for every ordered pair of vertices u, v

Digraph: directed graph

Dijkstra's Algorithm: algorithm to compute shortest paths from one vertex

Dilworth's Theorem: maximal number of pairwise incomparable elements equals minimum number of totally ordered subsets needed to cover all elements

Dinitz Conjecture: each bipartite graph G is $\Delta(G)$-list-edge-colorable

Direct product: Cartesian product

Directed graph: a set of vertices and a multiset of ordered pairs of vertices as edges

Directed walk, trail, path, cycle, etc.: same as without the adjective "directed" (the head
of an edge is the tail of the next edge)

Disc: in a surface of genus 0, the region bounded by a simple closed curve

Disconnected: a graph with more than one component

Disconnecting set: a set of edges whose deletion makes some vertex unreachable from
some other vertex

Disjoint union $G_1 + G_2$: the union of two graphs with disjoint vertex sets

Disjointness graph: complement of intersection graph

Distance $d(u, v)$: the minimum length of a (u, v)-path

Distance-preserving   embedding: mapping $f\colon V(G) \to V(H)$ so   that
$d_H(f(u), f(v)) = d_G(u, v)$.

Dodecahedron: planar graph with 20 vertices, 30 edges, and 12 faces

Domination number: the minimum size of a vertex dominating set

Dominating set: a set $S \subseteq V$ such that every vertex is in S or is adjacent to a vertex in S

Double jump: the phenomenon that the structure of the random graph in Model A is
markedly different for probability functions of the form c/n with $c < 1$, $c = 1$, and $c > 1$.

Double star: a tree with at most two vertices of degree > 1

Double triangle: $K_4 - e$

Double torus: the (orientable) surface with two handles

Doubly stochastic matrix: square matrix having identical sum in each row and column

Dual augmentation property (matroids): disjoint sets independent in a matroid and its
dual can be enlarged to a complementary base and cobase

Dual edge e^*: the edge of the dual graph G^* corresponding to each e of a planar graph G

Dual graph G^*: for a plane graph, the graph with a vertex for each region of G, in which
vertices are adjacent if the boundaries of the corresponding regions of G share an
edge (also makes sense for 2-cell embeddings on any surface)

Dual hereditary system (or matroid) M: the hereditary system whose bases are the complements of the bases of M

Dual problem: for a problem $\max c^t x$ such that $Ax \le b$ and $x \ge 0$, the dual is $\min y^T b$
such that $yA \ge c$ and $y \ge 0$

Duality gap: strict inequality between a pair of dual integer programs

Duplication of vertex x: adding x' with $N(x') = N(x)$

Ear: path between old vertices through new vertices

Ear decomposition: construction of G from a cycle by addition of ears

Eccentricity: for a vertex, the maximum distance to other vertices

Edge: 1) in a graph, a pair of vertices ($E(G)$ denotes the edge set); 2) in a hypergraph, a
subset of the vertex set

Edge-choosability: minimum k such that G is k-edge-choosable

k-edge-choosable: for all lists of size k assigned to edges of G, there exists a proper edge-
coloring that selects a color for each edge from its list

Edge-chromatic number χ': the minimum number of colors in a proper edge-coloring

k-edge-colorable: edge-chromatic number $\le k$

Edge-coloring: an assignment of labels to the edges

k-edge-connected: edge connectivity $\ge k$

Edge-connectivity $\kappa'(G)$: the minimum number of edges whose deletion disconnects G

Edge cover: a set of edges incident to all the vertices

Edge cut $[S, \bar{S}]$: the set of edges joining a vertex in S to a vertex not in S

Edge-reconstructible: a graph that can be determined (up to isomorphism) by knowing
the multiset of subgraphs obtained by deleting single edges

Edge-Reconstruction Conjecture: the conjecture that every graph with at least four edges
is edge-reconstructible

Edge-transitive: existence of a permutation for each pair $e, f \in E(G)$ that maps e to f

Eigenvalue: for a graph, an eigenvalue of the adjacency matrix

Eigenvector of A: a vector x such that $Ax = \lambda x$ for some constant λ

Elementary contraction: contraction

Elementary cycle: 1) boundary of a region in a plane graph; 2) used to mean (simple) cycle by some authors who use cycle to mean circuit

Elementary subdivision: replacement of an edge by a path of two edges connecting the endpoints of the original edge (also called "edge subdivision")

Embedding: a mapping of a graph into a surface, such that (the images of) its edges do not intersect except for shared endpoints

Empty graph: having no edges

Endpoint: 1) each member of an edge; 2) the first or last vertex of a path, trail, or walk

End-vertex: a vertex of degree 1

Equipartite: having part-sizes differing by at most one

Equitable coloring: having color classes differing in size by at most one

Equivalence relation: reflexive, symmetric, and transitive relation

Equivalence: as a graph, a union of disjoint cliques

Erdős number: distance from Erdős in the collaboration graph of mathematicians

Euler tour: Eulerian circuit

Eulerian circuit: a closed trail containing every edge

Eulerian (di)graph: a graph or digraph having an Eulerian circuit

Eulerian-extendible: a vertex from which every trail extends to an Eulerian trail

Eulerian trail: a trail containing every edge

Euler characteristic: for a surface of genus γ, $2 - 2\gamma$

Euler's formula: for any 2-cell embedding of a connected graph on a surface of genus γ, the formula $n - e + f = 2 : 2\gamma$, where f is the number of faces

Even cycle: cycle with an even number of edges (or vertices)

Even graph: graph with all vertex degrees even

Even pair: vertex pair x, y such that every chordless x, y-path has even length

Even triangle: triangle T such that every vertex has an even number of neighbors in T

Even vertex: vertex of even degree

Evolution: the model of generating random graphs by successively adding random edges

Expectation: (weighted) average

(n, k, c)-expander: bipartite graph with partite sets of size n and degree at most k such that each set S with at most half the vertices of the first partite set has at least $(1 + c(1 - |S|/n)|S|$ neighbors

Expansion: in a 3-regular graph, subdividing two edges and joining the two new vertices by an edge

Expansion Lemma: adding a vertex of degree k to a k-connected graph preserves k-connectedness

Expansive property: for a function σ on the subsets of a set, the requirement that $X \subseteq \sigma(X)$ for all X

Expectation: for a discrete random variable, $\Sigma k \mathrm{Prob}(X = k)$

Exterior region: the unbounded region in a plane graph

Face: a region of an embedding

Factor: a spanning subgraph

f-factor: a spanning subgraph with $d(v) = f(v)$

k-factor: a spanning k-regular subgraph

k-factorable: having a decomposition into k-factors

Factorization: an expression of G as the edge-disjoint union of spanning subgraphs

Fat triangle: a multigraph with three vertices and k copies of each pair as edges

x, U-fan: pairwise internally-disjoint paths from x to distinct vertices of U

Fary's Theorem: a planar graph has a straight-line embedding in the plane

k-factorization: a decomposition of a graph into k-factors

Feasible flow: a network flow satisfying edge-constraints and having net flow 0 at each internal vertex

Feasible solution: a choice of values for the variables that satisfies all the constraints in an optimization problem

Ferrers diagram: diagram of a partition of an integer, with λ_i positions in the ith row, where $\lambda_1 \geq \cdots \geq \lambda_k$ and $\Sigma \lambda_i = n$

Ferrers digraph: a digraph (loops allowed) with no x, y, z, w (not necessarily distinct) such that $x \to y$ and $z \to w$ but $z \not\to y$ and $x \not\to w$; equivalently, the successor sets or predecessor sets are ordered by inclusion; equivalently, the adjacency matrix has no 2 by 2 permutation submatrix.

Five Color Theorem: the theorem that planar graphs are 5-colorable

Flat: a closed set in a matroid

Flow: an assignment of values to variables for each arc of a network

Flower (in Edmonds' Blossom algorithm): consists of a stem (alternating path from an unsaturated vertex) and a blossom (odd cycle with a nearly-perfect matching)

Forcibly Hamiltonian: a degree sequence such that every simple graph with that degree sequence is Hamiltonian

Forest: a disjoint union of trees, an acyclic graph

Four Color Theorem: the theorem that planar graphs are 4-colorable

Fraternal orientation: an orientation such that two vertices are adjacent if they have a common successor

H-free: having no copy of H as an induced subgraph

Free matroid: matroid in which every set of elements is independent

Friendship Theorem: if every pair of people in a set have exactly one common friend in the set, then someone is everyone's friend

Fundamental cycle: for a spanning tree, a cycle formed by adding an edge to it

Gammoid: a matroid on E generated from vertex sets F, E in a digraph by letting independent sets be those that can be saturated by a set of disjoint paths starting in F

Generalized chromatic number: minimum number of classes needed to partition the vertices so that the subgraph induced by each color class has property **P**

Generalized Ramsey number $r(G_1, \ldots, G_k)$: the minimum n such that coloring the edges of K_n forces an i-monochromatic copy of G_i for some i

Generalized Petersen graph: the graph with vertices $\{u_1, \ldots, u_n\}$ and $\{v_1, \ldots, v_n\}$ and edges $\{u_i u_{i+1}\}$, $\{u_i v_i\}$, and $\{v_i v_{i+k}\}$, where addition is modulo n

Genus γ: 1) for a surface, the number of handles in its topological description 2) for a graph, the minimum genus surface on which it is embeddable

Geodesic: a shortest path between its endpoints

Geodetic: having the property that each pair of vertices u, v are the endpoints of a unique path of length $d(u, v)$

Girth g: the length of the shortest cycle in G

k-gon: in an embedding, a k-cycle bounding a region

Good algorithm: runs in polynomial time

Good characterization: characterization of a class that is checkable in polynomial time

Good coloring: often means proper coloring

Gossip problem: minimize the number of calls so that each vertex transmits to every other by an increasing path

Graceful labeling: an assignment of distinct integers to vertices such that 1) the integers are between 0 and $e(G)$, and 2) the differences between the labels at the endpoints of the edges yield the integers $1, \ldots, e(G)$

Graceful graph: a graph with a graceful labeling

Graceful tree: a tree with a graceful labeling

Graceful tree conjecture: that every tree has a graceful labeling

Graph: a collection of pairs of elements from some set

Graphic matroid $M(G)$: matroid whose independent sets are the acyclic subsets of $E(G)$

Graphic(al) sequence: a sequence of integers realizable as the degree sequence of a simple graph

Greedy algorithm: a fast non-backtracking algorithm to find a good feasible solution by iteratively making a heuristically good choice

Greedy coloring: with respect to some vertex ordering, color each vertex with the least-indexed color not already appearing among the neighbors of the vertex being colored

Grinberg condition: necessary condition for Hamiltonian cycles in planar graphs: summing (length−2) over the inside faces or over the outside faces yields the same total

Grötsch graph: the smallest triangle-free 4-chromatic graph

Grundy number: the maximum number of colors in an application of the greedy coloring algorithm

Hadwiger conjecture: a k-chromatic graph has a subgraph contractible to K_k (true for "almost all" graphs)

Hajós conjecture: a k-chromatic graph contains a homeomorph of K_k (false beyond $k = 5$)

Hall's theorem: a necessary and sufficient condition for the existence of a system of distinct representatives

Hamilton tour: Hamiltonian cycle

Hamiltonian: having a Hamiltonian cycle

Hamiltonian closure: graph obtained by successively adding edges between vertices whose degree-sum is as large as the number of vertices

Hamiltonian-connected: having a Hamiltonian path from each vertex to every other

Hamiltonian cycle: a cycle containing each vertex

Hamiltonian path: a path containing each vertex

Harary graphs: k-connected n-vertex graphs with minimal number of edges

Handle: addition to a surface to increase genus

Head: the second vertex of an edge in a digraph

Heawood's Formula: the chromatic number of a graph embedded on the oriented surface with γ handles is at most $\lfloor 1/2(7 + \sqrt{1 + 48\gamma}) \rfloor$.

Helly property: the property of the real line (or trees) that any collection of pairwise intersecting subsets has a common intersection point

Helly number: for some universe, the number k such that the sets in F have a common intersection if every k sets of F have a common intersection

Hereditary class: a class F for which every induced subgraph of a graph in F is also in F

Hereditary family: a family F of sets such that every subset of a member of F is in F

Hereditary system: a system consisting of a hereditary family and the alternative ways of specifying that family

Hole: a chordless cycle in a graph

Homeomorphic: obtainable from the same graph by subdivision of edges

Homogeneous: in Ramsey theory, a set whose colored pieces have the same color

Homomorphism: a map $f: V(G) \to V(H)$ that preserves adjacency

Huffman code: prefix-free encoding of data to minimize expected search time

Hungarian method: an algorithm for solving the assignment problem

Husimi tree: a graph in which every block is a clique

Hypercube: a Cartesian product of copies of K_2

Hypergraph: a generalization of graph in which edges may be any subset of the vertices

Hyperplane: in a matroid, a maximal closed proper subset of the ground set

Hypobase: in a matroid, a maximal set containing no base

Hypohamiltonian: a non-Hamiltonian graph whose vertex-deleted subgraphs are all Hamiltonian

Hypotraceable: a non-traceable graph whose vertex-deleted subgraphs are all traceable

Icosahedron: planar graph with 12 faces, 30 edges, and 20 vertices

Idempotence property: $\sigma^2(X) = \sigma(X)$ for all X

Identification: an operation replacing two vertices by a single vertex with the combined incidences (same as contraction if the vertices are adjacent)

Imperfect graph: having an induced subgraph with chromatic number exceeding clique number

Incidence matrix: 1) the 0,1-matrix in which entry (i, j) is 1 if and only if vertex j belongs to edge i (for a digraph, 1 if vertex j is the head of edge i, -1 if it is the tail, 0 otherwise); 2) more generally, the matrix of a membership relation

Incident: 1) vertex v and edge e are incident if $v \in e$; 2) edges are incident if they intersect

Inclusion-exclusion formula: number of objects outside $A_1, \ldots, A_n$ is $\Sigma_{S \in [n]} (-1)^{|S|} |\cap_{i \in S} A_i|$

Incomparability graph: the complement of a comparability graph

Incorporation property (matroids): $r(\sigma(X)) = r(X)$

In-degree: for a vertex in a directed graph, the number of edges of which it is the head

Independence number α: maximum size of an independent set of vertices

Independent domination number: minimum size of an independent dominating set

Independent set: a set of pairwise nonadjacent vertices (the complement of a clique)

Indicator variable: a random variable taking values in $\{0, 1\}$

Indifference graph: representable by assigning weights to vertices such that $u \leftrightarrow v$ if and only if $|f(u): f(v)| \le 1$

Induced circuit property (matroids): adding an element to an independent set creates at most one circuit

Induced sub(di)graph $G[A]$: the sub(di)graph on vertex set $A \subseteq V(G)$ obtained by taking A and all edges of G that join two vertices in A

Integer program: linear program plus requirement that variables be integer-valued

Integrality Theorem: in a network with integer edge capacities, there is an optimal flow expressible as units of flow along source/sink paths

Internal vertices: 1) for a path, the non-endpoints. 2) for a plane graph, the vertices that do not belong to the boundary of the exterior face

Interlacing Theorem: the eigenvalues $\{\lambda_i\}$ of G and $\{\mu_i\}$ of $G - x$ satisfy $\lambda_1 \ge \mu_1 \ge \lambda_2 \ge \cdots \ge \mu_n \ge \lambda_n$ for each vertex x

Internally-disjoint paths: paths intersecting only at endpoints

Intersection graph: for a collection of sets, the graph with a vertex for each set, in which vertices are adjacent if the sets intersect

Intersection number: minimum size of a set such that G is an intersection graph of subsets of it (equals minimum number of cliques covering edges)

Intersection of matroids: the hereditary system whose independent sets are the common independent sets in the matroids

Intersection representation: an assignment of a set S_v to each vertex v such that $u \leftrightarrow v$ if and only if $S_u \cap S_v \ne \varnothing$

Interval graph: a graph having an interval representation

Interval number: minimum t such that G has a t-interval representation

Interval representation: a collection of intervals whose intersection graph is G

t-Interval representation f: a mapping $f: V(G) \to \mathbf{R}$ such that each image consists of at most t intervals and $v \leftrightarrow w$ if and only if $f(v) \cap f(w) \ne \varnothing$

In-tree: a directed tree in which each edge is oriented toward the root

Involution: a permutation whose square is the identity

Isolated: vertex (or edge) incident to no (other) edge

Isomorphic decomposition: decomposition into isomorphic subgrpahs

Isometric embedding: a distance-preserving mapping of $V(G)$ into $V(H)$

Isomorphism: a bijection between vertex sets preserving adjacency

Isthmus: a cut-edge

Join $G \vee H$: the graph consisting of the disjoint union $G + H$ plus the edges $\{uv: u \in V(G), v \in V(H)\}$

Joined to: adjacent to

Junction: vertex of degree at least three

Kempe chain: in forbidding a minimal counterexample to the four color theorem, a path between two vertices of a cycle that alternates between two colors

Kernel: in a digraph, an independent dominating set

Kernel perfect: having a kernel in each induced subgraph

Kirchhoff's current law: net flow around a closed walk is 0

König-Egerváry Theorem: maximum matching equals minimum vertex cover for bipartite graphs

König's Other Theorem: maximum independence set equals minimum edge cover for bipartite graphs

Kronecker product: weak product

Krausz decomposition: edge clique covering using each vertex at most twice, leading to the graph for which this is the line graph

Kruskal's algorithm: grows a minimum weighted spanning tree by iteratively adding the cheapest edge in the graph that does not complete a cycle

Kuratowski subgraph: subdivision of K_5 or $K_{3,3}$

Kuratowski's Theorem: a graph is planar if and only if it has no subgraph homeomorphic to K_5 or $K_{3,3}$

Labeling: assignment of integers to vertices

Leaf: vertex of degree 1

Leaf block: a block containing only one cut-vertex

Leeway: amount by which flow can be augmented along an augmenting path

Length: the number of edges (counted with multiplicity, if necessary), the sum of the edge lengths

Lexicographic product $G[H]$: composition

Line: another name for edge

Line graph $L(G)$: the intersection graph of the edges of G, i.e. vertices correspond to edges of G and are adjacent if the corresponding edges intersect

Linear matroid: matroid whose independent sets are the sets of independent columns of some matrix over a field

Linear program: problem of optimizing a linear function with linear constraints

Link: edge

k-linked: a stronger condition than k-connected, in which for every choice of two k-tuples of vertices $(u_1, \ldots, u_k)$ and $(v_1, \ldots, v_k)$, there exists a set of k internally disjoint paths connecting corresponding vertices u_i, v_i.

List Coloring Conjecture: edge-choosability always equals edge-chromatic number

List chromatic index: edge-choosability

List chromatic number: choosability

Literal: a logical (Boolean) variable or its negation

Lobster: a tree having a path from which every vertex has distance at most 2

Local connectivity $\kappa(x, y)$: minimum number of vertices whose deletion separates a non-adjacent pair

Local path-multiplicity $\lambda(x, y)$: max number of pairwise internally-disjoint x, y-paths

Local Menger Theorem: $\kappa(x, y) = \lambda(x, y)$, also $\kappa'(x, y) = \lambda'(x, y)$

Local search: technique for solving optimization problems by successively making small changes in a feasible solution

Loop: an edge joining a vertex to itself

(n, k, c)-magnifier: n-vertex graph of maximum degree k in which each set S with at most half the vertices has at least $c|S|$ neighbors outside S

Markov's inequality: for a nonnegative random variable, $\mathrm{Prob}(X \geq t) \leq E(X)/t$

Martingale: sequence of random variables such that $E(X_i | X_0, \ldots, X_{i-1}) = X_{i-1}$

Matching: a 1-regular subgraph

b-matching: given a constraint vector b, a subgraph H with $d_H(v) \leq b(v)$ for all v

Matrix rounding: problem of converting the data and row/column sums in a matrix to nearest integers up or down such that row and column sums remain correct

Matrix-tree theorem: subtracting the adjacency matrix from the diagonal matrix of degrees, deleting a row and column, and taking the determinant yields the number of spanning trees

Matroid: a hereditary system satisfying any of many equivalent and useful properties

Matroid basis graph: graph whose vertex set is the collection of bases of a matroid, adjacent when their symmetric difference has two elements

Matroid Covering Theorem: the number of independent sets needed to cover the elements of a matroid is $\max_{X \subseteq E} \lceil |X|/r(X) \rceil$

Matroid Intersection Theorem: the maximum size of a common independent set in two matroids on E equals the minimum over $X \subseteq E$ of the rank of X in the first matroid plus the rank of $\overline{X}$ in the second matroid

Matroid Packing Theorem: the maximum number of pairwise disjoint bases in a matroid is $\min_{r(X) < r(E)} \lfloor (|E| - CA(X))/(r(E) - r(X)) \rfloor$

Matroid Union Theorem: the union of matroids $M_1, \ldots, M_k$ is a matroid with rank function $r(X) = \min_{Y \subseteq X}(|X - Y| + \Sigma r_i(Y))$

Max-flow Min-cut Theorem: maximum flow value equals minimum cut value

Maximal (instance) satisfying P: no proper superset or supergraph also has property P

Maximal clique: a maximal vertex set inducing a clique

Maximal path or trial: non-extendible path or trail

Maximal planar graph: equivalent to planar triangulation

Maximum Cardinality Search: an algorithm for recognizing chordal graphs

Maximum (instance) satisfying P: no larger set or graph also has property P

Maximum degree Δ: maximum of the vertex degrees

Maximum flow: a feasible network flow of maximum value, or the value itself

Maximum genus $\gamma_M(G)$: the maximum genus surface on which G has a 2-cell embedding

Metric: a real-valued symmetric nonnegative binary function that is 0 only when the arguments are equal and satisfies the triangle inequality

Metric representation: an isometric embedding into a Cartesian product

Menger's theorems: minimax characterizations of connectivity by number of pairwise internally-disjoint or edge-disjoint paths between pairs of vertices

Meyniel graph: graph in which every odd cycle of length at least 5 has at least two chords

Minimal (instance) satisfying P: no proper subset or subgraph also has property P

Minimal imperfect graph: every proper induced subgraph is perfect

Minimally 2-connected: any edge-deletion destroys 2-connectedness

Minimum (instance) satisfying P: no smaller set or graph also has property P

Minimum cut: a network cut having minimum value, or the value of such a cut

Minimum degree $\delta(G)$: minimum of the vertex degrees

Minimum spanning tree: spanning tree with minimum sum of edge weights

Minor: graph or matroid obtained by a sequence of deletions and contractions

Mixed graph: a graph model allowing directed and undirected edges

Möbius ladder: the graph obtained by adding to an even cycle the chords between vertex pairs at maximum distance on the cycle (can be drawn as a ladder with a twist)

Möbius strip: the non-orientable surface obtained by identifying two opposite sides of a rectangle using opposite orientation

rth-moment: expectation of X^r

Model A: probability distribution making the graphs with vertex set $[n]$ equally likely

Model B: probability distribution making the graphs with vertex set $[n]$ and m edges equally likely

Monochromatic: in a coloring, a set having all elements the same color

Monotone graph property: preserved under deletion of edges or vertices

Multigraph: in this book, same as graph, allowing multiple edges and loops (some authors forbid loops from multigraphs)

Multinomial coefficient: counts arrangements of fixed multiplicities of items; with k_i items of type i, there are $\binom{\Sigma k_i}{k_1 \cdots k_m} = \frac{(\Sigma k_i)!}{\Pi(k_i!)}$ arrangements in a row

Multiple edges: repeated pairs of vertices in the edge set

Multiple interval representation: intersection representation with a union of intervals for each vertex

Mutual reachability: an equivalence relation on vertices of a digraph

Nearest-insertion: TSP heuristic to grow a cycle

Nearest-neighbor: TSP heuristic to grow a path

Near-perfect matching: a matching of size $(n(G) - 1)/2$

Neighbors: the vertices adjacent to a given vertex (as a verb, "is adjacent to")

Neighborhood $N(v)$, $\overline{N}(v)$: if "open", same as adjacency set; if "closed", includes in addition the vertex itself

Net flow: at a vertex, the sum of flows on exiting edges minus the sum of flows on entering edges

Network: a directed graph with a distinguished initial vertex (set) and a distinguished terminal vertex (set), in which each edge is assigned a flow capacity and sometimes also a flow demand (lower bound)

Node: vertex, especially in network flow problems

NOHO: "no one hears hir own information", a gossiping condition

Nondeterministic algorithm: allowed to "guess" by having parallel computation paths

Nondeterministic polynomial algorithm: polynomial-time computation paths after guessing a polynomial number of bits

Nonorientable: a surface with only one side

Nontrivial graph: having at least one edge

Nonplanar: having no embedding in the plane

Normal product: strong product

NP: the class of problems solvable by nondeterministic polynomial algorithms

NP-complete: NP-hard and in NP

NP-hard: can provide a polynomial algorithm for any problem in NP

Null graph: graph having no vertices

Obstruction: forbidden substructure
Odd anti-hole: complement of an odd hole
Odd component: component with an odd number of vertices
Odd cycle: cycle with an odd number of edges (vertices)
Odd graph: the disjointness graph of the k-subsets of $[2k + 1]$
Odd hole: chordless odd cycle
Odd vertex: vertex of odd degree
Odd walk: walk of odd length
Open walk: walk in which the end-vertices differ
Optimal tour: a solution to the traveling salesman problem or Chinese postman problem
Order: the number of vertices, sometimes called "size" when no confusion is possible
Ordered multigraph: a multigraph with an order relation (usually linear) on the edges
Order-preserving property: requirement that $X \subseteq Y$ implies $\sigma(X) \subseteq \sigma(Y)$ for a function on
    the subsets of a set
Orientable surface: a surface with two distinct sides
Orientation: an assignment of order to each of the edge pairs in an undirected graph,
    making it a directed graph
Out-degree: for a vertex, the number of arcs of which it is the tail
Outerplanar graph: a planar graph embeddable in the plane so that all the vertices
    belong to the boundary of the exterior region
Outerplane graph: a particular embedding of an outerplanar graph

Packing number: the maximum number of something findable in a graph without violat-
    ing some condition P
Page: one of the outerplanar subgraphs in a book embedding
Pagenumber: minimum number of pages in a book embedding
Pan-connected: the condition of having, for every pair of vertices u, v, (u, v)-paths of all
    lengths at least $d(u, v)$
Pancylic: having cycles of all lengths at least 3
Parallel elements: non-loops in a matroid that form a set of rank 1
Parent: neighbor of a vertex along the path to the root in a rooted tree
Parity: odd or even
Parity subgraph of G: subgraph H such that $d_H(v) \equiv d_G(v) \bmod 2$ for all $v \in V(G)$
k-partite: coverable by k independent sets (same as k-colorable)
Partite set: block of a vertex partition, often called a "part" of a k-partite graph
Partition matroid: matroid induced by a partition of the ground set in which a set is inde-
    pendent if and only if it has at most one element from each block of the partition
Partitionable graph: graph with $aw + 1$ vertices such that each graph formed by deleting
    one vertex is colorable by w stable sets of size a and coverable by a cliques of size w
Part-sizes: sizes of the partite sets
Parts: partite sets
Path: an open walk with no repeated vertices
(u, v)-path: a path with u and v as endvertices
Path addition: a step in an ear decomposition
Path decomposition: expression of a graph as a union of edge-disjoint paths
p-critical graph: an imperfect graph whose proper induced subgraphs are all perfect
Pendant edge: incident with a pendant vertex
Pendant vertex: 1-valent vertex
α-perfect: $\alpha(H) = \theta(H)$ for evey induced subgraph H
β-perfect: $\alpha(H)\omega(H) \geq n(H)$ for every induced subgraph
γ-perfect: $\chi(H) = \omega(H)$ for evey induced subgraph H
Perfect elimination ordering: deletion order such that each deleted vertex is adjacent to
    the vertices of a clique in the remaining graph (also simplicial elimination ordering)
Perfect graph: γ-perfect - $\chi(H) = \omega(H)$ for evey induced subgraph H
Perfect Graph Theorem (PGT): a graph is perfect if and only if its complement is perfect
Perfect order: a vertex order yielding optimal greedy colorings for all subgraphs
Perfectly orderable graph: having a perfect order
Perfect matching: a 1-factor
Peripheral vertex: a vertex of maximum eccentricity
Permutation: a bijection from a finite set to itself

Permutation graph: representable by a permutation σ by $v_i \leftrightarrow v_j$ if and only if σ reverses the order of i and j

Permutation matrix: a 0,1-matrix having exactly one 1 in each row and column

Petersen graph: a graph disproving many reasonable conjectures, it is the disjointness graph of the 2-sets in a 5-element set

Pigeonhole principle: every set of numbers has one at least as large as the average

Pigeonhole property: a probability space contains a point where the value of a random variable is at least as large as its expectation

Planar: embeddable in the plane

Plane graph: a particular planar embedding of a planar graph

Plane tree: tree with fixed cyclic embedding order of edges at each vertex

Planted tree: rooted plane tree

Platonic solid: bounded regular polyhedron

Point: vertex

Poisson distribution: with mean μ, given by $\text{Prob}(X = k) = e^{-\mu} \mu^k / k!$

Polygonal curve: piecewise linear curve

Polygonal u, v-path: polygonal curve from u to v

Polyhedron: an intersection of half-spaces

Polytope: the convex hull of a set of vertices

Positional game: objective is seizing the positions of a winning set

Potential: a vertex labeling used in the dual to the min-cost flow problem

kth-power (G^k): 1) the graph with vertex set $V(G)$ in which $u \leftrightarrow v$ if and only if $d_G(u, v) \leq k$; 2) the graph with vertex set $V(G)$ in which $u \leftrightarrow v$ if and only if u and v are connected by a walk of length k in G

Predecessor: for v in a digraph, a vertex u with $u \rightarrow v$

Predecessor set: for v in a digraph, the set of precedessors

Prefix-free code: no code word is a prefix of another

Prim's algorithm: grows a minimum spanning tree by adding a leaf to the current tree in the cheapest way

Principal submatrix: submatrix using the same rows as columns

Product graph: obtained by combining factors in several possible ways - see Cartesian product, direct product, normal product, strong product, weak product

Product dimension: minimum number of cliques whose weak product contains G as an induced subgraph

Product representation: encoding of graph such that vertices are adjacent if and only if their codes differ in every coordinate

Proper coloring: 1) for vertices, a coloring in which no edge is monochromatic; 2) for edges, a coloring in which no intersecting edges get the same color

Proper subgraph of G: a subgraph not equal to G

Proper subset of S: a subset not equal to S

Proposal Algorithm: procedure for creating a stable matching

Prüfer code: for a labeled tree, a sequence of length $n - 2$ obtained by successively deleting the leaf with smallest label and recording its neighbor's label

Pseudograph: model to allow loops used by some authors who require that a multigraph has no loops

Quasi-random: a sequence of graphs behaving asymptotically like random graphs in one of several equivalent precise manners

Quasi-transitive orientation: an orientation in which $x \rightarrow y$ and $y \rightarrow z$ implies that x and z are adjacent but does not specify the orientation of that edge

Radius: the minimum eccentricity of the vertices

Ramsey number: the minimum number of vertices such that assigning colors to all pairs of those vertices produces a monochromatic clique of specified size (or a specified graph) in one of the colors

Random graph: a graph from a probability space, most often the space in which each labeled pair of vertices independently has probability p of adjacency; typically, $p = 1/2$ or p is a function of n

Random variable: a variable that takes on a value at each point in a probability space

Rank function: 1) for a $S \subseteq V(H)$, the largest size of an edge (or independent set) contained in S (applies to hypergraphs and matroids)

Reachability matrix: for a directed graph, the matrix in which entry i, j is 1 if there is a path from vertex i to vertex j, otherwise 0

Reciprocity theorem: a relationship between functions f and $\bar{f}$ for related counting problems, generally stating that $\bar{f}(n, k) = (-1)^n f(n, -k)$

Reconstructible: a graph determined (up to isomorphism) by its multiset of subgraphs obtainable by deleting a single vertex

Reconstruction Conjecture: the conjecture that all graphs with at least 3 vertices are reconstructible

Rectilinear crossing number: the minimum number of crossings in a drawing of the graph in the plane in which all edges appear as straight line segments

Reducible configuration: a configuration that cannot appear in a minimal 5-chromatic planar graph

Reflexive: 1) for a digraph, having a loop at every vertex; 2) for a binary relation R, having xRx for all x

Region: for an embedding of a graph on a surface, a maximal connected subset of the surface that does not contain any part of the graph

Regular: having all vertex degrees equal

Regular matroid: representable over every field

k-regular: having all vertex degrees equal to k

Representable matroid: linear matroid

Restriction martingale: martingale in which the value of successive variables is an expectation over a shrinking subset of the probability space

Rigid circuit graph: chordal graph

Robbins' Theorem: every 2-edge-connected graph has a strong orientation

Root: 1) a distinguished vertex; 2) for a branching, the vertex from which every other is reachable

Rooted plane tree: a tree with a distinguished root vertex so that children of each nonleaf have a specified left-to-right ordering in the plane

Rotation scheme: a description of a 2-cell embedding; a circular permutation of the edges at each vertex, giving their counter-clockwise order around the vertex

SATISFIABILITY: the original NP-complete problem

Satisfiable: formula having a "yes" answer in the SATISFIABILITY problem

Saturated arc: for a network flow, an arc in which the flow equals the capacity

Saturated vertex: 1) for a matching, a matched vertex; 2) for a b-matching, a vertex with $b(v)$ incident edges.

Score sequence: the sequence of outdegrees in a tournament

Second moment method: method for obtaining threshold functions

Self-complementary: isomorphic to the complement

Self-converse: isomorphic to the converse

Self-dual: isomorphic to the dual

Semi-strong perfect graph theorem: intermediate between the perfect graph theorem and strong perfect graph conjecture, concerns the P_4-structure of a graph

Semipath: an semiwalk in which each vertex appears at most once

Semiwalk: a sequence of edges (or adjacent vertices) in a directed graph such that each successive pair of edges are adjacent, without regard to the orientation of the edges

Separable: having a cut-vertex

Separating set: a vertex set whose deletion increases the number of components

Separator theorem: for a hereditary class of graphs, specifies a small function of n such that deleting at most that many vertices from a graph in the class splits the remaining vertices in a balanced way

k-ary sequence: list chosen from a set of size k

Series-parallel: a single edge between source and sink, or obtained recursively from two series-parallel graphs by identifying the sink of one with the source of the other (series construction) or by identifying the two sinks and the two sources (parallel construction)

k-set: set of size k

Shannon Switching game: a game played by the Spanner and the Cutter, one seeking to seize a set of elements spanning a specified element, the other seeking to prevent this

Shift graph: graph on the 2-subsets of $[n]$ having $\{i, j\}$ adjacent to $\{j, k\}$ when $i < j < k$

Signed (di)graph: special case of weighted (di)graph, assinging $+$ or $-$ to each edge

Simple: 1) for a graph, having no loops or multiple edges; 2) for a matroid, having no loops or parallel elements

Simplicial: 1) a vertex whose neighbors induce a clique; 2) a hypergraph such that every subset of any edge is also an edge; 3) a polytope such that every face is a simplex

Sink: the distinguished terminal vertex (set) in a network

Size: the number of edges, sometimes used for the number of vertices

Skew partition: a partition X, Y of $V(G)$ such that $G[X]$ and $\overline{G}[Y]$ are disconnected

f-soluble: having an edge weighting so that the sum of the weights incident to v is $f(v)$

Source: the distinguished initial vertex (set) in a network

Source/sink cut: a partition of the vertices of a network into sets S, T such that S contains the source and T contains the sink

Span function: the span of a set X in a hereditary system consists of X and the elements not in X that complete circuits with subsets of X

Spanning subgraph: contains each vertex

Spanning set: a set whose span (in a hereditary system on E) is E

Spectrum: the set of eigenvalues

Sphericity: minimum dimension in which G is the intersection graph of spheres

Split graph: a graph whose vertices can be covered by a clique and an independent set

Splittance: minimum number of edges to be added or deleted to make a split graph

Square of a graph: the second power

Squashed cube dimension: minimum length of the vectors in a squashed cube embedding

Squashed cube embedding: encodes vertices by $0, 1, *$-vectors such that distance between two vertices is the number of coordinates where one has 0 and the other has 1

Stability number: independence number

Stable matching: a matching having no instance of x and y each preferring the other to their current partner in the matching

r-staset: stable set of size r

Stable set: set of vertices inducing a subgraph with no edges

Star: the tree with at most one non-leaf; i.e., $K_{1,n-1}$

Star-cutset: vertex-cut inducing a subgraph containing a dominating vertex

Star-cutset Lemma: no p-critical graph has a star-cutset

s, t-numbering: a numbering of the vertices of a graph such that from each vertex there is a path to s through successively lower-indexed vertex and a path to t through successively higher-indexed vertices.

Steinitz exchange property: the property of span functions that if e is in the span of $X \cup f$ but not in the span of X, then f is in the span of $X \cup e$

Steinitz's Theorem: 3-connected planar graphs have only one embedding in the plane (more precisely, only one dual graph)

Strict: digraph with at most one copy of each ordered pair, and no loops

Strictly balanced: average vertex degree is strictly greater than average vertex degree of any subgraph

Strength: of a theorem, the fraction of the time the conclusion holds that the hypothesis also holds

Strong component: maximal strongly connected subdigraph

Strong orientation: orientation of G in which each vertex is reachable from every other

Strong Perfect Graph Conjecture (SPGC): the conjecture that a graph is perfect if and only if it has no odd hole or odd antihole

Strong absorption property (matroids): if $r(X \cup e) = r(X)$ for all $e \in Y$, then $r(X \cup Y) = r(X)$

Strong product $G_1 \cdot G_2$: a graph product with vertex set $V(G_1) \times V(G_2)$ and edge set $(u_1, v_1) \leftrightarrow (u_2, v_2)$ if $u_1 = u_2$ or $u_1 \leftrightarrow u_2$ and $v_1 = v_2$ or $v_1 \leftrightarrow v_2$

Strongly connected (or strong): digraph with each vertex reachable from all others

Strongly perfect: a graph in which some stable set meets every maximal clique

Strongly regular: a k-regular graph whose adjacent pairs have λ common neighbors, and whose nonadjacent pair have μ common neighbors

Subconstituent: the subgraph induced by a vertex neighborhood or by a vertex non-neighborhood

Subdigraph: a subgraph of a directed graph

Subdivision: 1) the operation of replacing an edge by a 2-edge path through a new vertex; 2) a graph obtained by a sequences of subdivisions

H-subdivision: a graph obtained from H by subdivisions

Subgraph: a graph whose vertices and edges all belong to G

k-subset: subset of size k

Submodular: function such that $r(X \cup Y) + r(X \cap Y) \le r(X) + r(Y)$ for all sets X, Y

Submodularity property: the property of having a submodular rank function

Subtree representation: assigns subtrees of a host tree to each vertex of a chordal graph so that vertices are adjacent if and only if the corresponding subtrees intersect

Successor: for u in a digraph, a vertex v with $u \to v$

Successor set: for u in a digraph, the set of successors

Sum: 1) for cycles and cocycles, taken modulo 2. 2) for a graph, the disjoin union. 3) for matroids on disjoint sets, the matroid on their union whose independent sets are all unions of an independent set from each

Supbase: a set of elements containing a base

Supergraph: a graph of which G is a subgraph

Superregular: a regular graph that is null or whose subconstituents are all superregular

Supply: source constraint in a transportation network

Sweep subgraph: the curbs available for sweeping in the streetsweeper problem

2-switch: a degree-preserving switch of two disjoint edges for two others

Symmetric: 1) for a graph, having a non-trivial automorphism; 2) for a digraph, having $u \to v$ iff $v \to u$; 3) for a binary relation R, having xRy iff yRx

Symmetric difference $A \triangle B$: the set of elements in exactly one of A and B

System of distinct representatives (SDR): from a collection of sets, a choice of one member from each set so that all the representatives are distinct

Szekeres-Wilf Theorem: $\chi(G) \le 1 + \max_{H \subseteq G} \delta(H)$

Tail: the first vertex of an edge in a digraph

Tait coloring: for a planar cubic graph, a proper 3-edge-coloring

Tarry's Algorithm: procedure for exploring a maze

Telephone problem: gossip problem

Telegraph problem: directed version of gossip problem with one-way transmissions

Tensor product: weak product

Terminal edge: a cut edge incident with a leaf

Ternary matroid: representable over the field with three elements

Thickness: the minimum number of planar graphs whose union is G

Threshold dimension: minimum number of threshold graphs whose union is G

Threshold function: a function t of n such that some property almost always or almost never occurs, depending on whether a particular parameter belongs to $o(t)$ or to $\omega(t)$.

Threshold graph: having a threshold t and a vertex weighting w such that $u \nleftrightarrow v$ iff $w(u) + w(v) \le t$; many other characterizations, including absence of a 2-switch and existence of a construction ordering by adding isolated or dominating vertices

Topological graph theory: the study of drawings of graphs on surfaces

Toroidal: 1) for a graph, having a 2-cell embedding on the torus; 2) for a topological parameter, the version using the torus in place of the plane (toroidal thickness, crossing number, etc.)

Torus: the (orientable) surface with one handle

Total coloring: a labeling of both the vertices and edges

Total Coloring Conjecture: every graph G has a proper total coloring using at most $\Delta(G) + 2$ colors

Total domination number: minimum number of vertices in a set S such that every vertex has a neighbor in S

Total interval number: minimum of the total number of intervals used to represent G as the intersection graph of unions of intervals on the real line

Total graph: the intersection graph of the sets in $V(G) \cup E(G)$, for some G

Totally unimodular: a matrix in which all square submatrices have determinant 0 or ± 1

Toughness: the minimum t such that $|S| \ge t \cdot c(G - S)$, where S is any vertex cut and $c(G - S)$ is the number of components of the subgraph obtained by deleting S

Tournament: an orientation of the complete graph

Trace: sum of the diagonal elements of a matrix

Traceable: having a Hamiltonian path

Trail: a walk in which no edge appears more than once

Trail-coverable: having a trail whose vertices form a vertex cover

Transitive: 1) for a digraph, uw must be an arc whenever uv and vw are arcs; 2) for a group action, such as the automorphism group acting on the vertice set, the existence of a group operator mapping each element to any other

Transitive closure: 1) for a digraph D, the digraph with $u \to w$ whenever there is a path from u to w in D; 2) for a relation R, the relation S with xSy whenever there is a sequence $x_0, \ldots, x_k$ with $x = x_0 R x_1 R \cdots R x_k = y$

Transitivity of dependence (matroids): $e \in \sigma(X)$ and $X \subseteq \sigma(Y)$ imply $e \in \sigma(Y)$

Transportation Problem: generalization of the assignment problem with supplies at each source and demands at each destination

Transportation constraints: supplies and demands

Transversal: a system of distinct representatives (this is the word used when the concept is generalized); also used for a system of representatives not necessarily distinct

Transversal matroid: a matroid whose elements are one partite set of a bipartite sets and whose independent sets are those that can be saturated by matchings

Traveling Salesman Problem (TSP): finding a minimum weight spanning cycle

Tree: a connected graph with no cycles

k-ary tree: rooted tree with at most k children at each internal node

Triangle: a cycle of length 3

Triangle inequality: $d(x, y) + d(y, z) \geq d(x, z)$

Triangle-free: not having K_3 as a subgraph

Triangular chord: chord of length two along a path or cycle

Triangulated: having no chordless cycle

Triangulation: a graph embedding on a surface such that every region is a 3-gon

Trivalent: having degree 3

Trivial graph: graph with no edges (some authors restrict to one vertex)

Turán graph: complete equipartite graph

Turán's theorem: charcterization of the complete equipartite r-partite graphs as the largest graphs of a given order with no $r + 1$-clique

Tutte polynomial: a generalization of the chromatic polynomial and of other polynomials

Tutte's Theorem: 1) for connectivity, characterization of 3-connected graphs by contractions to wheels; 2) for planar graphs, 3-connected planar graphs have embeddings with all bounded faces convex; 3) for factorization, a necessary and sufficient condition for the existence of an f-factor

Twins: vertices having the same neighborhood (false twins are adjacent vertices with the same closed neighborhoods)

Unavoidable set: a collection of configurations such that every graph in a specified class contains some configuration in the collection

Underlying graph: the graph obtained by igoring the orientation on the edges of a directed graph

Unicyclic: having exactly one cycle

k-uniform hypergraph: having each edge of size k

Uniform matroid $U_{k,n}$: a matroid on $[n]$ in which the independent sets are those of size at most k

Uniformity property (matroids): for all $X \subseteq E$, the maximal independent subsets of X have the same size

Union ($G_1 \cup G_2$): a graph whose vertex set is the union of the vertices in G_1 and G_2 and whose edge set is the union of the edges in G_1 and G_2 (written $G_1 + G_2$ if the vertex sets are disjoint)

Union of matroids: the union of matroids $M_1, \ldots, M_k$ is the hereditary system whose independent sets are $\{I_1 \cup \cdots \cup I_k : I_i \in \mathbf{I}_i\}$

Unimodular: for matrices, having determinant 0, +1, or -1

Unlabeled graph: informal term for isomorphism class

Unit-distance graph: the graph with vertex set $\mathbb{R}^2$ in which points are adjacent if the distance between them is 1

M-unsaturated: vertex not belonging to an edge of M

Upper embeddable: having a 2-cell embedding on a surface of genus $\lfloor (e(G) - n(G) + 1)/2 \rfloor$

Valence: degree
Vámos matroid: a particular nonlinear matroid with eight elements
Value of a flow: the net flow out of the source or into the sink
Variance: expected squared deviation from the mean
Vectorial matroid: linear matroid
Vertex: element of $V(G)$, the vertex set
Vertex chromatic number: chromatic number
Vertex connectivity: connectivity
Vertex cover: a set of vertices containing at least one endpoint of every edge
Vertex cut: a separating set of vertices
Vertex-critical: deletion of any vertex changes the parameter
Vertex cut: a separating set of vertices
Vertex multiplication: a replacement of vertices of G by independent sets such that copies of x and y are adjacent if and only if $xy \in E(G)$
Vertex partition: a partition of the vertex set
Vertex set $V(G)$: the set of elements on which the graph is defined
Vertex-transitive: for each pair $x, y \in V(G)$, some automorphism of G maps x to y
Vizing's Theorem: $\chi' \leq \Delta + 1$

Walk: a sequence of vertices and edges in a graph such that each vertex belongs to the edge before and after it (in a digraph, must follow arrows)
u, v-walk: a walk from u to v.
Weak elimination property: property of matrices that the union of distinct intersecting circuits contains a circuit that avoids a specified point in the intersection
Weak product $G_1 \otimes G_2$: a graph product with vertices $V(G_1) \times V(G_2)$, and edges $(u_1, v_1) \leftrightarrow (u_2, v_2)$ iff $u_1 \leftrightarrow u_2$ and $v_1 \leftrightarrow v_2$
Weakly connected: a directed graph whose underlying graph is connected
Weakly chordal: having no chordless cycle of length at least 5 in G or $\overline{G}$
Weight: a real number
Weighted: having an assignment of weights (to edges and/or vertices)
Wheel: a graph obtained by taking the join of a cycle and a single vertex
Well Ordering Property: every nonempty set (of natural numbers) has a least element
Whitney's 2-isomorphism Theorem: a characterization of the pairs of graphs whose cycle matroids are isomorphic
Wiener index: the sum of the pairwise distances between vertices

Glossary of Notation

Usage of non-alphabetic notation

$\leftrightarrow$	adjacency relation
$\rightarrow$	successor relation (digraph)
$\cong$	isomorphism relation
xRy	general relation
$\lfloor x \rfloor$	floor of number
$\lceil x \rceil$	ceiling of number
$[n]$	$\{1, \ldots, n\}$
$\lvert S \rvert$	cardinality of set
$\{x\colon P(x)\}$	set description
$\overline{X}$	complement of set
$\varnothing$	empty set
∞	infinity
uv	edge
$\overline{G}$	complement of graph G
G^*	(planar) dual
$G[S]$	subgraph of G induced by S
D^*	condensation of a digraph D
$[S, \overline{S}]$	edge cut
$[S, T]$	source-sink cut
$G - v$	deletion of vertex
$G - e$	deletion of edge
$G \cdot e$	contraction of edge
$G{\circ}x$	vertex duplication
$G{\circ}h$	vertex multiplication
$G + H$	disjoint union of graphs
$G \vee H$	join of graphs
$G \square H$	Cartesian product of graphs
$G \triangle H, A \triangle B$	symmetric difference
$A \times B$	Cartesian product of sets
$\cup$	union
$\cap$	intersection
$\binom{n}{k}$	binomial coefficient
$\binom{S}{r}$	r-subsets of S
$\binom{n}{n_1 \cdots n_k}$	mulitnomial coefficient
$a \equiv b(\bmod n)$	congruence relation
$\mathbf{1}_n$	n-vector with all entries 1
$Y \mid X$	conditional variable or event

Usage of Roman alphabet

$A(G)$	adjacency matrix
$\mathrm{Adj}\,A$	adjugate matrix
$B(G)$	bandwidth
$b(G)$	biparticity
$b(v)$	count of blocks containing v
$\mathbf{B}_M$	bases of matroid
$\mathbf{C}_M$	circuits of matroid
C_n	cycle with n vertices
C_n^d	circulant graph
$c(G)$	number of components
$c(G)$	circumference
$C(G)$	(Hamiltonian) closure of G
$c(xy)$	cost of an edge
$c(e)$	capacity of an edge
$cap(S, T)$	capacity of the cut $[S, T]$
$d_1, \ldots, d_n$	degree sequence
$d(v), d_G(v)$	degree of v in G
$d^+(v), d^-(v)$	out-degree, in-degree
D, D^+, D^-	diagonal matrices of degrees
D	digraph
$d(u, v)$	distance from u to v
$\mathrm{diam}\,G$	diameter
$\det A$	determinant
$\partial(v)$	demand at at a vertex
$E(G)$	edge set
$E(X)$	expectation of X
$e(G)$	size (number of edges)
$ex(n; H)$	generalized Turán number
$f^+(v), f^+(S)$	total flow on exiting edges
$f^-(v), f^-(S)$	total flow on entering edges
G^k	kth power of G
G	graph (or digraph)
G^p	random graph in Model A
$H_{k,n}$	Harary graph
$\mathbf{H}_M$	hypobases of matroid
$\mathbf{I}_M$	independent sets of matroid
I	identity matrix

J	matrix of all 1's	$\gamma(G)$	genus
K_n	clique with n vertices	$\Delta(G)$	maximum degree
$K_{r,s}$	complete bipartite graph	$\delta(G)$	minimum degree
$L(G)$	line graph	$\delta^+(G), \delta^-(G)$	minimum out-, in-degree
$l(D)$	maximum length of path in D	$\varepsilon_G(u)$	eccentricity of u in G
$l(F)$	length of a face	$\Theta(f)$	growth rate
$l(e)$	lower bound on flow in edge	$\theta(G)$	clique cover number
$\lg x$	logarithm base 2	$\theta'(G)$	intersection number
$\ln x$	natural logarithm	$\kappa(G)$	(vertex) connectivity
$M(G)$	incidence matrix of G	$\kappa(x, y)$	local connectivity
$M(G)$	cycle matroid of G	$\kappa'(G)$	edge-connectivity
M^*	dual hereditary system	$\kappa'(x, y)$	local edge-connectivity
$M.F$	contraction of M to F	$\kappa(r; G)$	local-global connectivity
$M \mid F$	restriction of M to F	$\lambda(G)$	path-multiplicity
$\mathbb{N}$	set of natural numbers	$\lambda'(G)$	edge-path-multiplicity
$N^+(v), N^-(v)$	out-, in-neighborhood	$\lambda(x, y)$	local path-multiplicity
$N(v) N_G(v)$	(open) neighborhood of v in G	$\lambda'(x, y)$	local edge-path-multiplicity
$N[v]$	closed neighborhood of v	$\lambda_1, \ldots, \lambda_n$	eigenvalues
$n(G)$	order (number of vertices)	$\mu_1, \ldots, \mu_n$	eigenvalues
O_k	odd graph	$\nu(G)$	crossing number
$O(f), o(f)$	growth rate	Π	product
$o(H)$	number of odd components	$\rho(G)$	maximum density
$P(A)$	probability of an event	Σ	summation
$P(i, j)$	v_i, v_j-portion of path P	σ, π, τ	permutation
P_n	path with n vertices	$\sigma(v)$	supply at a vertex
$\mathrm{pdim} G$	product dimension	σ_M	span function
$\mathrm{qdim} G$	squashed-cube dimension	$\tau(G)$	number of spanning trees
Q_k	k-dimensional hypercube	$\Upsilon(G)$	arboricity
$R(k, l)$	Ramsey number	$\phi(G; \lambda)$	characteristic polynomial
$R(G, H)$	graph Ramsey number	$\chi(G)$	chromatic number
$R(G), R_k(G)$	diagonal Ramsey number	$\chi'(G)$	edge-chromatic number
$\mathbb{R}$	set of real numbers	$\chi'(G)$	chromatic index
$\mathbb{R}^2$	$\mathbb{R} \times \mathbb{R}$	$\chi(G; k)$	chromatic polynomial
r_M	rank function	$\hat{\chi}'(G)$	edge-choosability
S_γ	surface with γ handles	$\hat{\chi}(G)$	choice number, choosability
$\mathbf{S}_M$	supbases of matroid	$\hat{\chi}(G)$	list chromatic number
$\mathrm{Spec}(G)$	spectrum	$\psi(G; \lambda)$	minimum polynomial
A^T	transpose of matrix	$\Omega(f), \omega(f)$	growth rate
T	tree, tournament	$\omega(G)$	clique number
$T_{n,r}$	Turán graph	ω_n	slowly growing function
$t_r(n)$	size of Turán graph		
$t(n)$	threshold probability function		
$U_{k,n}$	uniform matroid		
$u(e)$	upper bound on flow edge		
$val(f)$	value of a flow f		
$V(G)$	vertex set		
$W(G)$	Wiener index		
W_n	wheel with n vertices		
$w(e)$	weight of edge		
$\mathbf{Z}$	set of integers		
$\mathbf{Z}_p$	integers modulo p		

Usage of Greek alphabet

$\alpha(G)$	independence number
$\alpha'(G)$	maximum size of matching
$\beta(G)$	vertex cover number
$\beta'(G)$	edge cover number

References

Articles cited in the text that were yet to appear at the time of publication have been marked with the margin date of 1996 for ease of identification. The full listing of the citation still says "to appear".

[1972] Abbott H.L., Lower bounds for some Ramsey numbers. *Discrete Math.* **2** (1972), 289–293. [370]

[1982] Acharya B.D. and M. Las Vergnas, Hypergraphs with cyclomatic number zero, triangulated graphs, and an inequality. *J. Comb. Th. B* **33** (1982), 52–56. [297]

[1993] Ahuja R.K., T.L. Magnanti, and J.B. Orlin, *Network Flows.* Prentice Hall (1993). [75, 130, 158, 163, 172]

[1979] Aigner M., *Combinatorial Theory.* Springer-Verlag (1979). [328, 347]

[1984] Aigner M., *Graphentheorie. Eine Entwicklung aus dem 4-Farben Problem.* B.G. Teubner Verlagsgesellschaft (1984) (English translation by BCS Associates, 1987). [270]

[1983] Ajtai M., J. Komlós, and E. Szemerédi, Sorting in $c \log n$ parallel steps. *Combinatorica* **3** (1983), 1–19. [360, 445]

[1986a] Alon N., Eigenvalues, geometric expanders, sorting in rounds and Ramsey Theory. *Combinatorica* **6** (1986), 207–219. [445]

[1986b] Alon N., Eigenvalues and Expanders. *Combinatorica* **6** (1986), 83–96. [446]

[1990] Alon N., The maximum number of Hamiltonian paths in tournaments. *Combinatorica* **10** (1990), 319–324. [407]

[1993] Alon N., Restricted colorings of graphs. In *Surveys in Combinatorics, 1993.* London Math. Soc. Lect. Notes 187 Cambridge Univ. Press (1993) 1–33. [389]

[1992] Alon N. and M. Tarsi, Colorings and orientations of graphs. *Combinatorica* **12** (1992), 125–134. [387]

[1984] Alon N. and V.D. Milman, Eigenvalues, expanders and superconcentrators. In *Proc. 25th IEEE Symp. Found. Comp. Sci.,* Singer Island, FL. IEEE (1984) 320–322. [445, 446]

[1985] Alon N. and V.D. Milman, λ_1, isoperimetric inequalities for graphs and superconcentrators. *J. Comb. Theory (B)* **38** (1985), 73–88. [445]

[1985] Alon N. and Y. Egawa, Even edge colorings of a graph. *J. Comb. Theory (B)* **38** (1985), 93–94. [401]

[1992] Alon N., J.H. Spencer, *The Probabilistic Method.* Wiley (1992). [405, 408, 445]

[1977] Andersen L.D., On edge-colourings of graphs. *Math. Scand.* **40** (1977), 161–175. [217]

[1996] Ando K., A. Kaneko, and S. Gervacio, The bandwidth of a tree with k end vertices is at most $(k + 1)/2$. In *Proc. Intl. Conf. Comb..* Elsevier (1996) . [60]

474

[1976] Appel K. and W. Haken, Every planar map is four colorable. *Bull. Amer. Math. Soc.* **82** (1976), 711–712.   [270]

[1977] Appel K. and W. Haken, Every planar map is four colorable. Part I: Discharging. *Illinois J. Math.* **21** (1977), 429–490.   [270]

[1986] Appel K. and W. Haken, The four color proof suffices. *Math. Intelligencer* **8** (1986), 10–20.   [270]

[1989] Appel K. and W. Haken, *Every Planar Map Is Four Colorable, Contemporary Mathematics* **98**. Amer. Mathematical Society (1989).   [270]

[1977] Appel K., W. Haken, and J. Koch, Every planar map is four colorable. Part II: Reducibility. *Illinois J. Math.* **21** (1977), 491–567.   [270]

[1982] Ayel J., Hamiltonian cycles in particular k-partite graphs. *J. Comb. Theory (B)* **32** (1982), 223–228.   [229]

[1979] Babai L. and L. Kučera, Canonical labelling of graphs in linear average time. In *Proc. 20th IEEE Symp. Found. Comp. Sci.* Puerto Rico. (1979) 39–46.   [419]

[1980] Babai L., P. Erdős, and S.M. Selkow, Random graph isomorphisms. *SIAM J. Computing* **9** (1980), 628–635.   [418]

[1953] Bäbler F., Über eine spezielle Klasse Euler'scher Graphen. *Comment. Math. Helv.* **27** (1953), 81–100.   [95]

[1966] Bacharach M., Matrix rounding problems. *Management Sci.* **9** (1966), 732–742.   [168]

[1972] Baker B. and R. Shostak, Gossips and telephones. *Discrete Math.* **2** (1972), 191–193.   [384]

[1984] Batagelj V., Inductive classes of cubic graph. In *Finite and Infinite Sets.* (ed. A. Hajnal, L. Lovász, V.T. Sós), Proc. 6th Hung. Comb. Colloq. (Eger 1981) *Coll. Math. Soc. János Bolyai* **37**, Elsevier (1984) 89–101.   [48]

[1976] Bean D.R., Effective coloration. *J. Symbolic Logic* **41** (1976), 469–480.   [181]

[1965] Behzad M., *Graphs and their chromatic numbers.* Ph.D. Thesis, Michigan State University (1965).   [403]

[1971] Behzad M., The total chromatic number of a graph: A survey. In *Combinatorial Mathematics and its Applications.* (Proc. Conf. Oxford, 1969) Academic Press (1971) 1–8.   [403]

[1968] Beineke L.W., Derived graphs and digraphs. In *Beiträge zur Graphentheorie.* Tuebner (1968) 17–33.   [214]

[1959] Benzer S., On the topology of the genetic fine structure. *Proc. Nat. Acad. Sci. USA* **45** (1959), 1607–1620.   [298]

[1957] Berge C., Two theorems in graph theory. *Proc. Nat. Acad. Sci. U.S.A.* **43** (1957), 842–844.   [100]

[1958] Berge C., Sur le couplage maximum d'un graphe. *C.R. Acad. Sci. Paris* **247** (1958), 258–259.   [123]

[1960] Berge C., Les problèmes de coloration en théorie des graphes. *Publ. Inst. Statist. Univ. Paris* **9** (1960), 123–160.   [201]

[1961] Berge C., Färbung von Graphen, deren sämtliche bzw. deren ungerade Kreise starr sind. *Wiss. Z. Martin-Luther-Univ. Halle–Wittenberg Math.-Natur. Reihe* **10** (1961), 114.   [290]

[1973] Berge C., *Graphs and Hypergraphs.* North-Holland (1973) (translation and revision of *Graphes et Hypergraphes* (Dunod, 1970).   [177, 181]

[1984] Berge C. and P. Duchet, Strongly perfect graphs. In *Topics on Perfect Graphs.* (ed. C. Berge, V. Chvátal), *Annals Discr. Math.* **21** North-Holland (1984) 57–61.   [301]

[1984] Berge C. and V. Chvátal, *Topics on Perfect Graphs, Annals Discr. Math.* **21**. North-Holland (1984).   [288]

[1976] Bermond J.C., On Hamiltonian walks. In *Proc. Fifth Brit. Comb. Conf.*. (ed. C.St.J.A. Nash-Williams and J. Sheehan) Utilitas Math. (1976) 41–51.
                                                                                         [395, 396]

[1981] Bernstein P.A. and N. Goodman, Power of natural semijoins. *SIAM J. Computing* **10** (1981), 751–771.   [297]

[1988] Bertschi M. and B.A. Reed, Erratum: A note on even pairs. *Discrete Math.* **71** (1988), 187 (also B.A. Reed, A note on even pairs, *Discrete Math.* 65(1987), 317–318. [320]

[1994] Bhasker J., T. Samad, and D.B. West, Size, chromatic number, and connectivity. *Graphs and Combinatorics* **10** (1994), 209–213. [190]

[1993] Biggs N., *Algebraic Graph Theory (2nd ed.).* Cambridge University press (1993) (1st ed. 1974). [433, 447]

[1912] Birkhoff G.D., A determinant formula for the number of ways of coloring a map. *Ann. of Math.* **14** (1912), 42–46. [194]

[1913] Birkhoff G.D., The reducibility of maps. *Amer. J. Math.* **35** (1913), 114–128. [272, 283]

[1981] Bixby R.E., Matroids and operations research. In *Advanced techniques in practice of operations research.* (ed. H.J. Greenberg, F.H. Murphy, and S.H. Shaw) North-Holland (1981) 333–458. [328]

[1992] Björner A., M. Las Vergnas, B. Sturmfels, N. White, and G. Ziegler, *Oriented Matroids.* Cambridge University Press (1992). [347]

[1979] Bland R.G., H.-C. Huang and L.E. Trotter Jr., Graphical properties related to minimal imperfection. *Discrete Math.* **27** (1979), 11–22. [306, 308, 320]

[1979] Blass A. and F. Harary, Properties of almost all graphs and complexes. *J. Graph Theory* **3** (1979), 225–240. [430]

[1981a] Bollobás B., Threshold functions for small subgraphs. *Math. Proc. Camb. Phil. Soc* **90** (1981), 197–206. [415]

[1981b] Bollobás B., Degree sequences of random graphs. *Trans. Amer. Math. Soc.* **267** (1981), 41–52. [418, 420]

[1982] Bollobás B., Vertices of given degree in a random graph. *J. Graph Theory* **6** (1982), 147–155. [418]

[1985] Bollobás B., *Random Graphs.* Academic Press (1985). [405, 409]

[1986] Bollobás B., *Extremal Graph Theory with Emphasis on Probabilistic Methods.* (CBMS Regional Conference Series #62, American Math Society (1986) Chapter 9 (List Colorings). [387]

[1988] Bollobás B., The chromatic number of random graphs. *Combinatorica* **8** (1988), 49–55. [421, 427, 429]

[1985] Bollobás B. and A.J. Harris, List colorings of graphs. *Graphs and Combinatorics* **1** (1985), 115–127. [387]

[1976] Bollobás B. and P. Erdős, Cliques in random graphs. *Math. Proc. Camb. Phil. Soc.* **80** (1976), 419–427. [422]

[1990] Bòna M., Problem E3378. *Amer. Math. Monthly* **97** (1990), 240. [369]

[1969] Bondy J.A., Properties of graphs with constraints on degrees. *Stud. Sci. Math. Hung.* **4** (1969), 473–475. [142]

[1971] Bondy J.A., Pancyclic graphs I. *J. Comb. Theory (B)* **11** (1971), 80–84. [372]

[1971b] Bondy J.A., Large cycles in graphs. *Discr. Math.* **1** (1971), 121–132. [395]

[1972] Bondy J.A., Induced subsets. *J. Comb. Theory (B)* **12** (1972), 201–202. [63]

[1972] Bondy J.A., Variation on the Hamiltonian theme. *Canad. Math. Bull.* **15** (1972), 57–62. [229]

[1977] Bondy J.A. and C. Thomassen, A short proof of Meyniel's Theorem. *Discrete Math.* **19** (1977), 195–197. [399]

[1988] Bondy J.A. and M. Kouider, Hamiltonian cycles in regular 2-connected graphs. *J. Comb. Theory (B)* **44** (1988), 177–186. [224]

[1976] Bondy J.A. and U.S.R. Murty, *Graph Theory with Applications.* North Holland, New York (1976). [34, 39, 40, 59, 172, 264, 266]

[1976] Bondy J.A. and V. Chvátal, A method in graph theory. *Discrete Math.* **15** (1976), 111–136. [222]

[1976] Booth K.S. and G.S. Lueker, Testing for the consecutive ones property, interval graphs, and graph planarity using PQ-tree algorithms. *J. Comp. Syst. Sci.* **13** (1976), 335–379.   [265]

[1926] Borůvka O., Příspevěk k řešeníotázky otázky ekonomické stavby elektrovodních sítí. *Elektrotechnicky Obzor* **15** (1926), 153–154.   [74, 75]

[1974] Borodin O.V. and A.V. Kostochka, On an upper bound of the graph's chromatic number depending on the graph's degree and density. *J. Comb. Theory (B)* **16** (1974), 97–105.   [183]

[1966] Bosák J., *Hamiltonian lines in cubic graphs*. Talk presented to the International Seminar on Graph Theory and its Applications (Rome, July 5–9) (1966).   [285]

[1941] Brooks R.L., On colouring the nodes of a network. *Proc. Cambridge Phil. Soc.* **37** (1941), 194–197.   [179]

[1980] Buckingham M.A., Circle Graphs (also Ph.D Thesis, Courant 1981). Courant Computer Science Report 21 (1980).   [308]

[1983] Buckingham M.A. and M.C. Golumbic, Partitionable graphs, circle graphs, and the Berge strong perfect graph conjecture. *Discrete Math.* **44** (1983), 45–54.
                                                                [306, 310, 320]

[1981] Bumby R.T., A problem with telephones. *SIAM J. Alg. Disc. Meth.* **2** (1981), 13–19.   [386]

[1974] Buneman P., A characterization of rigid circuit graphs. *Discrete Math.* **9** (1974), 205–212.   [293]

[1982] Burlet M. and J.P. Uhry, Parity graphs. In *Bonn Workshop on Combinatorial Optimization*. (ed. A. Bachem, M. Grötschel, and B. Korte), *Annals Discr. Math.* **16** North-Holland (1982) 1–26.   [300, 318]

[1974] Burr S.A., Generalized Ramsey theory for graphs—a survey. In *Graphs and Combinatorics*. Springer (1974) 52–75.   [371]

[1981] Burr S.A., Ramsey numbers involving graphs with long suspended paths. *J. London Math. Soc. (2)* **24** (1981), 405–413.   [362]

[1983] Burr S.A., Diagonal Ramsey numbers for small graphs. *J. Graph Theory* **7** (1983), 57–69.   [361]

[1983] Burr S.A. and P. Erdős, Generalizations of a Ramsey-theoretic result of Chvátal. *J. Graph Theory* **7** (1983), 39–51.   [362]

[1975] Burr S.A., P. Erdős, and J.H. Spencer, Ramsey theorems for multiple copies of graphs. *Trans. Amer. Math. Soc.* **209** (1975), 87–99.   [363]

[1991] Cameron P.J. and J.H. van Lint, *Designs, Graphs, Codes, and their Links, London Math. Soc. Student Texts* 22. Cambridge Univ. Press (1991).   [449]

[1978] Catlin P.A., A bound on the chromatic number of a graph. *Discrete Math.* **22** (1978), 81–83.   [183]

[1979] Catlin P.A., Hajós' graph-coloring conjecture: variations and counterexamples. *J. Comb. Theory (B)* **26** (1979), 268–274.   [189, 193]

[1889] Cayley A., A theorem on trees. *Quart. J. Math.* **23** (1889), 276–378.   [63]

[1959] Chang S., The uniqueness and nonuniqueness of the triangular association scheme. *Sci. Record* **3** (1959), 604–613.   [218]

[1994a] Chappell G.G., *A weaker augmentation axiom*. unpublished (1994).   [348]

[1994b] Chappell G.G., *Matroid intersection and the Gallai-Milgram Theorem*. unpublished (1994).   [352]

[1995] Chappell G.G., *Pairwise balanced designs and hyperplanes*. unpublished (1995).
                                                                [351]

[1968] Chartrand G. and F. Harary, Graphs with prescribed connectivities. In *Theory of Graphs*. (ed. P. Erdős and G. Katona) Acad. Press (1968) 61–63.   [141]

[1986] Chartrand G. and L. Lesniak, *Graphs and Digraphs* (2nd ed.). Wadsworth (1986).
                                                                [95, 155, 264]

[1968] Chein M., Graphe régulièrement décomposable. *Rev. Française Info. Rech. Opér.* **2** (1968), 27–42.   [155]

[1975] Choudom S.A., K.R. Parthasarathy and G. Ravindra, Line-clique cover number of a graph. *Proc. Indian Nat. Sci. Acad.* **41** (1975), 289–293.   [401]

[1976] Christofides N., Worst-case analysis of a new heuristic for the traveling sales-man problem. Graduate School of Industrial Administration, Carnegie-Mellon University, Pittsburgh, PA (1976).   [237]

[1978] Chung F.R.K., On concentrators, superconcentrators, generalizers and nonblock-ing networks. *Bell Syst. Tech. J.* (1978), 1765–1777.   [445]

[1981] Chung F.R.K., On the decompositions of graphs. *SIAM J. Algeb. Disc. Meth.* **2** (1981), 1–12.   [375]

[1988] Chung F.R.K., Labellings of graphs. In *Selected Topics in Graph Theory, Vol. 3.* (ed. L.W. Beineke and R.J. Wilson) Acad. Press (1988) 151–168.   [367]

[1983] Chung F.R.K. and C.M. Grinstead, A survey of bounds for classical Ramsey num-bers. *J. Graph Theory* **7** (1983), 25–37.   [360]

[1993] Chung M.-S. and D.B. West, Large P_4-free graphs with bounded degree. *J. Graph Theory* **17** (1993), 109–116.   [40]

[1970] Chvátal V., The smallest triangle-free 4-chromatic 4-regular graph. *J. Comb. Theory* **9** (1970), 93–94.   [183]

[1972] Chvátal V., On Hamilton's ideals. *J. Comb. Th. B* **12** (1972), 163–168.   [223, 229]

[1975] Chvátal V., A combinatorial theorem in plane geometry. *J. Comb. Theory (B)* **18** (1975), 39–41.   [283]

[1976] Chvátal V., On the strong perfect graph conjecture. *J. Comb. Theory* **20** (1976), 139–141.   [312, 314, 320]

[1977] Chvátal V., Tree-complete graph Ramsey numbers. *J. Graph Theory* **1** (1977), 93.   [362]

[1984] Chvátal V., Perfectly ordered graphs. *Ann. Discrete Math.* **21** (1984), 63–65.   [301, 302, 319]

[1985] Chvátal V., Star-cutsets and perfect graphs. *J. Comb. Theory (B)* **39** (1985), 138–154.   [304, 319]

[1972] Chvátal V. and F. Harary, Generalized Ramsey theory for graphs, III. Small Off-Diagonal Numbers. *Pac. J. Math.* **41** (1972), 335–345.   [362]

[1973] Chvátal V. and F. Harary, Generalized Ramsey theory for graphs, I. Diagonal numbers. *Period. Math. Hungar.* **3** (1973), 115–124.   [429]

[1974] Chvátal V. and L. Lovász, Every directed graph has a semi-kernel. In *Hypergraph Seminar.* Proc. 1st Working Seminar (Columbus OH 1972) *Lect. Notes Math.* **411**, Springer (1974) 175.   [50]

[1988] Chvátal V. and N. Sbihi, Recognizing claw-free perfect graphs. *J. Comb. Theory (B)* **44** (1988), 154–176.   [312]

[1972] Chvátal V. and P. Erdős, A note on hamiltonian circuits. *Discrete Math.* **2** (1972), 111–113.   [225, 421]

[1979] Chvátal V., R.L. Graham, A.F. Perold, and S.H. Whitesides, Combinatorial de-signs related to the strong perfect graph conjecture. *Discrete Math.* **26** (1979), 83–92.   [307, 308, 319]

[1983] Chvátal V., V. Rödl, E. Szemerédi, W.T. Trotter Jr., The Ramsey numbers of a graph with bounded maximum degree. *J. Comb. Theory (B)* **34** (1983), 239–243.   [363]

[1975] Chvátalová J., Optimal labelling of a product of two paths. *Discrete Math.* **11** (1975), 249–253.   [373]

[1974] Clapham C.R.J., Hamiltonian arcs in self-complementary graphs. *Discrete Math.* **8** (1974), 251–255.   [229]

[1971] Cook S.A., The complexity of theorem-proving procedures. In *Proc. 3th ACM Symp. Theory of Comp.* Association for Computing Machinery (1971) 151–158.   [238]

[1970] Crapo H.H. and G.C. Rota, *On the Foundations of Combinatorial Theory: Combi-natorial Geometries* preliminary edition. M.I.T. Press (1970).   [328]

[1980] Cull P., Tours of graphs, digraphs, and sequential machines. *IEEE Trans. Comp.* **C29** (1980), 50–54. [97]

[1979] Cvetković D.M., M. Doob, and H. Sachs, *Spectra of Graphs.* Academic Press (1979). [433, 450]

[1971] de Werra D., Balanced schedules. *Information J.* **9** (1971), 230–237. [217]

[1964] Demoucron G., Y. Malgrange and R. Pertuiset, Graphes planaires: reconnaissance et construction des représentations planaires topologiques. *Rev. Française Recherche Opérationnelle* **8** (1964), 33–47. [265, 266]

[1947] Descartes B., A three colour problem. *Eureka* (1947), (soln. 1948). [185, 191]

[1954] Descartes B., Solution to advanced problem 4526 (Ungar). *Amer. Math. Monthly* **61** (1954), 352. [185, 191]

[1959] Dijkstra E.W., A note on two problems in connexion with graphs. *Numer. Math.* **1** (1959), 269–271. [76, 82]

[1952a] Dirac G.A., A property of 4-chromatic graphs and some remarks on critical graphs. *J. London Math. Soc.* **27** (1952), 85–92. [188, 192]

[1952b] Dirac G.A., Some theorems on abstract graphs. *Proc. Lond. Math. Soc.* **2** (1952), 69–81. [221, 226, 394, 421]

[1953] Dirac G.A., The structure of *k*-chromatic graphs. *Fund. Math.* **40** (1953), 42–55. [187]

[1960] Dirac G.A., In abstrakten Graphen vorhande vollständigene 4-Graphen und ihre Unterteilungen. *Math. Nachr.* **22** (1960), 61–85. [152, 156]

[1964] Dirac G.A., Homomorphism theorems for graphs. *Math. Ann.* **153** (1964), 69–80. [190]

[1965] Dirac G.A., Chromatic number and topological complete subgraphs. *Can. Math. Bull.* **8** (1965), 711–715. [189]

[1967] Dirac G.A., Minimally 2-connected graphs. *J. Reine Angew. Math.* **228** (1967), 204–216. [157]

[1954] Dirac G.A. and S. Schuster, A theorem of Kuratowski. *Nederl. Akad. Wetensch. Proc.* Ser. A **57** (1954), 343–348. [264]

[1917] Dudeney H.E., *Amusements in Mathematics.* Nelson (1917). [247]

[1917] Dziobek O., Eine Formel der Substitutionstheorie. *Sitzungsber. Berl. Math. G.* **17** (1917), 64–67. [72]

[1965a] Edmonds J., Paths, trees, and flowers. *Can. J. Math.* **17** (1965), 449–467. [127]

[1965b] Edmonds J., Minimum partition of a matroid into independent sets. *J. Res. Nat. Bur. Stand.* **69B** (1965), 67–72. [62, 328, 345, 346]

[1965c] Edmonds J., Lehman's switching game and a theorem of Tutte and Nash-Williams. *J. Res. Nat. Bur. Stand.* **69B** (1965), 73–77. [62, 328, 345, 346]

[1965d] Edmonds J., Maximum matchings and a polyhedron with 0,1-vertices. *J. Res. Nat. Bur. Standards* **69B** (1965), 125–130. [130]

[1970] Edmonds J., Submodular functions, matroids and certain polyhedra. In *Combinatorial structures and their applications.* Proc. Calgary Intl. Conf. 1969 Gordon and Breach (1970) 69–87. [340]

[1973] Edmonds J., Edge-disjoint branchings. In *Combinatorial Algorithms.* (Courant Comp. Sci. Symp. Monterey, CA 1972 - B. Rustin, ed.) Academic Press (1973) 91–96. [382]

[1965] Edmonds J. and D.R. Fulkerson, Transversals and matroid partition. *J. Res. Nat. Bur. Standards Sect. B* **69B** (1965), 147–153. [325, 344]

[1973] Edmonds J. and E. Johnson, Matching, Euler tours, and the Chinese postman. *Math. Programming* **5** (1973), 88–124. [92]

[1972] Edmonds J. and R.M. Karp, Theoretical improvements in algorithmic efficiency for network flow problems. *J. Assoc. Comp. Mach.* **19** (1972), 248–264. [163]

[1931] Egerváry E., On combinatorial properties of matrices (Hungarian with German summary). *Mat. Lapok* **38** (1931), 16–28. [103, 201, 341]

[1979] Eitner P.G., *The bandwidth of the complete multipartite graph.* Presentation at Toledo Symposium on Applications of Graph Theory (1979). [372]

[1956] Elias P., A. Feinstein and C.E. Shannon, Note on maximum flow through a network. *IRE Trans. on Information Theory* **IT-2** (1956), 117–119. [149]

[1996] Ellingham M.N. and L. Goddyn, List edge colourings of some 1-factorable multigraphs. *Combinatorica* (to appear). [389]

[1985] Enomoto B., B. Jackson, P. Katerinis, and A. Saito, Toughness and the existence of *k*-factors. *J. Graph Theory* **9** (1985), 87–95. [220]

[1947] Erdős P., Some remarks on the theory of graphs. *Bull. Amer. Math. Soc.* **53** (1947), 292–294. [361, 406]

[1959] Erdős P., Graph Theory and Probability. *Canad. J. Math.* **11** (1959), 34–38. [185, 407, 408]

[1962] Erdős P., Remarks on a paper of Pósa. *Magyar Tud. Akad. Mat. Kut. Int. Közl.* **7** (1962), 227–229. [229]

[1963] Erdős P., On a combinatorial problem, I. *Nordisk Mat. Tidskrift* **11** (1963), 5–10. [429]

[1979] Erdős P., Some old and new problems in various branches of combinatorics. *Congr. Num.* **23** (1979), 19–37. [387]

[1981] Erdős P., On the combinatorial problems I would most like to see solved. *Combinatorica* **1** (1981), 25–42. [182]

[1988a] Erdős P., Problem E3255. *Amer. Math. Monthly* **95** (1988), 259. [400]

[1988b] Erdős P., Problem E3284. *Amer. Math. Monthly* **95** (1988), 762. [401]

[1966] Erdős P. and A. Hajnal, On chromatic numbers of graphs and set systems. *Acta Math. Acad. Sci. Hung.* **17** (1966), 61–99. [183]

[1966] Erdős P. and A. Rényi, On the existence of a factor of degree one of a connected random graph. *Acta Math. Acad. Sci. Hung.* **17** (1966), 359–368. [418]

[1935] Erdős P. and G. Szekeres, A combinatorial problem in geometry. *Composito Math* **2** (1935), 464–470. [182, 354, 358, 369, 370]

[1977] Erdős P. and R.J. Wilson, On the chromatic index of almost all graphs. *J. Comb. Theory (B)* **23** (1977), 255–257. [419]

[1959] Erdős P. and T. Gallai, On maximal paths and circuits of graphs. *Acta Math. Acad. Sci. Hung.* **10** (1959), 337–356. [372, 394]

[1960] Erdős P. and T. Gallai, Graphs with prescribed degrees of vertices (Hungarian). *Mat. Lapok* **11** (1960), 264–274. [132]

[1961] Erdős P. and T. Gallai, On the minimal number of vertices representing the edges of a graph. *Publ. Math. Inst. Hung. Acad. Sci.* **6** (1961), 181–203. [132]

[1966] Erdős P., A. Goodman, and L. Pósa, The representation of graphs by set intersections. *Canad. J. Math.* **18** (1966), 106–112. [374]

[1979] Erdős P., A. Rubin, and H. Taylor, Choosability in graphs. In *Proc. West Coast Conf. Comb., Graph Th., Computing.* Congr. Num. 26 (1979) 125–157. [386, 389, 403]

[1962] Eršov A.P. and G.I. Kožuhin, Estimates of the chromatic number of connected graphs (Russian). *Dokl. Akad. Nauk. SSSR* **142** (1962), 270–273. [190]

[1736] Euler L., Solutio problematis ad geometriam situs pertinentis. *Comment. Academiae Sci. I. Petropolitanae* **8** (1736), 128–140. [85]

[1758] Euler L., Demonstratio Nonnullarum Insignium Proprietatum Quibus Solida Hedris Planis Inclusa Sunt Praedita. *Novi Comm. Acad. Sci. Imp. Petropol* **4** (1758), 140–160. [255]

[1975] Even S. and O. Kariv, An $O(n^{2.5})$ algorithm for maximum matching in general graphs. In *Proceedings of the 16th Annual Symposium on the Foundations of Computer Science.* (1975) 100–112. [130]

[1975] Even S. and R.E. Tarjan, Network flow and testing graph connectivity. *SIAM J. Computing* **4** (1975), 507–518. [120]

[1984] Fan G.-H., New sufficient conditions for cycles in graphs. *J. Comb. Theory (B)* **37** (1984), 221–227.   [397]

[1948] Fáry I., On the straight line representations of planar graphs. *Acta Sci. Math.* **11** (1948), 229–233.   [260]

[1988] Feng T., A short proof of a theorem about the circumference of a graph. *J. Comb. Theory (B)* **45** (1988), 373–375.   [397]

[1968] Finck H.-J., On the chromatic numbers of a graph and its complement. In *Theory of Graphs*. (Proc. Colloq., Tihany, 1966) Academic Press (1968) 99–113.   [182]

[1969] Finck H.J. and H. Sachs, Über eine von H.S. Wilf angegebene Schranke für die chromatische Zahl endlicher Graphen. *Math. Nachr* **39** (1969), 373–386.   [182]

[1985] Fishburn P.C., *Interval Orders and Interval Graphs*. Wiley (1985).   [318]

[1994] Fisher D.C., K.L. Collins, and L.B. Krompart, Problem 10406. *Amer. Math. Monthly* **101** (1994), 793.   [286]

[1978] Fisk S., A short proof of Chvátal's watchman theorem. *J. Comb. Theory (B)* **24** (1978), 374.   [283]

[1974] Fleischner H., The square of every two-connected graph is hamiltonian. *J. Comb. Theory (B)* **16** (1974), 29–34.   [228]

[1983] Fleischner H., Eulerian graphs. In *Selected Topics in Graph Theory* Vol. 2. (ed. L.W. Beineke and R.J. Wilson) Academic Press (1983) 17–54.   [97]

[1991] Fleischner H., A maze search algorithm which also produces Eulerian trails. In *Advances in Graph Theory*. (ed. V.R. Kulli) Vishwa International Publ. (1991) 195–201.   [97]

[1992] Fleischner H. and M. Stiebitz, A solution to a coloring problem of P. Erdős. *Discrete Math.* **101** (1992), 39–48.   [387]

[1990] Floyd R.W., Problem E3399. *Am. Math. Monthly* **97** (1990), 611–612.   [108]

[1956] Ford L.R. Jr. and D.R. Fulkerson, Maximal flow through a network. *Canad. J. Math.* **8** (1956), 399–404.   [149, 152, 162]

[1958] Ford L.R. Jr. and D.R. Fulkerson, Network flows and systems of representatives. *Canad. J. Math.* **10** (1958), 78–85.   [153, 342]

[1962] Ford L.R. Jr. and D.R. Fulkerson, *Flows in Networks*. Princeton University Press, Princeton (1962).   [94, 158]

[1973] Fournier J.-C., Colorations des arêtes d'un graphe. In *Colloque sur la Théorie des Graphes (Bruxelles 1973)*. *Cahiers Centre Études Rech. Opér.* **15** (1973) 311–314.   [217]

[1993] Frank A., Applications of submodular functions. In *Surveys in Combinatorics, 1993*. (ed. K. Walker) *Lond. Math. Soc. Lect. Notes* **187** Cambridge Univ. Press (1993) 85–136.   [154]

[1981] Frankl P. and R.M. Wilson, Intersection theorems with geometric consequences. *Combinatorica* **1** (1981), 357–368.   [361, 371]

[1985] Fraughnaugh (Jones) K., Minimum independence graphs with maximum degree four. In *Graphs and Applications*. (Proc. Boulder CO 1982) Wiley (1985) 221–230.
                                                                                        [283]

[1917] Frobenius G., Über zerlegbare Determinanten. *Sitzungsber. König. Preuss. Adad. Wiss.* **XVIII** (1917), 274–277.   [101]

[1971] Fulkerson D.R., Blocking and anti-blocking pairs of polyhedra. *Math. Programming* **1** (1971), 168–194.   [289]

[1965] Fulkerson D.R. and O.A. Gross, Incidence matrices and interval graphs. *Pac. J. Math.* **15** (1965), 835–855.   [205, 316]

[1981] Gabber O. and Z. Galil, Explicit construction of linear sized superconcentrators. *J. Comput. Systems Sci.* **22** (1981), 843–854.   [445]

[1975] Gabow H.N., An efficient implementation of Edmonds' algorithm for maximum matchings on graphs. *J. Assoc. Comp. Mach.* **23** (1975), 221–234.   [130]

[1990] Gabow H.N., Data structures for weighted matching and nearest common ancestors with linking. In *Proc 1st ACM-SIAM Symp. Disc. Algs.*. (San Francisco 1990) SIAM (1990) 434–443. [130]

[1989] Gabow H.N. and R.E. Tarjan, Faster scaling algorithms for general graph matching problems. Technical Report CU-CS-432-89 Dept. Computer Science, University of Colorado - Boulder (1989). [130]

[1986] Gabow H.N., Z. Galil, T. Spencer, and R.E. Tarjan, Efficient algorithms for finding minimum spanning trees in undirected and directed graphs. *Combinatorica* **6** (1986), 109–122. [75]

[1957] Gale D., A theorem on flows in networks. *Pac. J. Math.* **7** (1957), 1073–1082. [166, 167]

[1962] Gale D. and L.S. Shapley, College admissions and the stability of marriage. *Amer. Math. Monthly* **69** (1962), 9–15. [117, 121]

[1959] Gallai T., Über extreme Punkt- und Kantenmengen. *Ann. Univ. Sci. Budapest, Eötvös Sect. Math.* **2** (1959), 133–138. [104, 109]

[1962] Gallai T., Graphen mit triangulierbaren ungeraden Vielecken. *Magyar Tud. Akad. Mat. Kut. Int. Közl.* **7** (1962), 3–36. [300]

[1963] Gallai T., Kritische Graphen I. *Publ. Math. Inst. Hungar. Acad. Sci.* **8** (1963), 165–192. [188]

[1964] Gallai T., . *Magyar Tud. Akad. Mat. Kut. Int. Közl.* **8** (1964), 373–385. [191]

[1968] Gallai T., On directed paths and circuits. In *Theory of Graphs*. (ed. P. Erdős and G. Katona) Academic Press (1968) 115–118. [178]

[1960] Gallai T. and A.N. Milgram, Verallgemeinerung eines graphentheoretischen Satzes von Rédei. *Acta Sci. Math. Szeged* **21** (1960), 181–186. [352, 391]

[1995] Galvin F., The list chromatic index of a bipartite multigraph. *J. Comb. Theory (B)* **63** (1995), 153–158. [387, 388]

[1976] Garey M.R. and D.S. Johnson, The complexity of near-optimal graph colouring. *J. Assoc. Comp. Mach.* **23** (1976), 43–49. [421]

[1979] Garey M.R. and D.S. Johnson, *Computers and Intractability*. W.H. Freeman and Company, San Fransisco (1979). [234]

[1976] Garey M.R., D.S. Johnson, and L. Stockmeyer, Some simplified NP-complete graph problems. *Theor. Comp. Sci.* **1** (1976), 237–267. [244]

[1976] Garey M.R., D.S. Johnson, and R.E. Tarjan, . unpublished (1976). [245]

[1978] Garey M.R., R.L. Graham, D.S. Johnson, and D.E. Knuth, Complexity results for bandwidth minimization. *SIAM J. Appl. Math.* **34** (1978), 477–495. [366]

[1972] Gavril F., Algorithms for minimum coloring, maximum clique, minimum covering by cliques and maximum independent set of a chordal graph. *SIAM J. Computing* **1** (1972), 180–187. [316]

[1974] Gavril F., The intersection graphs of subtrees in trees are exactly the chordal graphs. *J. Comb. Theory (B)* **16** (1974), 47–56. [293]

[1994] Gavril F. and J. Urrutia, Intersection graphs of concatenable subtrees of graphs. *Discrete Appl. Math.* **52** (1994), 195–209. [316]

[1991] George J., *1-Factorizations of tensor products of graphs*. Ph.D. Thesis, Univ. of Illinois (Urbana-Champaign) (1991). [216]

[1989] Georges J.P., Non-Hamiltonian bicubic graphs. *J. Comb. Theory (B)* **46** (1989), 121–124. [220]

[1960] Ghouila-Houri A., Une condition suffisante d'existence d'un circuit Hamiltonien. *C. R. Adac. Sci. Paris* **156** (1960), 495–497. [226, 399]

[1985] Gibbons A., *Algorithmic Graph Theory*. Cambr. Univ. Press (1985). [92, 239]

[1959] Gilbert E.N., Random graphs. *Annals Math. Statist.* **30** (1959), 1141–1144. [410]

[1984] Giles R., L.E. Trotter Jr., and A.C. Tucker, The strong perfect graph theorem for a class of partitionable graphs. In *Topics on Perfect Graphs*. (ed. C. Berge and V. Chvátal) North-Holland (1984) 161–167. [313, 315]

[1963] Glicksman S., On the representation and enumeration of trees. *Proc. Camb. Phil. Soc.* **59** (1963), 509–517.   [71]

[1977] Goldberg M.K., Structure of multigraphs with restrictions on the chromatic class (Russian). *Metody Diskret. Analiz.* **30** (1977), 3–12.   [217]

[1984] Goldberg M.K., Edge-coloring of multigraphs: recoloring technique. *J. Graph Theory* **8** (1984), 123–137.   [217]

[1980] Golumbic M.C., *Algorithmic Graph Theory and Perfect Graphs.* Academic Press (1980).   [288, 308, 318]

[1984] Golumbic M.C., Algorithmic aspects of perfect graphs. In *Topics on perfect graphs.* (ed. C. Berge and V. Chvátal) North-Holland (1984) 301–323.   [295]

[1946] Good I.J., Normal recurring decimals. *J. London Math. Soc.* **21** (1946), 167–169.
                                                                                    [92, 95, 96]

[1988] Gould R.J., *Graph Theory.* Benjamin/Cummings (1988).   [265]

[1983] Gould R.J. and M.S. Jacobson, On the Ramsey number of trees versus graphs with large clique number. *J. Graph Theory* **7** (1983), 71–78.   [362]

[1992] Graham N. and F. Harary, Changing and unchanging the diameter of a hypercube. *Discrete Appl. Math.* **37-38** (1992), 265–274.   [354]

[1996] Graham N., R.C. Entringer and L.A. Székely, New tricks for old trees: maps and pigeonhole principle. (to appear).   [354, 369]

[1973] Graham R.L. and D.J. Kleitman, Increasing paths in edge ordered graphs. *Period. Math. Hungar.* **3** (1973), 141–148.   [355]

[1971] Graham R.L. and H.O. Pollak, On the addressing problem for loop switching. *Bell Sys. Tech. J.* **50** (1971), 2495–2519.   [379]

[1973] Graham R.L. and H.O. Pollak, On embedding graphs in squashed cubes. In *Graph Theory and Applications.* (Proc. 2nd Internat. Conf. Graphs Theory, Kalmazoo 1972), Lecture Notes in Math., vol. 303 Springer-Verlag (1973) 99–110.   [379]

[1980] Graham R.L., B.L. Rothschild, and J.H. Spencer, *Ramsey Theory.* Wiley (1980) 2nd ed. 1990.   [356]

[1968] Graver J.E. and J. Yackel, Some graph theoretic results associated with Ramsey's Theorem. *J. Comb. Theory* **4** (1968), 125–175.   [359, 360]

[1973] Greene C., A multiple exchange property for bases. *Proc. Amer. Math. Soc.* **39** (1973), 45–50.   [348]

[1975] Greene C. and G. Iba, Cayley's formula for multidimensional trees. *Discrete Math.* **13** (1975), 1–11.   [317]

[1978] Greenwell D.L., Odd cycles and perfect graphs. In *Theory and Applications of Graphs. Lect. Notes Math.* **642** Springer-Verlag (1978) 191–193.   [316]

[1973] Greenwell D.L. and H.V. Kronk, Uniquely line colorable graphs. *Canad. Math. Bull.* **16** (1973), 525–529.   [228]

[1955] Greenwood R.E. and A.M. Gleason, Combinatorial relations and chromatic graphs. *Canad. J. Math.* **7** (1955), 1–7.   [359]

[1991] Grigni M. and D. Peleg, Tight bounds on minimum broadcast networks. *SIAM J. Discr. Math.* **4** (1991), 207–222.   [403]

[1975] Grimmett G.R. and C.J.H. McDiarmid, On colouring random graphs. *Math. Proc. Camb. Phil. Soc.* **77** (1975), 313–324.   [422]

[1968] Grinberg E.J., Plane homogeneous graphs of degree three without hamiltonian circuits. *Latvian Math. Yearbook* **5** (1968), 51–58.   [275]

[1978] Grinstead C.M., *The strong perfect graph conjecture for a class of graphs.* Ph.D. Thesis, UCLA (1978).   [312]

[1981] Grinstead C.M., The strong perfect graph conjecture for toroidal graphs. *J. Comb. Theory (B)* **30** (1981), 70–74.   [312]

[1982] Grinstead C.M. and S.M. Roberts, On the Ramsey numbers $R(3, 8)$ and $R(3, 9)$. *J. Comb. Theory (B)* **33** (1982), 27–51.   [359]

[1959] Grötzsch H., Ein Dreifarbensatz fur dreikreisfree Netze auf der Kugel. *Wiss. Z. Martin-Luther-Univ., Halle-Wittenberg, Math.-Nat. Reihe* **8** (1959), 109–120. [283]

[1963] Grünbaum B. and T.S. Motzkin, The number of hexagons and the simplicity of geodesics on certain polyhedra. *Canad. J. Math.* **15** (1963), 744–751. [259]

[1962] Guan M., Graphic programming using odd and even points. *Chinese Math.* **1** (1962), 273–277. [91]

[1966] Gupta R.P., The chromatic index and the degree of a graph (Abstract 66T-429). *Not. Amer. Math. Soc.* **13** (1966), 719. [210, 217]

[1989] Gusfield D. and R.W. Irving, *The Stable Marriage Problem: Structure and Algorithms.* MIT Press (1989). [118]

[1969] Guy R.K., The decline and fall of Zarankiewicz's theorem. In *Proof Techniques in Graph Theory.* (ed. F. Harary) Acad. Press (1969) 63–69. [280]

[1970] Guy R.K., Sequences associated with a problem of Turán and other problems. In *Proc. Balatonfüred Combinatorics Conference, 1969.* Bolyai János Matematikai Tarsultat (1970) 553–569. [280, 287]

[1972] Guy R.K., Crossing numbers of graphs. In *Graph Theory and Applications.* Proc. Kalamazoo 1972 (ed. Y. Alavi et al) Springer Lect. Notes Math 303 (1972) 111–124. [279]

[1967] Guy R.K. and F. Harary, On the Möbius ladders. *Canad. Math. Bull.* **10** (1967), 493–496. [287]

[1980] Gyárfás A., E. Szemerédi, and Z. Tuza, Induced subtrees in graphs of large chromatic number. *Discrete Math.* **30** (1980), 235–244. [193]

[1979] Győri E. and A.V. Kostochka, On a problem of G.O.H. Katona and T. Tarján. *Acta Math. Acad. Sci. Hung.* **34** (1979), 321–327. [375]

[1943] Hadwiger H., Über eine Klassifikation der Streckenkomplexe. *Vierteljschr. Naturforsch. Ges. Zürich* **88** (1943), 133–142. [189, 336]

[1996] Häggkvist R. and J.C.M. Janssen, New bounds on the list-chromatic index of the complete graph and other simple graphs. (to appear). [387]

[1961] Hajós G., Über eine Konstruktion nicht *n*-färbbarer Graphen. *Wiss. Z. Martin-Luther-Univ. Halle-Wittenberg Math.-Natur. Reihe* **10** (1961), 116–117. [188, 192]

[1962] Hakimi S.L., On the realizability of a set of integers as degrees of the vertices of a graph. *SIAM J. Appl. Math.* **10** (1962), 496–506. [43, 48]

[1967] Halin R., Unterteilungen vollständiger Graphen in Graphen mit unendlicher chromatischer Zahl. *Abh. Math. Sem. Univ. Hamburg* **31** (1967), 156–165. [182]

[1969] Halin R., A theorem on *n*-connected graphs. *J. Comb. Theory* **7** (1969), 150–154. [157]

[1948] Hall M., Distinct representatives of subsets. *Bull. Amer. Math. Soc.* **54** (1948), 922. [101]

[1956] Hall M., An algorithm for distinct representatives. *Amer. Math. Monthly* **63** (1956), 716–717. [164]

[1935] Hall P., On representation of subsets. *J. Lond. Math. Soc.* **10** (1935), 26–30. [100]

[1981] Hammer P.L and B. Simeone, The splittance of a graph. *Combinatorica* **1** (1981), 275–284 (also Dept. of Comb. and Opt., Univ. of Waterloo, CORR 77-39 (1977). [49, 317]

[1981] Hammersley J., The friendship theorem and the love problem. In *Surveys in Combinatorics.* (ed. E.K. Lloyd) Cambr. Univ. Press (1981) 31–54. [449]

[1962a] Harary F., The maximum connectivity of a graph. *Proc. Nat. Acad. Sci. U.S.A.* **48** (1962), 1142–1146. [135, 141]

[1962b] Harary F., The determinant of the adjacency matrix of a graph. *SIAM review* **4** (1962), 202–210. [433]

[1969] Harary F., *Graph Theory.* Addison-Wesley, Reading MA (1969). [231, 264]

[1973] Harary F. and A.J. Schwenk, The number of caterpillars. *Discrete Math.* **6** (1973), 359–365. [73]

[1974] Harary F. and A.J. Schwenk, The communication problem on graphs and digraphs. *J. Franklin Inst.* **297** (1974), 491–495. [402]

[1965] Harary F. and C.St.J.A. Nash-Williams, On eulerian and hamiltonian graphs and line graphs. *Canad. Math. Bull.* **8** (1965), 701–710. [227]

[1973] Harary F. and E.M. Palmer, *Graphical Enumeration.* Academic Press (1973). [432]

[1966] Harary F. and G. Prins, The block-cutpoint-tree of a graph. *Publ. Math. Debrecen* **13** (1966), 103–107. [143]

[1993] Harary F. and P.C. Kainen, The cube of a path is maximal planar. *Bull. Inst. Combin. Appl.* **7** (1993), 55–56. [287]

[1964] Harary F. and Y. Kodama, On the genus of an n-connected graph. *Fund. Math.* **54** (1964), 7–13. [143]

[1977] Harary F., D.F. Hsu, and Z. Miller, The biparticity of a graph. *J. Graph Theory* **1** (1977), 131–133. [401]

[1966] Harper L.J., Optimal numberings and isoperimetric problems on graphs. *J. Comb. Theory* **1** (1966), 385–393. [366]

[1995a] Hartman C.M., *Some results on critical edges related to the strong perfect graph conjecture.* Talk presented to the 8th annual Cumberland conference on Combinatorics and Graph Theory (1995). [109]

[1995b] Hartman C.M., *A short proof of a theorem of Giles, Trotter, and Tucker.* unpublished note (1995). [313]

[1955] Havel V., A remark on the existence of finite graphs (Czech.). *Časopis Pěst. Mat* **80** (1955), 477–480. [43]

[1985] Hayward R.B., Weakly triangulated graphs. *J. Comb. Theory (B)* **39** (1985), 200–208. [305]

[1890] Heawood P.J., Map-colour theorem. *Q. J. Math.* **24** (1890), 332–339. [269, 270]

[1969] Heesch H., Untersuchungen zum Vierfarben problem. Num. 810/810a/810b B.I. Hochschulscripten. Bibliographisches Institut (1969). [271]

[1873] Hierholzer C., Über die Möglichkeit, einen Linienzug ohne Wiederholung und ohne Unterbrechung zu umfahren. *Math. Ann.* **6** (1873), 30–32. [85]

[1941] Hitchcock F.L., The distribution of a product from several sources to numerous facilities. *J. Math. Phys.* **20** (1941), 224–230. [94]

[1996] Hochberg R., C.J.H. McDiarmid, and M. Saks, On the bandwidth of triangulated cycles (Proc. 14th Brit. Comb. Conf. Keele 1993). *Discrete Math.* (to appear). [367, 368]

[1960] Hoffman A.J., On the exceptional case in the characterization of the arcs of a complete graph. *IBM J. Res. Dev.* **4** (1960), 487–496. [218]

[1963] Hoffman A.J., On the polynomial of a graph. *Amer. Math. Monthly* **70** (1963), 30–36. [442]

[1964] Hoffman A.J., On the line-graph of the complete bipartite graph. *Ann. Math. Statist.* **35** (1964), 883–885. [218]

[1993] Holton D.A. and J. Sheehan, *The Peterson Graph.* Cambridge Univ. Press (1993). [10]

[1981] Holyer I., The **NP**-completeness of edge-coloring. *SIAM J. Computing* **10** (1981), 718–720. [211, 419]

[1972] Holzmann C.A. and F. Harary, On the tree graph of a matroid. *SIAM J. Appl. Math.* **22** (1972), 187–193. [351]

[1974] Hopcroft J. and R.E. Tarjan, Efficient Planarity Testing. *J. Assoc. Comp. Mach.* **21** (1974), 549–568. [265]

[1975] Hopcroft J. and R.M. Karp, An $O(n^{2.5})$ algorithm for maximum matching in bipartite graphs. *SIAM J. Computing* **2** (1975), 225–231. [118, 119]

[1982] Horton J.D., On two-factors of bipartite regular graphs. *Discrete Math.* **41** (1982),
        35–41.   [220]

[1976] Huang H.-C., *Investigations on combinatorial optimization.* Ph.D. Thesis, School
        of Organization and Management, Yale University (1976).   [307]

[1952] Huffman D.A., A method for the construction of minimum redundancy codes.
        Proc. Inst. Rail. Engin. **40** (1952), 1098-1011.   [80]

[1973] Ingleton A.W. and M.J. Piff, Gammoids and transversal matroids. *J. Comb. Theory
        (B)* **15** (1973), 51–68.   [352]

[1980] Jackson B., Hamilton cycles in regular 2-connected graphs. *J. Comb. Theory (B)*
        29 (1980), 27–46.   [224]

[1991] Jacobson M.S., F.R. McMorris, H.M. Mulder, Tolerance Intersection Graphs. In
        Proceedings of the 1988 International Kalamazoo Graph Theory Conference. (ed.
        Y. Alavi, G. Chartrand, O.R. Oellerman and A.J. Schwenk) Wiley (1991) 705–724.
                                                                                        [318]

[1993] Janssen J.C.M., The Dinitz Problem is solved for rectangles. *Bull. Amer. Math.
        Soc.* **29** (1993), 243–249.   [387]

[1930] Jarník V., O jistém problému minimálnim. *Acta Societatis Scientiarum Natur.
        Moravicae* **6** (1930), 57–63.   [75, 82]

[1869] Jordan C., Sur les assemblages de lignes. *J. Reine Angew. Math.* **70** (1869), 185–
        190.   [55, 369]

[1965] Jung H.A., Anwendung einer Methode von K. Wagner bei Färbungsproblemen
        für Graphen. *Math. Ann.* **161** (1965), 325–326.   [189]

[1985] Jünger M., G. Reinelt, and W.R. Pulleyblank, On partitioning the edges of graphs
        into connected subgraphs. *J. Graph Theory* **9** (1985), 539–549.   [404]

[1996] Kahn J., Asymptotically good list colorings. *J. Comb. Th. A* (to appear).   [387]

[1967] Kalbfleisch J.G., Upper bounds for some Ramsey numbers. *J. Comb. Theory* **2**
        (1967), 35–42.   [359]

[1977] Kapoor S.F., A.D. Polimeni, and C.E. Wall, Degree sets for graphs. *Fund. Math.*
        95 (1977), 189–194.   [49]

[1995] Karger D.R., P.N. Klein, and R.E. Tarjan, A randomized linear-time algorithm to
        find minimum spanning trees. *J. Assoc. Comp. Mach.* **42** (1995), 321–328.   [75]

[1972] Karp R.M., Reducibility among combinatorial problems. In *Complexity of Com-
        puter Computations.* (ed. R.E. Miller and J.W. Thatcher) Plenum Press (1972)
        85–103.   [239, 240]

[1984] Kelmans A.K., A strengthening of the Kuratowski planarity criterion for 3-
        connected graphs. *Discrete Math.* **51** (1984), 215–220.   [265]

[1984] Kelmans A.K., Problem. In *Finite and Infinite Sets.* (ed. A. Hajnal, L. Lovász, V.T.
        Sós), Proc. 6th Hung. Comb. Colloq. (Eger 1981) *Coll. Math. Soc. János Bolyai*
        37, Elsevier (1984) 882.   [265]

[1879] Kempe A.B., On the geographical problem of four colours. *Amer. J. Math.* **2** (1879),
        193–200.   [270]

[1981] Kimble R.J. Jr. and A.J. Schwenk, On universal caterpillars. In *The theory and
        applications of graphs.* Wiley (1981) 437–447.   [73]

[1847] Kirchhoff G., Über die Auflösung der Gleichungen, auf welche man bei der Unter-
        suchung der linearen Verteilung galvanischer Ströme gefürht wird. *Ann. Phys.
        Chem.* **72** (1847), 497–508.   [65]

[1970] Kleitman D.J., The crossing number of $K_{5,n}$. *J. Comb. Theory* **9** (1970), 315–323.
                                                                                        [280]

[1980] Kleitman D.J. and J.B. Shearer, Further gossip problems. *Discrete Math.* **30**
        (1980), 151–156.   [386]

[1989] Klotz W., A constructive proof of Kuratowski's theorem. *Ars Combinatoria* **28**
        (1989), 51–54.   [267]

[1976] Knuth D.E., *Mariages Stables.* Les Pesses de l'Université de Montréal (1976).
                                                                                        [118]

[1916] König D., Über Graphen und ihre Anwendung auf Determinantentheorie und Mengenlehre. *Math. Ann.* **77** (1916), 453–465. [105, 201, 209]

[1931] König D., Graphen und Matrizen. *Math. Lapok* **38** (1931), 116–119. [103, 201, 341]

[1936] König D., *Theorie der endlichen und unendlichen Graphen.* Akademische Verlagsgesellschaft (1936) (reprinted Chelsea 1950). [97]

[1947] Koopmans T.C., Optimum utilization of the transportation system. In *Proceedings of the International Statistical Conference, Washington, DC.* (1947) also in Econometrica **17**(1949). [94]

[1991] Korte B., L. Lovász, and R. Schrader, *Greedoids.* Springer (1991). [347]

[1979] Kotzig A., 1-Factorizations of cartesian products of regular graphs. *J. Graph Theory* **3** (1979), 23–34. [216]

[1943] Krausz J., Démonstration nouvelle d'une théorème de Whitney sur les réseaux (Hungarian). *Mat. Fiz. Lapok* **50** (1943), 75–89. [212]

[1956] Kruskal J.B. Jr., On the shortest spanning subtree of a graph and the traveling salesman problem. *Proc. Am. Math. Soc.* **7** (1956), 48–50. [75]

[1955] Kuhn H.W., The Hungarian method for the assignment problem. *Naval Research Logistics Quarterly* **2** (1955), 83–97. [113]

[1986a] Kung J.P.S., *A Source Book in Matroid Theory.* Birkhäuser (1986). [347, 351]

[1986b] Kung J.P.S., Strong maps. In *Theory of Matroids.* (ed. N. White) Cambridge Univ. Press (1986) 224–252. [351]

[1930] Kuratowski K., Sur le problème des courbes gauches en topologie. *Fund. Math.* **15** (1930), 271–283. [259]

[1953] Landau H.G., On dominance relations and the structure of animal societies, III: The condition for score structure. *Bull. Math. Biophys.* **15** (1953), 143–148. [47, 49]

[1971] Las Vergnas M., Sur une propriété des arbres maximaux dans un graphe. *C.R. Acad. Sci. Paris Ser. A-B* **272** (1971), 1297–1300. [230]

[1975] Las Vergnas M., A note on matchings in graphs. *Cahiers Centre Etudes Recherche Opér.* **17** (1975), 257–260. [132]

[1976] Lawler E.L., *Combinatorial Optimization: Networks and Matroids.* Holt, Rinehart, and Winston (1976). [130]

[1978] Lawrence J., Covering the vertex set of a graph with subgraphs of smaller degree. *Discrete Math.* **21** (1978), 61–68. [183]

[1973] Lawrence S.L., Cycle-star Ramsey numbers. *Notices Amer. Math. Soc.* **20** (1973), A-420 (Notice #73T-157). [372]

[1957] Lazarson T., *Independence functions in algebra.* Thesis, Univ. of London (1957). [347]

[1966] Lederberg J., Systematics of organic molecules, graph topology and Hamiltonian circuits (Instrumentation Res. Lab. Rept.). Stanford Univ. 1040 (1966). [285]

[1964] Lehman A., A solution of the Shannon switching game. *J. Soc. Indust. Appl. Math.* **12** (1964), 687–725. [338, 345, 348]

[1974] Lehot P.G.H., An optimal algorithm to detect a line-graph and output its root graph. *J. Assoc. Comp. Mach.* **21** (1974), 569–575. [213]

[1962] Lekkerkerker C.G. and J.Ch. Boland, Representation of a finite graph by a set of intervals on the real line. *Fund. Math.* **51** (1962), 45–64. [318]

[1973] Lick D.R., Characterizations of *n*-connected and *n*-line-connected graphs. *J. Comb. Theory (B)* **14** (1973), 122–124. [156]

[1970] Lick D.R. and A.T. White, *k*-degenerate graphs. *Canad. J. Math.* **22** (1970), 1082–1096. [182]

[1973] Lin S. and B.W. Kernighan, An effective heuristic algorithm for the traveling-salesman problem. *Oper. Res.* **21** (1973), 498–516. [236]

[1976] Linial N., A lower bound for the circumference of a graph. *Discrete Math.* **15** (1976), 297–300. [395]

[1996] Liu J. and H. Zhou, Maximum induced matchings in graphs. *Discrete Math.* (to appear).   [108, 132]

[1995] Locke S.C., Problem 10447. *Amer. Math. Monthly* **102** (1995), 360.   [50]

[1966] Lovász L., On decomposition of graphs. *Stud. Sci. Math. Hung.* **1** (1966), 237–238.   [183]

[1968] Lovász L., On chromatic number of finite set-systems.. *Acta Math. Acad. Sci. Hung.* **19** (1968), 59–67.   [186, 408]

[1968] Lovász L., On covering of graphs. In *Theory of Graphs.* (Proc. Colloq., Tihany, 1966) Academic Press (1968) 231–236.   [392]

[1972a] Lovász L., Normal hypergraphs and the perfect graph conjecture. *Discrete Math.* **2** (1972), 253–267.   [291]

[1972b] Lovász L., A characterization of perfect graphs. *J. Comb. Theory (B)* **13** (1972), 95–98.   [291, 304, 305]

[1975] Lovász L., Three short proofs in graph theory. *J. Comb. Theory (B)* **19** (1975), 269–271.   [122, 178]

[1976] Lovász L., On two minimax theorems in graph. *J. Comb. Theory (B)* **21** (1976), 96–103.   [382]

[1979] Lovász L., *Combinatorial Problems and Exercises.* Akademiai Kiado and North-Holland (1979).   [25, 72, 156, 158, 372]

[1983] Lovász L., Perfect graphs. In *Selected Topics in Graph Theory* volume 2. (ed. L.W. Beineke and R.J. Wilson) Academic Press (1983) 55–87.   [300]

[1986] Lovász L. and M.D. Plummer, *Matching Theory (Annals Discr. Math.* **29***).* North Holland (1986).   [107, 342]

[1980] Lovász L., J. Nešetřil, and A. Pultr, On a product dimension of graphs. *J. Comb. Theory (B)* **28** (1980), 47–67.   [377, 401, 402]

[1994] Lu X., A Chvátal-Erdős type condition for Hamiltonian graphs. *J. Graph Theory* **18** (1994), 791–800.   [225]

[1986] Lubotzky A., R. Phillips, and P. Sarnak, Explicit expanders and the Ramanujan conjectures. In *Proc. 18th ACM Symp. Theory of Comp.* ACM Press (1986) 240–246.   [446]

[1921] Lucas E., *Récréations Mathématiques IV.* Paris (1921).   [88]

[1936] MacLane S., Some interpretations of abstract linear dependence in terms of projective geometry. *Amer. J. Math.* **58** (1936), 236–240.   [321, 333]

[1996] Maddox, Solution to Problem 6617. *Amer. Math. Monthly* (to appear).   [452]

[1991] Madej T., Bounds for the crossing number of the N-cube. *J. Graph Theory* **15** (1991), 81–97.   [287]

[1967] Mader W., Homomorphieeigenschaften und mittlere Kantendichte von Graphen. *Math. Ann.* **174** (1967), 265–268.   [189, 190]

[1971] Mader W., Minimale n-fach kantenzusammenhängende Graphen. *Math. Ann.* **191** (1971), 21–28.   [158]

[1978] Mader W., A reduction method for edge-connectivity in graphs. *Annals Discr. Math.* **3** (1978), 145–164.   [154, 158]

[1991] Mahadev N.V.R., F.S. Roberts, and P. Santhanakrishnan, 3-choosable complete bipartite graphs. DIMACS Tech. Report 91–62 (1991).   [386]

[1907] Mantel W., Problem 28, soln. by H. Gouwentak, W. Mantel, J. Teixeira de Mattes, F. Schuh and W.A. Wythoff. *Wiskundige Opgaven* **10** (1907), 60–61.   [33]

[1959] Marcus M. and R. Ree, Diagonals of doubly stochastic matrices. *Quart. J. Math.* **2** (1959), 295–302.   [108]

[1973] Margulis G.A., Explicit constructions of concentrators. *Problems of Inforamtion Transmission* **9** (1973), 325–332.   [445]

[1988] Margulis G.A., Explicit constructions of concentrators. *Problems of Information Transmission* **24** (1988), 39–46.   [446]

[1984] Markossian S.E. and I.A. Karapetian, On critically imperfect graphs. In *Prikladnaia Matematika.* (ed. R.N. Tonoian) Erevan Univ. (1984) .   [109]

[1972] Mason J.H., On a class of matroids arising from paths in graphs. *Proc. Lond. Math. Soc.(3)* **25** (1972), 55–74.   [351]

[1984] Matthews M.M. and D.P. Sumner, Hamiltonian results in $K_{1,3}$-free graphs. *J. Graph Theory* **8** (1984), 139–146.   [228]

[1968] Matula D.W., A min-max theorem for graphs with application to graph coloring. *SIAM Rev.* **10** (1968), 481–482.   [182]

[1972] Matula D.W., The employee party problem. *Notices A.M.S.* **19** (1972), A-382.   [420]

[1973] Matula D.W., An extension of Brooks' Theorem. Center for Numerical Analysis, University of Texas–Austin 69 (1973).   [180, 183]

[1980] Maurer S., The king chicken theorems. *Math. Mag.* **53** (1980), 67–80.   [47]

[1980] Maurer S., I. Rabinovitch, and W.T. Trotter Jr., Large minimal realizers of a partial order II. *Discrete Math.* **31** (1980), 297–314.   [50]

[1972] McDiarmid C.J.H., The solution of a timetabling problem. *J. Inst. Math. Applics.* **9** (1972), 23–34.   [217]

[1994] McGuinness S., The greedy clique decomposition of a graph. *J. Graph Theory* **18** (1994), 427–430.   [375]

[1992] McKay B.D. and K.M. Zhang, The value of the Ramsey number $R(3, 8)$. *J. Graph Theory* **16** (1992), 99–105.   [359]

[1991] McKay B.D. and S.P. Radziszowski, The first classical Ramsey number for hypergraphs is computed. In *Proc. 2nd ACM-SIAM Symp. on Discrete Alg..* (San Francisco 1991) (1991) 304–308.   [359]

[1995] McKay B.D. and S.P. Radziszowski, $R(4, 5) = 25$. *J. Graph Theory* **19** (1995), 309–322.   [359]

[1984] McKee T.A., Recharacterizing Eulerian: intimations of new duality. *Discrete Math.* **51** (1984), 327–242.   [95]

[1993] McKee T.A., How chordal graphs work. *Bull. ICA* **9** (1993), 27–39.   [297]

[1971] Melnikov L.S. and V.G. Vizing, Solution to Toft's problem (Russian). *Diskret. Analiz.* **19** (1971), 11–14.   [316]

[1927] Menger K., Zur allgemenen Kurventheorie. *Fund. Math.* **10** (1927), 95–115.   [152]

[1973] Meyniel H., Une condition suffisante d'existence d'un circuit Hamiltonien dans un graph oriente. *J. Comb. Theory (B)* **14** (1973), 137–147.   [226, 399]

[1976] Meyniel H., On the perfect graph conjecture. *Discrete Math.* **16** (1976), 339–342.   [300, 312, 320]

[1987] Meyniel H., A new property of critical imperfect graphs and some consequences. *Europ. J. Comb.* **8** (1987), 313–316.   [320]

[1980] Micali S. and V.V. Vazirani, an $O(\sqrt{|V|} \cdot |E|)$ algorithm for finding maximum matching in general graphs. In *Proceedings of the 21st Annual Symposium on the Foundations of Computer Science.* ACM (1980) 17–27.   [130]

[1981] Miller Z., The bandwidth of caterpillar graphs. In *Proc. Southeastern Conf.. Congressus Numerantium* 33 (1981) 235–252.   [372]

[1962] Minty G.J., A theorem on n-coloring the points of a linear graph. *Amer. Math. Monthly* **69** (1962), 623–624.   [183]

[1966] Minty G.J., On the axiomatic foundations of the theories of directed linear graphs, electrical networks and network programming. *J. Math. Mech.* **15** (1966), 485–520.   [350]

[1971] Mirsky L., *Transversal theory.* Academic Press (1971).   [342]

[1967] Mirsky L. and H. Perfect, Applications of the notion of independence to combinatorial analysis. *J. Comb. Theory* **2** (1967), 327–357.   [325]

[1963] Moon J.W., On the line-graph of the complete bigraph. *Ann. Math. Statis.* **34** (1963), 664–667.   [218]

[1965] Moon J.W., On a problem of Ore. *Math. Gaz.* **49** (1965), 40–41.   [229]

[1965] Moon J.W., On the diameter of a graph. *Michigan Math. J.* **12** (1965), 349–351.
[62]

[1966] Moon J.W., On subtournaments of a tournament. *Canad. Math. Bull.* **9** (1966), 297–301.   [231]

[1970] Moon J.W., *Counting Labeled Trees.* Canad. Math. Congress (1970).   [63]

[1969] Mowshowitz A., The group of a graph whose adjacency matrix has all distinct eigenvalues. In *Proof Techniques in Graph Theory.* Academic Press (1969) 109–110.   [452]

[1969] Mowshowitz A., The group of a graph whose adjacency matrix has all distinct eigenvalues. In *Proof Techniques in Graph Theory.* (ed. F. Harary) Acad. Press (1969) 109–110.   [452]

[1957] Munkres J., Algorithms for the assignment and transportation problems. *J. Soc. Indust. Appl. Math.* **5** (1957), 32–38.   [113]

[1955] Mycielski J., Sur le coloriage des graphes. *Coll. Math.* **3** (1955), 161–162.   [184]

[1972] Myers B.R. and R. Liu, A lower bound on the chromatic number of a graph. *Networks* **1** (1972), 273–277.   [182]

[1960] Nash-Williams C.St.J.A., On orientations, connectivity and odd-vertex-pairings in finite graphs. *Canad. J. Math.* **12** (1960), 555–567.   [154, 156]

[1961] Nash-Williams C.St.J.A., Edge-disjoint spanning trees in finite graphs. *J. London Math. Soc.* **36** (1961), 445–450.   [58, 62, 346, 353]

[1964] Nash-Williams C.St.J.A., Decomposition of finite graphs into forests. *J. London Math. Soc.* **39** (1964), 12.   [62, 346]

[1966] Nash-Williams C.St.J.A., An application of matroids to graph theory. In *Theory of Graphs.* (Intl. Sympos., Rome) Dunod (1966) 263–265.   [344]

[1988] Nemhauser G.L. and L.A. Wolsey, *Integer and combinatorial optimization.* Wiley (1988).   [328]

[1979] Nešetřil J. and V. Rödl, A short proof of the existence of highly chromatic hypergraphs without short cycles. *J. Comb. Theory (B)* **27** (1979), 225–227.   [186, 408]

[1991] Nilli A., On the second eigenvalue of a graph. *Discrete Math.* **91** (1991), 207–210.
[446]

[1956] Nordhaus E.A. and J.W. Gaddum, On complementary graphs. *Amer. Math. Monthly* **63** (1956), 175–177.   [182]

[1959] Norman R.Z. and M. Rabin, Algorithm for a minimal cover of a graph. *Proc. Amer. Math. Soc.* **10** (1959), 315–319.   [109]

[1996] O'Donnell P., The choice number of $K_{6,q}$. (to appear).   [386]

[1988] Olariu S., No antitwins in minimal imperfect graphs. *J. Comb. Theory (B)* **45** (1988), 255–257.   [319]

[1989] Olariu S., The strong perfect graph conjecture for pan-free graphs. *J. Comb. Theory (B)* **47** (1989), 187–191.   [312]

[1969] Olaru E., Über die Überdeckung von Graphen mit Cliquen. *Wiss. Z. Tech. Hochsch. Ilmenau* **15** (1969), 115–121.   [300]

[1951] Ore O., A problem regarding the tracing of graphs. *Elemente der Math.* **6** (1951), 49–53.   [95]

[1955] Ore O., Graphs and matching theorems. *Duke Math. J.* **22** (1955), 625–639.
[108, 342]

[1960] Ore O., Note on Hamilton circuits. *Amer. Math. Monthly* **67** (1960), 55.
[222, 395]

[1961] Ore O., Arc coverings of graphs. *Ann. Mat. Pura Appl* **55** (1961), 315–321.   [229]

[1963] Ore O., Hamiltonian connected graphs. *J. Math. Pures Appl.* **42** (1963), 21–27.
[229]

[1967a] Ore O., *The four-colour problem.* Academic Press (1967).   [217, 270]

[1967b] Ore O., On a graph theorem of Dirac. *J. Comb. Th.* **2** (1967), 35–42.   [230]

[1992] Oxley J.G., *Matroid Theory.* Oxford University Press (1992).   [347]

References 491

[1974] Padberg M.W., Perfect zero-one matrices. *Math. Programming* **6** (1974), 180–196.
[306, 308]

[1985] Palmer E.M., *Graphical Evolution: An Introduction to the Theory of Random Graphs*. Wiley (1985). [405, 415, 421, 430]

[1982] Papadimitriou C.H. and K. Steiglitz, *Combinatorial Optimization: Algorithms and Complexity*. Prentice Hall (1982). [163, 328]

[1976] Parthasarathy K.R. and G. Ravindra, The strong perfect graph conjecture is true for $K_{1,3}$-free graphs. *J. Comb. Theory (B)* **21** (1976), 212–223. [312, 315]

[2979] Parthasarathy K.R. and G. Ravindra, The validity of the strong perfect graph conjecture for $K_4 - e$-free graphs. *J. Comb. Theory (B)* **26** (2979), 98–100. [312]

[1984] Peck G.W., A new proof of a theorem of Graham and Pollak. *Discrete Math.* **49** (1984), 327–328. [440]

[1992] Peled U., Problem 10197. *Amer. Math. Monthly* **99** (1992), 162. [452]

[1969] Petersdorf M. and H. Sachs, Spektrum und Automorphismengruppe eines Graphen. In *Combinatorial Theory and its Applications, III*. North-Holland (1969) 891–907. [452]

[1891] Peterson J., Die Theorie der regulären Graphen. *Acta Math.* **15** (1891), 193–220. [124, 125]

[1973] Pinsker M., On the complexity of a concentrator. *7th International Teletraffic Conference* Stockholm (1973), 318/1–318/4. [445]

[1977] Pippenger N., Superconcentrators. *SIAM J. Computing* **6** (1977), 298–304. [445]

[1975] Plesnik J., Critical graphs of given diameter. *Acta Fac. Rerum Natur. Univ. Comenian. Math.* **30** (1975), 71–93. [142]

[1968] Plummer M.D., On minimal blocks. *Trans. Amer. Math. Soc.* **134** (1968), 85–94. [157]

[1957] Prim R.C., Shortest connection networks and some generalizations. *Bell Syst. Tech. J.* **36** (1957), 1389–1401. [82]

[1995] Pritikin D., A Prüfer-style bijection proving that $\tau(K_{n,n}) = n^{(}2n-2)$. In *Proc. 25th S.E. Conf. Comb. Graph Th. Comp. (1994)*. *Congr. Numer.* **104** (1995) 215–216. [71]

[1918] Prüfer H., Neuer Beweis eines Satzes über Permutationen. *Arch. Math. Phys.* **27** (1918), 742–744. [63]

[1957] Rado R., Note on independence functions. *Proc. Lond. Math. Soc.* **7** (1957), 300–320. [327]

[1930] Ramsey F.P., On a Problem of Formal Logic. *Proc. Lond. Math. Soc.* **30** (1930), 264–286. [355, 356]

[1982] Ravindra G., Meyniel graphs are strongly perfect. *J. Comb. Theory (B)* **33** (1982), 187–190. [300]

[1967] Ray-Chaudhuri D.K., Characterization of line graphs. *J. Comb. Theory* **3** (1967), 201–214. [215]

[1989] Recski A., *Matroid theory and its applications in electrical network theory and in statics*. Spinger-Verlag (1989). [347]

[1934] Rédei L., Ein kombinatorischer Satz. *Acta Litt. Szeged* **7** (1934), 39–43. [231]

[1946] Rees D., Note on a paper by I.J. Good. *J. London Math. Soc.* **21** (1946), 169–172. [96]

[1959] Rényi A, Some remarks on the theory of trees. *Magyar Tud. Akad. Mat. Kut. Int. Közl.* **4** (1959), 73–85. [71]

[1985] Reznick B., P. Tiwari, and D.B. West, Decompostition of product graphs into complete bipartite subgraphs. *Discrete Math.* **57** (1985), 179–183. [440]

[1985] Richards D. and A.L. Liestman, Finding cycles of a given length. *Annals Discr. Math.* **27** (1985), 249–256. [244]

[1964] Ringel G., Problem 25. In *Theory of Graphs and Its Applications (Proc. Symp. Smolenice 1963)*. Czech. Acad. Sci. (1964) 162. [69]

[1939] Robbins H.E, A theorem on graphs, with an application to a problem in traffic control. *Amer. Math. Monthly* **46** (1939), 281–283.   [148]

[1968] Roberts F.S., *Representations of Indifference relations.* Ph.D. Thesis, Department of Mathematics, Stanford Univ. (1968).   [318]

[1978] Roberts F.S., *Graph Theory and Its Applications to the Problems of Society (CBMS-NSF Monograph 29).* SIAM Publications (1978).   [93, 298]

[1996] Robertson N., D.P. Sanders, P.D. Seymour and R. Thomas, The four colour theorem. (to appear).   [274]

[1993] Robertson N., P.D. Seymour, and R. Thomas., Hadwiger's conjecture for K_6-free graphs. *Combinatorica* **13** (1993), 279–361.   [189]

[1976] Rose D., R.E. Tarjan, and G.S. Lueker, Algorithmic aspects of vertex elimination on directed graphs. *SIAM J. Computing* **5** (1976), 266–283.   [294]

[1971] Rosenfeld M., On the total coloring of certain graphs. *Israel J. Math* **9** (1971), 396–402.   [403]

[1964] Rota G.C., On the foundations of combinatorial theory I. *Z. Wahrsch.* **2** (1964), 340–368.   [328, 333]

[1991] Rotman J.J., Problem E3462. *Amer. Math. Monthly* **98** (1991), 645.   [96]

[1967] Roy B., Nombre chromatique et plus longs chemins d'un graphe. *Rev. Française Automat. Informat. Recherche Opérationelle sér. Rouge* **1** (1967), 127–132. [178]

[1985] Rucínski A. and A. Vince, Balanced graphs and the problem of subgraphs of random graphs. *Congr. Numer.* **49** (1985), 181–190.   [415]

[1957] Ryser H.J., Combinatorial properties of matrices of zeros and ones. *Canad. J. Math.* **9** (1957), 371–377.   [49, 167]

[1964] Ryser H.J., Matrices of zeros and ones in combinatorial mathematics. In *Recent Advances Matrix Theory.* (Madison, 1963) U. Wisc. Press (1964) 103–124.   [49]

[1977] Saaty T.L. and P.C. Kainen, *The Four-Color Problem.* McGraw-Hill (1977) (reprinted by Dover, 1986).   [270]

[1967] Sachs H., Über Teiler, Faktoren und characterische Polynome von Graphen II. *Wiss. Z. Techn. Hosch. Ilmenau* **13** (1967), 405–412.   [435]

[1967] Sachs H., Über Teiler, Faktoren und charakteristische Polynome von Graphen. I. *Wiss. Z. Techn. Hochsch. Ilmenau* **13** (1967), 405–412.   [435]

[1970] Sachs H., On the Berge conjecture concerning perfect graphs. In *Combinatorial Structures and Their Applications.* (ed. R. Guy, H. Hanani, N.W. Sauer, J. Schönheim) Gordon and Breach (1970) 377–384.   [300]

[1976] Sahni S. and T. Gonzalez, P-complete approximation problems. *J. Assoc. Comp. Mach.* **23** (1976), 555–565.   [236]

[1969] Schäuble M., Bemerkungen zur Kounstruktion dreikreisfreier k-chromatischer Graphen. *Wiss. Zeitschrift TH Ilmenau* **15** (1969), 59–63.   [191]

[1990] Schnyder W., Embedding planar graphs on the grid. In *Proc. 1st ACM-SIAM Sympos. Discrete Algorithm.* (1990) 138–148.   [264]

[1996] Schrijver A., Theory of Combinatorial Optimization. (to appear).   [328, 344]

[1966] Schwartz B.L., Possible winners in partially completed tournaments. *SIAM Review* **8** (1966), 302–308.   [165, 172]

[1973] Schwenk A.J., Almost all trees are cospectral. In *New Directions in the Theory of Graphs.* Academic Press (1973).   [435]

[1983] Schwenk A.J., Problem 6434. *Amer. Math. Monthly* **6** (1983), .   [451]

[1962] Scoins H.J., The number of trees with nodes of alternate parity. *Proc. Camb. Phil. Soc.* **58** (1962), 12–16.   [71]

[1974] Seinsche D., On a property of the class of n-colorable graphs. *J. Comb. Theory (B)* **16** (1974), 191–193.   [40, 316]

[1986] Seress Á., Quick gossiping without duplicate transmissions. *Graphs and Combinatorics* **2** (1986), 363–381 (also in *Combinatorial Mathematics* (Proc. 3rd Intl. Conf. in Combinatorics, New York 1985) (New York Acad. Sci 1989), 375–382). [402]

[1987] Seress Á., Gossips by conference calls. *Stud. Sci. Math. Hungar.* **22** (1987), 229–238. [402]

[1976] Seymour P.D., *A short proof of the matroid intersection theorem.* unpublished note (1976). [340]

[1948] Shannon C.E, A mathematical theory of communication. *Bell Syst. Tech. J.* **27** (1948), 379–423, 623–656. [82, 84]

[1949] Shannon C.E., A theorem on coloring the lines of a network. *J. Math. Phys.* **28** (1949), 148–151. [208, 216, 217]

[1994] Shende A.M. and B. Tesman, 3-Choosability of $K_{5,q}$. Computer Science Technical Report #94-9, Bucknell University (1994). [386]

[1988] Shibata T., On the tree representation of chordal graphs. *J. Graph Theory* **12** (1988), 421–428. [297]

[1981] Shmoys D.B., *Perfect graphs and the strong perfect graph conjecture.* B.S.E. Thesis, Princeton University (1981). [305]

[1959] Shrikhande S.S., The uniqueness of the L_2 association scheme. *Ann. Math. Statist.* **30** (1959), 781–798. [218]

[1977] Spencer J.H., Asymptotic lower bounds for Ramsey functions. *Discrete Math.* **20** (1977), 69–76. [371]

[1928] Sperner E., Neuer Beweis für die Invarianz der Dimensionszahl und des Gebietes. *Hamburger Abhand.* **6** (1928), 265–272. [364]

[1973] Stanley R.P., Acyclic orientations of graphs. *Discrete Math.* **5** (1973), 171–178. [202, 205]

[1974] Stanley R.P., Combinatorial reciprocity theorems. *Advances in Math.* **14** (1974), 194–253. [203]

[1951] Stein S.K., Convex maps. *Proc. Amer. Math. Soc.* **2** (1951), 464–466. [260]

[1993] Steinberg R., The state of the three color problem. In *Quo Vadis, Graph Theory?* (ed. J. Gimbel, J.W. Kennedy, L.V. Quintas) *Annals Discr. Math.* **55** (1993), 211–248 [283]

[1985] Stiebitz M., *Beiträge zur Theorie der färbungskritischen Graphen.* Dissertation zu Erlangung des akademischen Grades Dr.sc.nat., Technische Hochschule Ilmenau (1985). [192]

[1974a] Sumner D.P., Graphs with 1-factors. *Proc. Amer. Math. Soc.* **42** (1974), 8–12. [132]

[1974b] Sumner D.P., On Tutte's factorization theorem. In *Graphs and Combinatorics.* (ed. R. Bari and F. Harary), *Lecture Notes in Math.* **406** Springer-Verlag (1974) 350–355. [142]

[1991] Sun L., Two classes of perfect graphs. *J. Comb. Theory (B)* **53** (1991), 273–292 (also Tech. Report DCS-TR-228, Computer Science Dept., Rutgers Univ. 1988). [312]

[1982] Syslo M.M. and J. Zak, The bandwidth problem: critical subgraphs and the solution for caterpillars. In *Bonn Workshop on Combinatorial Optimization.* (Bonn, 1980) North-Holland (1982) 281–286. [372]

[1968] Szekeres G. and H.S. Wilf, An inequality for the chromatic number of a graph. *J. Comb. Theory* **4** (1968), 1–3. [177]

[1943] Szele T., Combinatorial investigations concerning complete directed graphs (Hungarian). *Mat. es Fiz. Lapok* **50** (1943), 223–236. [407]

[1978] Szemerédi E., Regular partitions of graphs. In *Problémes combinatoires et théorie des graphes.* Orsay C.N.R.S. (1978) 399–401. [364]

[1878] Tait P.G., On the colouring of maps, *Proc. Royal Soc. Edinburgh Sect. A* **10** (1878–1880), 501–503, 729 [274, 284]

[1984] Tanner R.M., Explicit construction of concentrators from generalized N-gons. *SIAM J. Algeb. Disc. Meth.* **5** (1984), 287–293.   [445]

[1975] Tarjan R.E., A good algorithm for edge-disjoint branching. *Info. Proc. Letters* **3** (1975), 51–53.   [383]

[1976] Tarjan R.E., *Maximum cardinality search and chordal graphs.* Lecture Notes from CS 259 (1976).   [295]

[1984] Tarjan R.E., A simple version of Karzanov's blocking flow algorithm. *Oper. Res. Letters* **2** (1984), 265–268.   [75]

[1984] Tarjan R.E. and M. Yannakakis, Simple linear-time algorithms to test chordality of graphs, test acyclicity of hypergraphs, and selectively reduce acyclic hypergraphs. *SIAM J. Computing* **13** (1984), 566–579.   [295, 316]

[1895] Tarry G., Le problème des labyrinthes. *Nouv. Ann. Math.* **14** (1895), 187–190.
                                                                                         [97]

[1974] Thomassen C., Some homeomorphism properties of graphs. *Math Nachr.* **64** (1974), 119–133.   [190]

[1980] Thomassen C., Planarity and duality of finite and infinite graphs. *J. Comb. Theory (B)* **29** (1980), 244–271.   [261, 263]

[1981] Thomassen C., Kuratowski's Theorem. *Journal of Graph Theory* **5** (1981), 225–241.   [263]

[1984] Thomassen C., A refinement of Kuratowski's theorem. *J. Comb. Theory (B)* **37** (1984), 245–253.   [264]

[1988] Thomassen C., Paths, circuits and subdivisions. In *Selected Topics in Graph Theory, 3.* (ed. L.W. Beineke and R.J. Wilson) Academic Press (1988) 97–132.
                                                                                   [189, 190]

[1994a] Thomassen C., Grötzsch's 3-Color Theorem. *J. Comb. Theory (B)* **62** (1994), 268–279.   [283]

[1994b] Thomassen C., Every planar graph is 5-choosable. *J. Comb. Theory (B)* **62** (1994), 180–181.   [389]

[1995] Thomassen C., 3-List-coloring planar graphs of girth 5. *J. Comb. Theory (B)* **64** (1995), 101–107.   [389]

[1974] Toft B., On critical subgraphs of colour-critical graphs. *Discrete Math.* **7** (1974), 377–392.   [192]

[1973] Toida S., Properties of an Euler graph. *J. Franklin Inst.* **295** (1973), 343–345.
                                                                                         [95]

[1971] Tomescu I., Le nombre maximal de colorations d'un graphe. *C. R. Acad. Sci. Paris* **A272** (1971), 1301–1303.   [204]

[1993] Tovey C.A. and R. Steinberg, Planar Ramsey numbers. *J. Comb. Theory (B)* **59** (1993), 288–296.   [283]

[1992] Truemper K., *Matroid decomposition.* Academic Press (1992).   [347]

[1973] Tucker A.C., The strong perfect graph conjecture for planar graphs. *Canad. J. Math.* **25** (1973), 103–114.   [312]

[1975] Tucker A.C., Coloring a family of circular arcs. *SIAM J. Appl. Math.* **3** (1975), 493–502.   [313]

[1976] Tucker A.C., A new applicable proof of the Euler circuit theorem. *Amer. Math. Monthly* **83** (1976), 638–640.   [95]

[1977] Tucker A.C., Critical perfect graphs and perfect 3-chromatic graphs. *J. Comb. Theory (B)* **23** (1977), 143–149.   [308, 310, 312]

[1976] Tucker A.C. and L. Bodin, A model for municipal street-sweeping operations. In *Case Studies in Applied Mathematics, Committee on the Undergraduate Program in Mathematics.* Math. Assoc. of Amer. (1976).   [93]

[1941] Turán P., Eine Extremalaufgabe aus der Graphentheorie. *Mat. Fiz Lapook* **48** (1941), 436–452.   [34]

[1946] Tutte W.T., On Hamiltonian circuits. *J. London Math. Soc.* **21** (1946), 98–101.
                                                                                        [276]

[1947] Tutte W.T., The factorization of linear graphs. *J. London Math. Soc.* **22** (1947), 107–111. [122]

[1948] Tutte W.T., The dissection of equilateral triangles into equilateral triangles. *Proc. Cambridge Philos. Soc.* **44** (1948), 463–482. [58, 68]

[1952] Tutte W.T., The factors of graphs. *Canad. J. Math.* **4** (1952), 314–328. [125, 132]

[1954] Tutte W.T., A short proof of the factor theorem for finite graphs. *Canad. J. Math.* **6** (1954), 347–352. [132]

[1958] Tutte W.T., A homotopy theorem for matroids, I, II. *Trans. Amer. Math. Soc.* **88** (1958), 144–174. [268, 350]

[1958] Tutte W.T., Matroids and graphs. *Trans. Amer. Math. Soc.* **88** (1958), 144–174. [264]

[1960] Tutte W.T., Convex representations of graphs. *Proc. Lond. Math. Soc.* **10** (1960), 304–320. [260, 263]

[1961] Tutte W.T., On the problem of decomposing a graph into n connected factors. *J. London Math. Soc.* **36** (1961), 221–230. [62, 346]

[1963] Tutte W.T., How to draw a graph. *Proc. Lond. Math. Soc.* **13** (1963), 743–767. [260, 263]

[1970] Tutte W.T., *Introduction to the Theory of Matroids.* American Elsevier (1970). [328, 347]

[1971] Tutte W.T., On the 2-factors of bicubic graphs. *Discrete Math.* **1** (1971), 203–208. [220]

[1980] Tverberg H., A proof of the Jordan Curve Theorem. *Bull. Lond. Math. Soc.* **12** (1980), 34–38. [249]

[1982] Tverberg H., On the decomposition of K_n into complete bipartite subragphs. *J. Graph Theory* **6** (1982), 493–494. [438]

[1951] van Aardenne-Ehrenfest T. and N.G. de Bruijn, Circuits and trees in oriented linear graphs. *Simon Stevin* **28** (1951), 203–217. [90]

[1937] van der Waerden B.L., *Moderne Algebra Vol. 1* Second ed. Spinger-Verlag (1937). [321, 328, 349]

[1965] van Rooij A. and H.S. Wilf, The interchange graphs of a finite graph. *Acta Math. Acad. Sci. Hung.* **16** (1965), 263–269. [213]

[1994] Vazirani V.V., A theory of alternating paths and blossoms for proving correctness of the $O(|V^{1/2}||E|)$ general graph matching algorithm. *Combinatorica* **14** (1994), 71–91. [130]

[1962] Vitaver L.M., Determination of minimal coloring of vertices of a graph by means of Boolean powers of the incidence matrix (Russian). *Dokl. Akad. Nauk. SSSR* **147** (1962), 758–759. [178]

[1964] Vizing V.G., On an estimate of the chromatic class of a p-graph. *Diskret. Analiz.* **3** (1964), 25–30. [210, 217, 387, 419]

[1965] Vizing V.G., Critical graphs with a given chromatic class. *Diskret. Analiz.* **5** (1965), 9–17. [210, 217]

[1976] Vizing V.G., Coloring the vertices of a graph in prescribed colors (Russian). *Diskret. Analiz.* **29** (1976), 3–10. [389]

[1996] Voigt M., List colourings of planar graphs. presented at "Paul Erdős Is Eighty" (Keszthely 1993) (to appear). [389]

[1936] Wagner K., Bemerkungen zum Vierfarbenproblem. *Jber. Deutsch. Math. Verein.* **46** (1936), 21–22. [260]

[1937] Wagner K., Über eine Eigenschaft der ebenen Komplexe. *Math. Ann.* **114** (1937), 570–590. [260, 336, 339]

[1980] Wagon S., A bound on the chromatic number of graphs without certain induced subgraphs. *J. Comb. Theory (B)* **29** (1980), 245–246. [191]

[1972] Walter J.R., *Representations of rigid cycle graphs.* Ph.D Thesis, Wayne State Univ. (1972). [293]

[1978] Walter J.R., Representations of chordal graphs as subtrees of a tree. *J. Graph Theory* **2** (1978), 265–267.   [293]

[1995] Wang J., D.B. West, and B. Yao, Maximum bandwidth under edge addition. *J. Comb. Theory* **20** (1995), 87–90.   [373]

[1976] Welsh D.J.A., *Matroid Theory.* Academic Press (1976).   [328, 343, 347, 349]

[1967] Welsh D.J.A. and M.B. Powell, An upper bound for the chromatic number of a graph and its application to timetabling problems. *Computer J.* **10** (1967), 85–87.
                                                                         [177]

[1982a] West D.B., A class of solutions to the gossip problem, I. *Discrete Math.* **39** (1982), 307–326.   [402]

[1982b] West D.B., Gossiping without duplicate transmissions. *SIAM J. Algeb. Disc. Meth.* **3** (1982), 418–419.   [402]

[1996] West D.B., The superregular graphs. *J. Graph Theory* (to appear).   [452]

[1973] White A.T., *Graphs, Groups and Surfaces.* North-Holland (1973).   [432]

[1977] White D.E. and S.G. Williamson, Recursive matching algorithms and linear orders on the subset lattice. *J. Comb. Theory (A)* **23** (1977), 117–127.   [108]

[1986] White N., *Theory of matroids.* Cambridge University Press (1986).   [347]

[1992] White N., *Matroid Applications.* Cambridge University Press (1992).   [347]

[1960] Whiting P.D. and J.A. Hillier, A method for finding the shortest route through a road network. *Operations Research Quart.* **11** (1960), 37–40.   [76]

[1932] Whitney H., Congruent graphs and the connectivity of graphs. *Amer. J. Math.* **54** (1932), 150–168.   [144, 146, 152, 218]

[1933a] Whitney H., Planar graphs. *Fund. Math.* **21** (1933), 73–84.   [339]

[1933b] Whitney H., 2-isomorphic graphs. *Amer. J. Math.* **55** (1933), 245–254.   [251, 340]

[1935] Whitney H., On the abstract properties of linear dependence. *Amer. J. Math.* **57** (1935), 509–533.   [321, 328, 334, 349]

[1967] Wilf H.S., The eigenvalues of a graph and its chromatic number. *J. London Math. Soc.* **42** (1967), 330–332.   [440]

[1986] Wilson R.J., An Eulerian trail through Königsberg. *J. Graph Theory* **10** (1986), 265–275.   [85]

[1990] Wilson R.J. and J.J. Watkins, *Graphs, an Introductory Approach.* Wiley (1990).
                                                                         [13]

[1983] Winkler P.M., Proof of the squashed cube conjecture. *Combinatorica* **3** (1983), 135–139.   [380]

[1992] Winkler P.M., Problem 27. *Discrete Math.* **101** (1992), 359–360.   [375]

[1972] Woodall D.R., Sufficient conditions for circuits in graphs. *Proc. Lond. Math. Soc.* **24** (1972), 739–755.   [394, 398]

[1982] Xia X.-G., Hamilton cycle in two sorts of Euler tour graph. *Acta Xin Xiang Normal Inst.* **2** (1982), 8–10.   [231]

[1954] Zarankiewicz K., On a problem of P. Turán concerning graphs. *Fund. Math.* **41** (1954), 137–145.   [280]

[1986] Zhang F.-J. and X.-F. Guo, Hamilton cycles in Euler tour graph. *J. Comb. Theory (B)* **40** (1986), 1–8.   [231]

[1985] Zhu Y.J., Z.H. Liu, and Z.G. Yu, An improvement of Jackson's result on Hamilton cycles in 2-connected regular graphs. In *Cycles in graphs.* (Burnaby, B.C., 1982) North-Holland (1985) 237–247.   [224]

[1949] Zykov A.A., On some properties of linear complexes (Russian). *Mat. Sbornik* **24** (1949), 163–188.   [191]

Author Index

This is an index of citations of publications, generally appearing in the text as "Author [year]". Citations with multiple authors are listed under each.

Abbott H.L. 370
Acharya B.D. 297
Ahuja R.K. 75, 130, 158, 163, 172
Aigner M. 270, 328, 347
Ajtai M. 360, 445
Alon N. 387, 389, 401, 405, 407, 408, 445, 446
Andersen L.D. 217
Ando K. 60
Appel K. 270
Ayel J. 229

Babai L. 418, 419
Bäbler F. 95
Bacharach M. 168
Baker B. 384
Barnette D. 285
Batagelj V. 48
Bean D.R. 181
Behzad M. 403
Beineke L.W. 214
Benzer S. 298
Berge C. 100, 123, 177, 181, 201, 288, 290, 301
Bermond J.C. 395, 396
Bernstein P.A. 297
Bertschi M. 320
Bhasker J. 190
Bialostocki A. 401
Biggs N. 433, 447
Birkhoff G.D. 194, 272, 283
Bixby R.E. 328
Björner A. 347
Bland R.G. 306, 308, 320
Blass A. 430
Bodin L. 93
Boland J.Ch. 318

Bollobás B. 387, 405, 409, 415, 418, 420, 421, 422, 427, 429
Bòna M. 369
Bondy J.A. 34, 39, 40, 59, 63, 142, 172, 222, 224, 229, 264, 266, 372, 388, 395, 399, 430
Booth K.S. 265
Boppana R. 388
Borůvka O. 74, 75
Borodin O.V. 183
Bosák J. 285
Brooks R.L. 179
Brozinsky 371
Buckingham M.A. 306, 308, 310, 320
Bumby R.T. 386
Buneman P. 293
Burlet M. 300, 318
Burr S.A. 361, 362, 363, 371

Cameron P.J. 449
Catlin P.A. 183, 189, 193
Cayley A. 63
Chang S. 218
Chappell G.G. 348, 351, 352
Chartrand G. 95, 141, 155, 264
Chein M. 155
Choudom S.A. 401
Christofides N. 237
Chung F.R.K. 360, 367, 375, 445
Chung M.-S. 40
Chungphaisan V. 131
Chvátal V. 50, 183, 222, 223, 225, 229, 283, 288, 301, 302, 304, 307, 308, 312, 314, 319, 320, 362, 363, 421, 429
Chvátalová J. 373
Clapham C.R.J. 229
Collins K.L. 286

Cook S.A. 238
Crapo H.H. 328
Cull P. 97
Cvetković D.M. 433, 450
de Bruijn N.G. 90
de Werra D. 217

Demoucron G. 265, 266
Descartes B. 185, 191
Dijkstra E.W. 76, 82
Dirac G.A. 152, 156, 157, 187, 188, 189, 190, 192, 221, 226, 264, 394, 401, 421
Doob M. 433, 450
Duchet P. 301
Dudeney H.E. 247
Dziobek O. 72

Edmonds J. 62, 92, 127, 130, 163, 325, 328, 340, 344, 345, 346, 382
Egawa Y. 401
Egerváry E. 103, 201, 341
Eitner P.G. 372
Elias P. 149
Ellingham M.N. 389
Enomoto B. 220
Entringer R.C. 354, 369
Erdős P. 132, 182, 183, 185, 225, 229, 354, 358, 361, 362, 363, 369, 370, 372, 374, 386, 387, 389, 394, 400, 401, 403, 406, 407, 408, 418, 419, 421, 422, 429
Eršov A.P. 190
Euler L. 85, 255
Even S. 120, 130

Fan G.-H. 397
Fáry I. 260
Feinstein A. 149
Feng T. 397
Finck H.-J. 182
Fishburn P.C. 318
Fisher D.C. 286
Fisk S. 283
Fleischner H. 97, 228, 387
Floyd R.W. 108
Ford L.R. Jr. 94, 149, 152, 153, 158, 162, 342
Fournier J.-C. 217
Frank A. 154
Frankl P. 361, 371
Fraughnaugh (Jones) K. 283
Frobenius G. 101
Fulkerson D.R. 94, 149, 152, 153, 158, 162, 205, 289, 316, 325, 342, 344

Gabber O. 445
Gabow H.N. 75, 130
Gaddum J.W. 182
Gale D. 117, 121, 166, 167
Galil Z. 75, 445
Gallai T. 104, 109, 132, 178, 188, 191, 300, 352, 372, 391, 394
Galvin F. 387, 388
Garey M.R. 234, 244, 245, 366, 421
Gavril F. 293, 316
George J. 216
Georges J.P. 220
Gervacio S. 60
Ghouila-Houri A. 226, 399
Gibbons A. 92, 239
Gilbert E.N. 410
Giles R. 313, 315
Gleason A.M. 359
Glicksman S. 71
Goddyn L. 389
Goldberg M.K. 217
Golumbic M.C. 288, 295, 306, 308, 310, 318, 320
Gonzalez T. 236
Good I.J. 92, 95, 96
Goodman A. 374
Goodman N. 297
Gould R.J. 265, 362
Graham N. 354, 369
Graham R.L. 307, 308, 319, 355, 356, 366, 379
Graver J.E. 359, 360
Greene C. 317, 348
Greenwell D.L. 228, 316
Greenwood R.E. 359
Grigni M. 403
Grimmett G.R. 422
Grinberg E.J. 275
Grinstead C.M. 312, 359, 360
Gross O.A. 205, 316
Grötzsch H. 283
Grünbaum B. 259
Guan M. 91
Guo X.-F. 231
Gupta R.P. 210, 217
Gusfield D. 118
Guy R.K. 279, 280, 287
Gyárfás A. 193
Győri E. 375

Hadwiger H. 189, 336
Häggkvist R. 387, 404
Hajnal A. 183
Hajós G. 188, 192
Haken W. 270
Hakimi S.L. 43, 48
Halin R. 157, 182
Hall M. 101, 164
Hall P. 100
Hammer P.L 49, 317
Hammersley J. 449
Harary F. 73, 135, 141, 143, 227, 231, 264, 287, 351, 354, 362, 401, 402, 429, 430, 432, 433
Harper L.J. 366
Harris A.J. 387

Hartman C.M. 109, 313
Havel V. 43
Hayward R.B. 305
Heawood P.J. 269, 270
Heesch H. 271
Hierholzer C. 85
Hillier J.A. 76
Hitchcock F.L. 94
Hochberg R. 367, 368
Hoffman A.J. 218, 442
Holton D.A. 10
Holyer I. 211, 419
Holzmann C.A. 351
Hopcroft J. 118, 119, 265
Horton J.D. 220
Hsu D.F. 401
Huang H.-C. 306, 307, 308, 320
Huffman D.A. 80

Iba G. 317
Ingleton A.W. 352
Irving R.W. 118
Isaak G.A. 107

Jackson B. 220, 224
Jacobson M.S. 318, 362
Janssen J.C.M. 387
Jarník V. 75, 82
Johnson D.S. 234, 244, 245, 366, 421
Johnson E. 92
Jordan C. 55, 369
Jung H.A. 189
Jünger M. 404

Kahn J. 387
Kainen P.C. 186, 270, 287
Kalbfleisch J.G. 359
Kaneko A. 60
Kapoor S.F. 49
Karapetian I.A. 109
Karger D.R. 75
Kariv O. 130
Karp R.M. 118, 119, 163, 239, 240, 431
Katerinis P. 220
Kelmans A.K. 265
Kempe A.B. 270
Kernighan B.W. 236
Kimble R.J. Jr. 73
Kirchhoff G. 63, 65
Klein P.N. 75
Kleitman D.J. 280, 355, 386
Klotz W. 267
Knuth D.E. 118, 366
Koch J. 270
Kodama Y. 143
Komlós J. 360, 445
König D. 97, 103, 105, 201, 209, 341
Koopmans T.C. 94
Korte B. 347
Kostochka A.V. 183, 375

Kotzig A. 216
Kouider M. 224
Kožuhin G.I. 190
Krausz J. 212
Krompart L.B. 286
Kronk H.V. 228
Kruskal J.B. Jr. 75
Kučera L. 419
Kuhn H.W. 113
Kung J.P.S. 347, 351
Kuratowski K. 259

Landau H.G. 47, 49
Las Vergnas M. 132, 230, 297, 347
Lawler E.L. 130
Lawrence J. 183
Lawrence S.L. 372
Lazarson T. 347
Lederberg J. 285
Lehman A. 338, 345, 348
Lehot P.G.H. 213
Lekkerkerker C.G. 318
Lesniak L. 95, 155, 264
Lick D.R. 156, 182
Liestman A.L. 244
Lin S. 236
Linial N. 395
Liu J. 108, 132
Liu R. 182
Liu Z.H. 224
Locke S.C. 50
Lovász L. 25, 50, 72, 107, 122, 156, 158,
    178, 183, 186, 291, 300, 304, 305, 342,
    347, 372, 377, 382, 392, 401, 402, 408
Lu X. 225
Lubotzky A. 446
Lucas E. 88
Lueker G.S. 265, 294

MacLane S. 321, 333
Maddox 452
Madej T. 287
Mader W. 154, 158, 189, 190
Magnanti T.L. 75, 130, 158, 163, 172
Mahadev N.V.R. 386
Malgrange Y. 265, 266
Mantel W. 33
Marcus M. 108
Margulis G.A. 445, 446
Markossian S.E. 109
Mason J.H. 351
Matthews M.M. 228
Matula D.W. 180, 182, 183, 420
Maurer S. 47, 50
McDiarmid C.J.H. 217, 367, 368, 422
McGuinness S. 375
McKay B.D. 359
McKee T.A. 95, 297
McMorris F.R. 318
Melnikov L.S. 316

Menger K. 152
Meyniel H. 226, 300, 312, 320, 399
Micali S. 130
Milgram A.N. 352, 391
Miller Z. 372, 401
Milman V.D. 445, 446
Minty G.J. 183, 350
Mirsky L. 325, 342
Moon J.W. 62, 63, 218, 229, 231
Motzkin T.S. 259
Mowshowitz A. 452
Mulder H.M. 318
Munkres J. 113
Murty U.S.R. 34, 39, 40, 59, 172, 264, 266
Mycielski J. 184
Myers B.R. 182

Nash-Williams C.St.J.A. 58, 62, 154, 156, 227, 230, 344, 346, 353
Nemhauser G.L. 328
Nešetřil J. 186, 377, 401, 402, 408
Nilli A. 446
Nishiura 371
Nordhaus E.A. 182
Norman R.Z. 109

O'Donnell P. 386
Olariu S. 312, 319
Olaru E. 300
Ore O. 95, 108, 217, 222, 229, 230, 270, 342, 395
Orlin J.B. 75, 130, 158, 163, 172
Oxley J.G. 347

Padberg M.W. 306, 308
Palmer E.M. 405, 415, 421, 430, 432
Papadimitriou C.H. 163, 328
Parthasarathy K.R. 312, 315, 401
Peck G.W. 440
Peled U. 452
Peleg D. 403
Perfect H. 325
Perold A.F. 307, 308, 319
Pertuiset R. 265, 266
Petersdorf M. 452
Peterson J. 124, 125
Phillips R. 446
Piff M.J. 352
Pinsker M. 445
Pippenger N. 445
Plesnik J. 142
Plummer M.D. 107, 157, 342
Polimeni A.D. 49
Pollak H.O. 379
Polya G. 63
Pósa L. 374
Powell M.B. 177
Prim R.C. 82
Prins G. 143

Pritikin D. 71
Prüfer H. 63
Pulleyblank W.R. 404
Pultr A. 377, 401, 402

Rabin M. 109
Rabinovitch I. 50
Rado R. 327
Radziszowski S.P. 359
Ramsey F.P. 355, 356
Ravindra G. 300, 312, 315, 401
Ray-Chaudhuri D.K. 215
Raynaud H. 372
Recski A. 347
Rédei L. 231
Ree R. 108
Reed B.A. 320
Rees D. 96
Reinelt G. 404
Rényi A. 63, 71, 418
Reznick B. 440
Richards D. 244
Ringel G. 69
Robbins H.E 148
Roberts F.S. 93, 298, 318, 386
Roberts S.M. 359
Robertson N. 189, 274
Rödl V. 186, 363, 408
Rose D. 294
Rosenfeld M. 403
Rota G.C. 328, 333
Rothschild B.L. 356
Rotman J.J. 96
Roy B. 178
Rubin A. 386, 389, 403
Rucínski A. 415
Ryser H.J. 49, 167

Saaty T.L. 270
Sachs H. 182, 300, 433, 435, 450, 452
Sahni S. 236
Saito A. 220
Saks M. 367, 368
Samad T. 190
Sanders D.P. 274
Santhanakrishnan P. 386
Sarnak P. 446
Sbihi N. 312
Schäuble M. 191
Schnyder W. 264
Schönheim J. 401
Schrader R. 347
Schrijver A. 328, 344
Schuster S. 264
Schwartz B.L. 165, 172
Schwenk A.J. 73, 402, 435, 451
Scoins H.J. 71
Seinsche D. 40, 316
Selkow S.M. 418
Seress Á. 402

Seymour P.D. 189, 274, 340
Shannon C.E. 82, 84, 149, 208, 216, 217
Shapley L.S. 117, 121
Shearer J.B. 386
Sheehan J. 10
Shende A.M. 386
Shibata T. 297
Shmoys D.B. 305
Shostak R. 384
Shrikhande S.S. 218
Siegel 388
Simeone B. 49, 317
Simonovits M. 430
Snevily H.S. 229
Spencer J.H. 356, 363, 371, 405, 408, 445
Spencer T. 75
Sperner E. 364
Stanley R.P. 202, 203, 205
Steiglitz K. 163, 328
Stein S.K. 260
Steinberg R. 283
Stiebitz M. 192, 387
Stockmeyer L. 244
Sturmfels B. 347
Sulanke 286
Sumner D.P. 132, 142, 228
Sun L. 312
Syslo M.M. 372
Székely L.A. 354, 369
Szekeres G. 177, 182, 354, 358, 369, 370
Szele T. 407
Szemerédi E. 193, 360, 363, 364, 445

Tait P.G. 274, 284
Tanner R.M. 445
Tarjan R.E. 75, 120, 130, 245, 265, 294, 295, 316, 383
Tarry G. 97
Tarsi M. 369, 387
Taylor H. 386, 389, 403
Tesman B. 386
Thomas R. 189, 274
Thomassen C. 189, 190, 261, 263, 264, 283, 389, 399
Tiwari P. 440
Toft B. 192
Toida S. 95
Tomescu I. 193, 204
Tovey C.A. 283
Trotter Jr. L.E. 306, 308, 313, 315, 320
Trotter Jr. W.T. 50, 363
Truemper K. 347
Tucker A.C. 93, 95, 308, 310, 312, 313, 315
Turán P. 34
Tutte W.T. 58, 62, 68, 122, 125, 132, 220, 260, 263, 264, 268, 276, 328, 346, 347, 350
Tuza Z. 193
Tverberg H. 249, 438

Uhry J.P. 300, 318
Urrutia J. 316

van Aardenne-Ehrenfest T. 90
van der Waerden B.L. 321, 328, 349
van Lint J.H. 449
van Rooij A. 213
Vazirani V.V. 130
Vince A. 415
Vitaver L.M. 178
Vizing V.G. 210, 217, 316, 387, 389, 419
Voigt M. 389

Wagner K. 260, 336, 339
Wagon S. 191
Wall C.E. 49
Walter J.R. 293
Wang J. 373
Watkins J.J. 13
Welsh D.J.A. 177, 328, 343, 347, 349
West D.B. 40, 190, 373, 402, 440, 452
White A.T. 182, 432
White D.E. 108
White N. 347
Whitesides S.H. 307, 308, 319
Whiting P.D. 76
Whitney H. 144, 146, 152, 218, 251, 321, 328, 334, 339, 340, 349
Wilf H.S. 177, 213, 440, 451
Williamson S.G. 108
Wilson R.J. 13, 85, 419
Wilson R.M. 361, 371
Winkler P.M. 375, 380
Wolk E.S. 25
Wolsey L.A. 328
Woodall D.R. 394, 398

Xia X.-G. 231

Yackel J. 359, 360
Yannakakis M. 295, 316
Yao B. 373
Yu Z.G. 224

Zak J. 372
Zarankiewicz K. 280
Zhang F.-J. 231
Zhang K.M. 359
Zhou H. 108, 132
Zhu Y.J. 224
Ziegler G. 347
Zykov A.A. 191

Subject Index

In this index, an italicized item may designate a definition of the concept and perhaps also basic results about it. An item in boldface designates important material such as a proof of the result named; it may also include a definition of the concept. This index does not include page references to the glossary, where an alphabetized list of definitions can be found.

absorption property 323
acquaintance relation 2, 11
acyclic *51-53*
acyclic orientation 202-203
adjacency 15
adjacency algebra *442*
adjacency matrix *5-14*, 24, 38, 366-368
adjacency relation *1*, 8
affine independence 350
algorithm 74, 76-78, 80-82, 89-91,
    109-121, 113-120, 127-130, 161-163,
    177, 246, 265-267, 295-296, 404-405
alteration principle 407-408
M-alternating path *99-102*, 115
alternating path 112-114, 127-130, 211
ancestor *79*
Anderson-Goldberg bound 217
antichain 391
antihole *312*, 314-315
antipodal vertex 354
antitwin *319*
approximation algorithm *235*
approximation scheme 235
arboricity 62, 346
archeological seriation 298
Art Gallery Theorem 283
articulation point *138*
k-ary tree *79*
aspect *321-322*
Assignment Problem 112, 115
asteroidal triple *318*
augmentation *210-211*
augmentation property 324, 325,

327-329, 337, 348
f-augmenting path *159-163*
augmenting path *99-102*, 113, 118-119,
    129-130, 132, 343
Augmenting Path Algorithm 109-112
automorphism *10*, 13, 37, 61, 418, 452
average degree 26, 37, 38, 60, 181
Azuma's Inequality **423-429**, 432

backtracking *78*
balanced graph *414-415*, 415, 430
bandwidth 364-368, *366-368*, 373-374
Barnette's Conjecture 277
barycenter *61*
base *321-322*, 338, 348, 350
base exchange property 323, **327-329**,
    334, 348
Baseball Elimination Problem 165, 172
basis step *16*
Berge's Theorem **100**, 127
Berge-Tutte Theorem 123-124
best possible *30*
biclique decomposition 440
bigraphic *49*, 167, 171
bijection 7, *28*, 63-64, 66
bin-packing 432
binary matroid *330*, 350
binary tree *79*, 84
Binet-Cauchy formula 68
binomial coefficient *9*, 37
binomial distribution 431
biparticity 30, *401*

bipartite graph *3*, 14, 21, 24, 27, 30, 36,
    38, 40, 47, 49, 83, **98-121**, 125, 167,
    174, 182, 187, 200, 209, 216, 219, 229,
    253-254, 257, 258, 277, 325, 330, 341,
    347, 352, 373, 387, 401, 429, 430,
    435-436, 450
bipartite multigraph 217
bipartition *21*, 227
Birkhoff diamond *272*
block **139-141**, 179, 188, 351
block-cutpoint graph *139-140*, 143
blossom *127-130*
bond *138*, 253, 335, 339
bond matroid *335*, 337, 338-340
bond space *432*
Bondy's Lemma **395-397**
bottleneck spanning tree 83
bouquet 282
boxicity *431*
branch vertex *259*, 262
r-branching *382-383*
branching *68*, 381-383
breadth-first search (BFS) **78**, 79,
    118-119, 174, 379-380
breadth-first search (BSF) 234
breadth-first tree 380
Bridg-it 58-59, 337
H-bridge *264*, 264-266, 268
bridge *138*
bridge (game) problem 39
broadcasting 402
Brooks' Theorem 178, **179**, 180, 183, 188,
    204, 216
Brouwer Fixed-Point Theorem 365
Bus Driver Problem *120*

cactus *143*, 181
canonical labeling algorithm *418*
capacity constraint *158-163*
Cartesian product 175, 181, 211, 215,
    228, 231, 287, 316, 367, 373, 401, 404,
    441, 450
Cartesian product of graphs 216
caterpillar *70*, 73, 317, 372
Cauchy-Binet formula 72
Cayley's Formula *63-65*, 71, 72, 317
Cayley-Hamilton Theorem 437
ceiling *29*
cell *364-365*
2-cell embedding *282*
center *54-55*, 61, 369
centroid 369
α, β-chain *270*
chain 391
characteristic polynomial **433-436**, 450
characterization 18, 42, 48, 49, 59, 70, 86,
    95, 122, 126, 143, 144-145, 148, 156,
    199, 212-215, 259-263, 283, 317, 374,
    376, 435-436

Chebyshev's Inequality 413, 431
children *79*, 84
Chinese Postman Problem 91, 96
choice function 386
choosability **386-390**, 403
f-choosable *387-388*
k-choosable *386*, **389-390**
chordal graph **198-201**, **293-297**, 298,
    300-301, 316, 317, 318, 403
chordless cycle *199*, 293
chords (conflicting) *250*
k-chromatic *173*, 177, 183, 184-190, 272
chromatic index *208*
CHROMATIC NUMBER 241
chromatic number *4*, **173-205**, 253, 269,
    288-292, 373, 408, 422, 426-429, 440
chromatic number (hypergraph) *429*
chromatic polynomial *194-198*, 202-205
Chvátal's condition 223
Chvátal's Conjecture (toughness) 220
Chvátal's graph 229
Chvátal-Erdös' condition 225
circle graph *312*, 320
circuit *85-97*, *321-322*, 324, 347, 349, 350
circulant graph *306-307*, 310, 313
circular-arc graph *312-313*, 320
circulation *169*, 172
circumference *218*, **394-400**
CIRCUMFERENCE 234
clause *239*
claw $K_{1,3}$ 37
claw-free 132, 213, 217, 312-315
Clebsch graph 449
clique *3*, 8, 49, 179, 188, 212, 288-292,
    303-304, 316, 359-361, 362, 376, 379,
    406
CLIQUE 241
clique cover number *200*, 288-292
clique covering 306-311, 374, 401
clique decomposition 374-375
clique identification 315
clique number 174-175, 208, 288-292,
    373, 405, 420
clique tree *296*
clique-vertex incidence matrix 292,
    298-299, 308, 317
closed curve *249*
closed ear decomposition *147*
closed neighborhood 313
closed set *333*, 349, 350
closed trail 23, 86
closed walk *14*, 23, 36
closure *222-223*
closure (Hamiltonian) 397, 431
closure function *333*
closure operator *333*, 349
cobase **333-336**
cocircuit **333-336**, 350
co-critical pair *311*
cocycle matroid *335*

coffee cup 280
co-graphs *316*
college admission 116
color class *4*
color-critical *173*, 185, 186-187, 190
3-COLORABILITY 239, 240
k-COLORABILITY 241, 245
2-COLORABILITY 234
K-COLORABILITY 234
k-colorable *173-175*, 177, 181, 183, 336
k-coloring *173*, 179, 195-196, 269, 356
4-coloring 271
coloring 207, 253, 306-311
column matroid *324*, 350
common independent set 340-345, 352
common system of distinct representatives (CSDR) 153-154, 171, 341-342
comparability graph *201*, 298, 300-301, *390*
compatible pair *205*
complement *3*, 10, 12, 20, 61, 229, 258, 443
complement reducible *316*
complete bipartite graph *3*, 8, 11, 32, 72, 215, 218, 386, 433, 440
complete graph *3*, 12, 30
complete matching *98*, 101, 108, 227, 378, 430
complete multipartite graph *33*, **39**, 372
completely labeled cell *364-365*
complexity 232-236
component *18-19*, 23, 24, 48, 52, 56, 60, 64, 86, 196-197, 219, 413, 434
component (giant) *417*
S-component *187-188*, 261, *264*, 283
composition *303*, *370*
concave sequence *369-370*
condensation *144*
conflict *264*
conflict graph *264-265*, 268
conjugate partition *167*
k-connected *133*, *148*, 215, 251-261, 264, 266, 277, 420
2-connected 141, 144-147, 155, 395-397
connected graph *5*, *15*, 18, 23, 24, 25, 29, 31, 36, 48, 51-53, 57, 60, 61, 138, 190, 204, 255, 256, 371, 410, 413
connection relation *15*, 18
CONNECTIVITY 234
connectivity **133-144**, 225, 285, 373, 420
connectivity (digraphs) 148
connector *367*
consecutive 1's property *298*, 317
conservation *158-163*
consistent rounding *168*, 172
construction procedure *293*
contracted vertex 195-196
contraction *65*, 66, 128-130, 189, 255, 257, 260, 261, *336-341*, 351
contradiction *20*

contrapositive *17*
converse *17*
convex combination 365, 372
convex embedding 261-264, 268
convex function 423
convex polygon 358, 369-370
convex representation *260*
convex sequence *369-370*
copy *8*, 27, 53
cospectral tree 450
cost 5, 112-114
Coupon Collector Problem 431
F-covering *374*
COVERING CIRCUIT 246
covering problem 30
k-critical *173-174*, 177, 177-180, 183, 186-187, 191-193
α-critical *246*
critical edge *109*, 109, *311*
critically connected 157-158
crossing *248*
crossing number **277-280**, 286, 287
crossover *395*
cut capacity *160*
cut-edge *18-19*, 31, 36, 54, 60, 87-88, 141, 251, 252
cut-vertex *18*, 20, 24, 25, 141, 143, 217, 259-261
cycle *5*, 8, 11, *14-15*, *14*, 19, 22, 23, 24, 37, 54, 59-62, 203, 225-226, 253, 254, 268, 321, 372, 391-393, 434, 450
cycle C_4 25, 245, 371, 401
cycle C_n 106
cycle matroid *322*, *322*, 323, 325, 331, 335, 337, 338-340, 347, 351
cycle space *432*

deBruijn cycle 92-93, 96
deBruijn graph 93
decision problem 232
F-decomposition *374*, *390*, **391-393**
decomposition 12, 56, 62, *69*, 72-73, 86-87, 94, 96, 131, 401, 403-404, 440
decomposition procedure *293*
decreasing path *369*
defect *342*
deficiency 132
degree *6*, *25*, 60, 85-88, 399, 418, 429
degree sequence *40-50*, 48, 60, 73, 142, 177, 190, 223, 224, 230, 251-259
degree set 49
degree-sum formula *26*, 252, 256
deletion 66, 255, 257, 259-261, 260, 336
deletion method 407-408, 429
deletion of edges 18
dense graph 410
density 362, 366-368, *415*
dependence (linear) 67
dependence properties 349

dependent edge *205*
dependent set 321, 349
depth-first search (DFS) **78**, 79, 140,
    379-380
descendant *79*
design (pairwise balanced) 351
determinant 67-68, 72, 433
diagonal Ramsey number *361*
DIAMETER 234
diameter *54-55*, 59, 61-62, 70, 142, 373,
    411, 439
digraph *2*, 46-47, 49, 68, 182, 216, 330,
    351, 352
Dijkstra's Algorithm **76-78**, 83
dilation *366-368*
Dilworth's Theorem 391, 403
k-dimensional cube *26-27*, 36-37
Dinitz Conjecture 387-389
Dirac's Theorem **221**, 396
directed graph 77, 183, 259
DIRECTED HAMILT. CYCLE 243
DIRECTED HAMILT. PATH 239, 242
Directed Matrix Tree Theorem *68*, 90
disconnected graph *15*, 38, 255
disconnecting set *136*
disjoint union *19*, 23, 29
disjointness graph *13*
distance 44, *51*, 54-58, 61, 174
distance-preserving embedding *378*
DNA chains 298
dodecahedron 227
dominating set *407*
dominating set algorithm 430
double jump *416*
double torus 280, 282
double triangle *213*, 214, 217
doubly stochastic matrix *106*
doubly-critical graph *191*
doughnut 280
drawing *248*
dual (matroid) **333-340**, 350
dual augmentation property 335
dual edge 251, 253, 257
dual graph 250-255, 274, 338-340,
    364-365
dual problem 102, 104, 111, 120, 160, 161
dual vertex 251, 253
duality gap 292

ear decomposition *146*, 147
eccentricity *54-55*
edge *1*, *2*
edge cover **104-105**, 109
edge cut *136-144*, 141, 142, 143, *148*, 187
edge set *1*
edge-choosability *387-390*
edge-chromatic number *208*, 215,
    217, 217, 284, 376
edge-coloring **207-212**, 215-217, 217, 274,

356
$\Delta(G)$-EDGE-COLORING 234
k-edge-connected *136*, *148*, 180, 183, 215,
    258, 353
k-edge-connected orientation 154
edge-connectivity **136-144**, *148*, 186-187
edge-transitive *13-14*
Edmonds' Blossom Algorithm **127-130**
Edmonds' Branching Thm. 382-383, 402
eigenvalue 379, **432-452**
eigenvector **432-452**
elementary subdivision *259*
embedding 259-261, 287
endpoints *1*, *5*, *14*, 397
entropy 82, 84
enumeration 9, 26-28, 37-39, 63-78, 90
equality subgraph *112-114*
equitable edge-coloring **217**
equivalence *17*, *376*
equivalence class *8*
equivalence relation *7*
erasure 31
Erdős-Faber-Lovász Conjecture 182
Erdős-Gallai condition 48, 132
Euler's Formula 255-257, 258, 267, 282,
    287
EULERIAN CIRCUIT 234, 238
Eulerian circuit **85-97**, 125, 207, 217,
    230, 231, 237
Eulerian digraph 88-97
Eulerian graph *85-97*, 143, 253-254, 283
Eulerian trail *87-88*
even cycle 100, 123
even graph 28, *85-88*
even pair *320*
even triangle *213*
even walk *17*
evolution of random graphs 415
excess matrix 114-115
existence method 405-408
expander graph 433, **444-446**, 451
expansion 31
Expansion Lemma 145
expansive property **331-332**
expectation *406*
extremal graph theory 373
extremal problem *30*, 34, 39-40, 134-135,
    373-404
extremality *22*, 25, 41, 47, 86, 122, 221,
    223, 231, 303, 399-400

face *248-250*, 253-254
face-coloring 274, 275
factor **121-132**
2-factor 125, 209-210
1-factor *121*, 125, 142
f-factor **125-127**, 132
k-factor *121*, 131
1-factorable *209*

1-factorization *209*
x, U-fan *152-153*
Fan Lemma 152-153, 156
Fary's Theorem 260, 264, 268
fat triangle 208
feasible flow *158-163*, 166, 168-170
feasible solution *292*
Ferrers diagram *167*
Five Color Theorem 269, 270
flat *333*, 349
Fleury's Algorithm 87-88
floor *29*
flow *158-163*
flower *128*
forbidden subgraph 213, 214, 214, 317
forbidden substructure 293
Ford-Fulkerson labeling algorithm
    161-163, 170-171, 235
Ford-Fulkerson Theorem 166, 171
forest *51*, 55-56, 61-63, 62-63, 254, 317,
    336
Four Color Problem 270, 284
Four Color Theorem 5, 189, 272-275, 275,
    277, 451
fraternal orientation *316*
H-free *32*
free matroid *329*
Friendship Theorem 433, **449**
fundamental set of circuits 349

Gallai's Theorem *104*, 109, 352
Gallai-Milgram Theorem 352
Gallai-Roy Theorem **178**, 182, 183, 205
gambler 424
games 58-59, 106, 107, 337-338
gammoid *351*
gap *395*
gas-water-electricity problem 247
generalized coloring parameters *180*
generalized partition matroid *343*
generalized Petersen graph *216*
GENUS 234
genus *280*, 281, 287
geometry 35
Gewirtz graph 449
Ghouila-Houri's Theorem **226**, 231
girth *25*, 37, *185*, 191, 258, 339, 403-404,
    408
k-gon 256
good algorithm *73*, 111
good characterization *234*
gossip problem 384-385
gossip scheme *384-385*
Grötzsch graph *184*, 192, 227
Grötzsch's Theorem 283
graceful labeling *69*
Graceful Tree Conjecture *69*, 73
graph transformation 125-127, 126,
    151-152, 157, 166-169, 187, 224

graphic matroid *322*, 330, 332, 347
graphic sequence *42-46*, 48, 132
greedy algorithm *74*, 235, 236, **327-329**,
    340, 347, 348, 349
greedy coloring *176-179*, 181-182, 201,
    209, 302, 422
greedy decomposition 375
grid 367, 373, 404
Grinberg graph 275, 285
Grinberg's condition **275-277**, 276, 277,
    285
growth rate *73*, *233*

Hadwiger's Conjecture 189, 274, 336, 417
Hajós' Conjecture 188, 417
Hajós' construction 192
half-edge 251
Hall's Condition 100-102, 108
Hall's Theorem 100-102, 106, 107, 132,
    153, 352
Hamiltonian closure *222-223*
Hamiltonian cycle **218-231**, 269, 275-277,
    285, 394, **399-400**, 415, 417, 421
HAMILTONIAN CYCLE 234, 238, 239,
    243-244, 245
Hamiltonian graph 219, 275, 373
Hamiltonian path *219*, 224, 276, 286, 407
HAMILTONIAN PATH 234, 243, 245,
    352
Hamiltonian plane graph 284
Hamiltonian-connected *229*
handle *280*
handshake problem 38
Harary graph 134-135, 141
Harper's bound 366-367, 367
Havel-Hakimi condition 43, 48
head *2*
head partition matroid *330*
Helly property 62, 293, 317
hereditary class *200*, 201
hereditary family *321-322*
hereditary system 320-353, *321-322*, 348
heuristic 246
heuristic algorithm 235-238
homogeneous set *356*
Hopcroft-Karp Algorithm **118-120**
Huffman coding **80-82**, 84
Hungarian Algorithm 113-114
hunter/farmer problem 108
hydrocarbon problem 59
hypercube *27*, 61, 106, 134, 215, 267,
    321-322, 354, 354, 367, 378
hypergraph *429*
hyperplane *333-334*, 334, 349, 350
hypobase *333-334*, 350

icosahedron 190, 284
idempotence property **332**

imperfect graph 205, **305-315**
in-degree *46*, 89-93
in-neighborhood *46*
in-tree *68*, 89-91, 385
incidence *6*
incidence matrix *6*, 12, 67
inclusion-exclusion principle 196-197
incorporation property **332**
increasing path *384-385*
increasing trail 355
independence number *103*, 175, 207, 225, 288-292
independence ratio 283
INDEPENDENT *k*-SET 234
independent set *3*, 4, 24, 38, 60, 174, 180, 202, 204, *321*, 324, 359-361, 406
INDEPENDENT SET 241
indicator variable *406*, 413-414
induced circuit property **327-329**, 348
induced subdigraph 201
induced subgraph *3*
induction *15*, 57
induction hypothesis *16*
induction step *16*
induction trap 31-33, 52
integer linear program *292*
Integrality Condition 448
Integrality Theorem *163*, 165
Interlacing Theorem 439
internally-disjoint paths *144*
intersection graph *293*, 316, 347
intersection number *374*
intersection representation *293*, *374*, 401
interval graph *176*, 200, 201, 205, **298-299**, 317, 317, 318
interval number *431*
interval representation *176*
intractable 234
involution 452
isolated vertex *26*, 413-414, 431
isometric embedding *378*
isomorphism *7-11*, 12-13, 418, 432
isomorphism class *8*, 11, 12, 61, 71, 73, *257*
isomorphism relation *8*
isomorphism testing 418
isthmus *138*

Jackson's Theorem 224
join *19*, 23, 141, 304
joined 15
Jordan Curve Theorem 249, 253, 255, 269, 270, 275

König's Other Theorem *105*, 352
König-Egerváry Theorem **103-105**, 107, 109, 156, 171, 201, 341, 347, 391, 403
Königsberg bridge problem 85

Kempe chain *270*, 273
kernel *387-389*
kernel perfect digraph *387-388*
king *47*, 50
knight's move *227*
Kotzig's Theorem 216
Krausz decomposition **212**, 212-213, *218*
Kruskal's Algorithm **74-75**, 83, 237
Kruskal's algorithm 296
Kuratowski subgraph 250, 256, **259-264**, *260*, 261, 268, 281
Kuratowski's Theorem **259-264**, 267, 268, 338-340

Lagrangian multiplier *436*
Las Vergnas' condition 230
lattice 321, 333
leaf *51-53*, 55, 56, 59-61, 63-64, *79*, 83
leaf block *140*
leeway *159*
length *14*
length (face) *252-253*
lexicographic product *370*
light bulb problem 452
line digraph 215
line graph 200, **206-218**, 218, 227, 293, 387, 401
ε-linear *183*
linear matroid *324*, 326, 332
linear programming 352
linearity of expectation *406*
list chromatic index *387*
list chromatic number *386*
list coloring **386-390**
List Coloring Conjecture 387
list edge-coloring *387-390*
k-list-colorable *386*
literal *239*
lobes 281
lobster *70*
local connectivity parameters *149-152*
local density bound 366-367
local search *236*
local-global connectivity 382-383
logical variable *239*
longest cycle 230
longest path 24, 157, 178, 394, 397
loop *2*, 60, 65, 231, 251, 282, *323*, 347, 350
loopless graph 48, 59, 59

Mader's Theorem 154, 158
magnifier graph *445-446*
map *4*
Markov's Inequality **411**, 412, 424, 428
Marriage Theorem *101*
martingale 423-429, *423*
Martingale Tail Inequality 423

matching **98-132**, 191, 207, 207-208, 215, 238, 341, 347, 352, 371, 373, 385, 450
3 – D MATCHING 352
mates 308, 310
0, 1-matrix 49
matrix rounding problem 168
Matrix Tree Theorem **66-68**, 72, 432
matroid 58, **320-353**
matroid basis graph *351*
matroid covering theorem 346
3-MATROID INTERSECTION 352
matroid intersection 383
Matroid Intersection Theorem **340-345**, 352, 391
matroid packing theorem 346
matroid sum *343*
matroid union **343-346**, 383
Matroid Union Theorem **343-346**, 352
Max-flow Min-cut Theorem **162-163**, 165, 172
maximal *23*
maximal clique 299, 300-301, 375
maximal outerplanar graph *254*
maximal planar graph 287
maximal trail 23, 86
maximum *23*
Maximum Cardinality Search (MCS) **295-296**, 316
maximum clique 314
maximum degree *25*, 37, 43, 44, 48, 176-179, 209, 210-211, 215-217, 418, 419, 440, 442
maximum density *415*
maximum flow **158-163**
MAXIMUM INDEPENDENT SET 232, 234, 235, 239
MAXIMUM MATCHING 234
maximum matching **99-100**
melting point 404
Menger's Theorem 164-165, 171, 207, 341, 382-383, 402
Menger's Theorems **149-158**
Meyniel graph *300-301*, 318, 318, 319, 320
Meyniel's Theorem 399, 404
middle-levels graph *108*
Min-cost Flow Problem 166
min-max relation *103*, 123, 289, 292, 349
minimal imperfect graph *303-304*
minimal vertex separator *199*, 317
minimum cut *161*
minimum degree 22, *25*, 29, 37, 48, 53, 134-135, 136, 142, 142, 177, 188, 192, 193, 221, 226, 229, 230, 231, 258, 313, 373, 385, 407, 418, 440
minimum polynomial *438*
MINIMUM SPANNING CYCLE 232
minimum spanning tree (MST) 74-75, 82-83, 83, 237
Minimum Spanning Tree Problem 232

MINIMUM VERTEX COVER 235
minor **336-340**, 350
Minty's Theorem 183
Model A *409*, 412, 413, 418
Model B *409*, 412, 413
monochromatic 363-364
monotone property *412*
monotone subsequence 354-355
mountain range problem 36
mourse problem 227
multigraph *2*, 42, 48, 59, 65, 85, 208, 209, 230, 251
multinomial coefficient *65*
multiple edges *2*
Mycielski's construction 184, 191

Nash-Williams' Orientation Theorem 154, 156, 158
nearest-insertion *236*, 245
nearest-neighbor *236*
necessity *21*
u, v-necklace *156*
neighbor *26*, 29
neighborhood *26*, 100-103, 106, 108, 271
net demand *166*
network *158*
network flow problems 158-172
node *158*
NODUP property 402
NOHO property 384-385, 402
non-Hamiltonian graph 276, 285
nondecreasing path *369*
nondeterministic algorithm *233*
nonplanar graph 250, 256, *259-261*, 264-267
nontrivial graph *18*, 85
Nordhaus-Gaddum Theorem 182
NP *233*
NP-complete *234*
NP-completeness 238-246, 269, 343, 352
NP-hard *234*, 405, 419

obstruction *301-303*
odd component *121*, 121-124
odd cycle *17*, 21, 30, 109, 179, 180, 183, 188, 289, 315, 316, 378
odd degree 26, 36, 56, 60, 364-365
Odd graph *25*
odd hole *312*, 314-315
odd triangle *213*
odd walk *17*
off-diagonal Ramsey number 371
One-Way Street Problem *148*
open set *248*
optimal edge-coloring *216*
optimization 73-84
optimization problem 373
order *26*

order-preserving property **331-332**
ordered multigraph *384-385*
Ore's Theorem **222**, 396, 404
(*m*, *n*)-orientable *172*
orientation *46*, 67, 95-96, 178
orthonormal 437
out-degree *46*, 89-93
out-neighborhood *46*
out-tree *68*, 89-91, 385
outerplanar graph *254*, 258, 268, 283
outerplane graph *254*

P *233*
p-critical graph *305-315*
P=NP *234*, 236
pair-disjoint *427*
parallel element *323*, 348, 350
parameter *25*
parent *79*
parity graph *300-301*, 318
Parity Lemma *132*
parity of a point *249*
parity subgraph *60*
partial transversal *325*
k-partite graph *4*, *21*, 174
partite set *4*
partition 21, 212
k-partition *356*
partition matroid *330*, 347
partitionable graph **306-315**, 432
path *5*, 8, *14-17*, 22, 58-59, 60, 379,
    391-393, 450
x, *y*-path 16, 18, 23, 53, 149-152,
    261-263, 300-301, 318
path P_4 9, 25, 37, 39, 131, 181, 318, 371
path P_3 11, 36, 94, 96, 215, 371
path addition *146*
pendant vertex *51*
γ-perfect 288-292, 305
α-perfect 288-292, 305
be-perfect *305*
perfect eliminaiton ordering *198*
perfect graph *184*, 200-202, **288-315**
Perfect Graph Theorem (PGT) 200,
    288-292, 390
perfect matching *98*, 277
perfect order *301-303*
perfectly orderable graph *301-303*, 319
performance ratio 181
permutation 7, 23, 24, 302, 429, 434
permutation matrix *106*
Petersen graph *10*, 12-13, 25, 28, 37, 47,
    62, 99, 106, 124, 142, 174, 178, 204,
    209-210, 211, 215, 216, 220, 228, 258,
    267, 277, 286, 401, 403, 451
Petersen's Theorem 124, 131, 216
pigeonhole principle 29-31, 38, 135,
    353-355, 366-368, 369
pigeonhole property *406-408*

PLANAR 3-COLORABILITY 243-244
planar embedding *248*, 252, 255
planar graph 5, 243-244, **247-287**, 269,
    321, 336-340, 338-340, 351, 389-390
planar map 253
planar multigraph 255
PLANARITY 234
planarity testing 264-267, 268
plane graph *248*, **249-259**
planted tree *79*
Platonic solid 257
Poisson distribution *414*
police cars 35
polygonal path *248-250*, 258
polygonal curve *248*, 249
polygonal representation 282
polyhedron (regular) 256
polynomial-time algorithm 232, 283
positional game *107*
*k*th power 228
Prüfer sequence *63-65*, 71, 317
predecessor 202
predecessor set *46*
pretzel 281
Prim's Algorithm *83*
principal submatrix *434*
product dimension **376-378**, 401, 402
product graph 175
product representation *376*
proper *k*-coloring *173*
proper edge-coloring *208*
proper interval graph 318
proper labeling *364-365*
Proposal Algorithm (Gale-Shapley)
    116-118, 121

quota 355-357

radius *54*, 61-62
Ramsey multiplicity 372
Ramsey number *356*, 359-361, 370-371,
    373, 406
Ramsey number (graph) 429, *361*
Ramsey Theory *353-373*
Ramsey's Theorem 355-359
random graph 177, 185, 404-432, *409*
random variable *406*
rank *321-322*, 433
rank function *321-322*, 323, 326, 334,
    337, 340-348, 348, 349
reciprocity 203
recurrence 66, 72, 84, 106, **195-196**, 205
reducible configuration *270*, *271-273*
reduction *238*
refinement of a matroid 349
region *248*
regular graph 25, 36, 209, 217, **441-444**
k-regular *25-27*, 36, 134-135, 179, 183,

224, 257, 277
3-regular 31, 38, 48, 131, 141
regular embedding *287*
relation *7*
Replacement Lemma 304
representable matroid *324*
Restricted Jordan Curve Theorem **249**
restriction *336-341*, 351
restriction martingale *425*
ring *271-273*, 283
Ringel's Conjecture *69*
road network 5
Robbins' Theorem 148
root *79*
rootable family 316
rooted plane tree *79*, 84
rooted tree *79*, 317

3-SAT *239*, 240, 246
SATISFIABILITY 238, 239, 246
satisfiable *238*
saturated vertex *98*, 105
scheduling 4
score sequence *46*, 50
Scrabble problem 84
search strategy 358
Second Moment Method **412-415**, 423
self-complementary *12*, 48
Semi-strong Perfect Graph Theorem 315
separable graph *138*
separating set *133*, *148*, 204, 261
x, y-separating set *149*
x, y-separator *199*
Shannon Switching Game *337-338*, 345
shift graph *181*
SHORTEST CYCLE 234
shortest path 22, 61, 144
simple curve *249*
simple digraph *2*
simple graph *1*, 6, 14, 28, 29, 42, 48, 60,
    173, 210-211, 211
simple hereditary system *323*
simple matroid 350
simple polygon 268
simplicial construction ordering 296
simplicial elimination ordering *198*, 205,
    294, 316
simplicial subdivision *364-365*, 367, 372
simplicial vertex *198*, 205
sink *158*
sink set *160*
skew partition *319*
snowplow problem 245
f-soluble *132*
sorting 445
source *158*
source set *160*
source/sink cut *160*
span function **331-332**, 348, 349

spanning cycle 207
spanning forest 83
spanning path 182, 224
spanning set *333*, 350, 352
spanning subgraph *51*, 178, 196-197, 325,
    376
spanning tree *51*, 53, 54, **58-75**, 172, 179,
    205, 257, 321-322, 336, 346, 346, 352,
    353, 354, 444
spanning trees 143
sparse graph 417
Spectral Theorem **437**, 439
spectrum *433*, 444, 448
Sperner's Lemma 364-368, 372
sphere (embeddings) 252
split graph *49*, *317*
Squashed Cube Conjecture *379*
squashed-cube dimension *378-381*, 380,
    402, 450
squashed-cube embedding *378-381*
stable matching 116-118, 121, 387-389
Stable Roommates Problem 121
stable set 49, *288-292*, 300-304, 347
stage of evolution *416*
star 105, 271
star $K_{1,n-1}$ *70*, 107
star-cutset *303-304*
Star-cutset Lemma 320
Star-Cutset Lemma 303-304
r-staset *427*
Steinitz exchange property **331-332**,
    333, 349
Steinitz's Theorem 351
stem *128*
Stirling's formula 361, 420
straight-line embedding 264, 268
Street-Sweeping Problem 93-94, 116
strength *421*
strict digraph *226*, 398-400
strong absorption property **327-329**
strong component *141*, 143, 144
strong digraph *88-91*, 141
strong duality 292
strong elimination property **331-332**,
    348
strong induction *16*
strong orientation 148
Strong Perfect Graph Conjecture (SPGC)
    289, 305, 309, **312-315**, 320, 417
strongly connected *88-91*
strongly perfect graph *300-301*, 319
strongly regular graph **446-449**
subconstituent *451*
subdivision *146*, *188*, 193, *259*
K_4-subdivision 188, 192
F-subdivision 189-190
subgraph *3*, 40, 47, 53, 57, 217
submodular function 157, 347
submodularity property **327-329**, 341,
    348, 351

subtree *79*
subtree representation *294*
subtrees 293, 316
successor set *46*
sufficiency *21*
sum *19*
supbase *333-334*, 350
n-superconcentrator *445*
supergraph 209
superregular graph *451*
supply/demand 94, *166*
2-switch *44-46*
switching 211
Sylvester's Law of Inertia **438**, 450
symmetric difference *99*, 118-119, 123, 143, 150, 320
symmetric digraph *68*
symmetric matrix 436
system of distinct representatives (SDR) *106*, *153*, 325
Szekeres-Wilf Theorem **177**, 182
Szemerédi Regularity Lemma 364

tail *2*
tail partition matroid *330*
Tait coloring *275*, 283
Tait's Conjecture 284
Tait's Theorem **274-275**, 284
Tarry's Algorithm 97
telegraph problem 402
telephone problem 402
tensor product 376
ternary matroid *330*
*k*th-moment *412*
thickness *286*
threshold 355-357, 404
threshold edge function *412*
threshold function **412-415**, 430
threshold probability function *412*
Tic-Tac-Toe 107
topological graph theory 247
topologically minimal *259*
toroidal embedding 281
toroidal graph 287
torus *280*, 281, 287
total coloring *403*
Total Coloring Conjecture 403
totally unimodular *72*
t-tough 220
toughness *220*
tournament *46*, 49, 50, 182, 231, 299, 407
trace 433
traffic light phasing 298
trail *14-15*, 87-88
transformation *238*
transitive digraph *390*
transitive orientation *201-202*, 299, 301
transitivity of dependence **332**
transportation constraint *166*

Transportation Problem 94, 116, 166
transposition 434
transversal matroid *324*, 325, 326, 342, 347, 351
transversal of a matrix *112*, 120, 121
Traveling Salesman Problem (TSP) 5, 232, 235-238, 245, 432
Traveller's Dodecahedron 218
k-tree *317*
tree *51-97*, 105, 108, 203, 317, 362, 369, 371, 403-404
triangle inequality 245
triangle-free *32*, 37-39, 183, 184, 190, 256, 283, 376, 400-401, 451
triangular grid 366-368, 367
o-triangulated graph *300-301*, 318
triangulation *256*, 271, 274, 283, 284
tripartite graph 287
trivial matroid *329*
Tucker's Algorithm 95
Turán problem 373
Turán graph 34, 39
Turán's Theorem **32-35**, 50, 182, 401
Tutte Conjecture 277
Tutte graph 276
Tutte's 1-factor Theorem *122-127*, 215
Tutte's condition 122-127, 132
Tutte's Theorem (planarity) 260-263
twin *319*
two-step method 407-408

unavoidable set *270*, 271-273
k-uniform hypergraph *429*
uniform matroid *329*, 347, 350
uniformity property **327-329**
union *19*, 60
unipathic digraph 50
uniquely *k*-edge-colorable 228
unit edge *164*
unit interval graph 318
unit-distance graph *181*, 259
unlabeled graph 8
unsaturated edge 110-111, 127-130
unsaturated vertex *98*, 99-101, 105, 109-111, 114-115, 127-130, 132

Vámos matroid 350
k-valent vertex *26*, 277
variance *412*
vectorial matroid *324*
vertex *1*
vertex-color-critical *186*, 192
vertex cover **102-105**, 110-111, 131, 246, 341
VERTEX COVER 241-243
vertex-*k*-critical subgraph 440
vertex cut *133*, *148*, 187
vertex duplication *289*, 320

vertex multiplication *289*, 303
vertex ordering 50
vertex separator 204
vertex set *1*
vertex k-split 156
vertex-transitive *10-11*, 14
Vizing's Theorem 208, **210-211**, 216-217,
   217, 377, 387, 419

walk *14-17*
u, v-walk *14*, 16, 19, 21, 23, 53
weak absorption property **327-329**, 348
weak augmentation property 348
weak dual *254*
weak duality 161, 292, 352
weak elimination property 324, 325, 326,
   327-329, 347, 348
weakly chordal graph *300-301*, 304, 319
weighted average 406
weighted cover *112-114*
weighted graph 4, *73*, 73-77, 91
weighted intersection graph *297*
weighted matching (bipartite) 111-116
Well Ordering Property *15*
wheel $K_1 \vee C_{n-1}$ 70, 72, 83, 156, 203
Whitney's 2-isomorphism Theorem
   251-259, 351
Wiener index *58*, 61
Woodall's Theorem 399, 404